The Bernard E. Harkness

SEEDLIST HANDBOOK

THE
BERNARD E. HARKNESS

SEEDLIST HANDBOOK

A Guide to the Plants Offered in the Major Plant Societies' Seed Exchanges

Compiled and Updated
by
MABEL G. HARKNESS
and
DEBORAH D'ANGELO

TIMBER PRESS
Portland, Oregon

ISBN 0-88192-059-2
Printed in Hong Kong

TIMBER PRESS
9999 SW Wilshire
Portland, Oregon 97225

CONTENTS

ACKNOWLEDGMENTS

The editors wish to acknowledge the help received from the following:

Audrey Harkness O'Connor, Ithaca, New York
Dr. Benjamin C. Blackburn, Gladstone, New Jersey
Letitia McKinney, Rochester, New York
Brian Mulligan, Kirkland, Washington
Peter Hypio, Ithaca, New York
William Dress, Ithaca, New York
Geoffrey Charlesworth, Sandisfield, Massachusetts
Nina Lambert, Ithaca, New York
William J. Hamilton, Jr., Ithaca, New York
Baldassare Mineo,, Medford, Oregon
Mark McDonough, Bellevue, Washington
Sallie D. Allen, Seattle, Washington
Judith Jones, Seattle, Washington
Rush Rhees Library, University of Rochester, New York (especially Karl Kabelac and Frances Carducci)
Library of the New York State Agricultural Experiment Station, Geneva
Panayoti Kelaidis, Denver, Colorado
Library of the New York Botanical Garden (especially Mrs. Lothian Lynas)
Wayne Morrison, Ovid, New York
Elizabeth Sheldon, Ithaca, New York
William Shipley, Rochester, New York
Mrs. William Flook, Wilmington, Delaware
Mr. and Mrs. John Gyer, Clark's Mills, New Jersey
and last, without whose patience and skill this work could not have been completed: Dorothy Densk of The Word Works, Geneva, New York

INTRODUCTION

Bernard Harkness spent a lifetime dedicated to horticulture. A degree from Cornell University in 1929 with a specialty in Ornamental Horticulture was followed by additional work at the Harvard School of Design in Landscape Architecture. This was the basis for his work on the Taconic Parkway in the state of New York, and his many years as taxonomist for the Monroe County, N.Y., Department of Parks, which includes the notable Highland Park of Rochester. His expertise in trees and shrubs is as well known as his knowledge of alpine plants. He was, for one term, head of the American Association of Arboreta and Botanical Gardens.

The original *Seedlist Handbook,* first published in 1974, grew from Bernard Harkness's interest in alpine gardening, coupled with a need to answer numerous letters and telephone calls for information, received during and following his 12 years as Seed Director for the American Rock Garden Society. As a charter member of this society, organized in 1935, and four years as its president, Mr. Harkness recognized the alpine gardener's need for a handy and compact guide to plants, with each entry having coded information about a plant, its place of origin, and a bibliographic reference for further investigation. This remains the purpose of this book.

The Seedlist Handbook had appeared in three editions by the end of 1980. Each edition was based on the previous several years' lists from the seed exchanges of the American Rock Garden Society, the Alpine Garden Society (England) and the Scottish Rock Garden Club. This fourth edition was compiled and edited from the previous three and includes new items from the three seedlists through 1985. It remains a dictionary for alpine gardeners, based on the guidelines and format developed for earlier editions.

H. Lincoln Foster wrote in the conclusion to the Foreword of the third edition: "Unpretentious as it is, just as was its author, *The Harkness Seedlist Handbook* is a veritable storehouse of dependable information, based on the prodigious industry and immaculate scholarship of Bernard Harkness."

This edition is dedicated to the memory of Bernard Harkness (1907–1980).

Mabel G. Harkness
Deborah D'Angelo

INTRODUCTION

Bernard Harkness spent a lifetime dedicated to horticulture. A degree from Cornell University in 1929 with a specialty in Ornamental Horticulture was followed by additional work at the Harvard School of Design in Landscape Architecture. This was the basis for his work on the Taconic Parkway in the state of New York and his many years as taxonomist for the Monroe County, N.Y., Department of Parks, which includes the notable Highland Park of Rochester. His expertise in trees and shrubs is as well known as his knowledge of alpine plants. He was, for one term, head of the American Association of Arboreta and Botanical Gardens.

The original *Seedlist Handbook*, first published in 1974, grew from Bernard Harkness's interest in alpine gardening, combined with a need to answer numerous letters and telephone calls for information received during and following his 12 years as Seed Director for the American Rock Garden Society. As a charter member of this society, organized in 1935, and four years as its president, Mr. Harkness recognized the alpine gardener's need for a handy and compact guide to plants, with each entry having coded information about a plant, its place of origin, and a bibliographic reference for further investigation. This remains the purpose of the book.

The Seedlist Handbook had appeared in three editions by the end of 1980. Each edition was based on the previous several years' lists from the seed exchanges of the American Rock Garden Society, the Alpine Garden Society (England) and the Scottish Rock Garden Club. This fourth edition was compiled and edited from the previous three and includes new items from the three seedlists through 1985. It remains a dictionary for alpine gardeners, based on the guidelines and format developed for earlier editions.

H. Lincoln Foster wrote in the conclusion to the Foreword of the third edition: "Unpretentious as it is, just as was its author, *The Harkness Seedlist Handbook* is a veritable storehouse of dependable information, based on the prodigious industry and immaculate scholarship of Bernard Harkness."

This edition is dedicated to the memory of Bernard Harkness (1907–1980).

Mabel G. Harkness
Deborah D'Angelo

KEY TO SYMBOLS

Some short-cuts are necessary to have the maximum of information in a single line description. The first section indicates the habit of the plant according to the following abbreviations:

A	- Annual	Gr	- Ornamental Grass
B	- Biennial	HH	- Half-hardy
Bb	- Bulb	lvs	- leaves
C	- Corm	P	- Perennial
Cl	- Climbing Vine	P gr. as A	- Tender perennial flowering in first year
F	- Fern	R	- Rock Garden Plant (mostly under 12")
fls	- flowers	Sh	- Shrub
frs	- fruits	X	- Tree
Gh	- Greenhouse Plant	Tu	- Tuber

Additional abbreviations - '________' - a cultivar

X - a known hybrid

* - obviously identified or determined by garden trial by B.E.H.

In the second space the approximate height of the plant is given.

In the third space the outstanding ornamental or distinguishing characteristic is given, usually by flower color.

In the fourth space the country or area of origin is given.

The numbers at the end of the line are keys to the Main References, pages 405-417, printed materials that give additional botanical or horticultural data. Numbers in parenthesis refer to Supplemental References, pages 419-428. Numbers underlined refer to illustrations; they will lead to both Main and Supplemental References. In some cases a footnote contains the bibliographic data.

Botanical sub-divisions below the species level exhibit no consistency in the listings of this handbook. Older authorities' preference for var. (variety) is reflected. Most European plants will be listed by the equivalent ssp. (sub-species) as appears in Flora Europaea. It is not the intent of this work to recommend any standardization of plants' names. In not a few instances seed of the same species is listed under two or more names.

ABELIA - Caprifoliaceae
chinensis / HH Sh / to 6' / white fls / c & e China 25

ABELMOSCHUS MALVACEAE
manihot / P gr, as A / 3-9' / fls yellow / China, India 95 / 25

ABIES - Pinaceae
balsamea f. Hudsonia / T /to 75'/ cones viol-purp./Lab.-W.Va., Ia. 25
borisii-regis / T / 100'+ / cones rare / Balkan Pen.
concolor compacta / T /100'/ bl-gr needles, dense br./Columbia-Mex. 25
fraseri / T / to 75' / cones purple / Alleghany Mts. 25
homolepis / T / 80' / lvs white-banded beneath / Japan 167 / 127
koreana / T / 20'+ / cones purplish / s Korea 127 / 36
koreana f. prostrata / 5'+ / horizontally branched 79
lasiocarpa / T / to 100' / purple cones / n North America 127 / 25
mariesii / T / 50' / needles white-banded / Japan 139
procera / T / 150' / purplish-brown cones / nw United States 127 / 25
spectabilis / T / 100' / violet-purple cones / Himalayas 127 / 25

ABRONIA - Nyctaginaceae
latifolia / P gr. as A / prostrate / lemon-yellow fls / w N.Am. 227 / 25

ABUTILON - Malvaceae
darwinii / Gh Sh / dwarf / fls orange-red / Brazil 28
X milleri / Gh Sh, 6'+ / orange bell-fls; lvs mottled yellow 180 / 139
X milleri 'Variegatum' - descriptive of the hybrid 139
ochsenii / HH Sh / 12' / fls blue-purplish / Chile 19 / 36
X suntense / HH Sh, 2-3' / bright mauve fls 110
theophrasti / A / 2-4' / yellow fls, ornamental pods / India 220
vitifolium - see CORYNABUTILON vitifolium 129 / 25
vitifolium 'Album" - white fls 36

ACACIA - Leguminosae
armata / Gh Sh / 10' / spiny plant, rich yellow fls / Australia 128
Baileyana / HH Sh / 18-20' /intense yellow fls/N.S. Wales, Australia 25
cavenia / HH Sh / to 20' / deep yellow fls / Chile 25
cyanophylla / HH sh / 18' / fls golden yellow / Australia 211 / 25
dealbata / HH t / 50' / yellow-gold fls in winter / Australia 208 / 25
farnesiana / HH Sh / 6-10' / deep yellow fls / Texas, Mexico 28
leprosa / Gh Sh / 5' / lvs willow-like / Australia 28
longifolia / Gh / 15'+ / fls yellow / Australia 36
myrtifolia / HH sh / 3-6' / yellow-cream fls / Australia 45 / 25
neriifolia / T / small / lvs long, fls golden balls/ N.S.Wales, Aust 109
podralyriifolia / Gh Sh / 10' / fls yellow / N.S. Wales, Australia 128
polybotria / Gh Sh / 15'+ / lvs pubescent, feathery / Australia 25
pravissima / Gh Sh / to 20' / yellow fls / se Australia (361) / 25
pulchella / HH Sh / 5-18' / intense yellow fls / w Australia 184 / 208
retinoides / HH sh / 20' / fls pale yellow / Tasmania 212 / 25
riceana / HH T / 30' / pale yellow fls / Tasmania (361) / 25
spectabilis / Gh Sh / 18'+ / golden ball fls / Australia 25
'spinescens' / Sh / to 3' / fls yellow; frs linear / e Australia 25
verticillata / HH Sh / tall / bright yellow fls / Australia (361) / 25

ACAENA - Rosaceae
adscendens - see A. affinis 293 / 118
affinis / R / 8-12" / blue-gray lvs; reddish heads / Antarctica 118
anserinifolia / R / creeping / lvs brownish-green/ N. Zealand 118 / 25
anserinifolia X inermis - cross by donor?
argentea / R / ground cover / gray lvs / Chile 96
buchananii / R / creeping / pale green lvs/ N. Zealand 118 / 25
caesiiglauca / R / creeping / blue-gray lvs / N. Zealand 118 / 268
glabra / R / 2' spread / green or reddish heads / N. Zealand 118 / 61
glauca - see A. caesiiglauca 25
glaucophylla - see A. magellanica 118
hirsutula / R / to 4" / lvs densely hairy, anth & stigma purp / N.Z. 15
inermis / R / dwarf / lvs purplish to greenish / N. Zealand 118 / 25
magellanica / R / creeping / glabrous lvs; anthers red / Patagonia 118 / 25
microphylla / R / mat / brownish lvs / N. Zealand 48 / 25
microphylla v. inermis - see A. inermis 118
microphylla X novae-zelandiae - cross by donor? *
myriophylla / R / ascending / silky lvs; green fls / Chile 25
novae-zelandiae / R / creeping / green lvs; purplish burrs / N. Z. 118 / 25
ovalifolia / HH R / sprawling / larger lvs / Peru 25
saccaticupula / R / 6" / purplish heads / s alps of N. Zealand 15
sanguisorbae - the Australian form of A. novae-zelandiae 118
sanguisorbae caesiiglauca - see A. caesiiglauca 268
sericea / HH R / 6" / silky lvs; globose heads / Mexico 81
splendens / HH Sh / 6" / lvs covered with long silky hairs / Chile 61

ACANTHOLIMON - Plumbaginaceae
albanicum - a listed name of no botanical standing 25
androsaceum / R / dense cushion / fls purple / s Balkans 240 / 289
araxanum / P / 6" / silver foliage; rosy flowers / Greece eastward 249
armenum / R / to 8" / pink & white fls / Asia Minor 25
aulietense / R / subhemispherical / pale pink fls / c Asia 10
creticum - see A. echinus v. creticum 25
diapensoides / R / cushion / pale pink fls / Afghanistan 96
echinus / R / cushion / purple fls / e Medit. region (3) / 25
echinus v. creticum - hairy lvs 25
glumaceum / R / 6" / corolla pink, limb white / Caucasus, Kurd. 93 / 25
gramineum - see A. aulietense 10
venustum / R / 6" / fls deep rose / Asia Minor 194 / 25

ACANTHOPANAX - Araliaceae
senticosus / Sh / to 15' / purplish-yellow fls / ne Asia 25
sessiliflorus / Sh / to 12' / fls purplish; frs black / China, Korea 25

ACANTHOPHYLLUM - Caryophyllaceae
microcephalum / Sh / 6-12" / white fls / Iran, Transcaucasia 77
pungens / R / 1' / reddish-pink fls / c Asia 25
spinosum - see A. pungens 25
verticillatum / Sh / to 10" / fls white / n Iran, Turkey 77

ACANTHUS - Acanthaceae
balcanicus / P / 3' / mauve fls / Balkans 279
caroli-alexandri - see A. spinosus 288
longifolius - see A. balcanicus 288

mollis / HH P / 4' / white fls / Mediterranean region 129 / 25
spinosissium - see A. spinosus 288
spinosus / P / 8-32" / reddish fls; spiny lvs / Italy, Balkans 154 / 288

ACARPHA
spatulata - see ACICARPHA spathulata 138

ACER - Aceraceae
buergerianum / T / 35' / ivy-like lvs / China 36
campbellii / HH T / 60-70' / 5 to 7-lobed lvs / e Himalayas 25
campestre / T / 35'+ / small tree / Europe, Near East, n Africa (4) / 36
cappadocicum / T / 70' / yellow fls / Caucasus, Asia Minor 36
circinatum / T / 20'+ / lvs have good fall color / w North America (342) / 36
crategifolium / T / to 30' / green & white bark / Japan 167 / 36
davidii / HH T / 30' / white-striped bark / China (5) / 36
davidii 'Forrest's Form' = 'George Forrest' - open tree in habit 36
dissectum - see A. palmatum 'Dissectum' 139
ginnala / T / 10'+ / may be shrubby / China, Manchuria 36
glabrum v. douglasii / T / 18'+ / lvs 3-lobed / w North America 294 / 139
glabrum v. torreyi / mostly 3-lobed leaf form / Sierra Mts., Calif. 25
grandidentatum - see A. saccharum
griseum / T / 25-45' / orange peeling bark / c China 269 / 139
grosseri / T / to 20' / green-striped bark / China 36
grosseri v. hersii / T to 20' / striped bark / Honan 36
hersii - see A. grosseri v. hersii 184 / 36
japonicum / T / 20' / purplish-red fls / Japan 36
japonicum 'Aconitifolium' / Sh / 8' / cut-leaved form 36
japonicum 'Aureum' - lvs pale golden-yellow 25
japonicum laciniatum - see A. japonicum 'Aconitifolium' 36
japonicum v. parsonsii - see A. japonicum 'Aconitifolium' 25
japonicum 'Vitifolium' - large lvs. richly colored in autumn 36
laxiflorum / T / to 25' / frs in lax racemes / w China 36
Maximowiczianum / T / 40-50' / autumn lvs yellow or red / Japan 25
miyabei / T / 30' / yellow fls / Japan 36
monspessulanum / T / 30'+ / reddish frs / s & c Europe, Near East 36
nikoense - see A. 'Maximowiczianum' 25
palmatum / T / to 20' / purplish-red lvs / Japan 167 / 36
palmatum 'Atropurpureum' / 15'+ / red lvs 36
palmatum 'Dissectum' / Sh / 4'+ / finely cut green lvs 269 / 36
palmatum 'Dissectum Atropurpureum' / Sh-T / 20'+/crimson lvs/ Japan 184
palmatum 'Dissectum Viridis' - see A. palmatum 'Dissectum' 36
palmatum v. heptalobum 'Osakazuki' - 7-lobed lvs, red in fall (41) / 36
palmatum 'Linearilobum Atropurpureum' - finely toothed purple lvs 36
pensylvanicum / T / 15-20'+/ fls yel, anther lemon yel / e N.A. 36 / 25
platanoides / T / 60' / greenish-yellow fls / Europe (41) / 36
platanoides 'Schwedleri' - lvs purplish in spring 139
rufinerve / T / 30' / blue-white young shoots / Japan 167 / 36
rufinerve f. albolimbatum - lvs white spotted 36
saccharinum / T / 90-130' / Silver Maple/ autumn fol yel & red/ N.A. 25
saccharum / T / 100' / Sugar Maple / e North America (342) / 36
spicatum / T / to 25' / often shrubby; red frs / e North America 22 / 36
syriacum / HH T / 10'+ / leathery lvs, evergreen / Syria, Cyprus 139
tataricum / T / 30' / distinctive foliage / se Europe, Asia Minor (414) / 36
tegmentosum ssp. grosseri - see A. grosseri 25
velutinum b. vanvolxemii / large tree with large lvs / Caucasus 222

ACERAS - Orchidaceae
anthropophorum / R / 6-16" / greenish-yellow fls / Eur., n Africa (69) / 85

ACERIPHYLLUM - Saxifragaceae
rossii / R / 6-9" / large glossy fls, lvs white / Korea 154

ACHILLEA - Compositae
abrotanoides / P / 6-9" / fls white / e Alps / 289
ageratifolia / R / 4" / fls white / n Greece (6) / 123
ageratifolia aizoon / R / to 1' / fls yellow / Balkan Pen.
ageratifolia v. serbica - see A. serbica 289
ageratum / P / 2' / yellow fls / s Europe 289
argentea - see A. clavenae 28
asplenifolia / P / 1½-3½' /pink to purp, rarely wht/Czech. to Jugo. / 289
atrata / R / 4-8" / fls white / Austrian Alps 93
barrelieri / R / to 6" / fls white / Italy, Apennines 289
brachyphylla - see A. monocephala 154
cartilaginea / P / 1'-4" / fls white/ USSR to E. Germany / 289
chamaemelifolia / R / 9" / white fls / France 314 / 206
chrysocoma / R / to 7" / fls light yellow / Albania, Macedonia 148 / 123
clavennae / R / 6" / white fls; gray lvs / e Alps 148 / 123
clavennae 'King Edward' - see A. X lewisii 123
clusiana / R / 8" / white fls /Alps 149 / 25
clypeolata (of hort.) / P / 2' / fls bright yellow heads / ? 263
coarctata - see A. compacta 262
compacta / R / 6" / tight head of small fls / Bulgaria 25
'Coronation Gold' - Achillea clypeolata X Achillea filipendulia 95
depressa / P / 1' / fls yellowish / ec Europe, Balkan Pen. / 289
erba-rotta / R / to 8" / white fls / sw Alps 148 / 289
erba-rotta ssp. rupestris / plant caespitose / Apennines 289
filipendula / P / 4-5' /fls yellow / Asia Minor, Caucasus 25
filipendulana 'Gold Plate' /P / 5' / flat heads of golden-yel fls 49 / 268
'Golden Plate' - see A. filipendulana 'Gold Plate' 268
herba-rotta - see A. erba-rotta 289
herba-rotta rupestris - see A. erba-rotta ssp. rupestris 289
huteri / R / 6" / white fls; silvery lvs / Switzerland 61
X jaborneggii / R / 6" / fls white; silky-green lvs / Europe 25
X kellereri / R / 6" / gray lvs; white fls / Bulgaria 123
X 'King Edward' - see A. X lewisii 123
X kolbiana - hybrid of A. clavennae x A. moschata 93
X lewisii / R / 6" / fls light yellow fading cream 289 / 25
millefolium / P / 3' / white fls / Eurasia, naturalized in US 308 / 25
millefolium v. alpicola / R / white fls / Alaska & s in Rockies 25
millefolium 'Rosea' / P / fls pink 25
monocephala /mats,moss like/ silver shoots; large wht fls; 6" stems 154
'Moonshine' / P / 18" / deep canary-yellow fls 49 / 135
moschata /R / to 6" / fls wht; lvs br grn, leaf lobes uncut / Alps 218 / 25
nana / R / to 8" / fls dull white / s Europe 25
X obristi - similar to A. huteri, but looser habit 154
odorata / R / 6"+ / white fls / Spain, France 314 / 206
oxyloba / R / 4-10" / fls white / Tyrolean Alps 218
oxyloba ssp. schurii / R / 8" / white daisies, Carpathians 289
'Pearl White fl. pl.' - see A. ptarmica 'The Pearl' 268
prichardii / R / 4" / white fls, free-flowering / ? 46

ptarmica / P / 2' / white fls; spreading plant / Eurasia 208 / 25
ptarmica 'The Pearl' / P / 2½' / pure white fls, double 129 / 268
pyrenaica / R / 6-10" / ptarmica-like fls / Spain, s France 314 / 96
rupestris - see A. erba-rotta ssp. rupestris 123 / 289
schurii - see A. oxyloba ssp. schurii 289
serbica / R / 4" / clustered white fls / Balkans 306 / 289
taygeta (of hort.) / P / 18-24" / light yellow corymbs 268
tomentosa / R / 6-12" / fls yellow; lvs whitish / n hemisphere 210 / 107
tomentosa v. aurea - deeper colored fls 24
tomentosa 'King Edward' - see A. X lewisii 46
tomentosa 'Moonlight' - fls lighter yellow 135
tomentosa 'Nana' - dwarfer form 25
umbellata / R / 6" / fls white; lvs woolly / Greece 289
X wilczekii / R / 6" / fls white; gray-woolly lvs 25

ACHLYS - Berberidaceae
triphylla / P / 16" / aromatic woods-plant / nw North America 229 / 78

ACHYROPHORUS
uniflorus - see HYPOCHAERIS uniflora 89

ACICARPHA - Calyceriaceae
scapigera / HH R / rosette / grn-white to pink fls / Argentina 64
spathulata / A / procumbent / spatulate lvs / Brazil, Argentina 138 / 81

ACIDANTHERA - Iridaceae
bicolor v. murielae - see GLADIOLUS callianthus 129 / 279
bicolor v. murielae / HH C / 2½-3½' / white fls / Ethiopia 267
murielae - see A. bicolor v. murielae 267
viridis / HH C / 1' / green fls / South Africa 61

ACINOS - Labiatae
alpinus / R / 12" / fls violet with white markings / c & s Europe 210 / 288
alpinus ssp. meridionalis / to 18" / s Europe 288
arvensis / A / to 16" / violet & white fls / most of Europe 148 / 288
triphylla / P / 16" / fragrant woods-plant / nw North America 78

ACIPHYLLA - Umbelliferae
anomala / HH P / to 18" / slender, tufted plant / New Zealand 15
aurea / HH P / to 3' / yellow-green tussocks / New Zealand 15
colensoi / HH P / to 3' / yellow-green tussocks / N. Z., in mts. 238 / 15
congesta / HH R / compact / fls on 1" stems / mts New Zealand 15
crenulata / HH P / to 2' / rather lax tufts / New Zealand 15
divisa / HH P / to 32" / tufted pl, invested with dead lvs / N. Z. 15
dobsonii / HH R / close-set rosettes / New Zealand 15
ferox / HH P / to 3' / stout tussocks / New Zealand, in mts. 238 / 15
glacialis / HH P / 18" / spectacular plant / N.S. Wales, Australia 109
glaucescens / HH P / to 6' / lvs glaucous, grayish-green/ N. Z. 15
gracilis / HH R / 8" / delicate plant / New Zealand 15
hectori / HH P / 2-4' / tall inflorescence / New Zealand 193 / (7)
hookeri / HH R / 1' / narrow paniculate inflorescence / N.Z. in mts 15
horrida / HH R / to 1' / thick coriaceous lvs / New Zealand, in mts. 238 / 15
kirkii / HH R / to 1' / rigid tufts; coriaceous lvs / N.Z., in mts. 15
lyallii / HH R / 1'+ / slender plant; lvs 1-pinnate / N.Z., in mts. 15
monroi / HH R / 10" / yellowish fls; crowded lvs / N.Z., in mts. 238 / 15

montana - see A. lyallii 15
'Otago' / HH R / 10"+ / golden yellow fls (360)
pinnatifida / HH R / 1'+ / flower bracts deep orange/ N.Z., in mts. 15
polita / HH R / 8" / plant small, tufted / New Zealand 15
scott-thomsonii / HH R / to 1' / hard pungent leaf-points / N.Z. (417)/ 15
similis / HH P / to 28" / fascicled umbels; large bracts / N.Z. 15
simplex / HH R / 2-4" / cream fls; bronze lvs / New Zealand (360)/ (7)
spedenii / HH R / to 10" / white fls; soft striate lvs / N.Z., in mts 15
squarrosa / HH P / 3' / sweet-scented brownish fls / New Zealand (9) / 238
subflabellata / HH P / 20" / dense rosettes / N. Zealand, in mts. 15
takahea / HH P / to 2' / inconspicuous pungent fls / New Zealand 15
townsonii / HH R / to 1' / slender, rather flaccid / N. Zealand 15
traillii v. cartilaginea / HH R / 8"/thick yellow leaf margins /N.Z. 15
trifoliolata / HH R / to 16" / lvs 1-pinnate; stems grooved / N.Z. (360)/ 15

ACONITUM - Ranunculaceae
altaicum / P / ? / fls dark violet / w Siberia 5
[1] amplexicaule / P / to 4' / shiny violet fls / Nepal
anthora / P / 12-24" / fls pale yellow / s Europe 148 / 286
[2] arendsii - probable hybrid
[2] arendsii bicolor - probably hybrid
X bicolor 'Bressingham Spire' / to 4'/ fls violet-blue 251 / 279
X cammarum 'Bressingham Spire' / 3' / fls deep violet 268
carmichaelii / P / 30-36" / fls deep blue-purple / China 75 / 129
carmichaelii v. wilsonii / 6' / amethyst-blue fls / e China 268
columbianum / P / 2-6' / fls deep blue / w North America 229 / 78
delphinifolium / P / 2-3' / fls deep blue / arctic Asia & N. Am. 29
delphinifolium paradoxum - included in the sp. 209
excelsum - see A. septentrionale 286
ferox / P / 6" to 6' / fls pale blue / Himalayas 107
firmum / P / 3-4' / fls blue to violet / c Europe 286
fischeri - see A. carmichaelii 93
grandidentatum - see A. senanense 165
grossedentatum / P / to 5' / large-toothed lvs / mid-Hondo, Sikoko (10)
hookeri / P / to 8" / lvs kidney sh; fls blue or purp/Sikkim, Tibet 61
japonicum / P / 2-6' / fls bluish-purple / Japan 25
japonicum v. montanum / blue-purple fls / 1' form of variable sp. *
judenbergensis - see A. variegatum 286
kusnezoffii / P / 3-5' / bright dark blue or white fls / e Siberia *
lamarckii / P / 3-6' / fls yellowish / s Europe, in mts. 148 / 286
luridum - see A. novoluridum 25
lycoctonum - see A. vulparia 286
maximum / P / 6' / fls pale blue / Kamtschatka 81
napellus firmum / P / divided lvs; fls blue-violet / c Europe 286
napellus nanum - see napellus firmum 256
[3] nepalensis / P / to 1' / fls blue / Nepal

---

[1] Notes from Royal Botanical Garden, Edinburgh, vol 27, p. 7, 1964.

[2] Originated Arends Nursery, Germany, William Dress, 1985, correspondence.

[3] William Dress, 1985, correspondence.

napellus / P / 3-4' / fls purple to blue / Eurasia 148 / 129
napellus 'Albus' - almost white fls 208
napellus 'Blue Sceptre' / 2½' / blue & white fls 49
napellus judenbergensis - see A. variegatum 286
napellus ssp. napellus / P / to 1' / fls violet-blue / w & wc Europe
napellus paradoxum - see A. delphinifolium 209
napellus tauricum - see A. tauricum 286
neomontanum - see A. napellus 286
novoluridum / P / 2½' / red-purple fls / e Himalayas 25
orientale / P / tall / yellowish white fls / Caucasus 25
palmatum / HH P / 2-5' / fls large, greenish-blue / Sikkim, Garwa 81
paniculatum / P / to 4' / fls violet-blue / s Europe 123
pulchellum / R / 6-9" / deep purple-blue / Yunnan 67
pyranaicum - see A. lamarckii 286
ranunculiflorum - see A. lamarckii 286
reclinatum / P / 3-6' / fls white / Alleghenies 263
sachaliensis / P / 2-3½' / many fls, bluish-purple / Hokkaido 200
senanense / P / 2-4' / fls bluish-purpl / Japan, in alpine region 165 / 200
septentrionale / P / 3-6' / fls dark violet / nc Europe 148 / 286
tauricum / P / to 3' / fls blue to violet / e Alps, Roumania 286
tauricum nanum - dwarf form of A. tauricum 286
uncinatum / P / 3-5' / fls blue; stem weak / ec United States 263
uncinatum ssp. muticum / P / 3'+/ blue fls; 5-lobed lvs/ Pa. to N.C. (492)
variegatum / P / 1-6' / fls blue to white, variegated / c Europe 286 / 25
variegatum v. gracile - 4-5' / fls blue to white / Hungary 292
volubile / P / 3', trailing / fls bluish-purple / Korea, Manchuria 200
vulparia / P / to 5' / fls yellow / c Europe 219 / 274
yamazakii - see A. senanense 165
yuparense / P / stems 15-20" / bluish purple / alpine slopes, N. Hem 200

ACORUS - Araceae
calamus / Aq P / 3' aromatic rhizome / Sweet Flag / Eurasia 99

ACTAEA - Ranunculaceae
alba - see A. pachypoda 99
arguta - see A. rubra ssp. arguta 7
arguta alba - see A. rubra f. neglecta 48
arguta neglecta - see A. rubra f. neglecta 48
arguta rubra - see A. rubra ssp. arguta *
asiatica / P / 16-28" / white fls; black frs / Far East 165 / 64
eburnea - see Actaea rubra f. neglecta 99
erythrocarpa / P / 1-2' / small red frs / ne Europe 286 / 200
pachypoda / P / 12-18" / dull white frs / ne North America 224 / 107
pachypoda alba / P / berries wht / Nov. Sco. - Ga., Minn., Mo. 25
pachypoda f. rubrocarpa - frs dull red 99
rubra / P / 18-24" / shining red frs / ne North America 279 / 107
rubra alba - see A. rubra f. neglecta *
rubra arguta / P / 1-2' / berries spherical, red / w N. America 25
rubra f. neglecta - frs glossy, ivory-white 279 / 99
spicata / P / 1-2' / frs. purplish-black / Eurasia 210 / 25
spicata 'Alba' - either A. spicata or A. pachypoda 25
spicata v. leucocarpa - see A. erythrocarpa 200
spicata nigra - see A. asiatica 200
spicata f. rosea - as above except frs. pinkish 96
spicata rubra - see A. rubra 25

ACTINEA
angustifolia - see HYMENOXYS scaposa 25
chenensis / Sh / ? / fls orange-yellow, Kiwi berry / New Zealand 25
grandiflora - see HYMENOXYS grandiflora 25
herbacea - see HYMENOXYS acaulis 25

ACTINELLA
acaulis - see HYMENOXIS acaulis 25
grandiflora - see HYMENOXYS grandiflora 25

ACTINIDIA ACTINIDIACEAE
arguta / Cl / high / fragrant fls; edible frs / Far East 36

ACTINOTUS - Umbelliferae
helianthi / HH B / 2' / woolly plant; cream fls / N.S. Wales, Astrl 194 / 45

ADENANDRA - Rutaceae
fragrans / HH Sh / 2-3' / rosy fls / Cape of Good Hope 28
umbellata / Gh Sh / 1-2' / fls pink / Cape of Good Hope 223 / 81
uniflora / Gh Sh / small / light pink fls; aromatic lvs / S Africa 223

ADENOCARPUS - Leguminosae
decorticans / T / to 25' / white bark; gold-yel. fls / Spain 25

ADENOPHORA - Campanulaceae
bulleyana / P / 3-4' / fls pale blue / w China 30
coelestis / P / 4" to 2' / fls white, blue, violet / Yunnan 67
confusa / P / 3' / deep blue fls / w China 36 / 25
divaricata/2-3'/fls blue, style longer than cor./Honshu, Kor. Manch. 200
farreri - see A. confusa 25
forrestii - see A. coelestis 67
himalayana / R / to 16" / fls blue-lavender / Himalayas 93
khasiana / P / 2' / fls violet / ne India (1)
koreana / P / to 3' / corolla blue / Korea 25
latifolia / P / 18" / blue fls / Dahuria 61
lilifolia / P / 1½' / fls pale blue / c Europe to Manchuria 182 / 25
nikoensis / R / 8-16" / fls blue / alps of Japan 164 / 200
nikoensis alba / P / to 1½' / white form, corolla bluish / Japan 25
nikoensis f. macrocalyx - high mt. form of the sp. 273
nikoensis nipponica - see A. nikoensis v. stenophylla 25
nikoensis v. petrophila / P / to 1½' / stems & lvs elongated/ Japan 200
nikoensis v. stenophylla - alpine var., calyx lobes entire 25
nikoensis v. linearifolia f. leucantha - white fls; narrow lvs *
ornata / P / 2-4' / deep blue fls, open trumpet / se Tibet, w China 30
pereskeaefolia - see A. latifolia 61
pereskiaefolia v. hererotricha/P/12-20"/dense flr raceme/Hok.,Hon. 200
pereskiaefolia v. heterotricha 'Alba' / white form of above 200
polymorpha / P / 2-4' / fls pale blue-violet / ne China 30
polymorpha tashiroi - see A. tashiroi
potaninii / P / weak-stemmed; fls violet-blue / w China 93 / 25
remotiflora / P / 28-40" / fls bluish / Far East 164 / 200
Takedae / P / to 2' / corolla violet blue / Japan 200 / 25
Takedai v. howozana / P / 2-6" / flowers smaller / Honshu 200

Takedai v. howozana 'Alba' - white form 200
tashiroi / R / 4-12" / fls blue-violet / Kyushu, s Korea 200
tetraphylla - see A. triphylla 200
triphylla / P / 16-40" / fls blue / Kyushu, Formosa, China 200
triphylla ssp. aperticampanulata - see A. triphylla v. japonica 200
triphylla v. hakusanensis / P / to 1½' / dense fls / Alpine Japan 200 / 25
triphylla v. japonica / 3' / pale bluish-violet fls / Japan 25
triphylla v. japonica 'Alba' / P / to 3' / wht form, 1" long/ Japan 25

ADENOSTYLES - Compositae
albifrons - see A. alliariae 93
alliariae / P / 2-6' / reddish-purple fls / Europe, in mts. 148 / 289
alpina / P / 12-20" / light purple fls / Alps 289
glabra - see A. alpina 148 / 289
leucophylla / R / 1' / pink-purple fls; whitish lvs / Alps 148 / 289

ADESMIA - Leguminosae
glomerata / R / scree plant / hairy lvs; small fls / Andes Mts. 64

ADIANTUM - Polypodiaceae
capillus-veneris / Gh F / 6-10" / evergreen fronds / s temp. zone (424) / 128
pedatum / F / 1-2' / for shade & moisture / n US, s Canada 147 / 123
pedatum v. aleuticum / 4-5" / glaucous / Aleutian Isls., Canada (250) / 159
pedatum f. imbricatum - compact form from Washington 25
pedatum minor - see A. pedatum v. aleuticum 279
trapeziforme f. pentdactylon - decurved lower lf. margin, tender 61

ADINA - Rubiaceae
rubella / HH Sh / small / pinkish-purple fls / se China 25

ADLUMIA - Fumariaceae
fungosa / A Cl / high / fls white to purplish / e US 224 / 313

ADONIS - Ranunculaceae
aestivalis / A / 4-16" / fls red / Europe 286
aestivalis ssp. parvifolia - smallish scarlet fls; annual of deserts 5
amurensis / R / to 1' / fls yellow / Far East 147 / 200
annua / A / 4-16" / scarlet fls / s Europe, sw Asia 211 / 25
brevistyla / R / to 1' / fls white, striped blue / Yunnan, Tibet 268
Chrysocyathus / R / 6-9" / fls red or yellow / Himalayas 25
distorta / R / to 8" / yellow fls / c Apennines, Italy 147
distorta tenore - Tenore is the author designation of the sp. *
pyrenaica / R / 12-16" / fls golden yellow / e Pyrenees 148 / 274
turkestanica / R / to 8" / fls intense yellow / c Asia 5
vernalis / R / to 8" / fls yellow / Europe 147 / 286

AEGILOPS GRAMINAE
ovata / A Gr / 4-12" / for dried decorations / Medit, reg. 61
squarrosa - see TRITICUM aegilops 160

AEGINETIA OROBANCHACEAE
indica / root parasite / short/ few pale-purple striate calyx/Japan 200
indica v. gracilis /R/to 1'/few lg fls, pale rose-purp/ s Asia,Jap. 200

AEONIUM - Crassulaceae
arboreum / Gh / 2-3' / fls bright yellow / sw Asia 77
canariense / Gh / to 2' / pale green fls / Canary Isls. 156
glandulosum / Gh / to 1' / fls golden yellow / Madeira 156
Haworthii /Sh/1-2'/bl-gr lvs, red-edge; fls yell., rose/Canary Isles 25
spathulatum v.cruentum / Gh Sh / 2' / yellow fls / Canary Isls. 25

AERVA - Amaranthaceae
tomentosa / Gh Sh / 1' / whole plant white-woolly / Israel (4)

AESCULUS - Hippocastanaceae
parviflora / Sh / to 15' / fls white / Georgia, Alabama 184 / 25
woerlitzersis / T / 50' / red fls / probable hybrid origin 222

AETHIONEMA - Cruciferae
antilibani - see A. grandiflorum 61
arabicum / A / 3-6" / minute lilac fls / se Europe, sw Asia 25
armenum / R / 5" / fls pink / e Anatolia, Transcaucasia 91 / 286
armenum 'Warley Rose' - see A. X warleyensis *
cordatum / R / to 8" / fls pink, white, yellow / Anatolia 77
coridifolium / R / 6-8" / fls pink / Lebanon, Turkey 182 / 268
creticum - see A. saxatile 286
diastrophis / R / to 10" / pale pink fls / Transcaucasia 268
edentulum / R / 8-12" / pink fls / Caucasus 6
graecum - see A. saxatile 268
grandiflorum / R / 10-18" / fls rose to pink / Russia, Iran 269 / 268
grandiflorum pulchellum / P / 12" / petals pink / Turkey, Iraq, Iran 25
iberideum / R / 6" / fls white / Turkey 184 / 268
kotchyi / R / dwarf / rich pink fls / e Europe, Middle East 154
oppositifolium / HH R / dwarf cushion / fls pink or lilac / Syria 77
ovalifolium - see A. saxatile 77
persicum - see A. grandiflorum 268
pulchellum / R / 6" / fls bright pink / Asia Minor 133
recurvum - see A. armenum 77
saxatile / R / 6" / fls white or pink / E Medit. reg. 148 / 268
saxatile f. gracile - form from Crete with larger fls 96
schistosum / R / 5" / fls pink, petals clawed / Cilician Taurus 77 / 25
speciosum / R / 6-10" / fls pink / Iraq, Iran 77
stylosum / R / 12" / fls pink, white, lilac / w Syria 268
theodorum - see A. iberideum 268
Thomasianum - differs in frs only from A. saxatile 286 / 25
'Warley Rose' - see A. X warleyensis 129 / *
X warleyensis / R / 3-6" / pale to deep rose fls 48 / 25

AGAPANTHUS - Liliaceae
africanus / HH P / to 2' / fls deep blue to blue-violet / Cape Prov. 128 / 268
campanulatus / HH P / 18" / fls blue or white / Natal 128 / 268
campanulatus ssp. patens / narrower lvs; smaller plant / Basutoland 268
comptonii / Gh P / to 4' / blue fls / e Cape Province 268
'Headbourne Hybrids' / 2-3' / hardy at Wisley / pale to violet-blue 129 / 268
'Headbourne Worthy' / 2-3' / hardier hybrid race 129
inapertus / HH P / to 6' / fls deep to violet-blue / Transvaal 268
'Mooreanus' / 15-18" / dark blue fls / A. campanulatus hybrid 121 / 268
orientalis / HH P / to 2'+/ fls funnel form, blue evergr/ S. Africa 25
orientalis albus - see A. praecox ssp. orientalis 268

'Peter Pan' / HH P / 18" / sky-blue fls 202
praecox ssp. minimus / HH P / 2' / pale blue fls / S. Africa 268
praecox ssp. orientalis / HH P / 4' / blue fls / S. Africa 268
praecox ssp. orientalis 'Albus' - fls white 268
umbellatus - see A. orientalis 25

AGASTACHE - Labiatae
anisata - see A. foeniculum 99
cana / P / to 2' / corolla pinkish / N. Mex. & w Texas 25
Foeniculum / P / 1-3' / fls blue; lvs anise-scented / n, c N. Am. 224 / 58
mexicana / HH P / 1-3' / fls bright pink / Arizona, Mexico 93
urticifolia / rel.to A Foeniculum/ corolla 3/8-5/8"/ B.C.,Mont,Cal. 25

AGATHOSMA - Rutaceae
rugosa / Gh Sh / 1-2' / white fls / Cape of Good Hope 81

AGAVE - Liliaceae
palmeri / HH P / large / greenish & purple fls / New Mex., & s-ward 227
parryi / HH P / ? / red & yellow fls / Ariz., N. Mex., Texas 226 / 227
schottii / Gh / lvs 1'; fls 6' / fls yellow / N. Mex., Mexico 227
utahensis / Gh / lvs 1'; fls 15' / fls yellow / Utah to Mexico 227
virginica - see MANFREDA virginica 99

AGERATUM - Compositae
Houstonianum /A/2½'/fls bl-lilac-lavender/c, s Mex., Guat., B.Hond. 25

AGONIS - Myrtaceae
flexuosa / HH T / 30' / fls white / Australia 93
linearifolia / HH SH / to 12' / white fls / w Australia 25

AGOSERIS - Compositae
aurantiaca / P / to 2' / fls burnt orange / New Mexico to Canada 227 / 25
glauca / P / to 2½ / fls yellow, aging pink / Rocky Mts. 229 / 25
glauca v. dasycephala / P / 4-29"/ yellow fls / Calif. & Washington 228

AGRIMONIA - Rosaceae
gyrosepala / P / 4-60" / small yellow fls / temp. US & Canada 263
pilosa / P / 2-5' / pale yellow fls / n Eurasia 287

AGROSTEMMA - Caryophylaceae
alpina - see LYCHNIS alpina 25
githago / A or B / 3'+ / fls magenta-purple / e Medit. region 229 / 99
githago 'Nana' - dwarf form

AILANTHUS - Simaroubaceae
altissima / T / 50'+ / Tree of Heaven (or Brooklyn) / n China 139

AJUGA - Labiatae
chamaepitys ssp. chia / R / to 1' /fls yellow marked red, purp./ Eu 288
genevensis / R / 6" / fls deep blue / Europe 148 / 123
genevensis 'Alba' - the rarely white fls 288
genevensis brockbankii / P / 6" / deep blue fls / Europe 184
pyramidalis / R / to 1' / fls pale violet-blue / Europe 288
pyramidalis ('Metallica') 'Crispa' / 5" / crinkled, colored lvs 210 / 25
reptans / R / 1' / fls blue, rarely pink or white / Europe 288

AKEBIA - Lardizabalaceae
quinata / Cl / 30'+ / fragrant red-purple fls / Far East 184 / 139
trifoliata / Cl / 25' / purple fls; violet frs / Far East 167 / 139

ALANGIUM - Alangiaceae
Chinense / HH Sh / 6'+ / small white fls / c China 139

ALBIZIA - Leguminosae
julibrissin / HH T / 30' / ball of pink stamens / Near East 187 / 36
lophantha / HH Sh or T / to 20' / fls sulphur-yellow / w Australia 139

ALBUCA - Liliaceae
altissima / Gh Bb / 18-24" / white petals, green backed / S. Africa 267
canadensis / HH Bb / to 3' / fls yellow & green / Cape of Good Hope 161 / 223
cooperi / Gh Bb / 6-18" / green & yellowish fls / Cape Peninsula 161
elwesii / Gh Bb / 12" / pale green fls / e Africa 111
humilis / HH Bb / 4" / fls white with markings / Lesotho (14) / 185
spiralis / Gh Bb / 9" / white fls / Cape of Good Hope 206
wakefieldii - see A. elwesii 111

ALCEA - Malvaceae
ficifolia / may be A. rosea if yellowish fls, or A. rugosa 25
pallida / Bb / to 6' / fls rose or lilac / se & c Europe 25
pisidica / P / tall / petals yellow, calyx striate/ Turkey 77
rosea / Bb/ to 10' / Hollyhock, white to pink-orange / Asia Minor 25
rugosa / B / to 10' / petals clr yel. or orng. yel./ Ukrane, s Rus. 25

ALCHEMILLA - Rosaceae
alpina / R / 8" / fls greenish; lvs silky / Europe 148 / 25
caucasica / dwarf to 4" / stems hairy; lvs kidney-shaped/Turkey 77
conglobata / P / to 14" / large lobes long incised / n U.S.S.R. 287
filicaulis ssp. vestita / to 16" / purp-red, hairy / nw Europe 287
glaucescens / P / to 8" / fls minute, yellowish / Europe 25
japonica / R / to 1' / fls green, lvs hairy / Japanese alps 200
major - see A. vulgaris 160
mollis / P / to 2½' / fls yellowish / e Carphathian Mts. 49 / 287
orbiculata - see A. conglobata 287
pubescens - see A. glaucescens 25
saxatilis / R / rhizomatous / palmate green lvs / s & c Europe 314 / 287
vulgaris / P / 1½' / yellowish fls / Europe 314 / 25
xanthochlora / P / 2' / fls yellowish / Europe 287

ALETES - Umbelliferae
acaulis / P / 2-14" / spiny lvs; fls yellow / Tex. to Col & N.M. 227

ALETRIS - Liliaceae
farinosa / P / 3' / tubular white fls / Maine to Texas 224 / 25
foliata / R / 8-15" / fls yellow-green / Japan, in mts. 273 / 200

ALISMA - Alismataceae
gramineum / Aq / on land, 4" / purplish to white fls / n hemisphere 99
lanceolatum / Aq / 1-3' / mauve fls; narrow lvs / Europe 301
plantago-aquatica / Aq / to 3' / lilac or roseate fls / Eurasia 138 / 245
plantago-aquatica v. americanum / all white fls / N. Am. 245

plantago-aquatica v. parviflorum / fls smaller / e N. Am. 245
subcordatum - see A. plantago-aquatica v. parviflorum 224 / 245
triviale - see A. plantago-aquatica v. americanum 224 / 25

ALKANNA - Boraginaceae
graeca / R / to 1' / bright yellow fls / Greece 211 / 288

ALLIARIA - Cruciferae
petiolata / B / to 4' / white fls; Garlic Mustard / w Europe 210 / 286

ALLIUM - Liliaceae
acuminatum / R / 6-12" / pink, lilac, purple fls / w N. Am. 229 / 267
affine - see A. vineale (3)
aflatunense / Bb / 2½-5' / light violet fls, darker nerve / c Asia 129 / 25
akaka / R / 1-4" / whitish fls; glaucous lvs / Iraq, Iran, Turkey (15) / 185
albidum / R / 4-12" / fls yellowish-white / Russia, Balkans 290
albopilosum - see A. christophii
album / P / 18" / pure white fls, hardier than A. neapol. / Italy 111
alleghaniense - see A. cernuum 25
altaicum / Bb / 1-2' / fls yellowish / c Asia 4
altissimum / Bb / 2-3½' / fls light violet / c Asia 4
1 amabile / R / 4-8" / fls rose to magenta-red/ ? A. mairei / w China 267
amblyophyllum - see A. platyspathum 4
amethystinum - see A. ampeloprasum 160
ampeloprasum / P / 16-32" / fls purple or rose / Europe, Asia 271
amphibolum / R / 4-8" / fls rose or rose-purple / Siberia 4
amplectens / P / 8-20" / fls white or pink/ w North America 227
angulosum / P / 10-20" / fls rosy-violet / c Europe to Siberia 4
anisopodium / R / 8-16" / rosy fls / Mongolia, China, Japan 4
ascalonicum - horticultural variant of A. cepa 268
atropurpureum / Bb / to 3' / very dark purple fls / s Europe 186 / 25
atrorubens / R / 3-6" / reddish-purple fls / Nevada, California 227
atroviolaceum / Bb / 2-3' / fls dark purple-violet / c Asia 4
beesianum / Bb / 18" / fls deep blue, edged white / China 267 / 25
bidwelliae - see A campanulatum 25
Bolanderi / R / to 8" / fls rose-pink / s Ore., n Cal. 25
brandegei / R / 4" / fls white / n Rocky Mts. 229
brevistylum / Bb / 1-2' / fls dark rose / Montana, Utah, Colorado 229 / 25
bucharicum / Bb / 1' / fls dirty rose to white / Afghanistan (16)
bulgaricum - see NECTAROSCORDUM siculum ssp. bulgaricum 290
caeruleum / Bb / to 2½' / blue fls / Asia 267 / 25
caesium / Bb / 6'-2" / fls azure blue / c Asia 4
callimischon / HH Bb / to 4" / fls whitish, pinkish / s Greece 185
calocephalum / Bb / to 12" / creamy-yellow fls / Middle East (15) / (16)
campanulatum / R / 8-12" / pale pink fls / California ranges 229 / 227
canadense / Bb/ to 2' / fls pinkish or all bulbils / e & c N. Am. 229 / 99
canadense v. canadense - see A. mutabile
2 canadense v. fraseri / fls lavender-pink / sw U.S.
canadense v. lavandulare - see A. canadense v. fraseri

---

1 Xu Jei-Mei, Flora Reipublicae Popularis Sinicae, Tomus 14, 1980.

2 Donovan Stuart Correll and M.C. Johnston, Manual of the Vascular Plants of Texas, 1970.

canadense v. mobilense - fls pink, no bulbils 25
[1] cardiostemon / P / 18-24" / many small deep purp. or red fls / Iran 121
carinatum / P / 16-20" / fls reddish-violet / Europe 210 / 4
carinatum ssp. pulchellum /umbel no bulbils; fls many/s Eur.-se Fr. 290
caspium / Bb / to 1' / fls greenish-violet / Russia 118
cepa - Onion of the vegetable garden 25
cepa v. ascalonicum - note A. ascalonicum reference in 25
cernuum / P / 12-18" / fls deep rose to purple / N. Am 224 / 129
cernuum 'Major' / 18" / deep pink selection *
cernuum oxyphilum - see A. oxyphilum 263
cernuum f. purpureum - several shades deeper in color *
chamaemoly / Bb / 4" / wht with red nerves / s Europe. n Africa 25
christophii / Bb / 6"-2½' / lilac fls / Iran, Asia Minor 129 / 25
cilicicum - see A. rotundum 4
condensatum / Bb / 1-2' / fls pale yellow / Siberia, Far East 4
cowanii - see A. album 111
crenulatum / R / 6-8" / pink fls / Oregon & n-ward 228 / 142
cupani / 4-12" / whitish or pink / Mediterranean to S. Balkan 290
cyaneum / R / 9-12" / fls violet or purplish-blue / China 267 / 25
cyaneum v. brachystemon - see A. kansuense 61
cyaneum macrostemon - see A. macrostemon 61
cyathophorum / P / to 18" / fls blue-violet / w China 121
cyathophorum v. farreri / 12" / fls reddish-purple / Kansu 267 / 25
cyrillii / R / 15" / fls white or rose / Italy to Asia Minor 25
decipiens / Bb / to 20" / fls pale rosy-violet / Caucasus, c Asia 4
delicatulatum / R / 6-18" / fls whitish to rose / w Siberia, c Asia 4
derderianum / R / 6" / stellate white fls, violet nerved / Iran 4
descendens - much confused name to be abandoned (18)
dichlamydeum / R / 4-12" / fls deep rose-purple / c California 228 / 25
dictuon / B / lvs 2; fls bright pink / Col. Co., Wash. 142
dioscoridis - see A. siculum 185
douglasii / R / 8-10" / fls rose-pink / Wash., Idaho, Oregon 121
drummondii / R / 6-8" / fls white to red, or yellow / sc US, Mexico 227
elatum / Bb / 2-3' / fls reddish-violet / c Asia (410)/ 4
ericetorum - see A. ochroleucum
falcatum - see A. platyspathum v. falcatum (4)
falcifolium / R / 3" / fls pinkish-purple / w N. Am. 228 / 267
farreri - see A. cyathophorum v. farreri 267
fetisowii / Bb / 16"-2' / fls rose / c Asia 4
fibrillum / R / to 4" / white to pale rose fls / Idaho 142 / 25
fimbriatum v. purdyi - see A. purdyi 25
fistulosum / P / 2-3' / Welsh Onion; fls yellowish-white / of cult (416)/ 25
flavescens / R / to 1' / fls pale yellowish / Europe, Asia 4
flavum / P / 1½-2' / bright yellow fls / Eurasia 148 / 25
flavum 'Minor' / 4" / found in s Europe & Caucasus (34) / 123
flavum v. minus - dwarf form 185
flavum 'Nanum' - dwarf form of A. flavum 185
flavum 'Pumilum Roseum' - dwarf mt. form of A. pulchellum 185
forrestii / R / 4-16" / fls claret-red to purple / w China, Tibet (351) / 67
frigidum / R / to 6" / fls reddish-yellow / Greece 290 / 25
fuscoviolacea / Bb / 12-28" / fls dark purple / Caucasus, Iran 4
galanthum / Bb / to 20" / fls white / c Asia 4

---

[1] Flora Iranica, 1971.

geyeri / Bb / to 20" / pink or rarely white fls / Texas to Alberta 226 / 25
giganteum / Bb / to 4' / fls lilac / Iran, c Asia 267
glaciale / R / 8-12" / rose fls in dense umbel / c Asia 4
goodingii / 14-18" / broad flat lvs; fls pink / Arizona 227
grandiflorum - should be attached to sp. name *
griffithianum / Bb / 1'+ / light carneous-red fls / c Asia 4
[1] haemanthoides / R / short-stemmed / stiff rose-purp. petals/Asia Min
haematochiton / Bb / to 1½' / deep purple or rose fls / s Calif. 227 / 25
heldreichii / Bb / to 1½' / large bright pink fls / Thessaly, Greece 121
hirtifolium / B / 3' / globe shaped heads, mauve-purple fls/ Iran 154
hookeri / P / 9-16" / white to greenish-white fls / India, China 121
humile / Bb / / dwarf white fls / Kashmir 154
huteri - a form of A. senescens 25
hyalinum / R / 6-12" / white or pinkish fls / Sierra Nevada, Calif. 227
hymenorrhizum / P / 1-3' / rose fls / w Siberia, c Asia 25
insubricum / R / 6-12" / purple campanulate fls / s Alps, on limestone 290
jajlae - see A. scordoprasum ssp. jajlae 290
jesdianum / Bb / 2' / fls small, starry, purple / Iran 121
kansuense - see A. sikkimense
karataviense / R / 6-9" / fls silvery-white / Turkestan 267 / 129
Kermesinum / B / 4-12" / fls red / se Alps 290
komarovianum - see A. thunbergii 166
kunthii / R / to 1' / white or pink fls / w Texas, se Arizona 227
ledebourianum / P / to 2' / fls rose / Siberia, Danuria 4
lemmonii / R / 5-8" / fls pale pink / California 121
leucanthum / Bb / 2-4' / white fls, green-nerved / Caucasus 4
loscosii - see A. sphaerocephalon v. A. loscosii (var, bulbils in umbel) 290
luteolum / Bb / 1-8" / many flowered, yellowish head / Greece 290
macranthum / P / 1-2' / fls dark purple / Sikkim, China 25
macrostemon / P / to 32" / fls pink, rose, purple / e Asia 166 / 258
macrum / R / 4-16" / fls white or rose / Blue Mts., Ore. & Wash. 229
mairei / R / 4-5" / fls pale pink / Yunnan 123
materculae / R / 4-12" / fls lilac, reddish-nerved / Caucasus 4
maximowiczii / P / 6-24" / lustrous rose fls / Far East 273 / 4
mirum / Bb / ? / wht to purp fls with dark stripe / Afghanistan 154
moly / Bb / 9-12" / fls bright yellow / sw Europe 212 / 129
moly v. bulbilliferum - rarely with bulbils, few fls 290
moly 'Bulbosum' - see A. moly v. bulbilliferum
montanum - see A. senescens v. montanum 314 / 4
monophyllum / R / dwarf / rose-violet fls / Kopet Dagh 4
multibulbosum - see A. nigrum 290
murrayanum - see A. unifolium
mutabile - see A. canadense v. mobilense (2)
narcissiflorum / R / 8-12" / fls deep rose / Italian Alps 148 / 267
narcissiflorum 'Grandiflorum' - large-flowered selection *
neapolitanum / HH Bb / to 18" / fls white /s Eu., n Africa, A.Minor 267 / 25
neapolitanum 'Grandiflorum' / 12" / fls white / larger 18
neriniflorum / P / ½' / rose-violet / Chusan, Mongolia 206
nigrum / P / 1-2' / fls white or pale lilac / s Europe 290
nigrum v. multibulbosum - with the bulb producing bulblets 181
nutans / P / 1-2' / fls rose to rosy-violet / Siberia 4
nuttallii - see A. drummondii 25

---

[1] Flora Iranica, 1971.

obliquum / P / 2-2½ / fls greenish-yellow / Europe, Asia 4
obtusum / R / 2" / fls greenish-white / Sierra Nevada, Calif 229 / 227
ochroleucum / R / 10-12" / fls yellow / s Europe 314 / 121
odorum - see A. ramosum 25
olympicum / R / 2-6" / fls pink or crimson / Turkey (14)
oreophilum / Bb / 3" / fls purple / Caucasus, Daghestan 96
oreophilum v. ostrowskianum - see A. ostrowskianum 25
oreophilum v. ostrowskianum 'Zwanenburg' - sel. for pink fls 185 / 129
oreoprasum / Bb / 15-18" / fls rose / c Asia 4
orientale / Bb / 15-18" / starry white or reddish fls / Asia Minor 121
ostrowskianum / Bb / 8-12" / rose / Turkestan 25
oviflorum - see A. macranthum 64
oxyphilum / P / 12-18" / fls white / Virginia & West Virginia 313
pallens / R / 4-12" / fls white or pink / s Europe 290
pallens purpureum - see A. paniculatum 29
paniculatum / P / 12-18" / fls many colors / Eurasia, n Africa 4
paradoxum / R / 8-12" / nodding white fls / Caucasus, n Iran 121
parryi / R / 3-8" / fls rose-purple / San Bernardino, Calif. 121
pedemontanum - see A. narcissiflorum 96
pendulinum / HH R / 6" / white fls / s Italy, Corsica 206
peninsulare / Bb / to 1' / fls deep rose-purple / California 227 / 25
perdulce / R / dwarf / fragrant rosy-purple fls / Neb. to Texas 25
persimile - see A. tolmiei v. persimile 142
petraeum / P / 1-1½' / fls pale yellow / c Asia 4
platycaule / R / 6" / fls deep rose / n Cal., s Oregon 228
platyphyllum - see A. tolmiei v. platyphyllum 142
platyspathum / Bb / 4-28" / fls rose / w Siberia, c Asia 4
platyspathum v. falcatum / 1'+ / globose head / rosy-lilac/Turkestan (5)
pleianthum / R / 4-6" / fls white / Oregon, Idaho 121
plummerae / HH P / 1-2' / fls white or pink / se Arizona, Mexico 227
polyastrum / Bb / 2' / fls magenta-rose / w China 67
polyphyllum / P / 15-24" / fls rose / c Asia 4
porrum - the garden Leek, distinct after centuries of cultivation 290
pseudoampeloprasum / Bb / 2' / fls rosy, purple nerved / Caucasus 4
pseudoflavum / R / 6-10" / yellow fls / Caucasus, Iran 4
pskemense / Bb / 2-2½' / fls wht, green nerve / c Asia, mts Turkey 4
pulchellum - see A. carinatum ssp. pulchellum
pulchellum 'Album' - fls white (15) / (16)
purdyi / R / to 1' / pale pink fls, darker nerve / California 25
pyrenaicum / Bb / 1½' / white fls, green nerved / Pyrenees 25
ramosum / P / 6-20" / fls white / Siberia 121
rosenbachianum / Bb / 2-3' / fls dark violet / c Asia (329)/ 4
roseum / Bb / 1½' / rose fls, rarely white / s Eu., n Africa 25
roseum 'Grandiflorum' - selection for larger flower 181
rotundum - see A. scorodoprasum
rubellum / Bb / 6-18"/ fls rose, purple nerve / Persia, w Kazanhstan 290
[1] sarawschanicum / Bb / 10-28" / fls rose-violet / Afg. & c Asia
sativum - Garlic of the vegetable garden 25
saxatile / P / 9-20" / fls pale rose, wht & purp./ c Europe, c Asia 314 / 25
schoenoprasum / R / 6-9" / fls pale lilac; Chives / n hemisphere (15) / 129
schoenoprasum 'Album' - white fls 93
schoenoprasum alvarense - a dwarf form

---

[1] *Flora Iranica*, 1971.

schoenoprasum v. sibiricum - lvs shorter than flowering stalk 25
schubertii / HH Bb / to 1' / fls whitish or rosy / e Medit. reg. 4
scorodoprasum / Bb / to 3' / fls deep purple / Eur., Asia Minor (330)/ 25
scorodoprasum ssp. jajlae / fls uniformly rose-violet/ Crimea, in mts 290
scorodoprasum ssp. rotundum/10-40"/out ptl purp, in pale, wh marg/n Eu 290
senescens / P / 1-2' / fls rose to violet / w Eu. to Manchuria 4
senescens 'Altai-Sayan' / 12" / fls rosy-violet / c Asia 4
senescens v. glaucum / 12" / fls rosy-violet / Far East 300 / 107
senescens 'Huteri' - name of no botanical standing; var. A. senescens 25
senescens v. montanum / 18-24" / fls light to dark rose / Europe 4
senescens petraeum - see A. petraeum 4
serratum / Bb / 6" / rose-pink fls / c California 228 / 25
sibiricum - see A. schoenoprasum 209
siculum - see NECTAROSCORDUM siculum 129 / 25
[1] sikkimense / R / 10" / fls purplish-blue / Sikkim 121
simillimum / R / 3-4" / fls pinkish-white / Idaho, Montana 229
siskiyouense / R / 1-3" / fls rose or paler / n Calif., s Oregon 228
sphaerocephalum / P / 1-2' / fls dark purple / Europe, Asia Minor 149 / 267
splendens / Bb / to 20" / fls bright rosy-lilac / e Asia 25
'Stars of Persia' - catalog name for A. christophi 25
stellatum / Bb / to 18" / lavender-pink fls / prairie states 224 / 25
stipitatum / P / 2-3' / fls lilac / c Asia (426)/ 4
strictum / Bb / to2' / rose fls, purple nerved / Eurasia 314 / 25
suaveolens / R / 8" / fls white, flushed purple / c & s Europe 314 / 121
subhirsutum / HH Bb / 8-15" / fls white / Medit. region 211 / 25
subhirsutum ssp. album - stamens exceeding perianth 181
subvillosum - see A. subhirsutum ssp. album 181
taeniopetalum / Bb / / dark rosy-violet, green nerve / c Asia *
tanguticum / R / 6-8" / fls bluish-lilac / w China 121
tataricum - see A. ramosum 25
tel-avivense / HH Bb / 6-10" / fls lilac / Israel, Syria 271
tenuifolium - see A. anisopodium 64
textile / R / to 1' / fls white or pinkish, dark nerved / w N. Am. 229 / 25
thunbergii / Bb / to 2' / purplish fls / Japan 166 / 25
tibeticum - see A. sikkimense
togasii / B / 10 18" / fls white / Shikoku 200
Tolmiei / Bb / 4-6" / light rose-purple / Idaho to Oregon 25
tolmiei v. persimile / 10" / fls pink, longer stamens / Idaho 142
tolmiei v. platyphyllum / 10" / fls pink / Oregon, Idaho 142
tribracteatum / R / 5" / fls pale rose / California 227
tricoccum / Bb / to 1' / fls white / e US 224 / 313
triquetrum / R / 9-12" / fls white or greenish-white / s Eu. 212 / 129
triquetrum pendulinum - see A. pendulinum 181
tuberosum / Bb / tp 20" / white fls, colored nerve / China, India 166 / 200
turkestanicum / Bb / 3'+ / rose fls / c Asia 4
tytthanthum / R / 6" / fls pale yellow, aging rose / c Asia 4
unifolium / P / 8-24" / fls rose, rarely white / w Coast Ranges, US (22) / 228
ursinum / Bb / 6-20" / fls white; weedy / n Eu., n Asia 210 / 93
validum / Bb / 1½-2½' / fls rose to near white / Wash., Calif. 228 / 25
vavilovii / Bb / to 3' / white fls, green nerved / n Iran 4
victorialis / P / 1-2' / fls greenish-white / c Eu., c Asia 148 / 121
violaceum / 2' / fls violet / Hungary 206

---

[1] Xu Jie-Mei, Flora Reipublicae Popularis Sinicae, Tomus 14, 1980.

1 viride / B / 2-5' / fls green
wallichianum / Bb / 18-36" / rose-violet fls / c Asia 25
wallichii - see A. wallichianum 25
watsonii / R / 3" / fls bright rose-purple / n Cal., s Ore. (3)
winklerianum / Bb / 6-12" / fls rosy-violet / c Asia 4
zebdanense / Bb / 1-2' / fls white / Lebanon, in mts. 267

ALNUS - Betulaceae
crispa / Sh / to 10' / lvs aromatic when young / Labrador to N. C. 25
crispa maximowiczii - see A. maximowiczii 222
incana / T / to 60' / male catkins to 4" / Asia, Eur, intro to N.A. 25
maximowiczii / Sh or T / to 30' / fine-toothed lvs / Japan, in mts 167 / 36
oregona - see A. rubra 36
rubra / T / to 60' / early catkins / w N. Am. 36
sinuata / Sh or T / to 40' / lvs lustrous green beneath / w N. Am. 36

ALOE - Liliaceae
aristata / Gh / 4" / fls reddish-yellow / South Africa 93 / 25
cryptopoda / Gh / stemless / scarlet fls / Rhodesia, Nyasaland 268
marlothii / Gh / to 3' / orange fls / Bechuanaland 268
millotii / Gh / to 10" / fls scarlet / Madagascar 156
saponaria / Gh / rosette of 8" / fls red, orange, yel. / Natal 268

ALONSOA - Scrophulariaceae
incisifolia / P gr. as A / 2' / scarlet fls / Chile 25 / 208
linearia / Sh / 2-3' / fls brick-red / Peru 25
Warcewiczii / P gr. as A / 18-24" / fls scarlet-orange / Peru 128 / 208

ALOPECURUS - Gramineae
lanatus / R Gr / 6-8" / silver-gray lvs / dry areas, c Asia 93
pratensis 'Aureau' / P Gr / 1-5' / lvs yellow / Eurasia 210 / 268

ALOPHIA - Iridaceae
amoena - see TRIFURCIA lahue ssp. caerulea (23)
drummondii / HH C / 1' / blue fls / prairies, Texas, Louisiana 25
2 lahue / lavender blue fls, ctr wht, blotched viol-bl / Andes
platensis / Gh C / light porcelain-blue fls / La Plata, Argentina 28
pulchella / Gh C / 9" / blue-purple fls / Brazil, Chile 25

ALSTROMERIA - Amaryllidaceae
angustifolia / HH P / 1'+ / fls pale rose; lvs grass-like / Chile (231)
aurantiaca / P / 3' / fls orange / Chile 171 / 129
aurantiaca 'Dover Orange' / 3' / orange-red fls 129
brasiliensis / Gh P / 3' / reddish-bronze fls / Brazil 111
chilensis - a confused name 25
'Dover Orange' - see A. aurantiaca 'Dover Orange' 129
haemantha / HH P / to 3' / fls deep red, yellow streak / Chile 25
haemantha 'Rosea' / HH P / to 3' / fls rose / Chile 25
hookeri / HH P / 1½' / pinkish fls / Chile 25
ligtu / HH R / to 1' / fls deep red / Chile 75 / 25

---

1 Flora Iranica, 1971.

2 Quarterly Bulletin of the Alpine Garden Society, 45:2, illus.

ligtu v. pulchra - with narrower and longer lvs 28
'Ligtu Hybrids' / HH P / 2' / fls pink, orange, yellow 129 / 268
patagonica / HH R / under 1' / fls intense yellow / s Chile (231)
pelegrina / Gr P / 1'+ / pale lilac fls / Peru 25 / 268
pelegrina 'Alba' - white fls 261 / 25
psittacina / Gh P / 3' / dark red fls / Brazil 25
pulchella / Gr P / 3' / wine-red fls tipped green / Brazil 111
pulchra / HH P / 18" / fls white to pink / Chile 25
pygmaea / HH R / 1" / orange fls / Patagonia 19
recumbens / HH R / 9-12" / maroon & yellow fls / Chile (26)
spathulata / HH R / rosettes / pink fls / Andes Mts. (27)
versicolor / HH P / yellow to orange fls / Chile 25
violacea / HH P / 5' / mauve fls, spotted / Chile (26)

ALTERNANTHERA - Amaranthaceae
amoena / HH P / 6" / foliage bedding plant / trop. Am. 93

ALTHAEA - Malvaceae
armeniaca / P / 6'+ / fls lilac-pink; lobed lvs / c & sw Asia 287
officinalis / P / 4' / pink / Europe 25
pallida - see ALCEA pallida 25
rosea / P / 5-8' / varii-colored fls; Hollyhock / Orient 149 / 129
rosea cv. 'Summer Carnival' - see ALCEA rosea
zebrina - see Malva sylvestris 'Zebrina' 300 / *

ALYSSOIDES - Cruciferae
graeca - slight botanical difference from A. utriculata 286
sinuata / P / to 18" / fls yellow / nw Balkan Pen., e Italy 25
utriculata / R / 15" / fls yellow / Alps, Balkan Pen. 148 / 286
utriculata ssp. graeca - see A. graeca 286

ALYSSUM - Cruciferae
alpestre / R / to 8" / fls pale yellow / c & w Alps 314 / 286
arduinii - see A.saxatile 286
argenteum - see A. murale 25
armenum / R / to 8" / fls straw-yellow / Caucasia, Anatolia 77
atlanticum / R / to 8" / large yellow fls / s & c Spain 286
bertolonii / R / 8-12" / fls yellow / Italy, Balkan Pen. 286
bertelonii ssp. scutarinum / slightly distinct / nw Balkans 286
borzeanum / R / 4-12" / whitish lvs / Black Sea coast 286
compactum - see A. minutum 286
condensatum / R / to 8" / pale lemon-yellow fls / Syria, Lebanon 96
corymbosum / P / 8-20" / yellow fls / w & s Balkan Pen. 286
cunefolium / R / to 10" / almost white lvs / s Europe, in mts. 286
cunefolium pirinicum / P / 2-3½"/ fls yel, small/ mts. s Eur. 286
desertorum / A / 4" / pale yellow fls / Eurasia 25
'Dudley Neville' - see A. saxatile 'Dudley Neville' 123
handelii / P / 2-6" / silver-gray or gray-grn; no fls/ n Greece 286
idaeum / R / trailing / fls soft yellow / Mt. Ida in Crete 96
'Markgrafii' / R / to 1' / yellow fls / Albania, Jugoslavia 25
minutum / A / 5" / fls yellow / s & e Europe 286
moellendorfianum / R / prostrate / yellow fls / Jugoslavia 25
montanum / R / to 10" / fls bright yellow / e & s Europe 148 / 286
'Mountain Gold' - see A. montanum 203
muelleri - close to, or part of A. montanum 64

murale / P / to 2' / yellow fls / e Europe 25
ovirense / R / 5"+ / golden-yellow fls / c Europe 25
petraeum / P / to 2' / yellow fls; inflated seed-pod / Balkans 286
procumbens - name of no botanical standing 25
propinquum / P / 1-2" / petals pale / Turkey 77
pyrenaicum - see PTILOTRICHUM pyrenaicum 286
repens / P / 8-10" / fls golden-yellow / se & sc Europe 286
reptans / P / to 2' / fls large orange / s Europe 25
rochellii - see A. repens 25
rostratum / A or B / to 2' / fls yellow / se Europe 286
saxatile / R / to 15" / fls yellow / s & se Europe 50 / 256
saxatile 'Basket of Gold' / R / 3"+ / fls yellow 184
saxatile 'Citrinum' - pale yellow fls 129
saxatile 'Compactum' - 6-9" 300 / 46
saxatile 'Dudley Neville' - light buff fls (230) / 123
saxatile 'Basket of Gold' - see Aurinia saxatilis 25
saxatile 'Silver Queen' - lemon-yellow fls; silvery lvs 300 / 135
saxatile 'Variegatum' - variegated white, yellow and green lvs 93
scardicum / R / 2-8" / non-flowering stems / Balkan Pen. 286
serpyllifolium / R / trailing / fls pale yellow / sw Europe 286
spinosum - see PTILOTRICHUM spinosum 286
spinosum 'Roseum' - see PTILOTRICHUM s. 'Roseum' *
stribrnyi / R / 3-8" / fls yellow / E. Balkans 286
tortuosum / P / 3-15" / fls yellow / ec & se Europe 286
troodii / Sh / little/ silvered lvs; small yellow fls / Cyprus 64
wierzbickii / P / 15-30" / fls rich yellow / c Danube Basin 286
wulfenianum / R / trailing / fls pale yellow / se Alps 148 / 286

AMARACUS - Labiatae
dictamnus - see ORGANUM dictamnus 25
pulchellus - see ORGANUM pulchellum 25

AMARANTHUS - Amaranthaceae
caudatus / A / 3'+ / crimson tassels / tropics 129 / 25

AMARYLLIS - Amaryllidaceae
belladonna / HH Bb/ 18" / rose-red to white fls / South Africa 128 / 25

AMBERBOA - Compositae
muricata / HH P / to 2' / pink to purple fls / Morocco 25

AMBROSIA - Compositae
chamissonis / HH R / mats to 1' / silvery lvs / coastal Cal. to B.C. 142

AMELANCHIER - Rosaceae
alnifolia / Sh / 3-20' / white fls / w N. Am. (325) / 25
alnifolia v. semiintegrifolia /longer petals, may be tree-like/ w US 142
bartramiana / Sh / 2-6' / pure wht fls; purple-black frs / ne N. Am. 36 / 99
canadensis / Sh / to 18' / fls white / e N. Am. 269 / 99
laevis / T / to 40' / white fls; young lvs reddish / e US 313 / 25
oligocarpa - see Amelanchier bartramiana
ovalis / Sh / to 8' / fls in erect racemes / Europe 25
pumila / Sh/ 3' / fls white / Rocky Mts. 294
stolonifera / Sh / to 5' / white fls / ne N. Am. 25

AMELLUS - Compositae
annus / A / 6" / fls deep purplish-blue / South Africa 283

AMIANTHIUM - Liliaceae
muscaetoxicum / Bb / 3' / fls white / se & sc US 224 / 25

AMMOBIUM - Compositae
alatum / A / 18" / silvery-white fls, yellow center / Australia 45 / 129

AMORPHA - Leguminosae
canescens / Sh / to 4' / fls dull purplish-blue / e & c N. Am. 224 / 36
fruticosa /Sh/20'/6" dark purp fls vary-blue wht/Sask., s Fla., Mex. 25
glabra / Sh / to 6' / fls blue / N.C. to Ga, & Ala. s-ward 25
nana / Sh / 2' / fls purple, fragrant / Minn. to Rocky Mts. 229 / 36

AMORPHOPHALLUS - Araceae
rivieri / Gh Tu / 4' / dark purplish fls / Viet Nam (332) / 25

AMPELOPSIS - Vitaceae
brevipedunculata / Cl / high / frs amethyst-blue / Far East 36

AMPHICARPAE - Leguminosae
bracteata / P / twining / fls pale lilac, small / e N. Am. 263

AMSONIA - Apocynaceae
angustifolia - see A. ciliata 160
ciliata / HH P / to 5' / pale blue fls / N. C., Fla., Texas 225
elliptica / P / 15-30" / blue fls / Japan, Korea, China 200
hubrichtii / P / to 5' / blue fls / Ozark Mts. (506)
illustris / P / 2-3' / pale blue fls / sc United States 259
salicifolia - variety of A. tabernaemontana 25
tabernaemontana / P / to 3½' / fls pale blue / se US 224 / 25

ANACAMPSEROS - Portulacaceae
pyramidalis / P / 9-18" / fls bright rose-pink / Purbeck, Ireland 301
rufescens / Gh / 3-4" / fls pink / Cape of Good Hope 156 / 93
telephiastrum / Gh / 6" / reddish fls / South Africa 25

ANACAMPTIS - Orchidaceae
pyranidalis / P / 9-18" / fls bright rose-pink / Pyrenees, Alps 148 / 85

ANACYCLUS - Compositus
clavatus / A / to 20" / fls yellow / Medit. to Portugal & w Africa 289
depressus / R / 3-6" / fls white, red reverse / Morocco 129 / 25
maroccanus / A / prostrate / fls white, purple reverse / Morocco 25
pyrethrum maroccanus - see A. maroccanus 64

ANAGALLIS - Primulaceae
arvensis / A or B / to 20" / blue, red or paler fls / Europe 224 / 288
arvensis ssp. caerulea - see A. foemina 288
arvensis ssp. phenicia - included in the sp. 288
caerulea - see A. foemina 288
foemina - like A. arvensis / fls blue / s, w & c Europe 194 / 288
linifolia - see A. monellii 288

linifolia 'Monellii' - see A. monellii, a Linnean sp. 129 / 288
monellii / P gr. as A / to 18" / fls blue & reddish / Medit. reg 212 / 25
monellii ssp. collina / bright red fls / n Africa, s Spain 194
monellii ssp. linifolia - lvs linear 25
tenella / P / trailing / fls pink, rarely wht / w Europe 288

ANAPHALIS - Compositae
alpicola / R / 4-8" / bracts brown-red & white / alps of Japan 273 / 200
margaritacea / P / 18" / fls pearly-white / N. Am., ne Asia 147 / 129
nubigena - see A. triplinervics v. monocephala
sinica / P / 8-14" / fls snow white / mts. of Japan 200
triplinervis / R / 9-12" / waxy-white fls / Tibet 284 / 129
triplinervis v. monocephala /P /8"/yel, involucre wht/Alp,Himalayas 25

ANARRHINUM - Scrophulariaceae
bellidifolium / B / 18-24" / bluish-purple fls / sw Europe 210 / 25

ANCHUSA - Boraginaceae
angustifolia - see A. officinalis 288
angustissima / R / 1' / fls true blue / Asia Minor, Armenia 107
azurea / P / 3-5' / fls blue-purple / Caucasus 129 / 25
azurea 'Dropmore' / P / 3-5' / fls blue purple / Caucasus 184
barrelieri / P / 2' / bright blue fls / Eurasia 25
caespitosa / R / 9" / fls deep, clear blue / Crete 123
capensis / B / 1-2' / corolla blue marg. red, white throat/S. Africa 25
italica - see A. azurea
officinalis / B or P / 2' / bright blue, purple fls / Eu, Asia Minor 210 / 25
officinalis 'Angustifolia' - narrow lvs 25
sempervirens - see Pentaglottis sempervirens 25
undulata / B or P / 8-10" / blue-violet or purple fls / Medit. reg 212 / 312

ANDRACHNE - Euphorbiaceae
phyllanthoides / Sh / to 3' / fls inconspicuous/Mo. to Ark. to Tex. 25

ANDROMEDA - Ericaceae
nana - see ARCTERICA nana 200
polifolia / Sh / to 18" / fls pink / n hemisphere 148 / 36
polifolia v compacta/Sh/1'/fls in umbels, evergr./Eur, n Asia, n Am. 25
polifolia v. glaucophylla / hairier lvs / N. Am 36
polifolia v. glaucophylla f. latifolia - lvs a bit wider 221
polifolia 'Grandiflora' - Japanese selection, deeper-colored fls 36
polifolia 'Grandiflora Compacta' / to 1' / coral-pink fls / Japan (413) / 36
polifolia 'Grandiflora Minima' / 2" / fls deep rose-pink 132

ANDROPOGON - Gramineae
gerardii 'Big Blue' / P-Gr / to 7' / purplish / N. America 25
scoparius - see SCHIZACHYRIUM scorium 25

ANDROSACE - Primulaceae
aeizoon / R / 6-12" / varying pinks / c Asia 96
aeizoon coccinea - see A. bulleyana
albana / B / 3-10" / fls large, white or pale pink / Caucasus 252 / 10
alpina / R / 1" / fls in shades of pink / European Alps 148 / 123
X aretioides / R / 1" / bright pink fls 132
argentea - see A. vandellii 288

armeniaca / R / 4" / fls white, cream, pink; eyed / Iran, Turkey		77
brevis / R / dense cushion / pink fls / s Alps	148 /	288
brigantiaca - see A. carnea v. brigantiaca		
bulleyana / 12" / pink to intense red / Yunnan, Szechuan, Kashmir	252 /	87
carinata / P / to 2½" / white / c mts & plains of U.S.		25
carnea / R / 1-3" / fls pink or white / European Alps	148 /	107
carnea 'Alba' - the rarely white form, yellow throat		252
carnea ssp. 'Alba' - entirely white fls		132
carnea ssp. brigantiaca / scape to 6" / fls white or pink / sw Alps	252 /	25
carnea ssp. carnea /mat 2"+ rosette, pnk, yel. throat/w Alps, e Pyr.		252
carnea v. halleri - see A. carnea ssp. rosea		288
carnea v. laggeri /wider shorter lvs; lg rose fls, gold eye/e Pyrenees		123
carnea X pyrenaica - see X 'Pink Gin'		
carnea ssp. rosea / open, lgr fls, rarely wht/ c France, Vosges Mts		252
carnea 'Rubra' - presumably deep pink fls		*
chaixii / A / to 1' / white or pink fls / se France, in mts.	148 /	288
chamaejasme / R / 2" / small white fls / n hemisphere	148 /	123
charpentieri / R / 1" / soft shell pink fls / Switzerland		132
chumbyi - see A. primuloides 'Chumbyi'		268
ciliata / R / 1" / fls pale to deep rose / wc Pyrenees	148 /	123
cylindrica / R / 2" / milky-white fls/ w Pyrenees	148 /	10
cylindrica X A. hirtella - leaf rosettes larger than A. hirtella		252
elongata / A / 1½-4" / fls white / c Germany east to ec Russia		288
foliosa / R / to 5" / fls pink, yellow eye / w Himalayas	252 /	25
glacialis - see A. alpina		132
globifera / small / pale pink fls / Kashmir		87
halleri - see A. carnea ssp. rosea		25
hausmannii / R / caespitose / white to pink fls, yellow throat/ Alps	148 /	288
hedraeantha / R / ht. variable / fls white to violet-pink / Balkans	252 /	96
helvetica / R / dense cushion / sessile white fls / European Alps	148 /	129
hirtella / R / short / single white fls / w Pyrenees	148 /	132
hookeriana / R / creeping / white to purplish fls / Himalayas		64
imbricata - see A. vandellii		252
jacquemontii - see A. villosa v. jacquemontii		252
lactea / R / 8" / fls snow-white / mts. of Europe	148 /	25
lactiflora / A / 6-10" / fls white / Asia		107
laggeri - see A. carnea ssp. laggeri		
lanuginosa / R / 3-4" / fls pale to lavender-pink / Himalayas	(644) /	129
lanuginosa Leichtlinii /P/ prostrate/fls wht-yel, crimson eye/Himal		25
lehmaniana / P / fls 2-4, wht with yel ctr /Japan, N. Kor. to Alaska		200
mairei - is it A. carnea spp. brigantiaca?		252
mathildae / R / 4" / fls white / Abruzzi mts., Italy	314 /	132
maxima / A / 6" / fls white or pink / Europe	314 /	288
microphylla - of gardens is A. sempervivoides		64
X 'Millstream' - probably A. carnea x pyrenaica		
montana - see DOUGLASIA montana		252
mucronifolia - if pink fls, is A. sempervivoides		252
multiflora - see Androsace vandellii		252
muscoidea / R / 1' / large white fls / Kashmir	252 /	123
muscoidea f. longiscapa / mauve-pink to lilac-blue fls/ Tibet, Nepal		252
obtusifolia / R / 4"+ / fls white / Alps to Balkans	252 /	123
occidentalis / A / to 3" / insignificant fls / c North America		259
X pedemontana / R / 1" / white fls, orange eye / Pyrenees		25
X 'Pink Gin' - cultivar of A. carnea x pyrenaica	(31) /	87

primuloides / R / 2-4" / fls flesh-pink / Himalayas 93 / 268
primuloides 'Chumbyi' - neat silky form 24
primuloides v. watkinsii - included in the sp. 268
puberulenta - see A. septentrionalis v. puberulenta 25
pubescens / R / 2" / fls milky-white / w Alps, Pyrenees 252 / 25
pyrenaica / R / cushion/ fls white / wc Pyrenees 252 / 129
X pyrenaica - see A. carnea v. laggeri 87
pyrenaica 'Pink Gin' - see A. X 'Pink Gin' 252
rotundifolia / R / to 6" / pink fls, fading white / nw Himalayas 252
'salicifolia' / A of cult. / 2" / large clear white fls 252
sarmentosa / R / to 4" / rose to carmine-red fls /Kashmir to w China 25
sarmentosa v. chumbyi - see A. primuloides 'Chumbyi' 24
sarmentosa v. watkinsii - half the size of typical sarmentosa 252 / 123
sarmentosa v. yunnanensis / 3-4" / slightly larger fls / Yunnan 129
sempervivoides / R / 3" / fls pink to pinkish-mauve /Kashmir, Punjab 107
sempervivoides microphyllum - see A. mucronifolia 87
septentrionalis / A / 6-10" / fls white / circumboreal 148 / 107
septentrionalis v. puberulenta / 6" / Rocky mts. to New Mexico 25
spinulifera / R / 5-6" / fls pale pink, red reverse / w China 87 / 123
strigillosa / R / 6" / pink, white fls, purplish-red reverse /Himal. 194 / 252
turczaninovii - see A. maxima 288
umbellata / A or B / white fls, fragrant / Japan, se Asia 164 / 200
vandellii / R / 2" / white fls, honey-scented / Alps, Pyrenees 148 / 274
villosa / R / 2-3" / fls white, pale pink / Europe 148 / 212
villosa v. arachnoidea / 2" / rounder & hairier lvs / Carpathians 81 / 129
villosa v. jacquemontii / under 2"/fls purp-red to purple /w Himal. 252
villosa taurica / distinct form/ fls pink, scarlet eye/ Crimean mts. 252
wulfeniana / R / 1" / fls pink / e Alps 148 / 132

ANDROSTEPHIUM - Liliaceae
caeruleum / Bb / 6-10" / blue fls / sw & c United States 226

ANDRYALA - Compositae
aghardii / HH R / 5-6" / fls lemon-yellow / Spain 212 / 127
lanata - see HIERACEUM lanatum 96

ANEMONE - Ranunculaceae
albana ssp. armena / R / to 6" / fls crm-yel, purp outside/Asia Min. 25
alpina - see PULSATILLA alpina 286
alpina sulphurea - see PULSATILLA alpina ssp. apiifolia 286
altaica / R / to 1' / fls white, veined blue / n Asia 286
apennina / R / 4-9" / fls sky-blue / s Europe 267 / 25
apennina 'Alba' - fls white 25
aurea - see PULSATILLA aurea 5
baicalensis / R / 6-9" / white fls / e Asia 93
baldensis / R / 4-5" / fls white / Alps, Jugoslavia 148 / 286
barbulata / R / 2-3" / fls white / China 96
biarmiensis / R / 8-16" / fls white, somtimes pink / ne Russia 286
biflora / R / 8" / red fls / Kashmir 96
blanda / Tu / to 10" / blue fls / se Europe 269 / 286
blanda 'Alba' - white-flowered *
blanda 'Atrocaerulea' - fls deep violet-blue 129 / 111
blanda 'Charmer' - rose-red fls 17
blanda 'Radar' - fls bright purplish-red (31) / 185
blanda 'Rosea' - bright pink fls 132

blanda v. scythinica / fls white or blue / n Turkestan 25
blanda 'White Splendour' - large white fls (32)
bucharica / R / 6" / fls crimson or yellow / Afghanistan 5
bungeana - see PULSATILLA bungeana 5
californica lutea - see A. drummondii 'Lutea' 196
californica rosea - see A. drummondii 'Rosea' 196
campanella - see PULSATILLA campanella 5
canadensis / P / 2' / white fls / N. Am. 224 / 107
caroliniana / R / 1' / fls pink, white, violet / se US 224 / 107
capensis / HH R / 1' / purplish fls / Cape of Good Hope 223
caucasica / resembling A. blanda / Caucacus 25
coronaria / HH Tu / to 18" / fls red, white, blue / Medit. region 267
coronaria 'De Caen' - single-flowered in wide color range 129
coronaria 'St. Brigid' - hort. strain, mixed colors, semi-dbl. fls 268
crinita / P / to 18" / white fls / Siberia 5
cylindrica / P / 2' / fls greenish-white / North America 224 / 107
decapetala / P / to 18" / fls pink to greenish-white / N. & S. Am. 308 / 25
deltoidea / R / 6-8" / white fls, sometimes colored / w N. Am. 228 / 107
demissa / R / 12-14" / white fls / Himalayas, w China 93
drummondii / R / 4-12" / white fls, usually tinged blue / w N. Am. 312 / 228
elongata / P / 1-2' / dull white, to 1" across / Himalayas 25
fannina / Gh / 2-5' / fragrant white fls / South Africa 61
fasciculata - see A. narcissiflora v. fasciculata 81
flaccida / R / 3-4"/ 1-3 cream white fls, 1" across / Japan 25
X fulgens / HH Tu/ to 1' / fls scarlet, Greece 267 / 129
X fulgens v. annulata - selection for larger fls 267
X fulgens v. annulata 'Grandiflora' - red fls, yellowish center 185
X fulgens v. multiflora - semi-double fls 267
X fulgens 'St. Bavo' - see A. pavonina 'St. Bavo' *
globosa - see A. multifida 25
graeca - see A. pavonina 61
halleri - see PULSATILLA halleri 286
heldreichii - see A. hortensis v. heldreichii (444) / 149
hepatica - see HEPATICA nobilis 25
hortensis / HH Tu / 10" / fls pale purplish / c Medit. region 267 / 286
hortensis v. heldreichii / 1'/ white fls, colored reverse/ Greek Ils. 149
hudsoniana - see A. multifida 25
hupehensis / P / to 3' / fls rose / c & w China 129 / 93
hupehensis 'Alba' - white fls *
hupehensis v. japonica / 2½' / fls light purple/ China 165 / 93
hupehensis 'September Sprite' - see A. hybrida 'September Sprite' 135
X hybrida / 5' / fls in various colors 208 / 25
X hybrida 'September Sprite' / 2-3' / fls rose-pink 135
japonica - see A. hupehensis v. japonica 93
japonica hupehensis - see A. hupehensis 93
[1] japonica 'Queen Charlotte' / 2½-3' / clear pink single fls
X lesseri / P / 1½-2' / fls rosy-purple 129
Leveillei / P / to 2' / fls white, lilac reverse / sw China 93
lutea - see A. palmata v. lutea 28
Lyallii / R / 6" / fls white or tinted blue, rose / nw N. Am. 228
magellanica / P / 6-18" / creamy-yellow fls / South America 107
magellanica 'Major' / 10" / sulphur-yellow fls 18

---

[1] Wayside Gardens Catalog, 1986.

montana v. australis - see PULSATILLA montana 286
montana rubra - see PULSATILLA rubra 286
multifida / P / to 2' / fls yellow or white / North America 228 / 25
multifida alba - see A. multifida v. richardsiana f. leucantha 99
multifida globosa - North American form of A. multifida 227
multifida f. polysepala - red doubled flower 99
multifida v. richardsiana /to 2'+/ brt red fls/Hudson Bay reg.,Que. 245
multifida v. richardsiana f. leucantha / 2'/ milky-wht fls/ nc N.Am. 99
narcissiflora / R / 6-16" / fls white, pink tinged / e Eu., Asia 148 / 286
narissiflora demissa - see A. demissa 93
narcissiflora v. fasciculata - fascicled, umbellate inflorescence 81
narcissiflora polyanthes - see A. polyanthes 93
narcissiflora v. sachalinensis / plants densly pubescent / Kuriles 200
nemerosa / R / 2-12" / fls white or flushed pink / n & c Europe 267 / 286
nemerosa 'Alba' - fls white 25
nemerosa 'Alba Plena' - double white fls *
nemerosa 'Allenii' - large blue fls 129 / 25
nemorosa v. Robinsoniana / sepals powder blue, yellowish outside 25
nemorosa 'Rosea' - fls pale rose-purple 25
nuttalliana - see PULSATILLA nuttalliana *
obtusiloba / R / 6-12" / blue to violet or cream / Himalayas 25
obtusiloba alba / R / white form / Himalayas 25
obtusiloba f. patula / R / mats / fls deep blue / Himalayas 123
occidentalis - see PULSATILLA occidentalis 78
oregana / R / 1' / fls blue, pink, white / Oregon, Washington 229 / 142
oregana v. adamsiana - white or pinkish fls 64
palmata / Tu / to 1' / fls golden-yellow / w Medit. region 212 / 25
palmata v. alba / 6" / satin white fls / s France 111
palmata v. lutea - has basal lvs lobed; fls yellow
parviflora / R / 3-15"/ fls white, tinged rose or blue / arctic reg 229 / 209
patens - see PULSATILLA patens
pavonina / Tu / 6-12" / fls scarlet, pink purple / s Europe 212 / 286
pavonina 'St. Bavo' - florist's strain 124 / 17
polyanthes / R / 6-10" / fls white / Himalayas 93
pratensis - see PULSATILLA pratensis 286
pratensis nigricans - see PULSATILLA pratensis ssp. nigricans 286
pseudo-altaica / P / 4-12" / fls wht or pale purp. / Japan 200
pseudo-altaica alba / P / 4-12" / fls white / Japan 200
pulsatilla - see PULSATILLA vulgaris 286
quinquefolia / R / to 1' / fls white / e & c N. Am. 224
ranunculoides / R / to 1' / fls yellow / Europe 147 / 286
reflexa / R / to 1' / fls greenish white / e Russia 286
riparia / P / 2' / milk-white fls / n North America 99
riparia f. inconspicua - with shorter sepals 99
rivularis / R / to 15" / fls white, blue reverse / India, China 182 / 129
rupestris / HH R / small / fls yellow, purple, white / Himalayas 64
rupicola / R / to 1' / fls white, pink reverse / Kashmir, Sikkim 107
slavica / R / similar to A. pulsatilla / Carpathian Mts. 25
'St. Bavo' - see A. pavonina 'St. Bavo' 17
stellata - see A. hortensis 25
sulphurea - see A. alpina ssp. sulphurea 25
sylvestris / R / 6-15" / fls white / c & e Europe 147 / 286
sylvestris 'Grandiflora' - fls large 25
sylvestris 'Macrantha' - fls larger, free-flowering 61
taurica - see PULSATILLA halleri ssp. taurica 286

tetrasepala / P / to 5' / white fls / w Himalayas 130 / 25
trifolia / R / 8-10" / fls white / s Europe 148 / 25
tschernjaewii / R/ 3-8"/ 1, 2 or 3 fls, pink or wht / Turk, Afghan 268
tuberosa / R / 4-12" / greenish-white & rose fls / Texas to Utah 226
vernalis - see PULSATILLA vernalis *
virginiana / P / 2-3' / fls white / US & Canada 224 / 107
vitifolia / P / 1-3' / white fls / Nepal 279 / 25

ANEMONELLA - Ranunculaceae
thalictroides / R / 6-10" / fls white to pink / ec US 224 / 107
thalictroides flora pleno - double flowered form of A. thalictroides

ANEMONOPSIS - Ranunculaceae
macrophylla / P / 15-30" / fls pale purple / Japan 279 / 200

ANETHUM - Umbelliferae
graveolens / A / 8-20" / the herb Dill / India, sw Asia 201 / 287

ANGELICA - Umbelliferae
acutiloba v. iwatensis / P / 4-6' / white fls / Honshu Japan 200
archangelica / B / 6-8' / tiny white fls / Europe, n Asia 148 / 93
atropurpurea / P / 4-5' / dark purple stem / e & c North America 263
grayi / P / 1-2' / reddish brown fls / Wyoming, Colo 229
montana / P / 3' / from mt. pastures / Alps of Jura, Cevennes 81
razulii / P / 6' / white or pinkish fls / Pyrenees 148 / 287
triquinata / P / 6' / white fls / Pennsylvania to N. Carolina 25

ANGUILLARIA - Liliaceae
dioica / HH Bb / 3" / white fls, purplish markings / Australia 45 / 271

ANIGOZANTHUS - Haemodoraceae
bicolor / Gh P / 1' / fls red, green, yellow / w Australia 45
flavidus / Gh P / 4' / fls red / sw Australia 194 / 25
humilus / Gh P / 9" / yellow & orange fls / w Australia 45
manglesii / Gh P / 3' / fls green, red at base / sw Australia 45 / 25
viridis / Gh P / to 2' / green fls, yellowish at base / w Australia 45 / 25

ANISOTOME - Umbelliferae
aromatica / HH R / to 1' / white fls / New Zealand 238
aromatica v. lanuginosa - upper leaf surfaces buff to white hairy 15
haastii / Gh P / 1-2' / white fls / montane New Zealand 15
imbricata / HH R / 2" / compact plant spreading by offsets / N. Z. 15
intermedia / HH R / 8-10" / white fls / New Zealand 15
lanuginosa - see A. aromtica v. lanuginosa *
pilifera / HH P / 1-2' / white fls / New Zealand (360) / 15

ANOIGANTHUS - Amaryllidaceae
brevifolius / HH Bb / 2-10" / bright yellow fls / South Africa 185
luteus - see A. brevifolius 185

ANOMALESIA - Iridaceae
cunonia / Gh / 6-18" / fls rosy-pink / coastal Cape Peninsula (13)

ANOMATHECA
cruenta - see LAPEYROUSIA laxa 93

ANOPTERUS - Escalloniaceae
glandulosus / HH T / 20'+ / fls white, tinged rose / Tasmania 36

ANTENNARIA - Compositae
alborosea / R / to 6" / fls pink to roseate / c Alaska, Yukon 213 / 245
alpina / R / 104" / fls white / arctic Europe, Asia & North America 229 / 93
aprica - see A. parvifolia 93
campestris - see A. neglecta v. campestris
carpatica / R / 6"+ / fls cream to purplish/ Alps, Pyrenees, Carpath. 148 / 289
dimorpha / R / 1-4" / without stolons / w US 229 / 142
dioica / R / 1-4" / fls white or pinkish / arctic Europe & Asia 148 / 93
dioica alborosea - see A. alborosea
dioica 'Hyperborea' - trade name for pink form 289 / 160
dioica 'Minima' - small form 123
dioica 'Minima' rubra / 2"/ tiniest form silvergreen lvs; round fls 249
dioica 'Nyewoods' / 3" / fls bright rose 46
dioica 'Rosea' - fls pink 123
dioica 'Rosea Minima' - see A. dioica 'Minima' 123
dioica 'Tomentosa' - woolly form 123
eucosma / R / 3" / brownish fls / Newfoundland 99
gaspensis - see A. neglecta v. gaspensis 25
howellii / P / 10-14" / plant grey-woolly; fls same / n Cal.-s Ore. 228
magellanica / R / 2" / bracts light brown / Straits of Magellan reg. 25
microphylla - see A. rosea v. nitida 245
neglecta / R / to 3" / fls whitish / ne & nc North America 259
neglecta v attenuata - lvs somewhat longer than v. gaspensis 25
neglecta v. campestris / 4" / fls whitish, North America 259
neglecta v. gaspensis /stolons short, leafy;lvs to 3/4"/Nfld,Que,Me 25
neodioica / R / mats / white to roseate fls / Newfoundland to Minn. 99
oxyphylla - see A. rosea 245
parvifolia / R / to 6" / fls rosy / c North America 229 / 93
parviflora - see A. parvifolia 93
plantaginifolia / P / 6-18" / fls grayish / North America 259
porsildii / R / to 2" / white to pink fls / e & w Greenland 209
pulcherrima / R / to 15" /fls gray-white /Can. to Utah, Col., Nfld. 229
racemosa / 2-16" / bract tips reddish brown / Can., n U.S. 229
rosea / R / to 8" / fls pink to roseate / North America 229 / 245
rosea v. nitida / R/ to 8" / fls paper-white to brownish / n Canada 245
sornborgeri / R / to 6" / whitish fls / e Greenland 209
suffrutescens / HH R / 2-6" / evergreen lvs / n Calif., s Oregon 107
tomentosa - see A. dioica 'Tomentosa' 25
vexillifera / P / mat to 6" / tiny white fls / Nfld & Que. 99

ANTHEMIS - Compositae
aizoon - see ACHILLEA ageratifolia 93
alpina - see ACHILLEA oxyloba 289
barrelieri - see ACHILLEA barrelieri 289
biebersteinii - see A. marschalliana 25
biebersteinii v. rudolphiana - see A. marschalliana 25
carpatica ssp. carpatica / lvs sparsely hairy, light green 289
compacta / R / neat / single white fls / s Alps, Pyrenees 96
cretica / R / to 1' / white fls / mts. of e Europe 77 / 289
cupaniana - see A. punctata ssp. cupaniana 129 / 289
hauschnechtii / A / 6-10" / white fls / Syria 61

marschalliana / R / to 1' / yellow fls / Caucasus 127 / 25
melanoloma ssp. trapezuntia / A / 8-18" / 5-6 white fls / Sov.Armen. 77
montana / R / 4-8" / white fls / Spain, Italy Balkans 148 / 107
montana ssp. carpatica - see A. carpatica ssp. carpatica
nobilis - see CHAMAEMELIUM nobile 25
punctata ssp. cupaniana / HH P/ 1-2' / white fls; silvery lvs/Sicily 184 / 289
rigescens / P / 18" / white daisies / mts. of Caucasus 64
rudolphiana - see A. marschalliana 25
sancti-johannis / P / 15" / golden daisies / Jugoslavia 129 / 46
tinctoria / P / 2½' / yellow daisies / Europe 269 / 129
tinctoria 'E. C. Buxton' / P / 2-3' / pale lemon fls 268
tinctoria 'Kelwayi' - bright yellow fls 25
tinctoria 'Kelways Var.' / deep yellow fls 28
tinctoria 'Moonlight' - pale yellow fls 127 / 135
triumfetti / P / 1-3' / fls white / mts. s Europe 289
zyghia - resembles A. melanoloma ssp. trapezuntia, soft wht hairs 77

ANTHERICUM - Liliaceae
algeriense - see A. liliago ssp. algeriense 160
liliago / P / 18" / fls white / s Europe 148 / 129
liliago ssp. algeriense - adaptive to varied sites 181
liliago 'Major' - larger in all parts 28
liliastrum 'Major' - see PARADISEA liliastrum 'Major' 28
ramosum / R / 12" / starlike fls, white / Europe 148 / 267

ANTHOLYZA - Iridaceae
ringens / Gh C / to 1' / red & pink fls / coastal South Africa 194 / 223

ANTHYLLIS - Leguminosae
alpestris - see A. vulneraria ssp. alpestris 287
cytisoides /Sh /2' / fls 1-3, corolla yel /s & e Spain, s France 287
Hermanniae / Sh / to 2' no more / fls yellow / Mediterranean 25
jacquinii - see A. montana v. jacquinii 36
maura - see A. vulneraria ssp. maura 287
montana / Sh / 4-12" / fls rose / s & se Europe 148 / 274
montana atrorubens purpurea /Sh/mat 1½-4"/fls rosy pink/Medit.alps 61
montana 'Carminea' - bright colored selection *
montana v. jacquinii / with paler fls / e part of range 36
montana 'Rosea' - pink fls 25
montana 'Rubra' - color selection, deeper color 36
polyphyllos - see A. vulneraria ssp. polyphylla *
vulneraria / R / to 1" / fls yellow / n Europe 148 / 287
vulneraria ssp. alpestris / 1' / fls pale yellow / in mts. 287
vulneraria ssp. carpatica / to 1' / pale yellow fls, fodder plant 287
vulneraria ssp. maura / 2' / fls usually red/ Portugal, Spain, Italy 287
vulneraria ssp. polyphylla / 1-3' / fls yellow / c & e Europe 287
vulneraria ssp. praepropera / 14" / fls red-purp / Medit. reg. 149 / 287
vulneraria ssp. pulchella / R / 1-8" / yel or red / Balkan Penn. 287
vulneraria ssp. pyranaica / 10" / fls pink to red 287
vulneraria v. 'Rubra' / good red fls / Great Britain 154

ANTIRRHINUM - Scrophulariaceae
asarina - see ASARINA procumbens 25
barrelieri / HH P / to 3' / pink fls yellow marked/ s Spain, Portugal 212 / 288

boissieri / HH P / to 2' / fls white or pink / c Spain, e Portugal 288
braun-blanquetii / HH P / to 4' / fls yellow / nw Spain, ne Portugal 288
'Floral Carpet' - HH P / 8" / bright colors / Mediterranean 129
glutinosum / HH R / 8-14" / fls yellowish-white, striped red / Spain 25
glutinosum roseum / P / prostrate / lvs rounded ends; fls pink/Spain 61
graniticum / HH P / 4' / fls pink or whitish / c & s Spain, ne Port. 212 / 288
grosii / HH Sh / decumbent / fls pale yellow / wc Spain 288
hispanicum / HH Sh / decumbent to 2' / fls white or pink / se Spain 212
'Magic Carpet' - see A. majus 'Magic Carpet' 268
majus / HH P / to 4' / purplish-pink fls / e Pyrenees, sc France 210 / 288
majus 'Magic Carpet' / HH A / 6" / creeping habit, mixed col. 268
meonanthum / HH P / to 4' / fls pale yellow / Spain, Portugal 288
molle / HH Sh / 16" / white or pale pink fls / c Pyrenees 274 / 288
orontium / A / 9-15" / rose-pink fls / Eurasia, escape in US 211 / 25
pulverulentum / HH Sh / dwarf / fls buff to white / e Spain 207
sempervirens / Sh / to 9" / fls white or cream / Pyrenees
siculum / HH P / to 2' / fls pale yellow, veined red / Sicily, Malta 288

APHANOSTEPHUS - Compositae
skirrhobasis / A / to 20" / white fls, reddish reverse / se US 225 / 25

APHYLLANTHES - Liliaceae
monspeliensis / HH R / 4-10" / blue fls / Spain, Italy, n Africa 148 / 211

APLECTUM - Orchidaceae
hyemale / P / to 2' / greenish or yellowish fls / n & c N. Am. 99

APOCYNUM - Apocynaceae
androsaemifolium / P / to 2½' / pink fls / e & c North America 224 / 99
pumilum / P / 8-20" / fls pinkish / North America 229 / 228

APONOGETON APONOGETONACEAE
distachys / aquatic / 3-4" / fls white / S. Africa 129

AQUILEGIA - Ranunculaceae
adantoides - see Aquilegia eximea, a subform of A. formosa 154
akitensis - see A. flabellata v. pumila 200
akitensis 'Alba' - see A. flabellata v. pumila 'Alba' *
akitensis kurilensis - see A. flabellata 'Nana' 132
akitensis nana - see A. flabellata 'Nana' 268
akitensis 'Val Ferrera Form' *
alpina / P / 6-36" / nodding bright blue fls / Alps, n Apennines 148 / 286
alpina 'Alba' - fls white 25
alpina 'Hensol Harebell' - see A. X 'Hensol Harebell' 268
alpina 'Hensol Hybrid' - see A. X 'Hensol Harebell' 207
amaliae / P / 8-12" / pale blue violet / Balkans 286
atrata / P / 1½-3' / dark purple-violet fls, nodding / Alps 148 / 286
atrata 'Major' / leaflets pubescent beneath/ Switzerland (39)
atropurpurea - see A. viridiflora 268
atroviolacea - see A. atrata 219
aurea / R / 4-16" / fls yellow, suberect / Bulgaria 286
barnebyi / syn. A. rubicunda /blue-gr lvs; pink & yel fls/ nw Colo. 154
bernardii / HH P / to 2' / pale blue fls, nodding / Corsica 25
bertolonii / R / 4-12" / blue-violet fls, nodding/ France, Italy 148 / 286
bertolonii 'Alba' - white fls *

bicolor - see A. siberica 268
'Biedermeier' / P / 22-25" / blue fls 49 / (40)
'Biedermeir' Alba - white form
brevistyla / P/ to 2½' / blue & yellowish-white fls/Alaska to S. Dak. 25
buergerana / P / to 2½' / fls purple & yellowish / Japan 165 / 25
buergeriana /P / 1-2' / brownish purp-yel /Japan, Korea, Manchuria 200
caerulea / P / to 2½' / fls blue & white / Rocky Mts. 229 / 25
caerulia 'Alba' - fls white 28
caerulea 'Candidissima' / 27-34" / white fls (40)
caerulea v. 'Daileyae' / P / to 2½" / Colorado 25
caerulea helenae - see A. X 'Helenae' 268
caerulea v. ochroleuca / to 2½' / sepals whitish / w Rocky Mts 25
caerulea 'Rosea' / fls pink or red, often double / probably hybrid 25
canadensis / P / 1-2' / fls bright red / e & c North America 224 / 107
canadensis v. alba - white form of species 132
canadensis v. Australis / P/ to 3' /pale grn lvs, nar sepals/Fl.-Tex. 226
canadensis f. flavescens - yellow fls 123
canadensis v. latiuscula / basal lvs 3-ternate / Neb. to Texas 25
canadensis 'Nana' - 12" strain 25
candidissima - see A. caerulea 'Candidissima' 25
caucasica - see A. olympica 160
cazorlensis / HH R / 6-10" / bright blue fls / s Spain 286
chaplinii / P / to 20" / fls yellow / Texas, New Mexico 227
chrysantha / P / 1-4' / fls yellow / N. Mex., Arizona, Mexico 227 / 268
chrysantha v. flavescens / yellow fls, tinged red, curved spurs 208
'Clematiflora' - spurless mutant of A. vulgaris 268
colchica / R / ? / fls blue or white / Caucasus, on limy rocks 5
'Crimson Star' / 2' / fls crimson & white 147 / 129
dichroa / P / 8-26" / fls blue tipped white / nw Spain, Portugal 286
discolor / R / 4-6" / fls blue & white / n Spain 50 / 286
'Dragon Fly Hybrid' / 18" / long-spurred, bright colors 282
ecalcarata - see SEMIAQUILEGIA simulatrix 24
'Edelweiss' - see A. vulgaris v. erecta 'Edelweiss' (40)
einseliana / R / 4-18" / fls nodding, blue-violet / Alps 148 / 286
elegantula / R / 4-16" / red fls tipped yellow / Rocky Mts. 229 / 25
eximia / HH P / 2-3' / red to scarlet fls / California 175 / 25
'Fairyland' - dwarf strain 279
flabellata / R / 6-18" / fls white, tinged violet-rose / Japan 268
flabellata 'Alba' - form with milky-white fls 132
flabellata v. kurilensis - see A. flabellata v. pumila 25
[1] flabellata 'Ministar' / 6" / sky blue & white spurless fls
flabellata 'Nana' / 3" / sepals mid-blue, petals light yellow 306 / 132
flabellata 'Nana-Alba' - white fls on the dwarf form 107 / 132
flabellata 'Nana Rosea' see A. flabellata v. pumila
flabellata v. pumila / 1' / 1-2-flowered / alps of Japan 165 / 200
flabellata v. pumila 'Alba' - white form *
flabellata v. pumila 'Rosea' - pink fls *
flavescens / P / 1-2' / yellow to pinkish fls / Rocky Mts. 229
formosa / P / 2-3' / fls red & yellow / w North America 229 / 268
formosa f. arctica - of dwarf habit and good temper 64
formosa schockleyi - see A. shockleyi
formosa v. truncata / to 40" / fls smaller / Calif. to Ore. 268

---

[1] Thompson and Morgan Seed Catalog, 1986, p. 28.

formosa v. wawawensis - see A. formosa 64
fragrans / P / 1½-2' / fls white or pale purple / Himalayas 21 / 268
glandulosa / R / 8-12" / fls bright blue / Altai Mts. 268
glandulosa v. jucunda / 10" / fls white (334) / 268
glauca - see A.fragrans
'Granny's Bonnet' - common name for A. vulgaris 'Flore Pleno' 129
grata / P / 6-18" / purplish violet / Jugoslavia 286
X helenae - A. flabellata x A. caerulea / blue & white fls 93 / 268
X 'Hensol Harebell' - A. alpina x A. vulgaris, seedlings vary 268
hirsutissima / R / 4-10" / fls pale blue to white / se France 286
japonica - see A. flabellata v. pumila 200
Jonesii / R / 3" / fls solitary, blue / Rocky Mts. 129 / 268
Jonesii X discolor *
Jonesii x saximontana *
jucunda - like A. glandulosa/sepals bl;petals wht;seeds shiny/Siber 25
kareliniana - see A. karelinii 5
karelinii / P / 8-32" / fls violet or dark red / c Asia 5
kitabelii / R / to 1' / fls red-violet to blue-violet / Italy 286
kurilense - see A. flabella v. pumila
lactiflora / P / 18" / fls white or tinged blue / Altai Mts. 268
latiuscula - see A. canadensis v. latiuscula 96
longissima / P / 2-3' / fls pale yellow / Texas, Mexico 227 / 129
'McKana Hybrids' / 2'+ / various colors 49 / (40)
micrantha / P / 12-18" / white, blue, reddish fls / Utah, Col. 229 / 25
nevadensis / P / 18-24" / dull pale blue fls / s Spain 268
nigricans / P / 1½-3' / fls purple, large / c & se Europe 286
nivalis / may be A. jucunda or A. fragrans/ lg blue fls / Kashmir 154
'Nora Barlow' / P / to 28" / dbl fls, comb red, pink & green
olympica / P / 2' / fls claret to purple / Caucasus 268
Ottonis /P / 12-18" / white & blue fls / Greece, Italy 149 / 25
oxysepala / P / 2½' / fls blue / Siberia 268
oxysepala v. yabeana / 8-18" / fls lilac-blue / Manchuria 5
parviflora / half size A. vulgaris / n.e. Asia 64
pubescens /R / 8-18" / fls cream to pink / s Sierra Nevadas, US 228
pumila (japonica) - see A. flabelleata v. pumila
pyrenaica /R / 8-16" / fls clear pale blue / Pyrenees 148 / 274
reuteri - see A. bertolonii 24
rubicunda - variety of A. elegantula 64
saximontana / R / 6" / fls blue & white / Rocky Mts. 312 / 107
schockleyi / P / to 30" / fls red & yellow / s Calif., Nevada 227
scopulorum / R / 6" / fls large, lavender to violet / nw N. Am. 132 / 268
scopulorum v. calcarea / petioles glandular-pubescent / Utah 132
scopulorum v. perplexans / pygmy 3" / red, blue, wht, yel / Nevada 154
sibirica / R / 1'+ / fls blue to claret / Siberia, Mongolia 25
'Snow Queen' - pure white with spurs
skinneri / HH P / 2' / fls orange & red / Mexico 268
thalictrifolia / R / 4-12" / blue-violet fls / s Tyrol (427) / 132
'Toy Town' / P / 12-15" / lavender, blue, purple fls *
transsilvanica / 6-18" / blue violet fls / Carpathians 286
triternata / R / 4-8" / yellow & red fls / Arizona, w Colorado 229 / 132
truncata - see A. formosa v. truncata
ullepitschii - see A. nigricans 286
viridiflora / P / 1-1½' / fls greenish / Siberia, China 169 / 268
vulgaris / P / 2' / fls blue, purple, white / Europe 210 / 129

vulgaris alba - see A. vulgaris v. 'Nivea' 28
vulgaris 'Alba Plena' - fls white, double 25
vulgaris atrata - see A. atrata *
vulgaris atrata 'Major ' - see A. atrata 'Major' *
vulgaris 'Atrorosea' - fls deep rose-pink 29
vulgaris 'Edelweiss' - see A. vulgaris v. erecta 'Edelweiss' (1)
vulgaris v. erecta 'Edelweiss' / 19-25" / fls white (40)
vulgaris 'Flore Pleno' - fls doubled 25
vulgaris 'Nivea' - fls white 25
vulgaris olympica - see A. olympica 268
vulgaris 'Stellata' - spurless mutant 64
[1] yabeana - see A. oxysepala v. yabeana

ARABIS - Cruciferae
aculeolata / P / 18-26" / fls bright pink or purplish / sw Oregon 228
albida - see A. Caucasica 286
albida rosea - see A X arendsii 'Rosabella' 208
albida variegata - see A. caucasica v. variegata 208
allionii / P / to 18" / white fls / Europe, in mts 286
alpestris / variable/ fls white, hairy; lf rosette/ Eur., Brit 154
alpina / R / 2-16" / fls white / much of Europe 148 / 286
alpina billardieri - see A. billardieri 6
alpina v. flavescens - see A. caucasica ssp. causasica 77
alpina 'Grandiflora' / 6"/ large white fls 283
alpina 'Rosea' - pink form 169
alpina variegata / R / 9" / yel-wht, evergreen, trailing / Switz. 206
androsacea / R / 3" / fls white / Turkey, in mts. 25
X arendsii 'Rosabella' / 6-8" / purple-pink fls 208
aubrietioides / R / to 6" / fls purplish-pink / Cilician Taurus 77
bellidifolia - see A. soyeri ssp. jacquinii 286
billardieri / R / to 1' / fls white / Asia Minor, Caucasus 6
blepharophylla / R / to 1' / deep pink fls / w North America 208 / 107
blepharophylla 'Alba' - white fls *
blepharophylla 'Frulingszauber' / R / 2-8"/ fls rosy purple 228
blepharophylla 'Spring Charm' - selected pink form 46
borealis / like sagittata / dry calcareaous slopes / n Russia 286
breweri / R / 2-8" / fls pink to red-purple / w U.S. 228
bryoides / P / 3/4-2 1/4" / fls 3-6 white / mts of s Balkan Pen. 25
caerulea / R / 2-6" / fls pale blue / Alps 148 / 286
carduchorum / R / to 3" / petals white / Turkey 25
caucasica / R / 8-14" / fls white / s Europe 147 / 286
caucasica ssp. causasica/ P / variable, ptls wht-dry yel / Balkans 77
[2] caucasica 'Compinkie' / 4" compact / rose-pink fls; slivergreen lvs
caucasica 'La Fraicheur' / R / 5-10" / pale to deep magenta fls *
caucasica 'Rosabella' - see A. X arendsii 'Rosabella' 208
caucasica 'Snowcap' / compact plant / white fls 241
caucasica 'Variegata' - lvs variegated yellow & green 208
ciliata - see A. hirsuta 286
corymbiflora / R, may be B / 1' / fls white / c Europe 314 / 286

[1] Bailey, Manual of Cultivated Plants. 1949

[2] Park Seed Catalog, 1986.

cypria / HH R / tufted / fls white or pale pink / Cyprus (28) / 81
Drummondii / B or P/ to 3' / fls white & pink/ Que s to Del. 25
ferdinandi-coburgii / R / to 1' / fls white / Bulgaria (41) / 286
ferdinandi-coburgii 'Variegata' - lvs yellow & green (41)
fruticosa / P / 1-3' / petals pink or purplish / Yellowstone Pk, US 229
grandiflora - see A. alpina 'Grandiflora' 25
halleri - see CARDAMINOPSIS halleri 286
hirsuta / P or A / to 2' / fls white / most of Europe 229 / 286
holboellii / B or P / 3' / white or pink fls / North America 228 / 25
jacquinii - see A. soyeri ssp. jacquinii 286
japonica - see A. stelleri v. japonica 200
X kellereri / R / 3" / fls large, white 132
kohleri / R / 2-16" / fls bright red to purple / sw Oregon 228
'La Fraicheur' - see A. caucasica 'La Fraicher' *
lemmonii / R / 2-16" / fls pink or rose-purple / nw North America 228
ludowitziana / ludoviciana - see SIBARA virginica 236
lusitanica / B / 1'+ / white fls / Portugal 286
lyallii / R / 4-12" / fls purple / nw North America 229 / 228
lyrata / B / to 1' / fls white / N. & S. America, n Asia 273 / 313
lucida 'Variegata' / R/ large rosette lvs, white to yellow border 96
montana - see A. hirsuta 81
muralis / R / to 1' / fls white, rarely pink / s & sc Europe 286
muralis f. rosea - pink form 149
nuttallii / R / 3-4" / white fls / Montana to Washington 229
oregana / P / 20" / purple fls / n California, s Oregon 25
petraea - see CARDAMINOPSIS petraea 286
platysperma / R / 4-12" / fls pink-purple to white / n Oregon 228
procurrens / R / to 1' / fls white / Balkan mts. 300 / 286
pumila / R / 2-6" / fls white / Alps, Apennines 148 / 286
purpurascens - see A. oregana 25
purpurea / R / 3" / fls purple / Cyprian Olympus 96
rosea - see A. muralis f. rosea
serrata v. japonica / R / 4-12" / white fls / alpine Honshu 200
serrata v. japonica f. fauriei - see A. serrata v. japonica 200
'Snowcap' - see A. caucasica 'Snowcap' 241
soyeri / P / 6-20" / white fls / Pyrenees 148 / 286
soyeri ssp. jacquinii / 6-18"/ fls white/ Alps 286
sparsiflora / P / 10-36" / pink or purple fls / nw US 229 / 228
'Spring Charm' - see A. blepharophylla 'Spring Charm' 300 / *
stelleri / R / 6-16" / fls straw-yellow or white / Far East 6
stelleri v. japonica / 8-16" / fls white / Japan, Korea 165 / 200
stricta / R / to 10" / fls yellowish / Alps, Pyrenees 286
X sturii / R / to 6" / fls large, white 24
subpinnatifida / R / 6-16" / fls lavender or purple / Calif., Oregon 228
sundermannii / cushion/ wht fls, short stem / 1912 Bulgarian hybrid 154
turrita / B or P / to 2½' / pale yellow fls / e & c Europe 314 / 25
verna / A / 3-8" / fls violet-purple / Medit. region 77
vochinensis / R / to 8" / fls white / se Alps 286
X wilczekii - A. bryoides X carduchorum / tufts / fls white 154 / 61

ARACHNOIDES - Polypodiaceae
aristatum 'Variegatum' / Gh F / 2' / ovate-triangular frond / Austrl 25
standishii / F / 18" / 3-pinnate leathery fronds / e Asia (424) / 25

ARALIA - Araliaceae

californica / HH P / to 10' / reddish-black frs / Ore. to Calif. 228 / 25
elata / T / to 30' / fls small, whitish / Japan, Korea 127 / 36
hispida / P / to 3' / globose black frs / e North America 224 / 25
nudicaulis / P / 8-16" / black frs on leafless stems / e & c N. Am. 22 / 224
racemosa / P / 2-10' / frs purple / ne & nc North America 224 / 313
spinosa / T / to 30' / frs red, turning black / se United States 139

ARAUJIA - Asclepiadaceae

sericifera / Gh Cl / vigorous / white fls, streaked maroon/s Brazil (317) / 25

ARBUTUS - Ericaceae

menziesii / T / to 100' / fls dull white / w North America 36
texana / HH T / to 30' / fls white or pink / Texas to Guatemala 294
unedo / HH T / 15-30' / white or pinkish fls / Medit. region 36

ARCTERICA nana - see PIERIS nana 25

ARCTOSTAPHYLOS - Ericaceae

alpina / Sh / fls white or pink, trailing / circum-subpolar 25
alpina erythrocarpa - see ARCTOUS alpinus v. ruber 36
alpina v. japonica / Sh / mats / black frs / Japan 25
glandulosa 'Cushingiana Repens' / 6-8" / gray lvs, white fls 25
glauca / HH Sh / 6-14" / white or pinkish fls / S. California 36
nevadensis / Sh / trailing / fls white or reddish / California 36
patula / HH Sh / to 6' / pink fls / Oregon, Arizona, Utah 131 / 25
Pringlei / Sh / to 10' / fls rose; fr. dk. red/Ariz. to Baja, Calif. 25
rubra / Sh / trailing / white fls; scarlet frs / n N. Am & Asia 299 / 99
tomentosa / Sh / to 5' / fls white / California, coast ranges 36
uva-ursi / Sh / trailing / white or pink fls / n hemisphere 148 / 36
viscida / HH Sh / to 9' / fls light pink; frs deep red / Calif. 222

ARCTOTIS - Compositae

acaulis / HH R / 6" / yellow fls, reddish reverse / South Africa 208 / 25
grandis - see A. stoechadifolia v. grandis 25
stoechadifolia v. grandis /P gr as A/ 2' /fls lavender-blue/ S. Afr. 184 / 208

ARCTOUS - Eriaceae

alpinus v. japonicus - see ARCTOSTAPHYLOS alpina v. japonica 25
alpinus v. ruber / Sh / 4" / fls white, frs red / n Hemisphere 36

ARDISIA - Myrsinaceae

crenata / Gh Sh / 6' / fls white or pink; frs red / Far East 167 / 25
crispa / Gh Sh / 8-16" / white fls; red frs / Far East 128 / 200
japonica / Gh Sh / 1½' / fls white; frs red / Japan, China (5) / 25

ARENARIA - Caryophyllaceae

aculeata / P / stems to 8" / mat forming, wht fls/ e Ore. to e Calif 25
aggregata / R / 2" / fls white / sw Europe 25
aggregata ssp. erinacea /P/ white fls/ e Spain, se France, n Port. 286
arctica / R / caespitose / fls white / arctic Asia & North America 209
armeriastrum - see A. armerina 286
armerina / R / to 8" / dense terminal white fls / s Spain 286
austriaca - see MINUARTIA austriaca 286
balearica / HH R / mats / solitary white fls / Balearic Isls. 212 / 25

bertolonii / R / 2-5" / white fls / c Medit. reg., on rocks 286
caespitosa - see MINUARTIA verna *
capillacea / P / to 1' / fls white / s & sc Europe 25
capillaris / R / to 8" /fls white / Eurasia 209
capillaris v. americana / R / 8"/ fls white / nw North America 229 / 142
caroliniana / R / to 8" / white fls in clusters / R.I. to Fla. 224 / 225
cephalotes / P / 8-20" / white fls / s Ukraine, Moldavia 286
ciliata / R / to 3" / white fls / mts. of c Europe 314 / 286
circassia - see MINUARTIA circassia 77
congesta / R / 4-12" / globular inflorescence; fls wht / Rocky Mts. 229 / 25
cretica - related to UNIVERIA of Greece 77
dedeana - is this the white-flowered DRABA of n Spain? *
dyris / short stemmed / white fls / Morocco 154
fendleri / R / tuft / white fls; dark anthers / Colorado 229 / 198
foliosa / R / 6" / white fls / Himalayas 25
glacialis - see A. gracilis 160
gracilis / R / to 3" / fls white / s & w Jugoslavia 286
graminifolia - see A. procera 286
grandiflora / R / 6" / fls white / c & s Europe 148 / 286
groenlandica / R / 6" / white fls / Greenland s to Tenn. 22 / 25
gypsophilioides / R / 1'+ / white fls / s & e Bulgaria, sw Asia 286
hookeri / R / to 12" / white fls / Rocky Mts 229 / 25
humifusa / P / mat to 1" / fls wht, pale purp. anthers / arctic Eu. 286
huteri / P / cushion / 2-5" / fls white / ne Italy 286
imbricata / R / 3" / fls white / alps of the Caucasus 81
juniperifolia - see A. grandiflora 96
kingii / R / to 8" / fls white / Great Basin of US 229 / 228
koriniana - see A. procera ssp. glabra 286
kotschyana / R / to 8" / mat-forming; fls white / Turkey 77
laricifolia - see MINUARTIA laricifolia 209
ledebouriana / R / to 10" / fls white / Turkey 77
lithops / P / cushion 2-32" / fls white / mts se Spain 286
longifolia / P / 8-16" / many fls / Russia, Ukraine 286
macrocarpa / R / mats, to 6" / solitary white fls / Alaska, Yukon 209
merckioides / P / 2-6" / fls wht; seeds red-brn / Honshu, Hokkaido 200
montana / R / to 1' / fls white / sw Europe 212 / 286
norvegica / R / 3" / white fls / Scandinavia, Scotland, Ireland 148 / 286
obtusiloba / R / 2" / fls white / Rocky Mts. 229 / 25
pinifolia - see MINUARTIA circassica 77
polaris / R / to 6" / fls violet-tinged / Siberia 286
procera / R / 8-16" / fls white / c & e Europe 25 / 286
procera ssp. glabra / to 16" / fls white / c & e Europe 286
pungens / P / to 8" / white fls / mts. s Spain & Morocco 25
purpurascens / R / to 4" / fls pale purplish to white / Pyrenees 148 / 286
purpurascens 'Elliott's Var.' - free flowering pink form 123
recurva / P / to 5" / white fls / mts. Portugal to Romania 25
rigida / P / to 16" / fls wht / s Ukraine to Romania & Bulgaria 25
Rossii / R / to 2" / fls white / Alaska s to Wash. & Col. 25
rotundifolia / R / to 8" / white fls / n Balkan Peninsula 286
rubella / P / to 8" / mat-forming / white fls / arctic-alpine 25
sajanensis - see MINUARTIA biflora (428) / 276
sedoides / P / cushion-forming / white fls / mts s Eu. & Scotland 25
serphyllifolia / A or B / to 12" / white fls / Europe 229 / 25
setacea - more densely clustered fls than A. verna 64
stellata / P / rosettes / white fls / mts. Greece & s Albania 25

striata kitabelii - see MINUARTIA laricifolia ssp. kitabelii 286
stricta / A or P / to 10" / white fls / e & c United States 25
tetraquetra / R / cushion / fls white / s & e Spain, Pyrenees 148 / 212
tmolia / R / to 3" / fls white / endemic to Turkey 77
Tweedyi / fine clusterheads / Colorado 64
verna - see MINUARTIA verna 286
verna aurea - see MINUARTIA verna 'Caespitosa Aurea' 46
verna caespitosa - see MINUARTIA verna 'Caespitosa' 123
verticillata - see ACANTHOPHYLLUM verticillatum 77

ARGEMONE - Papaveraceae
alba / A / 1-3' / fls white / southern states, US & s-ward 28
hispida - see A. platyceras v. hispida 78
mexicana / A / to 3' / yellow fls / West Indies (327) / 25
mexicana ochroleaca / P / 1-3' / prickly; fls yellow / Mex. & U.S. 226
munita / A or P / 2-5' / white fls / Calif to Baja, e to n Mexico 25
platyceras / A / 1-2½' / fls white or pale yellow / Mexico 80 / 25
platyceras v. hispida / B or P / 1½-4' / fls white / w US 78
polyanthemos / A or B / 3'+ / white fls / Rocky Mts., Texas 228 / 25
sanguinea / A / to 3' / fls white, lav., pink / s & w Texas 226

ARGYROLOBIUM - Leguminosae
zanonii / R / 10" / yellow fls / Albania & w-ward in s Europe 212 / 287

ARISAEMA - Araceae
amurense / Tu / to 2' / greenish fls / Far East 25
atrorubens - see A. triphyllum 246
candidissimum / HH Tu / to 1' / fls greenish-yellow / w China 279 / 129
consanguineum / HH Tu / to 3' / fls green & purple & white / e Asia 267
dracontium / Tu / to 1' / fls green / e North America 224 / 267
flavum / Tu / ? / fls purple & yellow / Himalayas 154
griffithii / Tu / 2' / spathe lined violet / Himalayas, Sikkim 267
'Jacquemontii' / Tu / grn, wht striped / nw Pak & Kashmir to Bhutan 25
japonicum / Tu / 12-20" / fls dark purple / Japan, Korea 70 / 200
ringens / Tu / to 10" / fls green or purple / Japan, China 166 / 200
robustum / Tu / 14" / spathe green, striped white / Japan, Korea 200
serratum / P / 1-2' / dark purple, white stripe / Honshu 200
sikokianum / Tu / to 8" / fls dark purple / Japan 194 / 200
speciosum / Tu / 1-2' / greenish purp. & violet / Nepal & Sikkim 184
tortuosum / HH Tu / 2' / green & purple spathe / w Himalayas 218 / 267
tortuosum v. helleborifolium / to 2' / deeper purple / India 267
triphyllum / Tu / 1'+ / fls green, purple, brown / e North America 224 / 267
triphyllum ssp. stewardsonii - strongly fluted tube 25
triphyllum cv. 'Zebrinum' - cultivar of Eastern 'Jack-in-the-Pulpit' 226
urashima / 12-20" / spathe bronze-purple / Japan 200
wallichianum / HH Tu / spathe to 8"/spathe striped purp. / Himalayas 25
yamatense v. sugimotoi / spathe grn, blade yellowish inside / Honshu 200

ARISARUM - Araceae
proboscideum / Tu / 3-4" / greenish fls / Apennines (336) / 123
simonorrhinum / Gh / ? / purplish fls / Algeria, Morocco 61
vulgare / Tu / 6-8" / pale green to mahogany fls / Medit. reg. 212 / 185

ARISTEA - Iridaceae
africana / Gh P / 6" / fls blue / Cape of Good Hope 238
'Ecklonii' / Gh P / 1½-3' / bright blue fls / tropical Africa 25
pusilla / Gh P / 1-3' / blue fls / Cape of Good Hope 206
spiralis / P / to 15" / very small lvs; fls white & green/ S. Africa 61
thyrsiflora / Gh P / 3-5' / bracts wht marg, perianth blue / S. Afr 25
woodii / Gh P / 18" / blue fls / South Africa 184

ARISTOLOCHIA - Aristolochiaceae
baetica / Gh Cl / to 15' / fls brownish to blackish purple / Spain 212 / 286
californica / HH Cl / to 10' / fls green & purple / California 228
chilensis / Gh Cl / 10' / purple & green fls / West Indies 206
clematitis / P / to 3½' / yellow fls / Europe 286
durior - see A. macrophylla 36
elegans / Gh Cl / high / purple, white & yellow fls / Brazil 128
macrophylla / Cl / high / fls purplish / Pennsylvania to Georgia 28
rotunda / P / 6-24" / yellow fls / s Europe 286
sempervirens / Cl/ evergreen to 15'/fls yel, purp, striped/ s Italy 286
serpentaria / P / to 2' / basal fls, greenish / e United States 226

ARISTOTELIA - Elaeocarpaceae
fruticosa / HH Sh / to 14' / rosy fls / New Zealand 238 / 36
peduncularis / HH Sh / 6' / fls white / Tasmania 36

ARMERIA - Plumbaginaceae
alliacea / P / 8-20" / fls purplish to white / w Europe 314 / 288
alliacea 'Alba' - the white phase 288
allioides - see A. alliacea 288
alpina - see A. maritima ssp. alpina 288
alpina ssp. balcana - see A. montana 61
arctica - see A. maritima ssp. sibirica 288
'Bees Ruby' - see A. pseudarmeria 'Bees Ruby' 46
berlengensis / sim. to A. mauritanica/ calyx tube hairy / Portugal 25
X 'Bloodstone' / 9" / fls almost red 268
'Brookside' - see A. maritima
bupleuroides - see A. alliacea 288
X caesalpina / 6" / fls pink (42)
caespitosa - see A. juniperifolia (429) / (42)
caespitosa 'Alba' - see A. juniperifolia 'Alba' *
caespitosa 'Bevans Var.' - see A. juniperifolia 'Bevans Var.' (41)
caespitosa 'Bloodstone' - see A. X 'Bloodstone' *
canescens / R / 10-14" / rose-pink fls / sc Europe (42)
canescens ssp. nebrodensis/3-10"/corolla pink-red-purp / c & e Medit. 288
cariensis / R / ?/ white or pink fls / ne Greece 288
cinerea / HH Sh / 10" / plant villous / n Portugal 288
corsica - see A. maritima 'Corsica' 24
elongata - see A. maritima v. elongata 208
filicaulis / R / tufts / pink or white fls / se Spain 212 / 96
'Formosa Hybrids' / 18" / deep carmine to terra-cotta fls 279
girardii / R / to 6" / pink fls / sc France on dolomite 288
girardii 'Alba' / white fls *
gussonei - see A. morisii 288
halleri - see A. maritima ssp. halleri 288
juncea - see A. giraldii 288
juncea f. alba - pure white form 25

juniperifolia / R / 6" / fls pale purplish or pink / Spain 212 / 25
juniperifolia 'Alba' / white fls *
juniperifolia 'Bevans Var.' / 2-4" / dark rose fls 268
juniperifolia 'Bloodstone' - see A. X 'Bloodstone' *
juniperifolia 'Six Hills' - see A. maritima 'Laucheana' 208
labradorica / R/to 1'/ pink to purple fls / Greenland, James Bay reg. 99
labradorica v. submutica / R / 1' / pink or purple fls / Gaspe & ne 99
latifolia - see A. pseudarmeria (42)
leucocephala / HH R / to 1' / fls white to rose / Corsica, Sardinia 25
littoralis / R / ? / white or pinkish fls / sw Spain, s Portugal 288
macrophylla - see A. pinifolia 25
majellanica - see A. maritima 25
majellensis - see A. canescens ssp. nebrodensis 288
maritima / R / 6" / fls in pink shades / coastal Eu. & N. Am. 147 / 129
maritima f. alba / 6-9" / fls white 300 / 268
maritima ssp. alpina / to 10" / fls deep reddish or white 288
maritima arctica - see A. maritima ssp. sibirica 288
maritima 'Corsica' / brick-red fls 24
maritima 'Corsica' f. alba - white form
maritima 'Dusseldorf Pride' / deepest red fls 48
maritima v. elongata - see A. maritima 208 / 25
maritima ssp. halleri / to 1' / fls bright pink to red / w & c Eu. 314 / 288
maritima ssp. labradorica - see A. maritima ssp. sibirica 288
maritima 'Laucheana' / 6" / bright pink fls / hort. selection 95 46
maritima 'Merlin' / soft pink fls 154
maritima 'Pride of Dusseldorf' - see A. maritima 'Dusseldorf Pride' *
maritima ssp. purpurea / 8-15" / fls purplish / Germany, Italy 288
maritima ssp. sibirica /to 8"/ fls pale pink/ Siberia, arctic N. Am. 288
maritima 'Splendens' - brilliant red fls 208
maritima 'Vindictive' / 6" / fls rosy-red 147 / 268
mauritanica / HH P / to 2' / pink fls / Spain & n Africa 25
montana - see A. alliacea 288
morisii / HH P / 8-16" / fls pale pink / Sardinia, Sicily 288
mulleri - see A. maritima ssp. halleri 288
pinifolia / P / 12-20" / downy and densely tufted / coastal Spain 96
plantaginea - see A. alliacea 288
pseudarmeria / HH Sh / to 10" / fls usually white / Portugal 212 / 288
pseudarmeria 'Bees Ruby' / to 18" / fls glistening pink 46
pungens / HH R / to 10" / densely spiny lvs; fls pale rose / s Port. 212
rigida - see A. plantaginea 288
rumelica / P / 10-14" / pinkish to red purple / c & se Balkan Pen. 288
scabra - see A. maritima 209
[1] setacea - see A. giraldii
sibirica - see A. maritima v. sibirica 25
'Six Hills' - see A. maritima 'Laucheana' 208
splendens - see A. maritima 'Laucheana' 25
splendens ssp. bigerrensis /Dw Sh / 2-6"/ corolla purp/ n & w Spain 288
undulata / H / to 20" / corolla usually white / Greece 288
vulgaris - see A. maritima v. purpurea 25
welwitschii / HH Sh / to 10" / fls pink to white / wc Portugal (485) / 288

[1] Lawrence, American Rock Garden Society Yearbook, 1941

ARNEBIA
echioides - see ECHIOIDES longiflorum 129 / 25

ARNICA - Compositae
acaulis / P / 5-36" / pappus whitish / Pa. & Del. to Fla. 99
alpina / R / under 1' / yellow fls / n Scandinavia 209 / 245
alpina ssp. angustifolia - peduncles glandular, circumpolar 245
alpina ssp. sornbergii / stem solitary to 3-flowered / e Canada 245
amplexicaulis / P / 1-3' / pale yellow fls / Calif., Alaska, Mont. 229
amplexicaulis piperi / silv. woolly lvs; fls yel /Cal. n to Alaska 229
angustifolia alpina - see A. alpina ssp. angustifolia 245
chamissonis / P / 8-40" / fls yellow / nw North America 227 / 228
chamissonis ssp. chamissonis/P/8-40"/fls yel;pappus tawny/Cal-Alaska 229
chamissonis ssp. foliosa / to 2' / lvs entire / nw North America 25
chionopappa / R / to 1' / fls yellow / e Quebec 96
cordifolia / P / to 18" / fls bright yellow / nw & c N. Am. 229 / 25
frigida / R / to 16" / yellow fls / Alaska, nw Canada, Siberia 13
latifolia / P / to 2' / fls yellow / Wyo., Col., Calif., Alaska 228 / 25
lessingii / R / to 1' / fls pale yellow to orange / Yukon, Alaska 209
longifolia / P / to 2' / yellow fls / Calif., Colorado, Canada 229
louiseana / R / 8" / fls pale yellow / w Canada 96
mollis / P / to 2' / fls yellow / North America 229 / 25
montana / R / 6" / fls deep orange / Europe, in mts. 148 / 132
nevadensis / 4-12" / yel. & orange fls / Cascade Mts. & Sierra Nev. 64 / 229
sachalinensis / P / 1-3' / yellow fls / Sakhalin (430) / 200
unalascensis / R / 6-14" / fls yellow / Japan 273 / 200
unalascensis v. tschonoskyi - corollas slightly pilose 200

ARONIA - Rosaceae
arbutifolia / Sh / 5-10' / fls white; frs red / e North America (337) / 36
arbutifolia 'Brilliantissima' - see A. arbutifolia f. macrocarpa 43
arbutifolia f. macrocarpa - hort. form with larger frs 43
atropurpurea - see A. prunifolia 221
melanocarpa / Sh / to 3' / fls white; frs black / e North America 222
prunifolia / Sh / to 9' / fls white, frs purple-black / e N. Am. 221

ARRHENATHERUM - Gramineae
elatius / P Gr / to 5' / pale green or purplish panicle / Europe 25
elatius v. bulbosum / 2-3' / may be weedy / Europe 25
tuberosum - see A. elatius v. bulbosum 25

ARTEMISIA - Compositae
absinthium / P / 3' / fls yellow; lvs silky / Europe 129
alpina / P / to 1' / corolla yellowish, few fls / Caucasus 289
arborescens / HH P / to 3½' / bright yellow fls / Medit. region 110 / 25
arctica / P / to 2' / greenish & dark brown fls / arctic Eurasia 209
assoana - see A. pedemontana (521) / 289
borealis / R / to 1' / silky gray lvs / circumboreal 209
campestris ssp. maritima / HH P / 2'+ / lvs fleshy, dk grn / Spain 212
chamaemelifolia / P / 12-10" / aromatic, yellow fls / Alps to Iran 77 / 289
douglasiana / P / tall / elliptic, unlobed lvs / Calif., Oregon 228
eriantha / R / 6" / silvery ferny lvs / s & c Alps 96
frigida / R / to 1' / silvery, candescent plant / w United States 28
genipi / R / to 4" / fls yellowish / European Alps 148 / 93
glacialis / R / 4" / yellow fls; silvery lvs / Alps 148 / 107

glomerata / R / to 6" / silky lvs / alps of Far East 200
kitadakensis / R / 8-12" / white-villous lvs / alps of Honshu 200
lactiflora / P / 4-5' / fls creamy-white / China, India 129
lanata / Sh / to 8" / yellow fls / c & s Europe, in mts. 50 / 61
latifolia - plant of the western Siberian lowland 266
laxa - see A. umbelliformis 289
ludoviciana / P /to 3' / lvs wht / Mich to Wash, s to Ark, Tex, Mex. 25
maritima - see A. campestris ssp. maritima 212
michauxiana / P / 18" / lvs green, white reverse / nw N. Am. 25
mutellina - see A. umbelliformis 314 / 289
nitida - see A. assoana 93 / 212
pedatifida / mat / fl spikes 3" / Wyoming 96
pedemontana / R / to 1' / yellowish fls / s Europe 289
petrosa - see A. eriantha 289
schmidtiana 'Nana' / R / 4" / bright silver lvs 46
scopulorum / Dwarf / silvery / Colorado 154
sinanensis / ? / 8-20" / silky pubescent / Honshu 200
spicata - see A. genipi 95
splendens / R / tufted / non-woolly heads / Caucasus 96
stelleriana / P / 8-24" / white-woolly lvs; yellow fls / ne Asia 164 / 200
tridentata / Sh / to 10' / silvery-gray-pubescent / w N. Am. (331) / 25
tridentata trifida - see A. tripartita 294
trifurcata / R / 6" / lvs densely silky / Hokkaido 200
tripartita / Sh / to 12' / fls yellow to brown / w North America 294
umbelliformis / Sh / to 10" / fls pale yellow / s & c Eu., in mts. 289

ARTHROPODIUM - Liliaceae
candidum / HH R / 10" / white starry fls / New Zealand 123
candidum v. maculatum - purplish-leaved (44)
candidum 'Purpureum' - see A. candidum v. maculatum *
cirrhatum / Gh / 3' / fls white / New Zealand (44) / 128
milleflorum / HH P / 2'+ / pale rosy mauve fls / se Australia 109
millefoliatum / R / 6-9" / white or soft lilac fls / Tasmania 154
minus / Gh / 18" / white fls / Australia 206
paniculatum / HH P / 18" / branching panicles / Tasmania 64

ARUM - Araceae
alpinum - see A. maculatum v. alpinum (45)
creticum / Gh Tu / 10-14" / fls pale green & whitish / Crete 128 / 25
dioscoridis / HH Tu / to 1' / fls grn mottled purp. / e Medit. reg. 149 / 267
hygrophilum / Gh Tu / 12" / pale green & purple fls / Syria 61
italicum / HH Tu / 18" / fls pale yellow; lvs veined white / s Eu. 149 / 267
italicum 'Marmoratum' - somewhat gray-marbled along veins (259) / 279
italicum ssp. neglectum /Tu/6-16"/spathe pale gr.-yel./w Eur.& Medit. 290
italicum 'Pictum' - narrower lvs marbled gray & cream 279
korolkowii / Tu / to 2' / green & white fls / c Asia 3
korolkowii f. immaculatum / spotless lvs; white & purple fls 28
maculatum / Tu / 12" / fls green, purple spotted / Eu., England 184 / 107
maculatum v. alpinum / spathe purple margined / Alps 25
1 maculatum 'Immaculatum' - unspotted; spathe wht inside / Aus. 25
nigrum / R / 6-8" / purplish-maroon spathes / se Europe (78) / 185
orientale / HH Tu / to 18"/similar to italicum / Jugo.-Turk-Turkmen 25
palaestinum / Tu / 2' / fls black purple / Israel 25
pictum / HH Tu / 10" / dark purple spathe / Corsica, Sardinia 25

1 Prime, Lords and Ladies, 1960.

ARUNCUS - Rosaceae
aethusifolius / P / 12" cream fls, gr Astilbe-like lvs / Korea 249
dioicus / P / to 6' / showy, white fls / Eurasia, North America 224 / 287
sylvester - see A. dioicus 129 / 287
vulgaris - see A. dioicus 287

ARUNDO
conspicua - see CORTADERIA richardii 268

ASARINA - Scrophulariaceae
antirrhinifolia / Gh Cl / 2-6' / fls purp. & yellow to white / Texas 226 / 25
erubescens / Gh Cl / to 10' / rose-pink fls / Mexico 128 / 25
lophospermum / like A. erubescens/ rose-purple fls / Mexico 25
procumbens / R / trailing / pale yellow fls / Pyrenees 148 / 274

ASARUM - Aristolochiaceae
canadense / R / to 1' / fls basal, reddish-purple / e North America 224 / 107
caudatum / R / 4" / fls brownish, appendaged / w North America 229 / 107
europeum / R / 6" / evergreen lvs; tiny purple fls / Europe 149 / 46
hartwegii / R / 6" / fls light reddish-brown / Sierra Nevadas, US (46) / 107
sieboldii / R / rhizomatous / large lvs; solitary fls / Honshu 200
virginicum / R / 7" / purple fls; evergreen lvs / Va. to Tenn. 224 / 25

ASCLEPIAS - Asclepiadaceae
curassavica / Gh P / 2' / orange-red fls / tropical America 227 / 128
exaltata / P / 1½-4½' / greenish & white fls / e US 224 / 99
incarnata / P / to 5' / rose-purple to flesh fls / e & c N. Am. 224 / 25
incarnata f. albiflora - whitish fls 99
incarnata 'Alba' - see A. incarnata f. albiflora *
incarnata ssp. pulchra / pubescent / Nova Scotia to N. Carolina 25
physocarpa / Gh P / 6' / fls white / South Africa 25
pulchra - see A. incarnata ssp. pulchra 28
purpurascens / P / 1-4' / fls deep rose / e United States 225
quadrifolia / P / 2' / fls pink to white / n & c US & Canada 107
solanoana / P / to 1' / flatish stems; fls rose purple / n Calif. 228
speciosa / P / 3' / fls large, purplish / Canada & w US 229 / 25
sullivantii /P/to 3'/corolla purp-rose/Ont.to Minn, s to Nebr.,Okla. 25
syriaca /P/3- 6'/fls purplish-rose/N.Bruns. to Man., s to Ga & Okla. 224 / 25
tuberosa / P / to 3' / orange fls, varying red, yellow / e & c N.Am. 49 / 25
verticillata / P / 2'+ / greenish-white fls / e & c North America 229 / 99
verticillata v. pumila / 10" / greenish-white fls / w United States 107

ASCYRUM - Guttiferae
hypericoides - see HYPERICUM hypericoides 25
stans / P / 1-3' / bright yellow fls / e United States 28

ASIMINA - Annonaceae
triloba / T or Sh / 10-40' / fls purplish / e & s U.S. 36

ASPARAGUS - Liliaceae
asparagoides / Gh Cl / high / red frs / South Africa 25
densiflorus 'Sprengeri' / Gh / small pinkish fls; br. red frs/ Natal 128
medeoloides - see A. asparagoides 25
plumosus - see A. setaceus 128

setaceus / Gh Cl / high / red frs / South Africa 128
sprengeri - see A. densiflorus 'Sprengeri' 128
verticillatus / Cl / to 15' / red-berried / Iran to Siberia 28

ASPERELLA
hystrix - see HYSTRIX patula 268

ASPERULA - Rubiaceae
aristata / Dw. Sh / 4-24" / variable fls / s Eur. to se Austria 289
arvensis / A / 4-22" / usually bluish violet / Europe 289
cynanchica / R / 9-12" / fls light pink / Europe, Asia 306 / 107
lilaciflora v. caespitosa / R / mats/ starry pink fls / e Medit. reg. 48 / 123
nitida / R / to 10" / fls pale purplish / mts. of c Greece 299
nitida ssp. puberula - see A. sintenisii 249
odorata - see GALIUM odoratum 25
orientalis / A / to 1' / blue fls / Eurasia (338) / 25
pontica / R / 2" / fls pink / ? 107
purpurea / R / cushion / few fls / s Eur to s Aust & nc Romania 289
rupicola / R / 4-6" / fls pink / sw Europe 289
sintenisii / R / 1" / lt. gr. cushion; pink tabular star fls/ Turkey 249
tinctoria / P / to 2' / fls white or reddish / Europe 25

ASPHODELINE - Liliaceae
brevicaulis / B / stems slender; lvs ascending; fls yel, veins grn 61
liburnica / P / 1-2' / fls yellow / Greece, Austria, Italy 182 / 25
lutea / P / to 4' / fragrant yellow fls / Medit. reg. 194 / 25
taurica / P / to 2' / white fls / Greece to Caucasus 25

ASPHODELUS - Liliaceae
acaulis / HH R / rosette / fls pale pink / Algeria 132 / 123
aestivus / P / 4½' / white fls, reddish veined / Medit. reg. 210 / 212
albus / P / 3' / white or pinkish fls / Portugal, Medit. reg. 148 / 212
cerasiferus / P / 4-5' / white fls / w Medit. reg. 279 / 25
fistulosus / B or HH P / 2' / fls white to pale pink / Canary Isls. 148 / 25
microcarpus - see A. aestivus 148
ramosus - see A. cerasiferus 25

ASPIDOTUS - Polypodiaceae
densa / F / to 4" / small rock-fern / nw North America 142

ASPLENIUM - Polypodiaceae
adiantum-nigrum ssp. anopteris - finely cut fronds 159
adulterinum / F / 1' / thin, light green blades / ec Eu., in mts. 286
antiquum / Gh F / 2-3' / coriaceous fronds, epiphytic / s Japan 272 / 200
X Asplenosarus X ebenoides/ F/ intergen.hyb, ASPLENIUM & CAMPTOSORUS 99
billottii / Gh F / low / lancolate blade / coastal England 142
ceterach - see CETERACH officinarum 142
cuneifolium / F / 8-40" / flabellate leaf-segments / s & c Europe 286
X ebenoides - see A. X Asplenosarus X ebenoides 99
flaccidum / Gh F / to 3', hanging / epiphytic / N. Zealand 159
fontanum / F / 3-4" / narrow, lanceolate fronds / c & s Europe 142
fontanum halleri - included in the sp. 286
forisense / F / 5-6" / linear-lanceolate blade / French Alps 142
incisum / F / to 1' / sterile fronds shorter / Far East 272 / 200
onopteris - see A. adiantum-nigrum ssp. anopteris *

pinnatifidum / F / 3-9" / evergreen fronds / c & s United States 191 / 99
platyneuron / F / 6-15" / evergreen fronds / Canada to S. Am. 191 / 99
richardii / F / 5-15" / finely cut frond / New Zealand 142
ruta-muraria / F / 1-3" / on walls and basic rocks / Europe 148 / 142
septentrionale / F / 2-5" / leathery frond / Europe 142
trichomanes / F / 3-5" / delicate fronds / n temperate zone 97 / 123
trichomanes 'Cristatum' - frond crested, partly true from spores (431) / 142
trichomanes 'Incisum' - narrow frond, deeply cut, true-breeding 142
viride / F / 2-6" / evergreen fronds / n Europe, n N. Am. 148 / 142

ASTELIA - Liliaceae
cockaynei / P / 16" / greenish fls / New Zealand 25
fragrans / Bb / to 8' / ripe berry brown-orange / New Zealand 193
linearis / R / acaulescent / large red fls / New Zealand
nervosa / P / to 2" / fragrant greenish fls / New Zealand 279 / 25
nervosa v. montana - included in the sp. 279
solandri / P / lge, densely tufted 2-5' / fls pale yel. / N.Z. 61

ASTER - Compositae
acuminatus / P / 2½' / fls white or purplish / e N. Am. 224 / 25
adscendens - see A. chiliensis ssp. adscendens 142
ageratoides v. viscidulus / p / 14-26" / fls blue-purple / Honshu 200
X alpellus / R / 1' / blue fls / A. Alpinus x A. amellus 25
X alpellus 'Triumph' / R / 9" / rays blue, disc orange 300 / 268
alpigenus / R / to 16" / fls violet to lavender / Calif. to Wash. 312 / 228
alpinus / R / 4-8" / fls mauve to rosy-purple / European Alps 148 / 129
alpinus 'Albus' - white form, rarely found with good fls 147 / 132
alpinus 'Beechwood' / lavender-blue fls/ one of the best forms 129
alpinus 'Berggarten' - see A. tongolensis 'Berggarten' 268
alpinus 'Coeruleus' / stems to 10" / ray fls blue / Switzerland 25
[1] alpinus 'Dark Beauty' / 12" / deep blue with gold eye
alpinus v. dolomiticus / 8" / near purple fls / Balkans 25
alpinus 'Goliath' / 10" / violet-blue fls 46
alpinus 'Pirinensis' - deep rose form 64
alpinus 'Roseus' / 6" / pale rose fls 25
alpinus v. speciosus / to 20" / fls dark violet / c Asia 25
alpinus 'Superbus' - large and showy form 46
alpinus 'Wargrave' - lilac-colored fls 29
alpinus v. wolfii / 1'+ / blue fls / Switzerland 25
amellus / P / 2-2½' / purple solitary fls / Italy 147 / 268
amellus 'Triumph' - see A. alpellus 'Triumph' 268
andinus / R / ? / cluster-headed dw. / alp. peaks, Mont. to Col. 107
anomalus / P / 1-4' / bracts bent back, fls colored/ Kans. & Okla. 229
apricus / solitary flowered / like A. foliaceus 64
asteroides / R / 2-6" / fls mauve / se Tibet, w China 268
asteroides 'Albus' - white fls *
azureus / P / to 3' / fls deep blue or blue-violet / e North America 123
batesii / P / to 2' / sky-blue fls / Nebraska prairies 236
'Beechwood' - see A. alpinus 'Beechwood' 46

---

1 Thompson and Morgan Catalog, 1986.

blakei / sim. to A. acuminatus except violet rays/Newfoundland-N.J. 224
brachytrichus / R / 1' / violet fls / China 25
bellidastrum / R / 1' / white fls / mts. of Europe 93
bigelovii - see MACHERANTHERA bigelovii 25
caucasicus / P / 2' / purple fls / Caucasus 25
chilensis ssp. adscendens /P/ 20-40"/ fls blue or pinkish/Rocky Mts. 142
conspicuus / P / 2' / violet fls / nw United States 229 / 25
corymbosus - see A. divaricatus 143
diplostephioides / P / 2½' / blue or pale purple fls / Himalayas 25
divaricatus / P / 3' / white fls / e United States 25
dolomiticus - see A. alpinus 'Dolomiticus' 46
dumosus 'Victor' / R / 9" / light blue fls 184
engelmannii / P / 2-4½' / fls white, aging pinkish / nw N. Am. 228 / 142
ericoides / P / 2-3' / fls white / e North America 28
ericoides 'Blue Star' / P / to 3' / pale blue / e North America 208
eriophorum / R / low / gray lvs, small daisies / Chile [1]
farreri / R / 9" / large violet-purple fls / w China, Tibet 123
farreri 'Berg Garten' - see A. tongolensis 'Berggarten' 268
fendleri / R / to 16" / bluish fls / plains of c United States 25
flaccidus / R / to 6" / blue or mauve fls / Asia 208 / 25
foliaceus / P / 1-2' / fls rose, blue, violet / n North America 229 / 228
foliaceus v. lyallii - involucral bracts much narrower / n Idaho 142
forrestii - see A. souliei 25
X frikartii / P / 2½' / fls sky-blue, orange disc 269 / 123
fruticosus - see FELICIA fruticosus 25
fuscescens / P / 2½' / fls pale purplish-blue / Yunnan 67
'Golden Sunshine' / P / 4' / bright golden-yellow fls 300
gormanii / P / 4-12" / lvs crowded; fls white or pink / N.C. - Ore. 229
hesperius / P / 3-5' / blue or pink fls / w North America 229 / 142
himalaicus / R / to 10" / purplish-blue fls / Himalayan reg. 25
junceus - see A. junciformis 99
junciformis / P / to 2' / fls purple, roseate, white / n N. Am. 224 / 99
'Knapsbury' - see A. tongolensis 'Napsbury' 268
kumleinii - see A. oblongifolius 259
laevis / P / to 3½' / fls blue or pale purple / n N. Am. 224 / 25
ledophyllus / P / 3' / fls lavender-purple / n Calif. to n Wash. 228 / 25
likiangensis - see A. asteroides 268
linariifolius / R / 1'+ / fls lavender / ne North America 224 / 107
linariifolius f. leucactis - the white form 99
linosyris / P / 2' / fls bright yellow / Europe 129
lipskyii / P / 18" / deep violet fls / Tibet 25
lowrieanus / P / to 6' / pale blue or violet fls / e N. Am. 224 / 99
luteus / P / to 2½' / golden yellow, rays canary yellow / France 25
maackii / P / 18-24" / blue fls / Honshu, Korea, Manchuria 164 / 200
macrophyllus / P / to 4' / pale blue or violet fls / e & c N. Am. 224 / 25
meritus - see A. sibiricus v. meritus 25
mongolicus / P / to 3' / ray fls bright lavender-blue / e Asia 25
montanus - see A. sibiricus 160
natalensis - see FELICIA rosulata 25
nemoralis / P / 2' / pink or lilac-purple fls / ne N. Am. 25
novae-angliae / P / to 6' / fls violet-purple or pink / e N. Am. 224 / 129
novae-angliae 'Roseus' - fls rose-pink 25

[1] G. Schenk, The Wild Garden, A Catalog, n.d.

novae-belgii 'Countess of Dudley' / R / 9-12" / clear pink fls 135
novae-belgii 'Lady in Blue' / 12-15" / veronica-violet fls 268
oblongifolius / P / 6-24" / fls violet-purple / c United States 224 / 259
oblongifolius v. orientis f. roseus / 12-58" / rosy fls / W. Va. 313
oreophilus / P / 18" / pale lilac to blue-mauve fls / w China 25
paludosus ssp. hemisphericus / P/to 1'/ fls pungent, viol./Ala.-Tex. 99
pappei - see FELICIA amoena 25
patens / P / to 3' / fls blue-purple or violet / e US 263
paucicapitatus / P / 18" / white fls / Olympic Pen., Vancouver Is. 312 / 25
peregrinus / R / 1'/ bluish-purple fls / North America 61
perrinensis - see A. alpinus 'Pirinensis' 64
petiolatus / HH R / prostrate / rose & yellow fls / S. Africa 61
porteri / P / 2-3' / white / Colorado 96 / 229
prenanthoides / P/ 1-3'/ lvs clasp stem; fls pale/ ne U.S. s to Va. 224
ptarmicoides / P / to 2' / fls white / e North America 224 / 99
pulchellus / R / 6"/ large purple-blue fl / Wyo. 96 / 154
purdomii - see A. flaccidus 25
pyrenaeus / P / to 3' / fls large, purple / ne Spain 148 / 96
schistosum - see A. lowrieanus 99
scopulorum / R / 6" / light violet fls / Montana to California 142
sedifolius ssp. trinervis / P / 1-4' / fls bluish, pinkish/ s France 314 / 142
sericeus / P / 8-14"/ violet-purple fls; silvery lvs / c US 259
sibiricus / P / to 16" / purple or blue fls / arctic Europe, N. Am. 13 / 25
sibiricus v. meritus / taller, heads more numerous / nw N. Am. 25
sikkimensis / P / 3-4' / fls purple-blue / Sikkim, Nepal 268
souliei / R / to 1' / blue or mauve fls / se Tibet, nw Yunnan 25
souliei v. limitanus / R/ 6-8"/ purp-viol fls / w China borderlands 123
spectabilis / P / 1½-2' / fls bright violet / Mass. to N. Carolina 312 / 129
steeleorum / R / to 15" / violet fls / Virginia & W. Virginia 313
subcaeruleus - see A. tongolensis 268
subintegerrimus - see A. sibiricus 289
subspicatus / P / 1-3' / purple fls / Aleutians to California 142
thomsonii / P / to 3' / fls clear lilac-blue / Nepal, Kashmir 96
tibeticus - either A. flaccidus or a form of A. alpinus 25
tongolensis / R / to 1' / fls pale blue / w China (605)
tongolensis 'Berggarten' / 1' / fls deeper blue 283
tongolensis 'Napsbury' / 18" / fls heliotrope-blue 268
trinervis - see A. sedifolius ssp. trinervis 289
trinervius v. viscidulus - see A. ageratioides v. viscidulus 200
tripolium / A / 10-22" / purplish fls / Far East 164 200
turbinellus / P/ 1-4'/stems long and wiry; fls colored/ se Kan-Okla 229
uliginosus / HH R / 2" / mauve fls / Basutoland (48)
umbellatus / P / to 8' / white fls / e North America 224 / 25
undulatus / P / 1-4' / fls pale blue, violet / e & c N. Am. 224
'Wargrave Pink' / 5" / large pink fls 16
1 'Wartburg Star' / P / 3' / deep violet-blue fls
yunnanensis / P / to 3' / fls blue-mauve / w China, se Tibet 268
yunnanensis 'Napsbury' - see A. tongolensis 'Napsbury' 268

ASTERANTHERA - Gesneriaceae
ovata /Gh Cl/?/ raspberry-red & yellow fls / Chile, in rain-forests 25

ASTERISCUS - Compositae
maritimus / HH R / to 8" / deep yellow fls / w Medit. reg. 212 / 289

---

1 Wayside Gardens Catalog, Spring 1985.

ASTILBE - Saxifragaceae
X arendsii 'Fanal' / 2' / dark crimson-red fls 184 / 279
biternata / P / to 6' / fls yellowish-white / Appalachian Mts. 25
chinensis / P / to 2' / fls white, rosy, purplish / Far East 129 25
chinensis v. davidii / 1-2' / fls rose-purple / China, Korea 200
chinenis v. pumila / 10" / raspberry-red fls 50 / 107
X crispa - a race of garden hybrids 25
X crispa 'Gnome' / 10" / pink fls 189
X crispa 'Perkeo' / 6-9" / bright pink fls (58) / 268
davidii - see A. chinensis v. davidii 200
'Fanal' - see A. X arendsii 'Fanal' 279
glaberrima - see A. japonica v. terrestris 200
glaberrima 'Saxatilis' - see A. japonica v. terrestris 'Saxatilis' 132
japonica v. terrestris / to 20" / fls white 200
japonica v. terrestris 'Saxatilis' / 3" / fls light purple, wht tips 132
microphylla / P / 12-32" / pale rose fls / Honshu, Kyushu 165 200
X 'Peach Blossom' - see A. X rosea 25
X rosea 'Peach Blossom' / best known cult./ fls pale peach-pink 25
rubra / P / to 2' / lflts oblique, numerous; fls rose / India 61
simplicifolia / R / to 1' / white fls / Honshu 165 / 200
tacquetii 'Superba' / 4' / rosy-purple fls; mahogany red stems 268
thunbergii / P / 2' / fls white, aging pink / Japan 165 25
thunbergii v. congesta - lower panicle branches compound 200
thunbergii v. fugisanensis / 2' / white fls / mts. of Honshu 200

ASTRAGALUS - Leguminosae
adsurgens / R / 1'+ / fls purple / Far East, w United States 165 / 200
alopecuroides / P / to 5' / yellow / Siberia & c Europe 25
alpinus / R / 4-6" / fls white & blue / European Alps, Scotland 210 123
ambiguus - see OXYTROPIS ambigua 287
angustifolius / R / 2-8" / white & purple fls / Balkan mts. 93 287
angustifolius ssp. angustifolius / racemes 3-8 fls / Balkan Pen. 287
austriacus / P / 4-24" / lflts 5-10 prs; fls blue & viol / Europe 287
austrinus / A or B / to 16" / fls pale purple-violet / Tex., Mex. 29
balearicus / HH R / caespitose / fls white / Balearic Isls. 287
caespitosus / R / small cushion / magenta-purple fls / nc N. Am. 64
calycosus / stems 1"-/ silvery, wht, pink, blue, purp/ sw Wyo & Col. 229
campestric ochroleuca - see OXYTROPIC nuriae 287
canadensis / P / 1-4' / fls greenish cream / e & c North America 99
caryocarpus - see A. crassicarpus 25
centralpinus / P / to 3'+ / yellow fls / sw Alps, Bulgaria 259
ceramicus /P/wiry stems spread 1-12"/ptls wht-flesh/ Ore, Mont, Col. 229
chamaeleuce /to 1½"/yellow, pink-purp., yellow eye / Wyo, Ut., Col. 229
Cottonianus / P / 5-12" / 10-20 fld, yellow / c Asia 7
crassicarpus / R / decumbent, 15" / fls violet-purple /Minn. to Tex. 229 / 25
danicus / P / 3-12" / lflts 6-13 prs; fls purp / Ireland, Russia 287
drummondii / P / to 2' / fls wht or crm, lilac tipped keel/ nw U.S. 229
exscapus /R / tufted / fls yellow / c Europe 314 / 287
falcatus / P / 16-32" / fls yellow / c & se Russia 287
filicaulis / A / 2-16" / fls 6-20, carolla pale viol. / c Sov. Asia 8
frigidus / R / 4-14" / yellow-white / n Europe, mts c Europe 287
glycyphyllos / P / 1-3', trailing / fls cream / Europe, Siberia 210 / 93
haarbachii / A to P / 2-20" / corolla yellow / e Bulg., ne Greece 287
hamosus / P / to 2' / corolla yellow / s Europe 287
inflexus / R / 12" / fls large, purple / nw United States 81

japonicus / P / 10-12" / calyx white-pubescent / Hokkaido 200
kentrophyta / R / variable hts / fls pink-purple / N. Dak to Calif. 229
lusitanicus / HH P / 18" / large white fls / Portugal, sw Spain 212
maritimus -like A. haarbachii / legume curved, tuberculate / Sardinia 287
melilotoides / P / 1½-3' / fls rose & violet / Far East 8
mexicanus (ex Kansas) - see A. mexicanus v. trichocalyx 99
mexicanus v. trichocalyx - decumbent / fls cream / calyx white 99
missouriensis / R / to 6" / fls pink-purple / Mont. to Okla. 229
mollissimus / P / 1-2' / dense racemes pink purp; lvs silky/c U.S. 229
monspessulanus / R / acaulescent / fls purplish-violet / s Eu. 148 / 287
nevadensis / R / 4-12" / yel. & orange / Cascade Mts. & Sierra Nev. 229
norvegicus / P / 8-16" / fls pale violet / arctic Europe 287
nutzotinensis / R / 4-6" / fls large, blue & white / Alaska 213
onobrychis / P / 2' , trailing / fls pale or dark violet / Europe 287
parnassi / lg tussocks/ corolla purp-pink, rarely yel/ s Balkan pen. 287
peduliflorus /R/8-20"/corolla yel./e&c Pyrenees,Alps,Carp., c Sweden 287
poterium - see A. balearicus 287
purpureus / R / 4-16" / purplish fls / s & w Europe 287
purshii / R / 2" / fls white or blue & white / Rocky Mts. 229 / 25
purshii v. glareosus / fls reddish-purple / Calif., Ore., B.C. 142
sachalinensis / R / to 7" / short racemes / Far East 8
scaberrimus / R / acaulescent / yellow fls / Far East 8
schelichovii / R / 8-16" / yellowish fls / betw. Yakutsk & Akhotsk 8
schizopterus / R / acaulescent / purple fls / Caucasus 8
sempervirens / R / to 16" / fls whitish to pale purplish / c Eu. 314 / 287
serbicus / P / tall / yellowish fls / Serbia 287
sericoleucus / P / stems prostrate to 2" / fls pink,purp/ nc U.S. 229
shiroumensis / R / to 1' / fls yellowish-white / Honshu alps 273 / 200
sinicus / B / 4-10" / reddish-purple fls / China, Japan, Taiwan 200
sirinicua / R / caespitose / fls yellowish & violet / Balkans 287
spatulatus / P / to 2" / Sask. to Neb., Col., Utah 25
Takhtadzhjanii / P/ stemless to 8"/ fls yel-grn/ Armeria-Transcauc. 8
tegetarioides / P/ prostrate thready stems; fls white / Ore & Wash 229
tennesseensis / R / 20", trailing / yellowish fls / mid-c US 225
utahensis /R /prostrate to 6"/ pink, purp., pale eye/ Ida, Wyo, Nev. 229
'Whitneyi' / low / yel, pink-wht, lilac / Sierra Nev, n coast ranges 228

ASTRANTIA - Umbelliferae
biebersteinii / R / to 1' / pink & white fls / Caucasus 107
carniolica / P / to 2' / fls whitish / se Alps 207 / 287
carniolica f. rubra / R / to 1' / fls dark red 47
carinthiaca - see A. major ssp. carinthiaca 287
gracilis - see A. carniolica 268
helleborifolia - see A. makima 25
major / P / 3'+ / fls white & purplish / s Europe 148 / 287
[1] major 'Margery Fish' / P / to 2' / tiny wht florets
major ssp. carinthiaca / with longer bracteoles / s Alps, nw Spain 287
maxima / P / 1-2' / fls pinkish / Caucasus 49 / 129
maxima alba / P / 1-2' / lvs 3-lobed; fls white / e Caucusus 61
minor / R / to 16" / fls whitish / Pyranees 148 / 287
minor variegata / P / 6-9"/ lvs 6-9 lobed, variegated; fls wht/ Eur. 61
pauciflora / P / to 16" / fls whitish / Italy 287
variegata - see A. minor variegata 61

---

[1] Wayside Gardens Catalog, Spring 1985.

ASYNEUMA - Campanulaceae
canescens / P / to 3' / pale lilac fls / Jugoslavia to Greece 25
canescens prenanthoides - see CAMPANULA prenanthoides 142
limonifolium / P / to 2' / blue fls / Jugoslavia to Turkey 25
lobelioides / P / to 2' / lilac-blue fls / Asia Minor 25

ATHAMANTA - Umbelliferae
cretensis / P / 2' / white fls / Spain to Jugoslavia 148 / 287
matthioli - see A. turbith 287
turbith / P / to 20" / white fls / nw Balkan Peninsula 287

ATHANASIA - Compositae
parviflora / Gh P / 2½' / fls yellow / Cape of Good Hope 206

ATHEROSPERMA - Atherospermataceae
moschatum / HH T / to 100' / creamy-white fls / Tasmania 36

ATHROTAXIS - Taxodiaceae
cupressoides / HH T / 20-40' / small close-pressed lvs / Tasmania 127
selaginoides / HH T / Tall /lvs & cones small/'King Billy Pine'/Tas. 109

ATHYRIUM - Polypodiaceae
crenulato-serrulatum /F/ 10-16"/ brown stripe, green pinnae/Far East 200
filix-femina / F / 1-4' / finely-divided fronds / temperate zones 22 / 159
filix-femina 'Cristatum' /F/lvs to 3'/rhiz. ascend./N.Am. s to Calif. 25
filix-femina 'Fieldiae' - a cruciate form 159
filix-femina 'Frizelliae' / 18" / pinnae bead-like / Ireland (424) / 159
[1] filix-femina 'Minutissima' / F / 6-10" / bi-pinnate, lanceolate
filix-femina 'Multifidum' - cultivar / N.Amer. to Calif., Eur origin 25
filix-femina 'Victoriae' / 3' / fronds cruciate and crestate (424) / 159
goeringianum 'Pictum' - see A. iseanum 'Pictum' 200
iseanum / F / 10" / mountain fern / Japan, Formosa 200
iseanum 'Pictum' / 2' / gray & green fronds 200
niponicum / F / creeping / Hokkaido, Honshu, China, Korea 200
pycnocarpon / F / to 2'+ / Glade-Fern / e & c North America (284) / 99
thelypterioides / F / 1-2½' / silvery indusia / e & c N. Am. 159 / 263

ATRAGENE
alpina - see CLEMATIS alpina 28

ATRAPHAXIS - Polygonaceae
buxifolia / Sh / to 2½' / fls pinkish white / Caucasus 36

ATRIPLEX - Chenopodiaceae
confertifolia / Sh / to 3' / lvs crowded entire 3/4" / e Ore to N.D. 25
hortensis 'Rubra' / A / 4-5' / whole plant reddish / Old World 129

ATROPA - Solanaceae
belladonna / P / 2-3' / dull purple-brown fls; frs poisonous / Eu. 148 / 25

AUBRIETA - Cruciferae
deltoidea / R / caespitose to trailing / fls red-purple / Greece 149 / 286

---

[1] The Rock Garden, vol. XVIII.

deltoidea 'Berachs White' - white fls *
deltoidea 'Dr. Mules' - deep violet-blue fls 48 / 123
deltoidea 'Eyrei' - large fls, deep violet 28
deltoidea 'Gurgedyke' - deep purple fls 46
deltoidea 'Leichtlinii' - pink fls in abundance 28
[1] deltoidea 'Mrs. Lloyd Edwards' - lilac-purple fls
deltoidea 'Purple Cascade' - blazing purple fls 300 / 282
deltoidea 'Purpurea' - horticult. variant of A. deltoides v. graeca 25
deltoidea 'Red Cascade' / 4" / reddish-purple fls 123
deltoidea 'Variegata' - lvs variegated 28
eyrii - see A. deltoidea 'Eyrei' 28
libanotica / R / short / fls variable color / Lebanon 154
olympica / P / prostrate/ petals violet; flattened frs / Turkey 77

AUCUBA - Cornaceae
japonica / HH Sh / 6-10' / purplish fls; scarlet frs / Japan 36

AULAX - Proteaceae
umbellata / Gh / 2' / yellow fls / Cape of Good Hope 206

AUREOLARIA - Scrophulariaceae
flava / P / to 4' / fls yellowish-orange / e & c United States 224 / 25
glauca - see GERARDIA flava 143
grandiflora / P / 20-40" / yellow fls / Wisconsin to Texas 224 / 236
laevigata / P / to 4½ / yellow fls / Pa., Ga., Tenn. 224
pedicularia - see GERARDIA pedicularia 263
virginica - see GERARDIA virginica 99

AURINIA
savatilis / P / mat / pale yellow fls / s & c Europe, Turkey 25
saxatilis citrina / P /mat / bright yellow fls / cultivar 25

AUSTROCEDRUS - Cupressaceae
chiliensis / T / to 60' / silvery-lined lvs; cones ½" long/ Chile 25

AVENA
candida - see HELICTOTRICHON sempervirens 268

AZALIA
procumbens - see LOISELEURIA procumbens

AZARA - Flacourtiaceae
gilliesii - see A. petiolaris 224
mycrophylaa / HH T / 15-30' / yellow fls, vanilla-scented / Chile 139
petiolaris / HH Sh / 15' / small yellow fls / Chile 224

AZORELLA - Umbelliferae
columnaris / HH R / close, hard, massive cushion / Peru 64

---

[1] Modic, Bulletin, American Rock Garden Society, 19:2.

BABIANA - Iridaceae
plicata / C / 4-12" / fls pale blue-lilac / sw Cape Province 185
pygmaea / C / to 2½", rarely 6"/yel, purp-maroon ctr / Cape Prov. 25
rubrocyanea / C / 6-8" / blue & scarlet fls / Cape Province 25
stricta / C / 6-10" / varii-colored fls / sw Cape Province 180 / 185
stricta v. rubrocyanea - see B. rubrocyanea 25
stricta v. sulphurea - fls creamy white, base stained blue 185
stricta v. villosa - deep crimson fls 267
stricta 'Zwanenburg Glory' - segments alternating blue & white 111
tubiflora / C / 1' / fls cr-white, red markings /coastal Cape Prov. 223 / 267
tubulosa tubiflora - see B. tubiflora *
'Zwanenburg Glory' - see B. stricta 'Zwanenburg Glory' 111

BACCHARIS - Compositae
halimifolia /S/to 12'/lvs toothed,gray-gr/coast marsh e N.& C. Amer. 25

BAECKEA - Myrtaceae
camphorosmae / HH Sh / 2' / fls white or pink / w Australia 61
ramosissima / HH Sh / low / pink fls / Australia 45
virgata / Gh Sh / 3' / fls white / Australia, New Caledonia 93

BAEOMETRA - Liliaceae
uniflora / HH Bb / 1' / fls red & orange-yellow / Cape of G. Hope 161 / 223

BAERIA
chrysostoma - see LASTHENIA chrysostoma 25

BAHIA - Compositae
oppositifolia / P / 4-12" / fls yellow / Mont., Col., to S. Dak. 229

BAILEYA - Compositae
multiradiata / R / 8-18" / fls yellow / sw United States, n Mexico 226

BALLOTA - Labiatae
nigra / P / to 3' / many fls, purple-white / Eur., n Africa, w Asia 25
pseudodictamnus / HH P / 12-20"/ fls purple & white/ s Aegean reg. 129 / 288

BALSAMORRHIZA - Compositae
dettoidea / P / to 3' / 2½-4" heads / B. C. to s Calif. 25
hookeri / R / 3-16" / yellow fls / Sierra Nevadas, S. Dakota, Col. 229 / 142
incana / P / 2' / yellow fls / nw US 229 / 25
rosea / R / 9" / fls golden-yel., aging roseate / e Cascades, Wash 229 / 142
sagittata / P / 8-32" / fls yellow / nw North America 229 / 228

BANKSIA - Proteaceae
baueri / HH Sh / to 10' / fls yellowish-brown / s Australia 45
marginata / Gh Sh / 6' / yellow fls / New South Wales 206
media / Gh Sh / 10-15' / fls golden / w Australia (215) / 38
meissneri / Gh Sh / 3' / linear lvs / w Australia 25
prionotes / Gh T / 15-25' / eleven inch lvs / Australia 28
repens / Gh Sh / 1' / yellow fls / Australia 206
solandri / Gh Sh / 6-12' / fls orange / w Australia 38
speciosa / Gh Sh / 4'+ / fls lemon-yellow / w Australia 45 / 25
tricuspis / Gh Sh / 8' / greenish-yellow fls / w Australia 38
victoriae / Gh Sh / 6-10' / orange fls / w Australia 38

BAPTISIA - Leguminosae
alba / P / to 3' / fls white / Virginia to Florida 225 / 25
australis / P / 3-4' / fls blue, occasionally white / midwest US 129 / 107
bracteata - see B. leucophaea 99
leucantha / P / to 4½' / white fls / c North America 224 / 282
leucophaea / P / to 2½' / fls cream / Michigan to Texas 224 / 25
perfoliata / HH P / 2'+ / yellow axillary fls / S. Carolina, Fla. 250
tinctoria / P / to 4' / fls bright yellow / Mass., Minn., Fla. 224 / 25

BARBAREA - Cruciferae
rupicola / HH P / 1½' / yellow fls / Corsica, Sardinia 286
vulgaris 'Variegata' / B / 18" / lvs cream-marbled 279

BARLIA - Orchidaceae
longibracteata - see B. robertiana 149
robertiana / P / 12-20" / reddish-violet fls / Medit. reg. 149

BARTSIA - Scrophulariaceae
alpina / A / to 8" / dull purple fls / n Europe 148 / 25

BAUHINIA - Leguminosae
variegata / HH T / 20-40' / fls in various colors / India, s China 187 / 25

BECKMANNIA - Gramineae
cruciformis - see B. eruciformis
eruciformis / P Gr / ? / bulbous stem base / Europe 268
syzigachne / A Gr / 1-3' / unusual inflorescence / N. Am., e Asia 61

BEGONIA - Begoniaceae
evansiana - see B. grandis 25
grandis / HH Tu / 2-3' / fls flesh-color / China, Japan 25
semperflorens 'Cinderella' - greenhouse plant 202
sutherlandii / Gh Tu / 1-2' / salmon-red fls / Natal 28

BELAMCANDA - Iridaceae
chinensis / P / 2-3' / orange fls, spotted red / e Asia 224 / 25
flabellata / P / 12-10" / lt yellow, orng. at base, fil. wht/ Jap. 25
punctata - see B. chinensis 200

BELLENDENA - Proteaceae
montana / HH Sh / 2' / white fls; scarlet frs / Tasmania (51)

BELLEVALIA - Liliaceae
albana / Bb / 6" / yellowish-violet fls / Caucasus 4
atroviolacea / Bb / 4-8" / fls dusky blue-violet / c Asia 185 / 4
ciliata / Bb / to 20" / fls purplish / s Europe, e Asia 181
dalmatica - see HYACINTHUS dalmaticus 123
dubia / Bb / to 10" / blue to white fls / European Medit. reg. 181
dubia hackellii - see B. hackellii 212
flexuosa / Bb / to 10" / white to purple fls / Near East 271
hackellii / HH Bb / 8-10" / fls bright blue / s Portugal 212
longistyla / Bb / 8-10" / rusty-purple & white fls / Caucasus 4
longipes / HH Bb / to 2' / purplish fls / Near East 271
paradoxa / Bb / 8" / fls dark blue / Caucasus, Iran (16) / 268

pycnantha / Bb / 8" / intense dark blue fls / e Turkey 185
romana / Bb / 8-16" / fls white or bluish / Medit. reg. 210 / 271
sarmatics / Bb / 10-20" / dingy-violet / s U.S.S.R., e Romania 290
saviczii / Bb / to 16" / fls white / c Asia 4
speciosa / Bb / 10-20" / yellow & brownish fls / Caucasus 4
spicata - see STRANGWEIA spicata
wilhelmsii / Bb / ? / fls brown, green banded / Caucasus 4

BELLIDASTRUM
michellii - see ASTER bellidastrum 93

BELLIS - Compositae
bellidioides - see BELLIUM bellidioides 289
perennis / R / 3-6" / fls pink, red, white / w Europe 138 / 129
perennis 'China Pink / to 5"/ small bright pink fls / Europe 208
perennis 'Longfellow' - dark pink fls 208
rotundifolia / HH R / 10" / ray florets purplish-red / s Spain 184 / 289
rotundifolia v. caerulescens / 2-3" / lt lavender fls / Atlas Mts. 123

BELLIUM - Compositae
bellidioides / R / 2" / miniature white daisies / Medit. reg. 212 / 25
minutum / A / 3" / white fls, purplish reverse / Asia Minor 25

BENSONIA
oregana - see BENSONIELLA oregona

BENSONIENELLA - Saxifragaceae
oregona / P / 12-16" / white / s Oregon 25

BERARDIA - Compositae
lanuginosa - see B. subacaulis 93
subacaulis / R / acaulescent / woolly lvs; fls not showy / w Alps 148 / 96

BERBERIS - Berberidaceae
aggregata / Sh / 3-5' / fls pale yellow; red bloomy frs / w China 36
amurensis / Sh / 10'+ / red frs / Manchuria 25
angulosa / Sh / 4'+ / orange-yellow fls; scarlet frs / Himalayas 36
aquifolium - see MAHONIA aquifolium 36
aristata / HH Sh / to 12' / evergreen; frs red / Nepal 25
Beaniana /Sh / to 8' / frs. dk red with heavy mauve bloom / w China 25
buxifolia / HH Sh / to 7' / evergreen; frs blue / s South America 208 / 25
buxifolia 'Nana' / 18" / rarely flowering 36
canadensis / Sh / 3-6' / spiny branches; yel fls; red fr / se U.S. 61 / 36
X carminea 'Fireflame' - lvs narrow; frs carmine 2
cauliata guhtgunica - see B. wilsonae guhtzunica 36
X chenaultii / evergreen Sh / 3'+ / lvs spiny-margined 139
chrysosphaera / Sh / dw. / blue-violet frs / s Tibet 36
concinna / HH Sh / to 3' / red frs / Sikkim 222
coxii / S to 7' / evergreen; bluish frs / upper Burma 36
cretica / HH Sh / low / yellow fls; black frs / mts. of Crete 36
darwinii / HH Sh / 6-12' / evergreen lvs; orange frs / Chile 269 / 36
dictyophylla / Sh / to 6' / pale yellow fls; red frs / w China 36
empetrifolia / HH Sh / 1½' / black frs; golden-yellow lvs / Chile 36
fendleri - see B. canadensis
floribunda / Sh / 15'+ / purple frs / Nepal 2

franchetiana v. macrobotrys / Sh / 4' / flgs. lge, to 3" / Yunnan 61
gagnepainii / Sh / 6' / bluish-black frs; evergreen lvs / w China 208 / 25
hookeri / HH Sh / to 6' / black-purple frs / Himalayas 222
irwinii corallina - see B. stenophylla v. corallina 2
julianae / Sh / 8-10' / evergreen lvs; blue-black frs / c China 36
koreana / Sh / to 6' / bright red frs / Korea 36
linearifolia / HH Sh / to 9' / orange-red frs / Chile 222
ludlowii v. deleica / HH Sh / 1-3' / frs red / Tibet, Yunnan 2
X meehanii / red frs / B. amurensis hybrid 222
morrisonensis /HH Sh / 6' / pale yellow fls; bright red frs / Taiwan 36
nervosa - see MAHONIA nervosa 139
repens - see MAHONIA repens 139
replicata / S / 4-5' / fls bright yel.; fruit red / sw Yunnan,China 36
sherriffii / Sh / 6' / lvs grey below; fruits, with grey bloom 184
X stenophyla / HH Sh / 8-10' / evergreen lvs; golden fls 36
X stenophylla 'Corallina' - fls reddish 2
X stenophylla 'Corallina Compacta' - less than 1' high 46 / 36
thunbergii / Sh / to 5' / fls reddish outside; frs red / Japan 208 / 25
thunbergii 'Atropurpurea' - lvs dark purple 184 / 25
thunbergii ' Atropurpurea Nana' / 2' / lvs brownish red 36
thunbergii 'Aurea' - lvs yellow 139
thunbergii 'Erecta' - upright habit 25
valdivana / HH Sh / 10' / long purple frs / Chile 36
verruculosa / Sh / to 6' / golden yellow fls / w China 139
vulgaris / Sh / 8' / fls yellow; frs red / Europe, n Africa, Asia 36
vulgaris v. purpurifolia - wine-purple lvs 184
wilsonae / Sh / 2-3' / frs soft, pinkish-red / w China 2
wilsonae 'Globosa' - dwarf globular form 139
wilsonae v. guhtzunica - fasciculate, globose frs. 36

BERGENIA - Saxifragaceae
'Ballawley' / HH P / to 20" / massive plant / purple fls (83)
beesiana - see B. purpurascens 268
ciliata f. ligulata / HH R / 9-12" / fls white aging red / se Himal. 268
cordifolia / P / to 15" / petals rose-pink / Siberia 147 / 268
crassifolia / R / 8-10" / fls reddish-pink / Siberia (339) / 135
delavayi - see B. purpurascens 268
purpurascens / R / to 9" / purple fls / Himalayas, China 184 / 129
X schmidtii / R / 8"+ / fls bright rose-pink 123 / 25
stracheyi / R / 1' / fls white becoming pink / Afganistan 218 / 268
stracheyi alba - included in the sp. 268

BERKHEYA - Compositae
macrocephala / HH P / 3' / yellow fls / Natal 184 / 279

BERTEROA - Cruciferae
incana / A or P / to 2' / white fls / Europe, naturalized in US 224 / 25

BESCHORNERIA - Agavaceae
yuccoides / P / to 4' / lvs red; fls green / Mexico 25

BESSERA - Liliaceae
elegans / Gh Bb / 1-2' / scarlet & white fls / Mexico 128

BESSEYA - Scrophulariaceae
- alpina / R / to 6" / fls violet-purple / Wyoming to New Mexico 229 / 25
- bullii - see WULFENIA bullii 99
- rubra / R / rosette / hoary, red-flushed lvs / Rocky Mts. 64
- wyomingensis / P/ 8-24"/plnt wht, hairy; fls pink-purp/Wyo, Ida, Nev 229

BETONICA
- divulsa - see STACHYS officinalia 288
- grandiflora - see STACHYS grandiflora 28
- grandiflora rosea - see STACHYS grandiflora 'Rosea' 28
- grandiflora superba - see STACHYS grandiflora 'Superba' 28
- macrantha - see STACHYS macrantha 25

BETULA - Betulaceae
- albo-sinensis / T / to 35' / pink & red bark / w China 139
- albo-sinensis septentrionalis / T/to 100'/ bark brn-orange/ w China 25
- apoiensis / Sh / 3' / many branched / Hokkaido 200
- aurata / Sh or T / 2' / toothed lvs, br. hairy 25
- costata / T / to 100' / bark papery, flaking / ne Asia 25
- Ermanii / T/ to 60'/ bark flaking, gray-wht to red /ne Asia, Japan 25
- glandulosa / Sh /to 6'/lvs pale, cones cylindr./n US, Can., Alaska 25
- humilis / Sh / 2-9' / glabrous, green lvs / Eu. & Asia, high lat. 36
- Jacquemontii / R / to 60' / bark white / Himalayas 25
- lenta / T / 70' / aromatic bark / e North America 36
- lutea / T / to 75' / aromatic twig bark / e United States 36 / 25
- Medwediewii / T / tall / catkins stalked, erect / Transcaucasus 61
- Michauxii / Sh / tiny, 4-24" / catkins very small / Labr.-Nfld 99
- Michauxii nana / 3-lobed bracts and longer catkins / Nova Scotia 99
- nana / Sh / 2-4' / shining dark green lvs / n Europe, n N. Am. 148 / 36
- papyrifera / T / 65' / whitest barked of the birches / N. Am. 36
- pendula 'Youngii' - weeping form of European White Birch 184 / 25
- platyphylla / T / to 60' / white bark / Korea, Manchuria 25
- platyphylla v. japonica / T / to 60' / brk white / Japan, China 25
- populifolia / T / 30' / smooth white bark / e North America 36
- pubescens v. carpatica /T /15'+ / densely br./ Iceland to Carpathia 132
- pumila / Sh / to 15' / frs 1" long / Newfoundland to N.J. & Minn. 25
- pumila v. glandulifera / 16"-9' / lvs glandular-warty / ne N. Am. 99
- tatewakiana / Sh / to 3' / catkins erect, 4-6" / Hokkaido, Japan 200
- tauschii - see B. platyphylla v. japonica 25
- utilis / HH T / to 60' / flaking, dark brown bark / Himalayas 218 / 268

BIARUM - Spathiflorae
- tenuifolium / Tu / 2-8" / fls purp, tinged green / Medit. & Port. 290

BIDENS - Compositae
- pilosa - see B. pilosa v. radiata 25
- pilosa v. radiata / A / to 5' / wht., yel., pinkish fls/pan-tropic 25

BILLARDIERA - Pittosporaceae
- cymosa / Cl / lvs dull, stalkless; fls pink or mauve / Australia 109
- longiflora / Gh Cl / 6' / fls greenish-yellow / Tasmania 36 / 25
- longiflora 'White Form' - white-fruiting form, true-seeding 36

BISCUTELLA - Cruciferae
coronopifolia / R / to 1' / yellow fls / France, Spain, Italy 286
frutescens / HH P / 20" / yellow fls; plant woolly / sw Spain 286
glacialis / R / to 6" / glabrous lvs; fls yellow / s Spain 286
laevigata / P / to 20" / yellow fls / s & e Europe 148 / 286

BLACKSTONIA - Gentianaceae
perfoliata / A / to 2' / fls yellow / w, s & e Europe 288

BLANDFORDIA - Liliaceae
grandiflora / HH P / 2' / fls red & yellow / N. South Wales, Astrl. 25
punicea / HH P / 2-3' / fls rich brown-red / Tasmania 194 / 25

BLECHNUM - Polypodiaceae
capense / Gh F / 1-3'/ suitable for house / Polynesia, S. Africa 128
discolor / HH F / 1-4' / pinnae closely-set, comblike / N. Zealand 159
discolor v. nudum / leathery frs / 1-3' / Australia 61
fluviatile / Gh F / 2½' / obtuse pinnae / N. A., Australia 25
Germainii / F/tiny sterile frds 2-3", fertile longer / Brazil, Chile 61
lanceolatum / Gh F / to 1' / leathery texture / Australia, Polynesia 28
patersonii / Gh F / to 10" / coriaceaous fronds / N.A., Pacific Isls. 15
penna-marina / HH F / 6' / dark green fronds / New Zealand 93 / 159
penna-marina 'Cristata' - frond apex crested 159
spicant / F / 9"+ / evergreen fronds / Alaska to Calif., Europe 97 / 25

BLEPHARIPAPPUS - Compositae
scaber / A / 1-3' / white fls / e Cascades, nw United States 229 / 142

BLEPHILIA - Labiatae
ciliata / P / 2' / fls bluish-purple / Vermont to Texas 224 / 25

BLETILLA - Orchidaceae
striata / HH P / to 2' / fls purple / China, Japan 129 / 25
striata 'Alba' - fls white 25
striata f. gebina - fls white 200

BLOOMERIA - Liliaceae
crocea / HH C / to 2' / fls orange-yellow / s coast of California 227 / 228
crocea v. aurea - slight botanical variation 25
humilis / HH C / to 6"/fls golden-yellow /San Luis Obispo Co, Calif. (52)

BLUMENBACHIA - Loasaceae
coronata / A / 1½' / white fls; stinging hairs / South America 28
laterita - see CAJOPHORA laterita 61 / 25

BOCCONIA - Papaveraceae
frutescens / Gh T / to 25' / purplish fls / American tropics 177 / 25

BOEA - Gesneriaceae
hygrometrica / R / tuft / color varies 64

BOENNINGHAUSENIA - Rutaceae
albiflora / P / 2' / fls white / Assam to c Japan 165 / 25

BOLANDRA - Saxifragaceae
oregana / P / 16-24" / fls purple / se Wash., ne Oregon 142 / 25

BOLAX - Umbelliferae
glebaria / R / 3" / yellow fls / Falkland Islands 147 / 107
glebaria 'Nana' / 1" / small form of B. glebaria 249

BOLTONIA - Compositae
asteroides / P / 2-8' / fls white to violet-purple / s & e US 28
asteroides v. latisquama - more showy fls, blue-violet 259
latisquama - see B. asteroides v. latisquama 259

BOMAREA - Alstroemeriaceae
caldasii / Gh Cl / twining / fls reddish-brown & yellow / n S. Am. 194 / 25
kalbreyeri - see B. caldasii 129 / 194
multiflora / Gh Cl / twining / reddish-gold fls / Venez., Columbia 93
salsilla / Gh Cl / twining / pink or red fls / Chile 28

BONGARDIA - Berberidaceae
Chrysogonum / HH P / 12-20" / yellow fls / Iran, Iraq, Syria 211 / 77

BORAGO - Boraginaceae
laxiflora - see B. pygmaea
officinalis / A / 1½-2' / blue or purple / Europe, n Africa 25
pygmaea / HH R / trailing / fls azure-blue / Corsica (418) / 288

BORDEREA
pyrenaica / Cl / stem to 8"/ lvs dk grn; fls sm greenish / Pyranees 290

BOSCHNIAKIA - Orobanchiaceae
hookeri / parasitic / 4" / yellow-purple fls / Calif. to Canada 228 / 142
strobilacea / parasitic /6-8"/red-brn bracts / Sierra Nev. to s Ore 25

BOSSIAEA - Papilidnaceae
prostrata / Sh/ prostrate/ lvs tiny, rarely ½"; fls br-yel/Austral 109

BOTHRIOSPERMUM - Boraginaceae
chinense / A or B / 10" / blue or white fls / n China 258

BOTRICHIUM - Ophioglossaceae
dissectum / F / to 10" / fronds bronzing in Autumn/ n & s N. Am. 191 / 99

BOTTIONEA - Liliaceae - see TRICHOPETALUM plumosom 25

BOUTELOUA - Gramineae
aristidoides / A Gr / to 1' / Needle Grama / Texas, South America 141
gracilis / P Gr / 8-20" / bristly, one-sided bracts / c & s US, Mex. 147 / 268
oligostachya - see B. gracilis 268

BOWIEA - Liliaceae
volubilis / Gh Bb / 5-15' / twining, leafless stem / S. Africa 93 / 25

BOYKINIA - Saxifragaceae
aconitifolia / P / 2' / fls small, creamy white / Va. to Ga. 93 / 279
elata / R / 2-6" / fls white / west coast ranges, United States 228

jamesii - see TELESONIX jamesii 129 / 25
jamesii v. heucheriformis - see TELESONIX jamesii v. heucheriformis 142
major / P / to 3' / white fls; lrge lvs / Oregon, California 229 / 25
occidentalis - see B. elata 25
rotundifolia / P / to 2' / white fl / California 25
tellimoides - see PELTOBOYKINIA tellimoides 184 / 200

BRACHYCHILUM - Zingiberaceae
horsfieldii / Gh P / 2'+ / white or yellowish fls / Java 25

BRACHYCHITON - Sterculiaceae
acerifolium / HH T / timber tree / red fls / Australia 28

BRACHYCOME - Compositae
aculeata / HH P / to 2' / white, blue, lilac fls / se Australia 25
iberidifolia / A / to 1½' / blue, white, rose fls / Australia 194 / 25
multifida / loose carpet or bushy plant/ lilac or pink / Australia 109
nivalis / HH R / to 1' / white fls / se Australia, Tasmania 161 / 25
nivalis v. alpina / basal lvs linear / New South Wales, Victoria 25
obovata / P / to 8" / lvs shrt; fls sm wht-lilac / Vict alps Austrl 109
rigidula / HH P / 15" / blue fls / se Australia, Tasmania 25
scapiformia / HH R / ? / lilac fls / Tasmania 64
scapigera / R / rosette / white fls / New South Wales, Australia 109

BRACHYGLOTTIS - Compositae
repanda / Gh T / 6-21' / fls whitish / New Zealand 15

BRASSICA - Cruciferae
olearacea ssp. robertiana - Wild Cabbage, pinnatifid term. leaf-lobe 286
repanda / P / to 18" / tufted plant; yellow fls / Spain, France 212

BRAVOA - Amaryllidaceae
geminiflora / HH Tu / 18" / coral-red fls / Mexico 111

BRAYA - Cruciferae
alpina / R / 4" / fls white / e Alps 314 / 286
humilis / R / 1-12" / fls white or purplish / Greenland to Alaska 99

BRICKELLIA - Compositae
grandiflora / P / 1-2' / fls greenish, yellowish-white / nw US 228

BRIGGSIA - Gesneriaceae
aurantiaca / Gr P / yellow, inside spotted, lined red-brown / Tibet 95
muscicola / R / stemless, fl stems 2" / yellow & purple fls / Tibet (398) / 286

BRIMEURA - Liliaceae
amethystina / Bb / 10" / light blue fls / Pyrenees 185 / 25
amethystina 'Alba' - white fls 25
fastigiata /HH Bb/ to 4"/pale blue fls / Corsica, Sardinia, Minorca 185

BRIZA - Gramineae
major - see B. maxima 25
maxima / A Gr / 1-1½' / elegant seedheads for drying / Medit. reg. 211 / 129
media / P Gr / to 2½' / spikelets usually purplish / Eurasia 141 / 268

minor / A Gr / 9-24" / shining white or purplish heads / w & s Eu. 141 / 268
subaristata / HH P Gr / purplish or green spikelets / Mex., S. Am. 268

BRODIAEA - Liliaceae
bridgesii - see TRITELEIA bridgesii 228 / 185
californica / C / 4-24" / fls violet-blue / c California 228 / 268
candida / C / 1-2' / fls white or bluish / California 28
capitata - see DICHELOSTEMMA pulchella 185
capitata alba - see DICHELOSTEMMA pulchella 'Alba' *
congesta / C / to 3' / blue-violet fls / California to Washington 196
coronaria / C / 3-9" / fls violet to lilac / B. C. to Calif. 229 / 268
coronaria v. macropoda / fl stem 0-2" / fls smaller 25
crocea - see TRITELEIA crocea 25
douglasii / C / ? / pale to deep blue fls / nw North America 142
elegans / C / to 16" / violet fls, rarely pink / Ore., Calif. 228 / 25
hendersonii - see TRITELEIA hendersonii 228 / 25
howellii / C / 15" / fls white to blue-purple / Oregon to B.C. 228
hyacinthina - see TRITELEIA hyacinthina 228 / 25
hyacinthina alba - included in the sp. *
ida-maia - see DICHLOSTEMMA ida-maia 228 / 25
ixioides - see TRITELEIA ixioides 186 / 25
ixioides 'Splendens' - see TRITELEIA ixioides v. scabra 25
lactea - see TRITELEIA hyacinthina 185
laxa - see TRITELEIA laxa 228 / 25
laxa candida alba - see B. candida 28
laxa 'Queen fabiola' - see B. fabuloia
leachiae - see TRITELEIA hendersonii v. leachiae 25
[1] leichtlinii - see IPHEION brevipes
lugens - see TRITELEIA lugens *
lutea - see TRITELEIA ixioides 228 / 185
lutea 'Splendens' - see TRITELEIA ixioides 'Splendens' *
minor / C / to 1' / violet fls / California 228 / 25
multiflora - see DICHLOSTEMMA multiflorum 228 / 25
peduncularis - see TRITELEIA peduncularis 228 / 25
pulchella - see DICHLOSTEMMA pulchellum 228 / 25
pulchella alba - see DICHELOSTEMMA pulchella 'Alba' *
purdyi - see B. minor 25
stellaris / P / to 16" / perianth violet / California 25
terrestris / Bb / 1" / fls yellowish / c Calif., coastal 28
X tubergenii - see TRITELEIA X tubergenii 185
volubilis - see DICHLOSTEMMA volubile 228 / 25

BROMUS - Gramineae
briziformia / A Gr / 1-2' / decorative spikelets / sw & c Asia 141 / 268
lanceolatus / A Gr / to 2' / leaf blades flat, 8-20 fls / Medit 25
macrostachys - see B. lanceolatus 25
secalinus / A or B Gr / to 4' / less fragile inforescence / Eurasia 166 / 268
squarrosus / A or B Gr / to 1½' /green or purplish fls/ Medit. reg. 268
tectorum / A Gr / to 2' / shining green, purplish fls / Medit. reg. 73 / 268

---

[1] Traub & Moldenke, Herbertia, 1955.

BROUSSONETIA - Moraceae
papyrifera / T / to 50' / catkin-bearing / Far East 167 / 36

BRUCKENTHALIA - Ericaceae
spiculifolia / Sh / 10" / pink fls / s Eu., Asia Minor 36 / 25
spiculifolia X gaulnettya - by donor?

BRUGMANSIA - Solanaceae
suaveolens / Gh Sh / 6-15' / white fls / se Brazil 25

BRUNNERA - Borginaceae
macrophylla / P / 2½' / blue fls / Caucasus, w Siberia 129 / 25

BRUNONIA - Brunoniaceae
australis / A small / cornflower-blue fls / inland Australia 25 / 45

BRUNSVIGIA - Amaryllidaceae
radulosa / Gh Bb / 12-20" / red or pink fls / South Africa

BRYANTHUS - Ericaceae
gmelinii / S / trailing / rosy pink / ne Asia, Japan to Berring Str. 95 / 36

BRYONIA - Cucurbitaceae
dioica / Tu Cl / 6-12' / fls greenish-white / Europe 268

BUCKLEYA - Santalaceae
distichophylla /Sh/ to 12'/fls greenish, drupes yel-gr/N. Car., Tenn 25

BUDDLEIA - Loganaceae
alternifolia / Sh / 15' / fls bright lilac-purple / Kansu 36
Colvilei / HH Sh / to 30' / purple or crimson fls / Himalayas 194 / 25
davidii / Sh / to 15' / fls lilac, orange eye / China; nat. Calif. 25
globosa / Gh Sh / to 15' / orange fls / Chile, Peru 129 / 25

BUGLOSSOIDES - Boraginaceae
purpureocaerulea / P / to 2' / fls purple turning blue / Europe 149 / 25

BULBINE - Liliaceae
annua - of cult. is B. semi-barbata 268
asphodeloides / Gh P / 1'+ / yellow fls / Cape of Good Hope 223
bulbosa / HH R / 9" / yellow fls / Australia, Tasmania 45
semi-barbata / HH P / 2' / fls yellow / Australia 25

BULBINELLA - Liliaceae
angustifolia / HH P / ? /fl yellow,foliage ages to coppery-red/ N.Z. 193
floribunda / HH P / 3' / creamy-white fls / Table Mt., S. Africa 111
gibbsii / Gh B / 12" / yellow turning red / New Zealand 193
hookeri / HH P / 2-3' / fls bright yellow / New Zealand 182 / 96
robusta - see B. floribunda 25

BULBOCODIUM - Liliaceae
vernum / Bb / 4" / reddish violet-purple fls / Europe, in mts. 212 / 267

BULBOSTYLIS - Cyperaceae
capillaris / A / to 14" / purplish spikelet / North America 53 / 99

BUPHTHALMUM - Compositae
salicifolium / P / to 2½' / yellow fls / c Europe 129 / 25
speciosum - see TELEKIA speciosa 184 / 25

BUPLEURUM - Umbelliferae
angulosum / R / 6-16" / petals yellow / Pyrenees 314 / 287
aureum / P / 1-2' yellow involucres; coriaceous lvs / Siberia 81
candollii / HH P / to 3' / narrow lvs / Himalayas 25
falcatum / P / to 3' / petals yellow / Europe 314 / 287
fruticosum / HH Sh / 6' / evergreen, leathery lvs / s Europe 129 / 25
lancifolium / A / 6-30" / yellowish-green fls / s Europe 287
longifolium / P / to 4' / fls yellow or purplish / c Europe 314 / 287
longiradiatum v. skikotanense / P/ 8-12"/lvs thick, fls yel/Hokkaido 200
ranunculoides / P / 2'/yellowish fls / c & s Eu., in mts., nw N. Am 219 / 287
ranunculoides canalese / P / to 2'/ fls yel. & purp. / c & sw Eur. 25
ranunculoides ssp. gramineum - with involute, linear lvs 287
rotundifolium / A / 1½-3½' / purplish lvs; yellow-green fls / Eu. 287
stellatum / R / 6-16" / yellowish fls / Alps, Corsica 148 / 287
trirodiatum /P/ 2-4"/basal lvs clasp; fls yel/Siberia,Sakh.Kamchatka 200

BURCHARDIA - Lilliaceae
umbellata / HH P / 2' / fls white to pinkish / Australia 45 / 61

BUROMUS - Butomaceae
umbellatus / P Aq / 2-4' / fls rose-pink / Eurasia, intro e N. Am. 147 / 129

BURSARIA - Pittosporaceae
spinosa / HH Sh / 8-15' / white fls / Tasmania, N.S. Wales, Astrl. 28

BURTONIA - Papilionaceae
scabra / HH Sh / 6-12" / fls yellow / sw Australia 238

BUXUS - Buxaceae
microphylla v. koreana / Sh / 2' / dense habit / Korea, China 36
sempervirens / T / to 30' / evergreen lvs / Eurasia 148 / 222

CACALIA - Compositae
adenostyles / P / 2-3'/ fls whitish / woods in mts., Japan 200

CACCINIA - Boraginaceae
crassifolia / R / ? / bluish fls / Iran 204

CAESALPINIA - Leguminosae
gilliesii / HH T / to 15' / rich yellow fls / Argentina 36
japonica / Sh / 3'+ / fls yellow, red stamens / Far East 167

CAIOPHORA - see CAJOPHORA

CAJOPHORA - Loasaceae
cirsiifolia / R / 4-16" / lg. cup fls, orange-red / Peru 64
lateritia / A Cl / to 20' / fls orange-red / Argentina 93 / 25

CALAMINTHA - Labiatae
alpina - see ACONOS alpina 25
grandiflora / P / 18" / lilac-pink fls / s Europe 210 / 279
officinalis / P / to 1½' / fls purplish / Europe 25
sylvatica ssp. ascendens / P / 1-2' / pink or lilac fls / Europe 288

CALAMPELIS
scabra - see ECCREMOCARPUS scabra 25

CALANDRINIA - Portulacaceae
caespitosa / R / 1-2" / white fls, tinged pink / Patagonia 64
ciliata / A / 1'+ / crimson fls / w N. Am., S. Am. 227 / 142
discolor / A or P / to 2' / light purple to violet / Chile 95 / 206
Feltoni / white to magenta fls / Falklands 64
grandiflora / P gr. as A / 1-3' / fls light purple / Chile 25
sericea / P gr. as A / 6"+ / white to red fls / South America (36) / 64
umbellata / P gr. as A / 4-6" / fls bright crimson / Peru 182 / 129

CALANTHE - Orchidaceae
discolor / Gr / to 16" / red, white, spotted pink / Japan 25
discolor v. bicolor / fls larger, not full expanding / Kyushu, Jap. 200
torifera / Gr / 12-20" / brownish-green, tinted purple / Japan 200
tricarinata - see C. torifera 200

CALCEOLARIA - Scrophulariaceae
acutifolia / R / 4" / fls golden-yellow / Patagonia 132
biflora / R / 10-12" / yellow fls / Chile 50 / 123
chelidonioides / P / 2' / golden-yellow fls / Peru 93
corymbosa / P / 1-2' / yellow with purple spots & stripes / Chile 95
crenatiflora / P / to 2½' / yellow fls / Chile 25
darwinii / R / 2-3" / orange-yellow & white fls / Patagonia 194 / 129
dentata - see C. X fruticohybrida 28
falklandica / R / 6" / yellow fls, purple spotted / Falkland Isls. 96
filicaulis / R / 4" / pale yellow fls, red spotted / Patagonia (79) / 123
Fothergillii /P/creeps;sm crwded lvs; fls yel,red spots/Pata, Falk. 61
fruticohybrida / Gh Sh / yellow or orange fls 28
integrifolia / Gh Sh / to 6' / fls yellow to red-brown / Chile 128 / 25
lagunae-blancae - similar to C. acutifolia 64
lanceolata - similar to C. acutifolia 64
mexicana / A / 10" / yellow fls / Mexico, South America 93
pavonii / P / 2' / yellow fls / Peru 128 / 206
picta / Gh / slender / purplish-pink fls / ? (486) / 64
pinifolia / Sh / dense bush / small yellow fls / alps of S. Am. (31) / 64
pinnata / A / to 3' / pale yellow fls / Peru, Chile, Bolivia 25
polyrhiza / R / 5" / yellow fls, purple spots / Patagonia (28) / 132
prichardii / R / leafy mat / yellow fls / Chile 64
scabiosifolia - see C. tripartita 25
tenella / R / 4" / fls clear yellow, red spots / Chile 132 / 123
tripartita / A / 2' / fls pale yellow / Colombia to Chile 25
volkmannii / R / 1' / large fls / South America, in mts. 54

CALENDULA - Compositae
arvensis / A / to 1' / yellow / c Eur. & Medit. 25
suffruticosa / P gr. as A / 2'+ / yellow or orange fls / w Medit.reg (46)

CALICOTOME - Leguminosae
spinosa / HH Sh / 3' / solitary fls / Spain, Jugoslavia 211 / 95

CALLA - Araceae
palustris / P Aq / to 1' / white spathe / Eu., N. Am., n Asia 147 / 107

CALLIANTHEMUM - Ranunculaceae
anemonoides / R / 3-10" / fls pink to white / ne Alps 314 / 286
coriandrifolium / R / 8" / fls white / Alps, Carpathians 286
kernerianum / R / to 4" / pink to white fls / s Alps (340) / 286
rutifolium / R / 8" / white fls, reddish nectaries / Tyrol (3) / 25

CALLICARPA - Verbenaceae
bodnieri v. giraldii / Sh / 8' / lilac fls & frs / s China 110 / 36
dichotoma / Sh / 7' / pink fls, violet frs/ Japan, China 200
giraldiana - see C. bodnieri v. giraldii 36
japonica f. leucocarpa / 7' / pink fls, white frs / Japan 221

CALLIRHOE - Malvaceae
digitata / P / 1-4' / fls white to purple-red / Kansas to Texas 224 / 25
involucrata / P / trailing / fls cherry-red / Minn. to Texas 224 / 107
papaver / HH R / trailing / crimson fls / Florida to Texas 225 / 25

CALLISTEMON - Myrtaceae
citrinus / HH Sh / to 20' / dark red stamens / Australia 187 / 177
linearis / HH Sh / to 9' / red fls / New South Wales 25
macropunctatus / HH Sh / to 9' / fls red / s Australia 25
rigidus / Gh Sh / 8' / red fls / Australia (432) / 36
salignus / Gh T / 30' / fls white to red / Australia 36
speciosus / Gh Sh / to 15' / pale yellow fls / se Australia 128 / 45
violaceus / Gh Sh / 6-8' / mauve-pink fls / Australia 184

CALLITRIS - Pinaceae
preissii / HH T / timber tree / stunted in coastal areas / Austrl. 127
rhomboidea / Gh T / 40' / conifer / Australia 127
robusta - see C. preissii

CALLUNA - Ericaceae
vulgaris 'Dainty Bess' / 4" / pink fls 291
vulgaris 'Tenuis' / 6" / fls scarlet-purple 291

CALACEDRUS - Cupressaceae
decurrens / T / 100-150' / columnar evergreen / w North America 127

CALOCEPHALUS - Compositae
brownii / HH Sh / to 3' / yellow fls / coastal Australia 25

CALOCHILUS - Orchidaceae
campestris / Gh Tu / small / greenish-yellow fls / Australia 45

CALOCHORTUS - Liliaceae
albus / C / 1-2' / fls pearly white / California 227 / 267
albus f. rosea - see C. amoenus 28
albus v. rubellus - variants with pinkish fls 185
albus rubescens - see V. albus v. rubellus *

amabilis / C / 3-20" / bright yellow fls / coastal n California 240 / 228
ambiguus / C / to 20" / fls pinkish to bluish-gray / Ariz., Utah 227 / 73
amoenus / C / 8-20" / rose fls / s half of Sierra Nevadas, Calif. 227 / 228
barbatus / HH C / 2' / lilac or light purple fls / Mexico 185 / 25
Bonplandianus / Gh C / 3' / fls yellow & purple / Mexico 28
bruneaunis / like nutllii / median green stripe on petals 229
caeruleus / C / 3" / fls lilac & blue / Sierra Nevada, Calif. 132
catalinae / HH C / 2' / fls white, tinged purplish / Calif., coastal 228 / 25
clavatus / C / to 3' / fls yellow / California 227 / 267
coeruleus / B / to 6" / bluish / w side of Sierra Nev., n Calif. 25
elegans / C / 6" / greenish-white fls / Idaho, Oregon, Wash. 229 / 25
elegans v. nanus - see C. lyallii 28
greenei / C / 4-12" / purplish fls / nc Calif., sw Arizona 228
gunnisonii / C / 1½' / white to purple fls / Mont. to N. Mex. 229 / 25
howellii / C / 8-12" / yellowish-white fls / sw Oregon 228
invenustus / B / 6-20" / white to dull lav to purp / California 25
kennedyi / HH C / to 8" / fls vermilion to orange / se Calif., Nev. 227 / 25
Leichtlinii / B / 16"+ / white tinged smoky blue / w Nev., n Calif. 228 / 25
luteus / C / 8-20" / fls yellow / California 185 / 267
luteus v. oculatus - see C. vestae 132
lyallii / C / to 20" / white or purple-tinged fls/ Wash., s B.C. 229 / 25
macrocarpus / C / 15" / purple fls, green banded / B.C. to Calif. 229 / 111
maweanus - see C. caeruleus 132
monophyllus / C / 3-8" / yellow fls / Sacrementa Valley, Calif. 228
nanus - see C. elegans v. nanus 28
nitidus / C / 18" / fls white, marked purple / California 184 / 111
nudus / C / 2-4" / fls greenish-white or pale lilac / California 28
nuttallii / C / 1-2' / fls white, blotched / N. Dakota & s-ward 227 / 267
palmeri / HH C / 1-2' / white to lavender fls / s California 25
plummerae / C / ? / fls pink to rose / s Calif 25
pulchellus / C / 8-16" / canary-yellow fls / c California 28
purpureus - see C. Bonplandianus 28
splendens / C / 1-2' / fls pale lavender / California 227 / 267
subalpinus / C / to 16" / yellowish-white fls / s Wash. to c Ore. 142
superbus / C / 2' / white, yellowish, lavender fls / Calif. 228 / 25
tolmiei / C / 4-16" / cream or white fls / Oregon, California 228
uniflorus / C / 1-1½' / fls lilac-pink / California 185 129
venustus / C / 2' / fls variable, all blotched / n California 185 267
venustus 'El Dorado Strain' / varii-colored fls / Eldorado Co. (433) / 28
vestae / C / to 2' / fls white to purple, blotched / n Calif. 132
weedii / Gh C / to 18" / fls lemon-yellow or orange / S. Calif. 227 / 267

CALOPOGON - Orchidaceae
pulchellus / C / 12-18" / fls magenta-crimson / e North America 28

CALOSCORDUM - Liliaceae
neriniflorum / HH Bb / 8" / bright rose fls / Pamirs, n China 185

CALOTROPIS - Asclepiadaceae
procera / Gh Sh / 10' / fls white & purple / India 28

CALTHA - Ranunculaceae
appendiculata / R / 2" / narrow-sepalled fls / Southern Andes Mts. 64
biflora / R / 1' / white fls / Alaska to California 229 / 142

biflora rotundifolia - see C. leptosepala v. rotundifolia 78
howellii - see C. biflora 142
introloba / R / of alpine streams / white fls / New South Wales (59) (123)
laeta / P / to 2' / varies slightly from Marsh Marigold / Europe 286
leptosepala / R / 1' / white, greenish, bluish-tinged fls/nw N. Am. 312 / 142
leptosepala v. rotundifolia - large orbicular-leaved 78
novae-zealandiae / R / 6" / fls pale yellow / New Zealand 96
palustris / R / 1' / fls golden-yellow / Europe, North America 148 / 129
palustris 'Alba' - single white form (54) / 129
palustris v. barthei / 10" / large yellow fls / Japan 200
radicans / smaller in all parts than C. palustris / Scotland, n Eng. 61

CALYCANTHUS - Calycanthaceae
floridus / Sh / to 10' / fls dark reddish-brown / Va. to Fla. 36 / 25
occidentalis / HH Sh / to 12' / light reddish-brown fls / Calif. 175 / 25

CALYDOREA - Iridaceae
nuda / dwarf Bb / 4" / blue / Uruguay 185
speciosa / HH Bb / ? / violet fls / Chile (55) / 64

CALYPSO - Orchidaceae
bulbosa / R / 6-9" / fls purple & pink & yellow / Me. to Wash., Eu. 229 / 107

CALYPTRIDIUM - Portulacaceae
umbellatum / A or P / to 10" / fls white or pink / nw N. Am. 229 / 25

CALYPTROSTIGMA
middendorffianum - see WEIGELA middendorffianum (56)

CALYSTEGIA - Convolvulaceae
sepium / Cl / 3-10' / fls white, insidius weed / Eu. 288
soldanella / R / creeping / fls white or purplish / temp zone shores 164 / 15
tuguriorum / P / creeping / white fls / New Zealand 238

CALYTRIX - Myrtaceae
tetragona / Gh Sh / 5' / blue fls / Australia, Tasmania 45 / 25

CAMASSIA - Liliaceae
angusta / Bb / 6-16" / fls lavender to pale purple / Illinois-Texas 259
azurea - see C. quamash v. azurea 142
cusickii / Bb / to 3' / pale blue to blue-violet fls / ne Oregon 129 / 25
esculenta - see C. quamash 93
fraseri - see C. scilloides 25
howellii / Bb / to 2' / fls pale purple / s Oregon 28
leichtlinii / Bb / to 4' / fls cream to white / Douglas Co., Ore. 228 / 25
leichtlinii 'Alba' - see C. leichtlinii ssp. suksdorfii 'Alba' 25
leichtlinii v. angusta - see C. angusta 259
leichtlinii v. azurea - pale bluish-violet fls, Washington 142
leichtlinii ssp. suksdorfii / fls lt. to bright blue/ B.C. to n Cal. 73 / 25
leichtlinii ssp. suksdorfii 'Alba' - white fls 25
leichtlinii ssp. suksdorfii 'Orion' - garden selection, deep blue 279
quamash / Bb / to 2½' / fls white, pale blue / nw North America 129 / 25
quamash 'Alba' - white fls *
quamash 'Orion' - see C. leichtlinii ssp. suksdorfii 'Orion' 279
scilloides / Bb / 18" / fls pale blue / c United States 224 / 267

CAMELLIA - Theaceae
'Donation' - see C. X williamsii 'Donation' 139
japonica 'Alba' / HH T / 25'+ / white fls 28
X williamsii 'Donation' / Gh / orchid-pink, semi-double fls 139

CAMPANULA - Campanulaceae
abietina - see C. patula ssp. abietina 289
affinis / B / to 2½' / white to pink or pale violet / Spain 25
aggregata / P / 2' blue fls / Bavaria 206
aizoon / B / 12-18" / purple, up to 5 fls per stem / s Greece 74
alaskana - C. rotundifolia alaskana 25
alata / Gh P / 5' / pink or white fls / Algeria, Tunisia 25
alliariifolia / P / 16-40" / fls creamy-white / Russia to Turkey 279 / 30
allionii - see C. alpestris 289
allionii 'Rosea' - pinker form *
alpestris / R / 102" / fls light purple-blue / sw Alps 314 / 289
alpestris 'Alba' - white forms are not uncommon 74
alpina / R / 10" / fls pale to dark blue / Austria, Italy 148 / 107
alpina 'Alba' - white fls *
americana / A / 1½-6' / fls light blue / e North America 99
americana 'Alba' - white form, botanical record not seen *
anchusiflora / B / 9-12" / blue tubular fls / e Greece 289
arvatica / R / 3" / violet to light purple fls / n Spain 107
arvatica 'Alba' - white stars for the blue 132
atlantis / Gh / procumbent / fls heliotrope-purple / Morocco 268
aucheri / R / 4-5" / fls violet-purple / n Iran, Armenia 90 / 30
autraniana / R / 8" / fls purple-blue / w Caucasus 74 / 30
'Avalon' - see raineri X turbinata 74
barbata / P / 4-18" / fls light lilac-blue / Alps, Carpathians 148 / 107
barbata 'Alba' - white fls 132 / 30
bellidifolia / R / 6" / violet fls / c Caucasus 50 / 93
betulifolia / R / 6" / white fls, flushed pink / Armenia 74 / 107
X 'Birch Hybrid' / 6" / purple-blue fls 46
bononiensis / B / 1½-3' / fls light purplish-blue / e Eu., sw Asia 74 / 30
'Brantwood' - see C. latifolia 'Brantwood' 268
X burghaltii / P / 2' / satiny gray-blue fls 129
caespitosa / R / 4-8" / fls deep lilac to white / c Europe 148 / 30
calaminthifolia / HH R / 6", decumbent / blue fls / Aegean reg. 74 / 289
carnica / R / 8-14" / fls lilac to rose-blue / s Alps 289
carniolica - see C. thyrsoides ssp. carniolica 74
carpatica / R / 6-12" / fls in shades of blue / Carpathians, e Eu. 147 / 129
carpatica 'Alba' - a white form 49 / 30
carpatica 'Blue Clips - stable 9" strain 282
carpatica 'Blue Moonlight / 6" / fls light blue 46
carpatica 'Chewton Joy' / 8" / smoky-blue fls 46
carpatica 'Ditton Blue' - good mid-blue 184
carpatica 'Hannah' / 6" / small white fls / freely produced 46
carpatica 'Jewel' - see C. carpatica v. turbinata 'Jewel' 251
carpatica 'Pallida' - see C. carpatica v. turbinata 'Pallida' 283
carpatica 'Riverslea' - large flat fls, dark blue 96
carpatica 'Turbinata' / 4" / flatter fls, light blue 46
carpatica v. turbinata 'Alba' - fls white *
carpatica v. turbinata 'Jewel' - clear blue fls 251

carpatica v. turbinata 'Pallida' - fls pale blue			282
carpatica v. turbinata 'Pseudoraineri' / 4" / deep violet fls			46
carpatica 'Wheatley Violet' / 6" / violet-blue fls	48	/	46
carpatica 'White Clips' - counterpart of 'Blue Clips'			283
cashmeriana / HH R / 5-6" / tubular blue fls / Kashmir, Afghanistan	74	/	123
caucasica - see C. sibirica			74
celsii / B / to 1' / lilac or blue-lilac fls / se Greece	(487)	/	289
cenisia / R / 2-3" / fls dark to bright violet-blue / Alps	148	/	30
cephallenica / R / trailing / blue fls / Cephalonia, Greece			25
cephalotes - see C. glomerata			74
cervicaria / B / 2-2½' / fls violet-blue / c & n Europe	(341)	/	30
'Chamissonis' / R / 4" / blue fls, darker nerves / Far East			25
[1] charadze / P / 12-16" / violet fls / Caucasus			12
X 'Chewton Joy' - see C. carpatica 'Chewton Joy'			46
cochlearifolia / R / 2-4" / blue or mauve fls / Alps	148	/	129
cochlearifolia 'Alba - white form			129
cochlearifolia 'Miranda' - pale china-blue fls			25
cochlearifolia 'Miss Willmott' - selected color form, RHS Award 1915			74
cochlearifolia 'Warleyensis' - double-flowered			24
collina / R / 1' / fls deep purple / Caucasus	74	/	107
collina 'Alba' - white fls			*
X 'Covadonga' / R / 6" / bright blue-violet fls			46
[2] cripsa / B / ? / fls pale blue / Caucasus			12
dasyantha - see C. chamissonis			200
'Ditton Blue' - see C. carpatica 'Ditton Blue'			184
divaricata / P / 1-3' / pale blue fls / Md., Kentucky, Ga., in mts	224	/	25
'E.K. Toogood' / 9" / light blue fls			46
elatines / R / to 6" / fls blue or white / nw Italy	74	/	289
elatines v. elatinoides - see C. elatinoides			289
elatines v. fenestrellata - C. fenestrellata			289
elatines v. garganica - see C. garganica			289
elatines v. istriaca - tomentose lvs; corolla to 3/4" long			30
elatinoides / R / 4-6" / velutinous plant / Italy	148	/	289
ephesia / P / 1-1½ / deep blue fls, whitish lvs / w Turkey	(58)	/	30
erinus / A / 1' / reddish-violet to white fls / n Medit. reg.			30
excisa / R / 4-6" / fls violet / Swiss & Austrian Alps	74	/	30
excisa X arvatica - see C. X kewensis			74
farinosa - syn. for glomerata			74
fenestrellata / R / 6-8" / blue fls & pollen / Jugoslavia, Albania			289
finitima - see C. betulifolia			77
formanekiana / B / 8-12" / white or light blue fls / Greece, Albania	30	/	107
fragilis / HH R / trailing / fls deep blue / s Italy	74	/	132
garganica / R / to 6" / blue fls; yellow pollen / w Greece, se Italy	268	/	30
garganica hirsuta - see C. elatines v. garganica			30
gentianoides - see PLATYCODON grandiforum			74
X 'G. F. Wilson' / 4" / violet-purple fls			46
'Glandore' / large mound, small bright blue fls			46
glomerata / P / 2' / fls violet-blue to white / Europe, Siberia	212	/	107
glomerata v. acaulis / R / 3-6" / fls blue-violet			107
glomerata 'Alba' - white-flowering form of the type			160
glomerata v. dahurica / 2'+ / fls variable / ne Manchuria	147	/	30
glomerata f. lilacina - lilac-flowered selection			160

[1] Shishkin & Bobrov, Flora U.S.S.R., Vol. 24, 1972 (trans.)

[2] English, Bulletin American Rock Garden Society, 17:3.

glomerata 'Nana' - see C. glomerata v. acaulis		30
grossekii / P / 2-2½' / fls violet / Hungary	74 /	30
X hallii / R / 4" / white fls		46
'Hannah' - see C. carpatica 'Hannah'		46
hawkinsiana / HH R / 4-5" / fls blue-violet / Greece, Albania	74 /	30
hercegovina / R / 4-5" / fls blue-lilac / Jugoslavia	74 /	30
hercegovina 'Alba' - white flowered		*
hercegovina 'Nana' - compact, erect deep lilac fls	(488) /	132
herminii / R / lax cushion / blue-violet fls / s Portugal, Spain		212
heterophylla / R / 4-8" / blue fls / Kikladhes, Greece	74	289
hondoensis / stems to 3' / hanging tubular fls, red-purp / Japan		74
incurva / P / 16-20" / fls pale violet to mauve / Greece	74 /	30
isophylla / Gh / trailing / violet-blue fls / Italy	30 /	132
isophylla 'Alba' - fls white	128 /	30
isophylla 'Mayi' - grayish-pubescent lvs	(398) /	25
X jenkinsae / R / to 1' / white fls		25
X jenkinsonii - see C. X 'Spetchley'		74
Kemulariae / P / to 1' / fls pale mauve / Transcaucasus	25 /	74
X kewensis / runs underground / 3" stems; clusters violet fls		74
kladniana - C. rotundifolia / Tatras mts.		74
kolenatiana / B / 15"+ / fls violet-blue / Caucasus	30 /	25
lactiflora / P / 5-6' / fls white to milky-blue / Caucasus	182 /	129
lactiflora 'Alba' / 4' / white fls		129
lactiflora 'Lodden Anna' / to 4' / fls flesh pink		268
lanata / R / trailing / fls ivory to milk-white / Bulgaria	74 /	30
lanceolata / 18-20" / similar to C. rhomboidalis / Pyrenees		74
lasiocarpa / R / 2-6" / fls blue / n Japan, Alaska to B.C.	164 /	200
lasiocarpa ' Alba' - attractive albino		74
latifolia / P / 4-5' / fls bluish-purple / Europe to Kashmir	147 /	129
latifolia 'Alba' / 4' / white fls	129 /	25
latifolia 'Brantwood' / 4' / violet-purple fls		129
latifolia 'Gloaming' - silvery lavender-blue		268
latifolia 'Macrantha' - taller, bright purple fls, larger	147 /	135
latifolia 'Macrantha Alba' - white form		
latiloba - see C. persicifolia ssp. sessiliflora		30
leutweinii - see C. incurva		74
lingulata / B / 12-16" / fls violet in close head / Balkans		30
linifolia - see C. carnica		289
longifolia - see C. speciosa		30
longistyla / B / 1½' / fls deep blue or amethyst / Caucasus	30	25
lourica / R / 2" / purple fls / Iran		96
macrorhiza / R / 1' / fls violet-blue / s France	(489) /	25
makaschvilii / P / 10-20" / pink fls / Caucasus		12
marchesettii / R / 6-16" / fls violet / nw Jugoslavia		189
medium / B / to 3' / fls white to blue / s Europe	184 /	25
medium 'Calycanthema' - Cup and Saucer Caterbury Bels	(418) /	25
meyeriana / prostrate to 4" / violet fls / alps of Caucasus		12
mirabilis / B / 1' / fls pale lilac / w Caucasus	74 /	30
modesta - see C. caespitosa		74
moesiaca / B / 1½-2' / fls lilac-blue / Balkans		30
mollis / R / to 6" / blue fls / s & se Spain	212 /	289
'Molly Pinsett' / 9" / light mauve-blue fls		46
morettiana / R / 2-3" / solitary lilac fls / Tyrols, Italy	148 /	30
morettiana alba / 1-2" / white fls / s Tyrol, Dalmatia		123
myrtifolia - see TRACHELIUM myrtifolia		74

'Mrs. G. F. Wilson' - see C. X 'G. F. Wilson'		*
muralis - see C. portenschlagiana		107
nitida - see C. persicifolia 'Planiflora'		96
nitida 'Alba' - see C. persicifolia 'Planiflora Alba'		96
ochroleuca / P / 2½' / fls pale ochre-yellow / Caucasus		25
olympica / R / creeping / fls pale violet-blue / Mt. Olympus, Greece		25
orbelica / 2½" / lilac blue / Balkans		74
oreadum / R / similar to C. rupicola / smooth-edged lvs		154
orphanidea / B / to 6" / fls pale to deep violet / ne Greece	74 /	30
parryi / R / to 1' / violet-blue to blue-purple fls / Rocky Mts.	229 /	30
patula / B / 2' / large blue-violet fls / Europe	182 /	279
patula ssp. abietina / P / 6-20" / light blue fls / Balkans		25
pelviformis / B / 1-2' / lilac fls / Crete	74 /	30
peregrina / Gh B / 2' / blue fls / Cape of Good Hope		206
persicifolia / P / 1-3' / blue fls / Europe incl. Britain	148 /	129
persicifolia 'Alba' - white form	147 /	30
persicifolia 'Grandiflora' - larger fls		30
persicifolia 'Planiflora ' / 9" / otherwise typical		268
persicifolia 'Planiflora Alba' / 9" / fls white		268
persicifolia ssp. sessiliflora / fls nearly sessile / Balkans		289
persicifolia 'Telham Beauty' / 3½' / fls pale china-blue		268
petrophila / R / 2-5" / pale lilac fls / e Caucasus	132 /	30
phyctidocalyx - see C. persisifolia		25
'Phyllis Elliott' - see C. X kewensis		74
pilosa / R / 3" / deep purple to bluish fls / Alaska, n Japan		74
pilosa dasyantha - see C. dasyantha		30
pilosa 'Superba' / 3" / blue fls		123
piperi / R / 3" / fls clear deep lavender / Olympics, Wash	312 /	107
[1] piperi 'Sovereigniana' - the white form		
planiflora - see C. persicifolia 'Planiflora'	(434)	268
planiflora 'Alba' - see C. persicifolia 'Planiflora Alba'		268
pontica / P / 2' / calyx white, warty / Transcauc.		74
portenschlagiana / R / 2-4" / fls light bluish-mauve / e Europe	147 /	129
portenschlagiana 'Major' - a gross form?		96
poscharskyana / R / trailing / pale lavender fls / nw Jugoslavia	207 /	107
poscharskyana 'Alba' - white flowered		*
poscharskyana 'Stella' - clear bright blue fls		280
prenanthodes / HH P / 12-32" / blue fls / California	228 /	142
primulaefolia / P / 2-3' / fls blue-purple / c Portugal	74 /	30
'Profusion' / 4" / bright blue fls		46
pseudoraineri - see C. carpatica v. turbinata 'Pseudoraineri'		46
pulla / R / 3" / fls deep rich purple / e Europe	148 /	123
pulla 'Alba' - white fls		*
pulla 'Miss Willmott' - see C. cochlearifolia 'Miss Willmott'		*
X pulloides / R / 8" / large fls, blue-purple		25
punctata / P / 16-32" / fls rose-purple / Japan	164 /	200
punctata 'Alba' - dull white fls		30
punctata v. hondoensis / corolla deeper colored, spotted / Honshu		200
punctata v. microdonta / fls smaller, paler / seashores of Honshu		200
pusilla - see C. cochlearifolia		107
pyramidalis / P / to 5' / white to pale blue / sc Europe	30 /	25
raddeana / R / 6-16" / fls deep lavender / Caucasus	74 /	107

[1] English, Bulletin American Rock Garden Society, 17:3.

raineri / R / 2-4" / light blue fls / c Alps 74 / 148
raineri x turbinata / dwarf fl. of raineri 74
ramosissima / A / to 1' / blue, whitish at center / Greece, Italy 25
rapunculoides / P / 1-3' / deep purple-blue fls; weedy / Europe 148 / 123
rapunculoides 'Alba' - white form of above weedy sp. 30
rapunculus / B / 2-3' / lilac fls / Europe, w Asia, n. Africa 107
recta / R / 8-16" / turnip-like root / Pyrenees 314 / 289
recurva / R / 10" / silvery-blue fls / origin not determined 30
Reiseri / B / 9-12" / violet / s Greece 25
rhomboidalis / P / 2' / bluish-purple fls / Europe 148 / 25
rigidipila / R / 6-12" / blue fls / Ethiopia 61
rotundifolia / P / 1-1½' / fls blue-violet / n Europe, n N. Am. 148 / 107
rotundifolia alaskana - merged with the sp. 25
rotundifolia 'Alba' - white fls 25
rotundifolia v. lancifolia - stem-lvs petioled, northern form 99
rotundifolia 'Olympica' - of rock garden stature 25
rupestris / HH R / 4-6" / blue-lilac fls / ne Greece 194 / 289
rupestris anchusaeflora - see C. anchusiflora 289
rupicola / HH R / 4" / bluish-purple fls / se Greece 74 / 289
sarmatica / R / 1' / soft gray-blue fls / Caucasus 74 / 107
sarmentosa - see C. rigidipila 268
sartorii / R / 6-9" / glistening white fls / Andrews Isls., Greece 74 / 107
saxatilis / R / 3-4" / lilac fls / Crete 74 / 30
saxifraga / R / 5-6" / fls purplish-blue / Caucasus, Asia Minor 194 / 123
scabrella / R / 2-5" / fls bright blue-lilac / Calif. to Wash. (490) / 30
scheuchzeri / R / 10" / dark blue-violet fls / Europe, in mts. 210 / 289
scheuchzeri 'Covadonga' - see X 'Covadonga' 184 / 46
scouleri / R / 3-12" / blue fls / Alaska to n California 228 / 25
serrata / R / 8-16" / fls mostly solitary / Carpathians 289
Shetleri / R / 2" / 2 prs teeth each leaf; fls blue / c Calif. 228
sibirica / B / to 18" / fls violet-blue / Eurasia (491) / 25
sibirica v. divergens - spreading branches, inflated fls 74
sibirica ssp. taurica / numerous stems / Crimea 289
X 'Spetchley' / F / 1' / snow-white fls 74
speciosa / B / pyramidal inflorescence / blue-violet fls / Pyrenees 74 / 25
spicata / B / 6-30" / blue-violet fls / Switzerland, Italy 30
X stansfieldii / R / 4-6" / fls violet-purple 30
X 'Stella' - see C. porscharskyana 'Stella' 280
stevenii / R / 4-15" / lilac fls / Caucasus to Siberia 74 / 30
takhtadzhianii / R / low / blue fls / Caucasus, on limestone 12
taurica - see C. sibirica ssp. taurica 289
thessala / R / decumbent / pale violet fls / Greece 25
thyrsoides / R / 6-12" / yellow fls / European Alps 148 / 123
thyrsoides ssp. carniolica / P / 2' / yellow fls / Alps 74 / 74
tomentosa - see C. celsii 289
tommasiniana / R / 4-14 / fls pale lilac / Jugoslavia 75 / 107
trachelium / P / 2-3' / fls lilac to white / Eurasia 210 / 30
trachelium 'Alba' - the white form 30
trautvetteri / R / 4-12" / lilac fls / e Caucasus 25
tridentata / R / 5-6" / purplish-blue fls / Caucasus 74 / 123
tridentata v. stenophylla / 4" / fls rose-purple / alpine 30
tubulosa / B / 8-15" / lavender-blue fls / Crete 74 / 30
turbinata - see C. carpatica 'Turbinata' 25
turbinata alba - see C. carpatica v. turbinata 'Alba' 30
X van houttei / P / 18" / fls indigo-blue or violet 30

velutina - see C. lanata 25
versicolor / P / 2' / fls light violet / Greece, c Italy 149 / 30
vidalii / HH P / 1-2' / fls white or cream / Azores 194 / 30
waldsteiniana / R / 4-10" / lavender fls / Jugoslavia 18 / 30
wanneri - see SYMPHYANDRA 'Wanneri' 25
warlevyensis - see C. cochlearifolia 'Warlevyensis' 24
X wockii / R / 2-5" / lavender fls (435) / 30
zoysii / R / 3-4" / fls pale lilac, tubular / e Alps 74 30

CAMPSIS - Bignoniaceae
grandiflora / Sh / creeping / corolla scarlet / China 25
radicans / Cl / 30' / fls orange to scarlet / Pa., Fla., Texas 75 / 222

CAMPTOSORUS - Polypodiaceae
rhizophyllus / F / 6" / Walking Fern / nc North America (424) / 313

CANARINA - Campanulaceae
canariensis/ P/ 6-8'/corolla yel., red, purp. br. lines/ Canary Isl. 25

CANTUA - Polemoniaceae
buxifolia / Gh Sh / 6-15' / rich red fls / Peruvian Andes 36

CAPPARIS - Capparaceae
spinosa / Sh gr. as A / to 5' / edible fl buds: Capers/ Medit.reg 210 / 25

CARAGANA - Leguminosae
arborescens / Sh / 15'+ / yellow fls / Siberia, Mongolia 36
aurantiaca / Sh / 4' / fls orange-yellow / c Asia 36
pekinensis / Sh / 6' / light yellow fls / ne China 222
pygmaea / Sh / 3-4' fls yellow / Caucasus, Siberia 36
pygmaea 'Pendula' - presumably refers to the sp., high grafted 28
traganthoides / Sh / 1-1½' / fls yellow / Tibet, n China 36

CARALLUMA - Asclepiadaceae
europea / Gh / to 6" / fls brownish-red / n Africa, s Spain 93

CARDAMINE - Cruciferae
alpina - see C. bellidifolia ssp. alpina 286
asarifolia / R / 8-16" / white fls / Pyrenees, Alps 148 / 286
bellidifolia / P / ½-3" / white / Arctic or sub-Arctic Eur. 286
bellidifolia ssp. alpina / 1-4" / fls white / Alps, Pyrenees 286
bulbifera / P / 1-2' / fls pale purple / Europe 286
bulbosa / P / 6-10" / white fls / e & n North America 224 / 99
enneaphyllos /P/8-12"/pendant fls, pale yel or wht/Carpath.-s Italy 286
heptaphylla / P / 1-2' / white, pink, purple fls / w & c Europe 93 / 286
pentaphyllos / P / 1-2' / fls white, pale purple / w & c Eu., in mts. 49 / 286
pratensis / P / 1-2' / fls white, violet tinged / Europe 286
pulcherrima v. tenella / R / 4-8" / fls pink / B.C. to California 142
trifolia / P / 8-12" / white or pink / c Europe to c Italy 286

CARDAMINOPSIS - Cruciferae
arenosa / A / 2-20" / fls white to lilac / Europe to s-w 286
halleri / P / 4-20" / fls white or lilac / mts. of c Eu. 286
neglecta / R / 2-8" / purple fls / Carpathians 286
petraea / R / 4-12" / fls white or purplish / n & c Europe 182 / 286

CARDIOCRINUM - Liliaceae		
cordatum / Bb / to 6' / creamy white fls / Japan	166 /	200
cordatum v. glehnii / Bb / 3'+ / greenish-white fls / Japan		166
cordatum v. yunnanense - see C. giganteum v. yunnanense		52
giganteum / Bb / to 10' / white lilies / Himalayas to Tibet	100 /	267
giganteum v. yunnanense / smaller, no red markings / China, Burma	129 /	267
glehnii - see C. cordatum v. glehnii		166
yunnanense - see C. cordatum v. yunnanense		267
CARDUNCELLUS - Compositae		
mitissimus /R/ ½-8"/ blue, purp, mauve fls/ c Spain, s-w-c France	289 /	(60)
pinnatus / R / sessile / purplish fls / n Africa		123
[1] pinnatus f. acaulis / R / sessile, silvery lvs / Atlas Mts.		
rhaponticoides / R / sessile / fls light violet / n Africa	194 /	268
CARDUUS - Compositae		
carlinoides / clustered pink fls, spiny stalks, narrow lvs		77
kerneri / P / 3' / fls rosy-purple / Bulgaria		10
nutans / B / 3'+ / fls purplish / Africa, Europe, Caucasus	210 /	77
CAREX - Cyperaceae		
atrata / P / 6-20" / Black Sedge / ne Eu., nw Asia, Greenland	148 /	209
aurea / P / to 10" / orange frs, drying brownish / N. Am.	53 /	99
baldensis / R / tufted, to 8" / Mt. Baldo Sedge / c Europe	148 /	268
bergrothii / P / 5-12" / green w. chartreuse / wet lands / w Russia		290
bootiana / P / 12-20" / lustrous, coriaceous lvs / Far East		200
buchananii / P / ? / reddish foliage / New Zealand		268
comans / P / 1½' / grass-like edging plant / New Zealand		268
crinita / P / to 3' / dark purple fls / ne N. Am.	53 /	99
cyperoides / A or B / to 1' / lt. green frs / Japan, Siberia, Eu.		200
dissitiflora / P / 16-30" / lf. blades soft bright green / Japan		184
flacca / P / 2' / purplish scales / Eu., escape in US	210 /	99
flava / P / to 2½' / yellow-green lvs & frs / e & c N. Am.	53 /	99
grayi / P / to 3' / inflated gray-green lvs / e & c N. Am.	147 /	99
japonica /P/ 8-16"/ spikes pale grn, many fls / Japan, Korea, China		200
mertensii / P / to 3' / dense, caespitose / Calif. to Yukon, Mont.		196
mitrata v. mitrata /P/4-12"/lf yel-gr, basal sh. yel-br./Japan, Kor.		200
morrowii / P / 1-2' / flat, thick, lustrous green lvs / Japan		25
morrowii 'Variegata' / HH P / 8-16" / white-striped lvs / Japan	76 /	200
muskingumensis / P / 1-3' / large greenish spikes / e & c N. Am.	53 /	99
nigra tornata / tiny / blackish-brown, very variable species/ n Eur.		290
oahuensis bootiana - see C. bootiana		200
pendula / P / 2-5' / tufted, spikes pendulous / Eurasia	182 /	25
pseudo-cyperus / Aq P / to 3' / stout plant / e N. Am., Eurasia	210 /	99
secta / Gr / 2-3' / terminal tuft of long grass-like lvs / N.Z.		61
stricta / P / 1½-3'+ / abundant in wet acidic lands / e & c N. Am.	53 /	99
vulpinoidea / P / 12-30" / reddish-brown / c Europe, not in N. Am.		290
CARISSA - Apocynaceae		
grandiflora / Sh /to 18'/ fls white; fr. 2", scarlet, edible/S. Afr.		25

[1] Anonymous, Quarterly Bulletin Alpine Garden Society, 35:4.

CARLINA - Compositae
acanthifolia / R / stemless / fls white to yellowish / Eurasia 314 / 25
acaulis / R / 2" / fls in heads, silky, shiny white / Alps 148 / 123
acaulis caulescens - se C. acaulis ssp. simplex 289
acaulis ssp. simplex - 2' stem 289
sicula / ? / ? / large flower heads / Sicily, Crete 61
vulgaris / B / 6-15" / purple fls / Europe 194 / 301

CARMICHAELIA - Leguminosae
australis / HH Sh / 3-12' / fls pale purple / New Zealand (436) / 36
compacta / HH Sh / 3' / purple & white fls / New Zealand 15
cunninghamii / HH Sh / to 9' / white & purple fls / New Zealand 15
Enysii / HH Sh / 6-12" / fls bright violet / S. Island, N. Z. 238 / 36
Enysii compacta - see C. ensyii v. nana *
Enysii v. nana - to 1'+ in height 132
Kirkii / Sh / 6-13" / fls white-cream, purp. vein / New Zealand 15
monroi / HH Sh / 6" / white fls, purple veined / New Zealand 15
odorata / HH Sh / 10' / fragrant fls, lilac-rose / New Zealand 36
petriei / HH Sh / to 6' / whitish & purplish fls / New Zealand 15
uniflora / HH Sh / 3" / purple & white fls / New Zealand 15
Williamsii / Sh / to 13" / yellowish, purple vein fls / N. Zealand 15

CARPENTERIA - Philadelphiaceae
californica / HH Sh / 6-15' / fls white / California 269 / 36

CARPHEPHORUS - Compositae
bellidifolius / HH P / 1-3' / rose-purple fls / Fla. to N. Carolina 225 / 250

CARPINUS - Betulaceae
betulus / T / 35'+ / gray, fluted bark / Europe, Asia Minor 210 / 139
caroliniana / T / to 40' / American Hornbeam / e North America (342) / 36

CARPOCERAS
cilicium - see THLASPI cilicium 77

CARTHAMUS - Compositae
tinctorius / A / to 3' /fls orange-yellow, Safflower/Eurasia, Calif. 25

CARUM - Umbelliferae
carvi / B / to 5' / Caraway / Europe, adventive in US 148 / 99

CARYOPTERIS - Verbenaceae
X clandonensis / Sh / 2' / fls bright blue 269 / 36
incana / Sh / to 5' / violet-blue fls / Far East 25

CASSIA - Leguminosae
artemisioides / HH Sh / 4' / fls yellow / temperate Australia 45
corymbosa / HH Sh / 5-6' / rich yellow fls / Argentina 128 / 36
covesii / HH Sh / 2' / yellow fls / New Mexico, s & Baja Calif 227
didymobotrya / HH Sh / 10' / yellow fls / Africa 25
eremophila / HH Sh / Bushy / yellow fls / Australia 25
fasciculata / P / 3' / yellow fls / Mass., Fla., N. Mex. 224 / 226
hebecarpa / P / 3'+ / yellow fls / ne & nc United States 224 / 99
marilandica / P / 4' / fls rich yellow / e & c United States 224 / 208
mexicana / HH Sh / 6-10' / petioles and branches velvety / Mexico 81

nicticans / P / 2' / yellow fls / Mass., Kan., Fla., Texas 224
occidentalis/3'/styptic weed, pilose; fr 4½"/Va.,s-Fla, Mex.-S.Am. 25
odorata / Sh / to 6' / 6-10 prs of lvs; fls yellow / Vic., s Austrl 109
roemeriana / P / to 2' / fls yellow / Texas, N. Mex., Okla. 308 / 226
stipulacea / HH Sh / 3' / fls yellow / Chile 206

CASSINIA - Compositae
aculeata / HH Sh / bushy / yellow fls / Australia 45
fulvida / HH Sh / 4' / white fls; yellowish lvs / New Zealand 36
vauvilliersii / HH Sh / 2-6' / white fls / New Zealand 238 / 36

CASSIOPE - Ericaceae
X 'Badenoch' / 4" / white fls 139
[1] X Edinburgh / Sh / 1' / white fls
fastigiata / Sh / 6-12" / tufted, much branched, fls white/Himalayas 25
hypnoides / Sh / 2" / bright pink fls / circumpolar 22 / 36
lycopodioides / Sh / 1-3" / fls white / Japan, Alaska 129 / 36
lycopodioides 'Beatrice Lilley' / 1" / dwarf cultivar (61)
X 'Medusa' / twisted growth / white fls 19
mertensiana / Sh / to 1' / fls white to pinkish / Idaho to Alaska 228 / 25
X 'Muirhead' / 6" / white fls 110 / 132
X 'Randle Cooke' / 6" / white fls 139
selaginoides / Sh / 2-10" / glistening white fls / Himalayas, China 132 / 36
stelleriana / Sh / 3" / fls creamy-white, pinkish / B.C.- Japan 36
tetragona / Sh / to 10" / fls white, tinged red / circumpolar 148 / 209
wardii / Sh / to 8" / white fls / Himalayas 132 / 36

CASTILLEJA - Scrophulariaceae
applegatei / P / 8-20" / fls bright red / California, Oregon 228
californica / HH P / 2½' / scarlet fls / coastal California 25
coccinea / A or B / 2' / scarlet fls / N.H. to Okla, Manitoba 224 / 25
exilis / A or B / ? / ? / Mont., Wash., Col., Nevada 235
hispida/P/1½'/fls scarlet, rarely crimson, yell./Wash., Ore., Mont. 25
indivisa / A / 8-16" / red fls / Texas, Oklahoma 226
[2] integra /P/ 1½'/vermil-ornge fls, part root parasite/ sw U.S.,Mex.
lindheimeri / R / 1' / fls rose to brick red / Texas, Colorado 29
linariaefolia / P / 1-3' / orange-red to red fls / w U.S. 229 / 227
miniata / P / 26-32" / red & green fls / nw United States 229 / 228
oreophila / P / 15" / deep rose-purple fls / w Canada 61
parviflora / R / 6-12" / white pink, deep rose fls / Olymp., Casc. 13 / 228
rhexifolia / P / 1' / fl crimson, purple, unbranched / New Mexico 227
septentrionalis / P/ to 1½' /pale yellowish to purplish fls/ N. Am. 25
sulphurea - see C. septentrionalis 25
thompsonii / P / to 6" / hairy or bristly, yellow fls / c Washington 229

CASUARINA - Casuarinaceae
distyla / HH Sh / diffusely branched / globular cones / Tasmania (418) / 25
equisetifolia / HH T / 70' / tropical hedge plant / Malaysia 25

---

[1] Quarterly Bulletin, Alpine Garden Society, V.52, #1.

[2] Plants of the Southwest Seed Catalog, Santa Fe, NM, 1985.

CATALPA - Martyniaceae
bignonioides / T / 25-50' / fls white, yel. & purple sp / e U.S. 36

CATANANCHE - Compositae
caespitosa/R/rosettes gray lvs; sh. stemmed fls, yell./Atlas Mts. 154
coerulea / P / 2-3' / fls deep mauve / s Europe 212 / 129
coerulea 'Alba' - white fls 28
lutea / A / 3-4" / yellow / Aegean to nw Italy 289

CATHCARTIA - Papaveraceae
villosa / R / 8-12" / yellow fls / Nepal, Sikkim, Bhutan 93

CAULOPHYLLUM - Berberidaceae
robustum / P / 16-32" / yellow fls / Japan 165 25
thalictroides / P / 2' / blue frs / e North America 224 / 107

CAUTLEYA - Zingiberaceae
gracilis / HH P / to 16" / yellow fls / Himalayas 25
lutea - see C. gracilis 25
spicata / P / to 6" / bracts red, corolla lip yellow / Himalayas 25

CEANOTHUS - Rhamnaceae
americanus / Sh / 3' / fls dull white / e & c United States 36
dentatus / HH Sh / low / fls bright blue / California 36
impressus / HH Sh / 5' / deep blue fls / s California 182 36
integerrimus / HH Sh / 10' / fls white or pale blue / California 36
prostratus / HH Sh / few inches / fls blue / Calif. to Wash. 131 / 36
pumilus / Sh / small-leaved extreme of C. prostratus/ Siskiyous 36
repens - see C. thyrsiflorus v. repens 25
sanguineus / HH Sh / 10' / white fls; red branches / Ore., Wash. 25
thyrsiflorus prostratus - see C. thyrsiflorus v. repens 36
thyrsiflorus v. repens / prostrate / fls pale blue / California 110 / 36
velutinus / HH Sh / 8-10' / fls dull white / California 36

CEDRELA - Meliaceae
sinensis / T / to 50'/ white fls; woody frs / China 222

CEDRONELLA - Labiatae
canariensis / Gh Sh / 3-5' / lilac to violet fls / Canary Isls. 25
Foeniculum - see AGASTACHE Foeniculum 99
mexicana / HH P / 1-3' / bright pink fls / Mexico in mts., Ariz. 28
triphylla - see CEDRONELLA canariensis 25

CELASTRUS - Celastraceae
scandens / Cl / high / orange frs / e United States 75 36

CELMISIA - Compositae
allanii / R / small / gray lvs / New Zealand (58) / 64
alpina / R / 4" / gray-green to white-tomentose lvs / New Zealand 238 / 15
angustifolia / R / dense tuft / lvs silvery-white reverse / N. Z. 25
argentea / R / 8", close cushions / grayish-white lvs / N. Zealand 64 / 15
argentea spectabilis - see CELMISIA spectabilis 15
armstrongii / P / 6-24" / white daisies / New Zealand 64
asteliaefolia - see CELMISIA longifolia 25
bellidioides / R / 1" mat / shining green lvs / New Zealand 238 / 123

bonplandii / R / to 1' / lvs white, satiny beneath / New Zealand 15
brevifolia / R / cushion / near-woody branches / New Zealand (58) / 64
compacta / Sh / prostrate / buff tomentose lvs / New Zealand 15
cordatifolia v. similis / R / 5" / lvs soft white to grayish/ N.Z. 15
coriacea / P / 10-18" / white fls / South Island, New Zealand 132 / 64
coriacea v. semicordata / R / to 12" / scape slender / New Zealand 15
coriacea v. stricta - narrower silvery lvs, more rigid plant 15
dallii / R / 8-12" / viscid plant / New Zealand, in mts. 15
densiflora / P / 1-2' / white fls / South Island, New Zealand 238 / 64
discolor / Sh / prostrate / white, satiny leaf-reverse / N. Zealand 15
discolor v. ampla / R / 14" / stems stout, ribs hairy / New Zealand 15
dubia / R / 4-10" / white, hairy plant / New Zealand, in mts. 15
durietzii / Sh / 8" / diverse leaf-forms, often white-tomentose/N.Z. 15
glandulosa / R / 6" / glandular bright green lvs / N. Zealand 238 / 15
glandulosa v. latifolia - broader leaf 238 / 15
gracilenta / R / low tufted / silvery lvs; fl stems to 16" / N.Z. 238 15
gracilenta v. latifolia - included in the sp. 15
graminifolia / R / 4-8" / lvs densely felted beneath / N. Zealand 238 / 15
haastii / R / compact / slender-stemmed daisies / New Zealand 64
hectori / R / prostrate / silvery lvs / South Island, N.Z. (492) / 64
hieracifolia / R / 8" / reverse of lvs buff-woolly / N. Zealand 64
hieracifolia v. oblonga - smaller plant 15
holosericea / P / 1-2' / reverse of lvs white / South Island, N.Z. 96
hookeri / HH P / 1-2' / 4-5" daisies / N.Z., low altitudes 194 / 64
incana / R / small / gray lvs / New Zealand 64
insignis / R / 12" / narrow-leaved / New Zealand 64
lanceolata / P / 2' / large lvs / New Zealand 238 / 64
laricifolia / R / 6" / thin white tomentum on leaf-reverse / N.Z. 15
lateralis / R / prostrate / coriaceous lvs / New Zealand, in mts. 15
lateralis v. villosa - lvs with glandular hairs 15
lindsayi / HH R / 8" / crowded, leathery lvs / New Zealand 182 / 96
longifolia / HH P / 18" / white fls / Tasmania, w Australia 45 / 25
lyallii / R / 6-12" / stiff lvs / New Zealand, in mts. 96
mackaui / P / 1-2' / green-leaved / New Zealand 96
major v. brevis / dwarfed form / silvery-white lvs / New Zealand 238 / 15
monroi / R / to 15" / near to C. coriacea / New Zealand 96
morganii/R/14"/tufted, if lower surface white satiny tomentum/N. Z. 15
novae zelandiae/R/7-8"/shrubby, ray fls long, narrow, recurved/N. Z. 15
petiolata / R / 8" / purplish fls / New Zealand 129 / 25
petriei / P / 2' / fierce foliage / New Zealand 64
polyvena / HH R / to 6" / subcoriaceous lvs / Stewart Is., N.Z. 15
prorepens / R / low / sticky lvs, 3" long / New Zealand 64
X pseudolyallii - less rigid lvs, more woolly 15
ramulosa / R / 1' / semi-shrubby / New Zealand 268
ramulosa v. tuberculata - lvs with minute papillae 15
rigida / R / 1' / tawny-woolly lvs; good fls / New Zealand 64
saxifraga / HH R / low / pointed silvery lvs / Tasmania 64
semicordata - see Celmisia coriacea v. semicordata 15
sericophylla / P / ? / silky lvs / Victorian Mts., Australia 109
sessiliflora / R / 4" / densely tufted / New Zealand (62) / 268
spectabilis / R / 10"/ green lvs, felted beneath / New Zealand 129 / 15
spectabilis 'Albo-Marginata' - conspicuous woolly margins 15
spectabilis 'Angustifolia' - smaller in all parts 15
spectabilis 'Magnifica' / 14" / robust variety (58) / 15
stricta - see C. coriacea v. stricta 15

traversii / P / 18" / leaf-reverse fawn-velvet / South Island, N.Z. 238 / 96
verbascifolia / P / 12-16" / tomentose plant / New Zealand 238 / 15
viscosa / R / 6" / stumpy-rayed fls / New Zealand 64
walkeri / R / creeping / large fls / New Zealand (58) / 64
webbii / Sh / 4-8" / felted gray-green lvs / New Zealand 123

CELSIA - Scrophulariaceae
acaulis - see VERBASCUM acaule 129 / 288
acaulis 'Alba' - see VERBASCUM acaulis 'Alba' *
arcturus - see VERBASCUM arcturus 288
bugulifolia - see VERBASCUM bugulifolium 288
cretica - see VERBASCUM creticum 288
roripifolia - see VERBASCUM roripifolia 288
suwarowiana / R / ? / yellow fls / Iran 204

CELTIS - Ulmaceae
occidentalis / T / to 120' / orange to purple frs / e N. Am. (342) / 25

CENTAUREA - Compositae
achtarovii / R / 4" / violet & blue fls / sw Bulgaria (368) / 289
aegialophila / HH R / short / purple fls / Crete 289
alpestris - see CENTAUREA scabiosa ssp. 25
americana / A / to 6' / fls rose / Missouri to Mexico 308 / 25
atropurpurea / P / 4' / intense crimson-purple fls / Hungary 279
bella / R / 8-12" / rosy fls / Caucasus (133) / 93
cineraria / P / 2-3' / purple / rocks w coast of Italy 289
clementei / P / robust / yellow fls; white-woolly lvs / Spain 212
collina / R / 8-12" / yellow-purple / sw Europe 289
conifera - see LEUZEA conifera 28
corymbosa / R / 4-12" / purple fls / s France 289
cretica - see C. aegialophila 289
cynaroides / P / stout / fls purple / Pyrenees, Canary Isls. 93
cyanus / A / 2' / usually blue fls, weedy / Europe, Near East 229 / 25
cyanus 'Nana Compact' / A / 1' / mixed colored fls 283
cyanus 'Polka Dot' / A / 15" / mixed colors (Bachelor's Buttons) 282
dealbata / P / 18" / purple fls / Caucasus 300 / 206
dealbata 'Steenbergii' / P / 2½' / fls white & purplish-red 129 / 25
hololeuca / P / ? / solitary golden fls / ? 64
interbaceus - see CHEIROLOPHUS interbaceus 289
jacea / P / 4' / rose-purple fls / Eurasia; weedy in U.S. 224 / 25
kerneriana / R / 10" / fls reddish / Bulgaria 64
kotschyana / P / to 3' / fls dark purple / Balkans 25
macrocephala / P / 3' / yellow fls; brown bracts / Caucasus 147 / 279
maculosa / B or P/to 4' /fls pink to pale purple / Eu., weedy in US 224 / 25
montana / P / 1½' / fls purple / Europe 148 / 129
nana compacta - see C. cyanus 'Nana Compacta' 283
nervosa - see C. uniflora ssp. nervosa 289
nigra / P / 1-2' / fls rose-purple / Europe 28
orientalis / P / 3' / straw-yellow fls / s Russia 93
paniculata / B / 16-32" / purple fls / sw Europe 314 / 289
plumosa - see C. uniflora 289
pulcherrima / P / 20-32" / rosy fls / Caucasus 93
pulchra / P / 1-2' / fls deep pink / Kashmir 47
rhapontica - see LEUZEA rhapontica 289
rhenana - see C. stoebe ssp. rhenana 262

rupestris / P / 2½' / fls yellow or orange / se Europe 29
scabiosa / P / to 5' / fls purple / Eurasia 148 / 25
scabiosa ssp. alpestris / 2½' / unbranched, fls purple-red/ c Eur. 25
sempervirens mauritanica - see CHEIROLOPHUS sempervirens 289
simplicicaulis / R / to 12" / fls lilac-pink / Armenia 24
spinosa / HH P / 2' / purple fls / Crete 206
stoebe ssp. rhenana / ? / ? / ? / Bulgaria 262
stricta / R / 12" / blue fls / Hungary 206
thracica / P / to 24" / yellow / e Balkan Pen., se Roumania 289
uniflora / R / 6-8" / fls violet; lvs whitish / c Europe 314 / 289
uniflora ssp. nervosa / R / 6" / fls violet; green lvs / sw Alps 289
vallesiaca / Biennial / 10"-3' / fls pink / sw Alps 289

CENTAUREA - Compositae
vochinensis / P / 2' / rose-purple fls / Me. to Ont., s to Va., Mo. 99

CENTAURIUM - Gentianaceae
caespitosum - see C. chloodes 81 / 288
capitatum - see C. erythraea 288
chloodes / A / 2' / fls rose-red / France 93
diffusum - see C. scilloides 288
Erythraea / B / to 20" / fls pink to pinkish-purple / Europe 210 / 288
Erythraea ssp. turcicum - scabrid-margined, narrower lvs 288
floribundum / A / 4-20" / pink fls / California 228
littorale / B / to 10" / pink fls / seasides of n Europe 288
littorale ssp. uliginosum / B / 16" / pinkish fls / ec & c Europe 288
massonii - see C. scilloides 64
minus - see C. erythraea 25
pulchellum / A / to 8" / rosy-purple fls / Europe 288
scilloides / A / tufted /deep rose-pink fls / Eu., including Britain 123
umbellatum - see C. erythraea 268
vulgare - see C. erythraea ssp. tauricum 288
uliginosum - see C. littorale ssp. uliginosum 288

CENTRANTHUS - Valerianaceae
angustifolius / P / 12-32" / pink fls / France, Italy, Swiss Alps 210 / 289
ruber / P / 12-32" / red, pink, white fls / Medit. region 210 / 289

CEPHALANTHERA - Orchidaceae
damasonianum / P / to 2' / white or cream fls / Eu., Asia Minor 85

CEPHALANTHUS - Rubiaceae
occidentalis / Sh / 3-6' / creamy-white fls / e North America 93 / 36

CEPHALARIA - Dipsaceae
alpina / P / 5-6' / sulphur-yellow fls / European Alps 148 / 25
alpina 'Nana' / P / 6'+ / dwarf form, yellow / sw Alps 290
gigantea / P / to 6' / fls cream or yellow / Caucasus 25
leucantha / P / 16-32" / creamy-white fls / s Europe 212 / 93
radiata / P / 4' / fls yellow to white / Hungary 25
scabra / HH P / 3' / white fls / South Africa 25
tatarica - see C. gigantea 268

CEPHALOTAXUS - Taxaceae
drupacea - see C. harringtoniana v. drupacea 127

harringtoniana v. drupacea / Sh / to 10' / evergreen / Japan 127
harringtoniana v. sinensis / acuminate lvs / China 36
sinensis - see C. harringtoniana v. sinensis 36

CERASTIUM - Cruciferae
alpinum / R / 2-8" / fls white / n Europe, s in mts. 148 / 286
alpinum ssp. lanatum - plant grayish to whitish lanate 286
alpinum tomentosum - see C. tomentosum 28
arcticum / R / to 6" / white fls / nw Europe 314 / 286
arcticum ssp. edmonstonii / R /compact/lvs dark green/Shetland Isls. 286
arvense ssp. lerchenfeldianum / R / to 1' / Carpathians 286
arvence ssp. thomasii / R / caespitose / lvs sessile / c Italy 286
beeringianum / R / 1-12" / wht fls; lvs blunt / Asian arctic 209
bialynickii - see C. beeringianum
biebersteinii / R / to 1' / white-lanate lvs, invasive / Crimea 286
candidissimum / R / to 1' / white to yellowish lanate / Greece 286
carinthiacum / R / 8" / white fls; shining lvs / Alps, Carpathians 314 / 286
cerastoides / R / mats / bifid white fls / n Arctic reg., s in mts. 148 / 286
chlorifolium / A / 8-14" / petals white / Turkey 77
lanatum - see C. alpinum ssp. lanatum 286
latifolium / R / to 4" / white fls / Alps, Apennines 286
lerchenfeldianum - see C. arvense ssp. lerchenfeldianum 286
maximum / R / 8-16" / white fls / n Asia, n North America 286
purpurascens / R / 4" / fls white / Asia Minor 28
pyrenaicum / R / to 4" / fls white / e Pyrenees 148 / 286
thomasii - see C. arvense ssp. thomasii 286
tomentosum / R / 6-10" / lvs white-woolly / mts. of Italy 148 / 25
uniflorum / R / 4" / lvs soft bright green / Alps, Carpathians 148 / 286

CERATOPETALUM - Cunoniaceae
gummiferum / Gh T / 40' / fls white, rose, yellow / Australia 28

CERATOSTIGMA - Plumbaginaceae
plumbaginoides / R / to 1' / fls deep blue / China 28
willmottianum / HH Sh / 2-4' / bright blue fls / w Szechuan 36

CERCIDOPHYLLUM - Cercidiphyllaceae
japonicum / T / 100' / good autumn color / Japan, China 36

CERCIS - Leguminosae
canadensis / T / to 40' / fls pale rose / e & c United States 60 / 36
canadensis f. alba - fls white 25
canadensis 'Withers Pink (Charm)' - soft pink fls (63)
chinensis / Sh / 18' / pink fls / China 300 / 36
occidentalis / HH Sh / 15' / fls reddish-purple / California 175 / 139
siliquastrum / HH T / 40' / fls purplish-rose / Europe 269 / 36

CERCOCARPUS - Rosaceae
betulifolius - see C. betuloides 36
betuloides / HH T / 20' / plumy style, no petals / Ore., Calif. 36
ledifolius / HH T / 40' / evergreen lvs / Oregon to New Mexico 131 / 36
montanus / HH Sh / to 10' / frs long-tailed / Oregon, California 36

CERINTHE - Boraginaceae
glabra / P or Bi / 6-20" / lvs smooth; fls yel, red spots/ c&s Eur. 288
major / A / 1-2' / yellow & purple fls / Medit. reg. 210 / 25

minor / A or B / 6-24" / fls yellow / c & e Europe 148 / 288

CEROPEGIA - Asclepiadaceae
woodii / Gh / trailing stems; small purple fls / South Africa 156 / 128

CESTRUM - Solanaceae
Newellii / Sh / to 10' / bright crimson / tropical Americas 25
Parqui / Sh / 15'+ / green-white to gr-yellow; fr blue / s S. Am. 25

CETERACH - Polypodiaceae
dalhouseae / Gh F / to 8" / rounded lobes / nw Himalayas 61
officinarum / F / to 10" / dry-rock fern / Europe 148 / 286

CHAENACTIS - Compositae
alpina / R / 4" / a pink daisy / Sierra Nevadas, U.S. 107
Douglasii / B / 18" / white to pink / B.C to Mont., Calif., Ariz. 25

CHAENOMELES - Rosaceae
japonia / Sh / to 3' / fls orange to blood-red / Japan 208 / 36
japonica v. alpina - dwarf plant 36
lagenaria - see C. speciosa 36
maulei - see C. japonica 36
speciosa / Sh / to 10' / fls scarlet to blood-red / China 208 / 36
Moerlooski / fls white, striped rose-pink, single 268 / 25

CHAENORRHINUM - Scrophulariaceae
glareosum / R / 6" / fls mauve-lipped / Spain 46
macropodium / R / 6" / blue-violet fls / Spain 96
minus / A / 1' / lilac / Europe, naturalized N. America 25
origanifolium / R / 6" / fls rose-purple / Pyrenees 148 / 96
villosum / R / to 1' / lilac & yellow fls / s Spain, sw France 288

CHAEROPHYLLUM - Umbelliferae
aureum / P / 5' / white fls / c & s Europe 77 / 286

CHAMAEBATIA - Rosaceae
foliolosa / HH Sh / 2-3' / fls white / California 36

CHAMAECYPARIS - Pinaceae
'Fletcheri' - see C. lawsoniana 'Fletcheri' 25
funebris / HH T / to 60' / drooping branches / China 25
lawsoniana / T / to 100' / branchlets frond-like / nw U.S. 127 / 25
lawsoniana 'Columnaris' / to 20' / narrowly conical 79
lawsoniana 'Erecta Virdis' / densely branched column 79
lawsoniana 'Fletcheri' - juvenile lvs, dense column 304 / 25
lawsoniana 'Forsteckensis' - low form, congested branches 23 / 25
lawsoniana v. minima glauca / rounded mound of dark green / N. Am. 107 / 25
nootkatensis 'Lutea' / to 100' / young shoots yellow 36
nootkatensis 'Pendula' - erect trunk, branches hang vertically 36
obtusa / T / to 120' / bark reddish-brown / Japan 167 / 25
obtusa 'Nana' / dwarf form of C. obtusa 200
thyoides / T / 50' / columnar, branches fan-shaped / e N. Am. 36
thyoides 'Andeleyensis' / 18' / nar. columnar, dk bl-gr foliage 139

CHAMAECYPARISSUS
nana - see SANTOLINA chamaecyparissus 'Nana' *

CHAMAECYTISUS - Leguminosae
albus / Sh / 12-32" / white fls / c & se Europe 287
austriacus / Sh / 6-28" / deep yellow fls / ec & se Europe 287
banaticus / Sh / 12-32" / pale yellow fls / ec Europe, ne Balkans 287
demissus / Sh / 4" / yellow fls / Mt. Olympus, Greece *
hirsutus / Sh / 3' / fls yellow or pinkish-yellow / c & e Europe 133 / 287
purpureus / Sh / to 1' / fls lilac-pink to purplish / s & se Alps 287
purpureus 'Incarnatus' - bluish-pink fls? *
pygmaeus / Sh / to 6" / yellow fls / e Balkans (64) / 287
ratisbonensis / Sh / 1-1½' / fls yellow, orange spots / c Europe 287
supinus / Sh / to 2' / yellow fls / c & s Europe 212 / 287

CHAMAEDAPHNE - Ericaceae
calyculata / Sh / 2-3' / white fls / e N. Am., Europe, Asia 299 / 36
calyculata v. angustifolia/Sh/4½'/ narrower, wavy-edged lvs/Eur. 95
calyculata nana / Sh / to 1' / fls white / n Eur., n Asia, N. Am. 25

CHAMAELIRIUM - Liliaceae
luteum / P / 1-4' / fls white / e & c North America 224 / 25

CHAMAEMELUM - Compositae
nobile / R / to 1' / white fls / w Europe, Azores 210 / 25

CHAMAENERION
caucasicum - see EPILOBIUM colchicum 77
latifolium - see EPILOBIUM latifolium 287

CHAMAEPERICLYMENUM
suecica - see CORNUS suecica 287

CHAMAEPEUCE
casabonae - see CIRSIUM casa bonae 93
diacantha - see CIRSIUM diacanthum 93

CHAMAESCILLA - Liliasceae
corymbosa / B / 3-10" / lvs basal limp; fls clear blue / se Austrl. 109

CHAMAESENNA - Leguminosae
didymobotrya / HH Sh / ? / large yellow fls / Africa 25

CHAMAESPARTIUM - Leguminosae
sagittale / P / 4-20" / yellow fls / c Europe 212 / 287
sagittale ssp. delphinensis / procumbent / smaller fls / France 133 / 287

CHAMORCHIS - Orchidaceae
alpina / R / to 5" / green, purple, yellow fls / n Eu., Alps 314 / 85

CHARIEIS - Compositae
heterophylla / A / 6-12" / blue fls / South Africa 182 / 25

CHASMANTHE - Iridaceae
aethiopica / Gh C / 4' / orange-red fls / South Africa 161 / 25

floribunda / Gh C / 2' / orange fls / Cape Province 184 / 111

CHASMANTHIUM - Gramineae
latifolium / P Gr / to 5' / Wild Oats of woodlands / Pa. to N. Mex. 25

CHEILANTHES - Polypodiaceae
argentea / F / 4" / white farina; reddish rachis / Siberia 159
feeii / F / 12" / fronds densely woolly beneath / Wis. to Calif. 25
fragrans / HH F / 4-5" / violet-scented fronds / s Eu., coastal 184 / 159
gracillima / F / 4" / minute bead-like pinnae / Idaho, California 28
hispanica / HH F / to 6" / of rocky cliffs / Spain, Portugal 286
lanosa / F / 6-8"/ white or light brown woolly reverse / Conn. 159
marantae / F / 6-10" / evergreen; rusty red scales / s Europe 159
siliquosa / F / to 1' / continuous revolute indusium / n N. Am. 99

CHEIRANTHUS - Cruciferae
allionii - see ERYSIMUM perofskianum 129
cheiri / P / 3' / fls yellow / Greece & its islands 286
linifolius - see ERYSIMUM linifolium 2
mutabilis / Gh / 3' / yellow or purple fls / Madeira 206
scoparius / Gh Sh / 2-3' / purple fls / Canary Islands 268
semperflorens / Gh Sh / 9-24" / pale lilac & white fls / Morocco 268
senoneri - included in C. cheiri 286

CHEIROLOPHUS - Compositae
interbaceus / HH P / 1-2' / purple fls / e Spain, s France 289
sempervirens / HH P / 1-2' / purple fls / c & s Portugal 289

CHELIDONIUM - Papaveraceae
japonicum / P / 12-15" / yellow fls / Japan 200
majus / B or P / to 4' / yellow fls / Eurasia 224 / 25
majus 'Flore Pleno' / P / 2' / double fls 279
majus v. laciniatum (Bowles Var.) - single fls, cut lvs 286

CHELONE - Scrophulariaceae
barbata - see PENSTEMON barbatus 25
glabra / P / 2-3' / white fls / e & c North America 22 / 107
glabra v. elatior / fls purple or deep rose/ e & c United States 99
glabra rosea - see C. glabra v. elatior 99
lyonii / P / 3-4' / fls clear pink / s United States 224 / 135
obliqua / P / ? / fls deep rose / sc United States 28

CHENOPODIUM - Chenopodiaceae
bonus henricus / P / to 2½' / potherb; fls incous / Eur., nat. U.S. 25
Botrys / A / to 2' / aromatic / Eur., Asia, Africa, native U.S. 25
foliosum / A / to 3' / fleshy red calyx / Alps, Spain 210 / 286

CHERLERA
sedoides - see MINUARTIA sedoides 286

CHIASTOPHYLLUM - Crassulaceae
caucasus - see C. oppositifolium 184
oppositifolium / R / 10" / small golden-yellow fls / Caucasus 147 / 24
oppositifolium 'Cotyledon' - see C. oppositifolium 25
simplificifolia / R / 2-12" /tiny yellow fls / Caucasus 95

CHILENA
subgibbosa - see NEOPORTERIA subgibbosa 25

CHILIOTRICHUM - Compositae
ameIloides - see C. diffusum 139
amelloides diffusum - see C. diffusum 139
diffusum / HH Sh / 3-5' / evergreen lvs; white daisies / s S. Am. 139
linearis / Sh / 20' / fls white-pink-purple, mottled / sw U.S., Mex. 25
rosmarinifolium - see C. diffusum 139

CHILOPSIS - Bignoniaceae
linearis / Sh / to 10' / fls wht-pink-purp mottled/ sw U.S. to Mex. 25

CHIMAPHILA - Ericaceae
maculata / R / 6" / waxy white fls / e North America 224 / 107
menziesii / R / 2-6" / white fls / ns North America 229 / 228
umbellata / R / 5" / fls white or roseate / Asia, Europe 164 / 200
umbellata v. cisatlantica / lvs conspicuously veined beneath/ e N.Am 25
umbellata v. occidentalis / coarser plant to 1' / w N. Am., n Mich. 99

CHIMONANTHUS - Calycanthaceae
fragrans - see C. praecox 36
praecox / Sh / to 10' / yellow fls / China 36 / 25
praecox 'Grandiflorus' - has showier yellow fls 36
praecox 'Luteus' - unstained waxy-yellow fls (418) / 139

CHIONANTHUS - Oleaceae
retusus / Sh / 15'+ / snow-white fringy fls / China 36
virginicus / Sh / to 30'/ feathery white fls / N.J. to Fla. 300 / 222

CHIONODOXA - Liliaceae
cretica / Bb / 3-8" / fls blue / Mts. Kriti 290
gigantea - see C. luciliae 'Gigantea 25
gigantea 'Alba' - good white form 267
Luciliae / Bb / 6" / blue & white fls / Turkey 185 / 129
Luciliae 'Alba' - white form 267
Luciliae 'Gigantea' - fls larger, more numerous 25
Luciliae 'Pink Giant' / 6" / pink form 129 / 267
Luciliae 'Rosea' / 5" / fls lilac-pink / white eye 17
Luciliae 'Tmoli' - late blooming 25
sardensis / Bb / to 8" / fls porcelain-blue / Turkey 267 / 129
Siehei / Bb / 8-12" / 15 fls per stalk /se Asia Min, Crete 61
tmoli - see C. luciliae 'Tmoli' 25

CHIONOGRAPHIS - Liliaceae
japonica / R / 1' / white fls in spikes / Japan 166 / 25
japonica v. hisauchiana / R / white fls / Honshu 200
koidzumiana / R / to 8" / fls greenish / Japan 166 / 25

CHIONOPHILA - Scrophulariaceae
jamesii / R / to 6" / fls greenish-white to cream / Col., Wyo. 229 / 25
tweedyi / R / 10" / pale lavender fls / Idaho, Montana 229 / 142

CHLORA
perfoliata - see BLACKSTONIA perfoliata 288

CHLORAEA - Orchidaceae
magellanica / HH P / 18'/ wht fls, black veins & lip / Magellan Str. 121

CHLORANTHUS - Chloranthaceae
japonicus / Sh / to 12" / fls white / woods of Japan 200
serratus / P / 12-20" / small white fls / Japan, China 200

CHLORIS - Graminaea
virgata / A / 12-20" / green, dull purple / Japan 200

CHLOROGALUM - Liliaceae
pomeridianum / HH P / 2-9' / fls white / California, Oregon 28

CHLOROPHYTUM - Liliaceae
capense v. variegatum - see C. comosum v. variegatum 128
comosum v. variegatum / Gh / 12"/ lvs green/ ivory stripes / S. Afr. 129 / 128

CHORDOSPARTIUM - Leguminosae
stevensonii / Gh T / to 20' / purple fls / New Zealand 238 / 25

CHORIZEMA - Leguminosae
cordatum / Gh Sh / 2-4' / orange to red fls / Australia (418) / 128

CHRYSANTHEMUM - Compositae
alpinum / R / 6" / glistening white daisies / Alps of Europe 148 / 107
alpinum v. tomentosum / ash-gray down / Corsica 96
anserinifolium / P / 18" / loose corymb of white fls / A. Minor 25
arcticum / R / 3-15" / large white fls / arctic Eurasia 28
argenteum / R / 4-6" / solitary white fl; lvs white hairs / w Asia 95
aricerinifolium / R / 6" / green lvs; single fls / ? 249
atlanticum / HH R / 3-4" / fls white, flushed pink / Morocco 25
atratum / R / to 1' / white daisies / European Alps 148 / 93
burnatii / R / tufted / white daisies / French Maritime Alps 25
catananche / HH R / 4" / crimson-backed white daisies / Morocco 194 / 132
cineariifolium / P / 15" / white fls / Jugoslavia 25
coccineum / P / 1-2' / varii-colored fls / sw Asia 147 / 25
coronarium / A / 3-4' / light yellow to white fls / Medit. region 28
corymbosum / P / 1-4' / white fls / Caucasus 25
densum / Gh P / to 16" / woolly lvs; white fls / Syria 25
densum f. armenum - see TANACETUM haradjanii 129
discoideum / P /16-24"/ligules yellow when present/sw Fr., nw Italy 289
frutescens / Gh / 3' / white or lemon yellow fls / Canary Island 28
haradjani - see TANACETUM haradjani 268
hispanicum / R / 2-10" / white, yellow, purple fls / alps of Spain 96
hosmariense / HH R / 6" / white fls / Morocco 152 / 268
leucanthemum / P / to 3' / Oxeye Daisy, weedy / Eurasia 224 / 25
leucanthemum ssp. montanum - ???, c Europe 262
mawii / HH R / to 18" / pale pink daisies / Morocco 208 / 123
maximum - see C. X superbum 25
nipponicum / P / 2' / fls bright white / Japan 28
pallasianum japonicum - see C. rupestre 200
parthenium / P / to 3' / white fls / se Europe to Caucasus 224 / 25
parthenium 'Aureum' - lvs golden-yellow 25
parthenium 'Golden Ball' / to 18" / yellow double fls *
parthenium 'Lemon Ball' / 12-18" / light yellow fls 208

ptarmiciflorum / Gh P / to 2' / white-tomentose lvs / Canary Islands 25
roseum - see C. coccineum 28
rupestra / R / to 7" / silvery lvs; rayless fls / Honshu, in mts. 200
segetum / A / 1-1½' / golden yellow fls / Eurasia, Africa 28
X superbum / P / 3' / Shasta Daisy 25
tomentosum - see C. alpinum tomentosum
vulgare / P / 2'+ / fls yellow / Far East 200
weyrichii / R / 4-6" / pink ray florets / Kamtchatka 268

CHRYSOBRACTRON
hookeri - see BULBINELLA hookeri 96

CHRYSOGONUM - Compositae
virginianum / R / 6-9" / fls yellowish-gold / e North America 49 / 129

CHRYSOPSIS - Compositae
camporum - see C. villosa var. comporum 259
falcata / R / 1' / golden daisies / Mass to New Jersey 224 / 107
graminifolia / P / 2'+ / yellow fls / La., Fla., to Va. 99
mariana / P / 1-2' / golden-yellow fls / N.Y. to Florida 224 / 107
pinifolia - see PITYOPSIS pinifolia 25
villosa / P / 6-40" / yellow fls / Wis., B.C., California 228 / 25
villosa v. camporum - fls larger than type 259
villosa v. prostrata - sprawls in a low mound 107
villosa v. rutteri - in the tallest range, later fls 28

CHRYSOSPLENIUM - Saxifragaceae
alternifolium / P / 2-4" / fls yel./ Japan, n China, Siberia, N. A. 200
fauriei Kiotense / P / ? / fls yellow / Japan (Honshu) 200
japonicum / P / 4-8" / fls pale green / Japan 200

CHRYSOTHAMNUS - Compositae
nauseosus / Sh / 9"-9' / golden yellow fls / Great Basin 294
paniculatus / HH Sh / small to large / rubber source / sw U.S. 294

CHUSQUEA - Gramineae
cummingii / Gh Sh / 6-10' / Bamboo / Andes Mts., Chile 36
quila / Gh Cl / 40' / evergreen bamboo / Chile 268

CICERBITA - Compositae
alpina / P / 20"-7' / ligules blue; achenes linear / e & c Europe 289
plumieri / P / 2-4' / ligules lube; achenes flat / w Europe 289

CICHORIUM - Compositae
intybus / P / 1-8' / sky-blue fls / Medit. region; weedy in U.S. 220
spinosum / P / to 8" / ligules blue, rarely pink or white / Medit. 289

CIMICIFUGA - Ranunculaceae
acerifolia - see C. acerina *
acerina / P / 20-48" / fls white / Japan 165 / 200
americana / P / 5' / white fls / N.Y. and s-wards 207 / 107
dahurica / P / 6' / fls white / c Asia 147 / 93
elata / P / 3-7' / white or pinkish fls / w Washington, nw Oregon 228
foetida / P / 3-6' / sepals greenish / se Europe, Siberia 25
japonica / P / 24-32" / fls white / Honshu 165 / 200

japonica v. acerina - see C. acerina 200
racemosa / P / 3-8' / long racemes / Mass. to Missouri 129 / 25
simplex / P / 18-40" / white fls / Japan, Kamchatka 165 / 200

CIRCAEA - Onagraceae
alpina / P / to 2' / open clustered fls, purplish / N. Am., Eurasia 25
lutetiana / P / 1-3' / small white fls / e N. Am., in wet woods 28
mollis / P / 8-20" / reddish stems / Far East 165 / 200
quadrisulcata / R / 1'+ / white fls / Far East 200

CIRSIUM - Compositae
acaule / R / to 6" / purple fls / Europe 210 / 289
canovirens / P / 20-40" / fls yellowish / Oregon, Montana, Wyoming 229
casabonae / P / 3' / fls pale purple / s Europe 93
coulteri / HH B / dark crimson fls / California 228 / 25
diacantha / B / to 3' / purplish fls / Asia Minor 93 / 25
eriophorum / P / to 4½' / blue-violet fls / c Europe, Russia 93
foliosum / B / 3' / fls lavender & rose to white / w U.S. in mts. 229 / 25
helenioides / P / 3'+ / purple fls / Europe 314 / 289
japonicum / P / to 3' / fls purplish to rose / Japan 164 / 200
occidentale / HH B / to 3' / purplish-red fls / coastal c Calif. 228 / 25
pectinellum v. alpinum / P / 10" / fls purplish / alpine Hokkaido 200
Pitcheri / P / stem to 31" / fls cream / shores of Great Lakes 99
purpuratum / P / 20-40" / purplish fls / mts. of Honshu 164 / 200
rivulare / P / 3'+ / purple fls / c Europe 314 / 289
rivulare 'Atrosanguinea' - reddish-purple fls *
scopulorum / P / to 2' / fls yellowish-white to white / mts. nw U.S. 25
spinosissimum / P / 16-20" / yellowish-white fls / Alps 148 / 93
spinosum / P / ? / cobwebby stems / along shore, s Japan 200
undulatum / Bi / 8-30" / waxy lvs; fls purp / Vt., NY w to Sask. 99
virginianum / B / 3'+ / fls purplish / New Jersey to Florida 225 / 99

CISSUS - Vitaceae
striata / Gh Cl / small / evergreen lvs / Chile, Brazil 25

CISTUS - Cistaceae
albanicus / Sh / 10" / fls white / c Albania 287
albidus / HH Sh / 5' / fls pale rosy-lilac / sw Europe, n Africa 212 / 36
algarvensis - see HALIMUM ocymoides 25
X canescens 'Albus' - derived from C. crispus 36
clusii / HH Sh / to 3' / fls white / s Spain, se Italy 212 / 25
crispus / HH Sh / 2' / fls purplish-red / sw Europe, n Africa 212 / 36
X cyprius / HH Sh / 6-8' / fls white, red-blotched 269 / 36
X cyprius v. albiflorus - fls without basal blotch 36
X florentinus / HH Sh / white fls 110 / 25
heterophyllus / Sh / to 3½' / fls purple, pink, much branched / se Spain 287
hirsutus - see CISTUS psilosepalus 25
incanus / HH Sh / 3' / fls purplish-pink / s Europe 210 / 287
incanus 'Albus' - fls white *
incanus ssp. corsicus - sepals less hairy 287
incanus ssp. creticus / lvs undulate-crispate / Aegean region 149 / 287
incanus ssp. tauricus - included in the sp. 25
ladanifer / HH Sh / 3-5' / fls white, red blotched / s Eu., n. Afr. 212 / 36
ladanifer v. albiflorus - pure white fls 36
ladaniferus albus - see C. ladanifer v. albiflorus 36

ladaniferus f. maculatus - see C. landanifer 25
laurifolius / HH Sh / 6-8' / white fls / sw Europe 212 / 36
Libanotis - see C. clusii 25
X loretii - C. ladanifer X C. monspeliensis 110 / 36
X lusitanicus / HH Sh / 2'+ / white fls, blotched at base 36
monspeliensis / HH Sh / 3' / fls white / s Eu., n Africa 212 / 36
X obtusifolius / Sh / 1½' / white / Medit. 25
palhinhae / HH Sh / to 2' / satin-white fls / Algarve, Portugal 212 / 36
parviflorus / HH Sh / 3' / fls clear rose-pink / e Medit. region 149 / 36
populifolius / HH Sh / to 5' / white fls / Spain, Portugal 212 / 287
populifolius lasiocalyx - see C. populifolius ssp. major 287
populifolius ssp. major - narrower lvs, sepals hairy 287
populifolius v. narbonnensis/Sh/3-7'/shorter stalked wht fls/sw Fr. 36
pouzzolzi - see C. varius 287
psilosepalus / HH Sh / 3'+ / fls white / Portugal, w Spain 287
X purpureus / Sh / 4' / petals purple, yellow base, maroon blotches 25
rosmarinifolius - see C. clusii 25
salvifolius / HH Sh / 2' / fls white, yellow at base / Medit. reg. 212 / 36
salvifolius v. prostratus - 1' / white fls, yellow stained 36
'Silver Pink' / 2' / fls clear pink 134 / 24
x skanbergii / Sh / 3' / blue-gr lvs, pale pink fls / Med., Greece 139
symphytifolius / HH Sh / 2-6' / fls purplish-pink / Canary Isls. 36
tauricus v. albiflorus - see C. incanus 287
tauricus f. typicus - see C. incanus 287
varius / Sh / to 1½' / fls white / s France 287
villosus - see C. incanus 287
villosus ssp. creticus / Sh / to 40' / fls purple, pink / s Europe 287

CITRULLUS - Cucurbitaceae
colocynthis / HH Cl / high / pale yellow fls / India 268

CLADIUM - Cyperaceae
sinclairii / P / to 3' / lg. choc. brown inflorescences / N. Zealand 124

CLADOTHAMNUS - Ericaceae
pyroliflorus / Sh / 3-6' / fls rosy / Alaska, B.C. 184 / 36

CLADRASTIS - Leguminosae
lutea / T / 50' / white, fragrant fls / se U.S. 36 / 25

CLARKIA - Onagraceae
amoena / A / to 3' / pink to lavender fls / coastal n California 228 / 25
biloba / A / to 3' /sepals pk-gr., pet. lav-pk. flecked red / Calif. 25
concinna / A / 8-16" / petals deep bright pink / California 25
pulchella / A / 20" / pink-lavender / Rocky Mts. to Pacific Coast 25
purpurea ssp. quadrivulnera / A /to 20"/lav.-purp, red fls/ nw U.S. 196
quadrivulnera - see C. purpurea ssp. quadrivulnera 196
rhomboidea / A / 3' / fls rose-purple / nw N. Am. 229 / 142
rubicunda v. blasdalei / A / 1-3' / rose-pink to lavender fls/ Calif 196
xantiana / A / 3' / lavender fls / s California 227 / 228

CLAYTONIA - Portulacaceae
australasica / HH R / creeping / fls white to rose / N.Z., Austrl. 193 / 15
caroliniana / R / 1' / pink to white, marked pink / e N. America 25
lanceolata / R / 6" / fls pink or yellowish / w N.Am. 228

megarrhiza / R / dense tuft / fls white, rose, pink / nw N. Am. (337) / 229
megarrhiza v. nivalis - fls deep rose, Washington (444) / 25
nivalis - see C. megarrhiza v. nivalis 25
rosea / P / to 6" / fls pink / N.M., Ariz. n to Wyo. & Utah 227
sibirica - see MONTIA sibirica 25
umbellata / R / 5" / fls rose or white / c & n Sierra Nevada Mts. 228
virginica / R / 4-8" / fls white, pink striped / e & c U.S. 224 / 107

CLEMATIS - Ranunculaceae
addisonii / P / to 3' / fls reddish to purplish / w Virginia 99
aethusifolia / Cl / to 6' / pale yellow fls / China, Manchuria 25
afoliata / HH Cl / 3'+ / fls greenish-white / New Zealand 238 / 36
albicoma / P / 2' / fls purplish / Virginia, West Virginia 313 / 99
albicoma v. coactilis - with felty lvs 99
alpina / Cl / 8' / blue fls / n Europe, n Asia, s in mts. 148 / 36
alpina 'Alba' - see C. Alpina v. sibirica 28
alpina 'Columbine' - soft lavender-blue fls (326) / (66)
alpina v. ochotensis / fls violet-blue / Korea, Japan e Siberia 25
alpina 'Frances Rivis' - fls more brightly colored 36
alpina 'Pamela Jackson' - rich, deep azure fls 25
alpina 'Rosea Fl. Pl.' - double rose fls *
alpina v. sibirica / white or yellowish-white fls/ n Eu. to Siberia (438) / 25
alpina 'White Moth' - see C. macropetala 'White Moth' (66)
apiifolia / Cl / 12-15' / fls dull white / China, Japan 36
aristata / HH Cl / high / white fls / Australia 25
armandii / HH Cl / high / fls creamy white / c & w China 36
australis / HH Cl / high / white of pale yellow fls / New Zealand 25
balearica / Cl / ? / greenish-yellow, spotted red / Corsica, Minorca 25
baldwinii / HH P / 1-2' / fls blue or lavender / Florida 225
Beadlei / Cl / ? / fls on bracted pedicels / Mts. Tenn. to Ga. 25
'Bees Jubilee' - mauve-pink large flowered cultivar 268
X 'Blue Bird' / Cl / 6'+ / semi-double blue-purple fls (66)
brachiata / Gh Cl / woody / fls cream or white / South Africa 25
calycina - see C. cirrhosa v. balearica 36
campaniflora / HH Cl / to 20' / fls white, tinged violet / Portugal 36
chrysocoma / HH Sh / 6-8' / fls white, tinged pink / Yunnan 61 / 36
chiisanensis / Sh / prostrate / yellowish fls / Korea 222
cirrhosa / HH Cl / 10' / fls yellowish-white / Medit. region 212 / 25
cirrhosa v. balearica - less hardy, dark green lvs, finely cut (67) / 36
coactilis - see C. albicoma v. coactilis 99
coccinea - see C. texensis 28
columbiana / Cl / 5' / fls violet-blue or white / B.C., Col., Ore. 229 / 25
columbiana v. columbiana / Cl /?/ lflts entire; fls blue/ Mo.to Wyo. 142
columbiana v. dissecta - see C. occidentalis v. dissecta 25
columbiana v. tenuiloba / P / 3' / fls violet-blue / Mont.-S. Dak. 229
'Crimson King' / Cl / large wine-red fls 93
'Crimson Star' / Cl / garnet-red fls 300
crispa / Cl / to 8' / fls bluish-purple / se United States 82 / 36
cylindrica - see C. crispa 93
Davidiana - see C. heracleaefolia v. Davidiana 25
dioscoreifolia / Cl / high / fls white / Korea 93
dioscoreifolia v. robusta - see C. maximowicziana 200
drummondii / HH Cl / straggler / fls white / Texas, Arizona 226 / 25
douglasii / P / 2' / fls pale purple, deeper within / Mont., N. Mex. 28
douglasii v. scottii - see C. scottii 25

douglasii v. scottii 'Rosea' - see C. scottii 'Rosea'		*
'Dutchess of Albany' - see C. texensis 'Dutchess of Albany'		268
X eriostemon / Sh / 8' / blue fls		25
X eriostemon 'Henderson' / 6-8' / deep bluish-purple fls	49 /	36
'Ernest Markham' - see C. viticella 'Ernest Markham'		93
'Etoile Rose' - see C. texensis 'Etoile Rose'		268
fargesii - see C. Potaninii		25
flammula / HH Cl / 10' / fragrant white fls / s Europe	149 /	36
foetida / HH Cl / ? / fragrant yellow fls / New Zealand		25
forsteri / Gh Cl / stout / white fls / New Zealand		25
fremontii / P / 2' / thick purple sepals / Mo. to Col.		28
fremontii v. riehlii / R / to 15" / dull pale purple fls / Missouri	259 /	(68)
fusca / Cl / to 15' / violet, brown-pubescent ext. / ne Asia		25
gentianoides / HH R / creeping / white fls / Tasmania	(493) /	25
glauca v. akebioides / slender climber / yellow fls / w China		28
glaucophylla / Cl / low / fls reddish-purple / sc US	225 /	99
grata / Cl / ? woody; fls cream-colored / Himalayas		25
'Gypsy Queen' / dark purple / large fls		93
'Hagley' / Cl / to 6' / large shell-pink fls		268
'Henryi' - probably is C. lawsoniana 'Henryi'	129 /	25
heracleaefolia / P / 3' / fls light blue / China		207
heracleaefolia v. Davidiana / 2' / fls light violet-blue / n China	129 /	25
hirsutissima / P / 8-28" / fls purplish brown / w North America	(343) /	36
indivisa / HH Cl / high / fls white / New Zealand		28
integrifolia / P / 3' / fls violet or blue / Eurasia	(343) /	36
integrifolia hendersonii - see C. X eriostemon 'Henderson'		93
X jackmannii / Cl / 10' / blue-purple to reddish fls	208 /	25
Jackmanni 'Purpurea Superba' / Cl tp 10' / fls dk viol-purp/Cultivar		25
Jackmanii 'Superba' - see C. Jackmanni 'Purpurea Superba'		
koreana / Sh / prostrate / red, yellow, violet fls / Korea		25
lanuginosa 'Alba' / Cl / 6' / white fls		25
lanuginosa v. candida - see C. lanuginosa 'Alba'		*
lanuginosa 'Nelly Moser' - see C. 'Nelly Moser'		268
lasiantha / Cl / high / fls white / California		28
X lawsoniana 'Henryi' - large cream-white fls		25
ligusticifolia / Cl / 20' / fls white / w N. Am.	229 /	36
macropetala / Cl / 6' / fls violet-blue / China, Siberia	269 /	36
macropetala 'Alba' - see C. macropetala 'Ballet Blanc'		(66)
macropetala 'Ballet Blanc' - double white fls		(66)
macropetala 'Blue Bird' - see C. X 'Blue bird'		(66)
macropetala 'Blue Lagoon' - see C. macropetala 'Lagoon'		(66)
macropetala 'Lagoon' - deep lavender fls, semi-double		(66)
macropetala 'Maidwell Hall' - sepals rose, suffused with purp at base		25
macropetala markhamii - see C. macropetala 'Markhams Pink'		268
macropetala 'Markhams Pink' - fls rose-colored		268
1 macropetala f. rosea - purplish-pink fls		
mandshurica - see C. recta v. mandshurica		25
marata / HH Cl / small / greenish-yellow fls / New Zealand		15
'Marie Boisselot' - eye catching large white blooms		184
maximowicziana / Cl / to 30' / white fls / Japan		36
microphylla / HH Cl / tall / white fls / Australia	194 /	81
'Mme Edouard Andre' / Cl / velvety-red fls, large		93

1 Pringle, Baileya, 19:2

'Mme. Le Coulture' - see C. 'Marie Boisselot'
montana / Cl / 20' / large white fls / Himalayas 269 / 36
montana 'Elizabeth' / larger, fragrant, light pink blooms 95
montana 'Rosea' / Cl / 25' / fls rose-pink 25
montana v. rubens - lvs purplish, fls rose or pink 129 / 25
montana 'Tetrarose' - chromosome doubled; purplish-pink fls 110 / 36
'Mrs. Cholmondeley' / Cl / large pale blue fls 268
napaulensis/ Cl / 30'/evergreen, creamy yel. fls/ n India, sw China 36
'Nellie Moser' / Cl / large pale mauve-pink fls, carmine border 300 / 268
[1] occidentalis / Cl / to 10' / fls pinkish-violet / Va. to Ontario
occidentalis v. dissecta / P / 2'+/fls reddish-violet/ Wash. in mts 25
ochotensis / Cl / ? / fls purplish / Japan & Siberia 200
ochotensis v. japonica / Cl / ? / fls purplish/Japan, Korea, Siberia 200
ochroleuca / P / 1-1½' / yellow or purplish fls / N.Y. to Ga. 225 / 25
orientalis / Cl / to 20' / fls yellow / Caucasus, Iran, Manchuria 129 / 36
orientalis 'Orange Peel' - thick lemon-yellow sepals 100
orientalis 'Sherriff's Var.' - see C. orientalis 'Orange Peel' 100
paniculata / HH Cl / high / white fls / New Zealand 238 / 36
patens 'Alba' / Cl / to 12' / fls violet, 3-5" across / China 25
Peterae / Cl / woody / many fls white or yellowish / China 25
Pitcheri / Cl / high / fls solitary, dull purplish / sc N. Amer. 25
Potaninii / Cl . to 20' / fls white, 2" across / China 25
pseudoaplina - see C. columbiana 25
quadribracteolata / HH R / trailing / fls dull light purple / N.Z. 193 / 15
'Ramona' / Cl / large lavender-blue fls (69)
recta / P / to 5' / fragrant white fls / s Europe 129 / 25
recta v. mandchurica / may be decumbent / e Asia 25
recta 'Purpurea' - purple lvs early in season 25
rehderiana / HH Cl / 25' / primrose-yellow fls / w China 36
reticulata/Cl/?/yellowish outside, violet inside/S.C. to Fla & Tex. 25
'Rosy O'Grady' / Cl / 6'+ / semi-double pink fls (66)
Scottii / P / to 2' / fls purple / S. Dak., Neb., Idaho, Colo. (439) / 25
Scottii 'Rosea ' - lighter color form 25
serratifolia / Cl / 10' / yellow fls / Korea 36
songarica / Cl / to 5' / fls yellowish-white / s Siberia 36
spooneri 'Rosea' - see C. chrysocoma 25
stans / P / to 3'+ / fls pale purple-blue / Honshu 165 / 200
tangutica / Cl / to 15' / fls rich yellow / c Asia 269 / 36
tangutica v. obtusiuscula - free-flowering and strong-growing 139
[1] tenuiloba - see C. columbiana v. tenuiloba
texensis / Cl / high / fls in shades of red / Texas 226 / 36
texensis 'Duchess of Albany' - pink fls 268
texensis 'Etoile Rose' - silvery-pink & cherry-red fls 110 / 268
'The President' / Cl / large purple-blue fls 268
tosanensis / P / scandent / fls creamy-white / Honshu 165 / 268
versicolor / Cl / to 12' / dull purple or bluish-lavender fls / c US 25
verticillaris - see C. occidentalis 25
verticillaris v. columbiana - see C. columbiana
'Ville de Lyon' / Cl / carmine-red fls 268
viorna / Cl / to 10' / fls dull reddish-purple / e U.S. 224 / 36
virginiana / Cl / 20' / fls dull white / e N.Am. 224 / 36
vitalba / Cl / to 40' / fls dull white / Europe (418) / 36

---

[1] Pringle, Brittonia, 23:4.

viticaulis / P / 2' / purplish fls / Bath Co., Va. (221) / 99
viticella / Cl / 12' / fls blue-purple or rosy purple / s Europe 208 / 36
viticella 'Ernest Markham' - dark petunia-red fls 93
viticella 'Lady Betty Balfour' - velvety purple fls 268

CLEOME - Capparaceae
hasslerana / A / 5' / dark pink fls, fading whitish / South America (418) / 25
lutea / A / 5' / fls yellow / nw United States 227 / 25
patens 'Lasurstern' / Cl / 12' / dark blue bls 93
petriei / HH Cl / high / greenish-yellow fls / New Zealand 15
'Pink Queen' / A / 3½' / apple pink fls 282
pitcheri / Cl / 12' / fls purplish-blue / c United States 224 / 36
pseudalpina - see C. columbiana 227 / 25
serrulata / A / 2-6' / rose or white fls / c United States 229 / 28
spinosa - see C. hasslerana 25
spinosa 'Helen Campbell' / A / 3' / white fls 283

CLERODENDRON - Verbenaceae
glabrum / Gh Sh / 15' / white or pink fls / South Africa 25
trichotomum / Sh / to 10' / fls white, fragrant / Japan 208 / 25
ugandense / Gh Sh / to 10' / blue & crimson fls / tropical Africa 25

CLETHRA - Clethraceae
acuminata / Sh / to 15' / white fls / Va., Ga., Ala 60 / 25
alnifolia / Sh / 9' / fls white / e North America 36
alnifolia 'Rosea' - buds deep pink 36
barbinervis / Sh / to 10' / fls white / Japan 129 / 36

CLIANTHUS - Leguminosae
formosus / Gh Sh / procumbent / fls red / w Australia 129
puniceus / Gh Sh / 6' / crimson fls / New Zealand 128 / 25
puniceus 'Albus' - white form 76 / 139

CLIFTONIA - Cyrillaceae
monophylla / HH T / 30'+ / fls white or pinkish / se United States 138 / 222

CLINOPODIUM - Labiatae
chinense / P / 10-30" / fls pale rose-purple / e Asia 25

CLINTONIA - Liliaceae
alpina / R / ? / "rather more choice" Farrer / Tibet, Nepal 162
andrewsiana / P / to 1½' / fls pink, nodding / Oregon, California (70) / 107
borealis / P / 1-2' / greenish fls; blue frs / ne North America 22 / 107
udensis / P / 8-28" / white fls; blackish frs / Far East 166 / 200
umbellata / R / 15" / white fls; black frs / ne North America 182 / 107
umbellulata / P / 1½' / spotted white fls / N.Y. to Ga. 224 / 25
uniflora / R / 6" / white fls / California to Alaska 229 / 107

CLITORIA - Leguminosae
mariana / Cl / to 3' / fls lilac / N.J. to Fla. 25
ternatea / Gh A Cl / 15' / bright blue fls / India 128

CLIVIA - Amaryllidaceae
gardenii / Gh / 12-20" / yellow fls / Natal, Transvaal 93
nobilis / Gh / 12" / red & yellow fls / South Africa 128 / 25

CNEORUM - Cneoraceae
tricoccum / HH Sh / 4' / deep yellow fls / s Europe 212 / 25

CNICUS - Compositae
acaulis - see CIRSIUM acaule *
benedictus / A / to 2' / yellow fls; white-veined lvs / c & se Eu. 229 / 289
casabonae - see CIRSIUM casabonae 93
diacantha - see CIRSIUM diacantha 184 / 25
spinosissimum - see CIRSIUM spinosissimum 93

COCCULUS - Menispermaceae
carolinus / HH Cl / 12' / red frs / Virginia to Texas 28
trilobus / Sh / scandent / small yellow fls / temperate Asia 70 / 25

COCHLEARIA - Cruciferae
alpina / B / to 20" / fls white or mauve / Britain 286
danica / A or B / to 8" / fls white or purplish / w & n Europe 210 / 286
glastifolia / A / to 50" / white fls / Spain, Italy 286
officinalis / A or B / 12" / small white fls / Europe 314 / 25

CODONOPSIS - Campanulaceae
bulleyana / R / low mat / water-blue fls / Yunnan (440) / 64
cardiophylla / P / 18" / white or flushed-blue fls / China, Tibet 25
clematidea / R / 1'+ / fls pale blue / mts. of Asia 30 / 107
convolvulacea / P / 2-3' / blue fls / Himalayas, w China 100 / 123
convolvulacea v. forrestii / fls nearly 3" across / Yunnan 67
cordifolia / HH P / twining / greenish-yellow fls / Yunnan 25
dicentrifolia / P / 2' / violet-blue fls / Nepal (71) / 268
Handeliana/HH P/to 3'/corolla gr., yell. fls, purp. streak / w China 30 / 25
lanceolata / P / twining to 3' / light blue or lilac fls / Far East 164 / 25
meleagris / P / 2' / yellowish fls / sw China (494) / 107
meleagris subscaposa/Cl/2'/crm to blue-choc.,veins, purp-yel./Yunnan 25
mollis / P / to 3' / blue fls, purple throat / s Tibet (440) / 25
ovata / R / 6-12" / fls china-blue / Himalayas 194 / 129
pilosula / P / 6', twining / pale greenish fls / ne Asia 25
rotundifolia / P / 3' / fls gray-blue to purplish / Kashmir (440) / 30
rotundifolia v. angustifolia / fls dull green / se Tibet (71)
subsimplex - little branched, yell. or whitish-blue bells 64
tangshen / P / 3'+ / greenish fls / w China 93
tubulosa / HH R / 15", twining / yellowish-green fls / sw China 25
ussuriensis / P / twining / purplish fls / Manchuria 164 / 30
vinciflora - see C. convolvulacea (440) / 25
viridiflora / P / twining to 4' / fls yellowish-green / w China 30

COELOGLOSSUM - Orchidaceae
viride / Tu / 4-14" / fls greenish-yellow / northern hemisphere 148 / 85

COIX - Gramineae
lachrymi-jobi / A Gr / 2-5' / shining white to black frs / tropics 141 / 268

COLCHICUM - Liliaceae
alpinum / C / 1" / rosy-lilac fls / Alps of Europe 148 / 111
agrippinum / B / 4-6" / fls checkered purple / not known in wild 25
atropurpureum / C / to 10" / deep reddish-purple fls / ? 185

autumnale / C / 4-8" / fls soft rosy-lilac / Europe	148 /	129
autumnale 'Album' - fls white		25
autumnale alpinum - see C. alpinum		185
autumnale f. minor - small form, rosy-mauve fls		132
autumnale 'The Giant' - well-named		17
boissieri / C / 2" / deeply colored fls / Greece	149 /	64
bornmulleri - see C. speciosum v. bornmulleri		267
byzantinum / Tu / 4" / rose-lilac fls, dotted / Turkey		17
cilicicum / C / 12" / rosy-lilac fls, autumnal / Asia Minor	129 /	25
corsicum / C / 4-6" / lilac-rose fls / Corsica		132
creticum / C / small / dark rosy-lilac fls / Crete		52
cupanii / C / to 6" / purplish-pink, anthers black / Medit.		290
'Disraeli' - magenta-purple fls		268
X 'The Giant' / fls warm rosy lilac, white throat		267
giganteum / C / 16" / rosy-purple fls, autumnal / Asia Minor	(73) /	25
hungaricum/C/8"/fls purp-pink to pink, white / Hungary, Macedonia		290
hydrophilum / Tu / ? / pink fls / Taurus Mts., Asia Minor		52
illyricum - confused name in Colchicum		52
kesselringii / C / 6" / white fls, striped / s Russia, Afghanistan	(378) /	(67)
kotschyi / C / short / fls white to pale pink / Armenia, Iran		185
libanoticum / B / ? / pale pink or white, 2" across / Mts. of Siberia		52
'Lilac Wonder' - free-flowering, rosy violet		267
lingulatum / C / small / pink fls / Greece		185
lusitanum / HH C / 14" / purplish-pink fls / Portugal		268
luteum / C / 1' / fls yellow / Himalayas		25
macrophyllum / HH C / to 20" / pale pink or lilac fls / Crete	149 /	185
neapolitanum / C / 6" / pale chequered fls, autumnal / Italy		185
parnassicum - like C. lingulatum, but fls fewer & larger		290
sibthorpii / C / 6" / rosy-lilac fls / Greece	149 /	111
speciosum / C / 10" / variable purple shades / Asia Minor	267 /	129
speciosum 'Album' - fls white	267 /	25
speciosum 'Atrorubens' - rich purplish-crimson fls		267
speciosum v. bornmulleri / 6" / lilac fls / Turkey	267 /	17
speciosum giganteum - see C. giganteum		185
speciosum illyricum - see C. giganteum		25
speciosum 'Maximum' - rosy-purple fls		52
speciosum rubrum / dark purple form, slightly paler than atrorubens		52
speciosum 'Violet Queen' - deep purplish-violet		267
stevenii / HH C / 3" / rosy-mauve fls / Israel	(496) /	111
szovitsii / C / 9" / whitish-pink fls / n Iran, Caucasus		4
triphyllum / C / 5-6" / purplish-pink, anther dark		290
troodii / HH C / 8" / white or pale pink fls / Troodos Range, Cyprus		185
umbrosum / C / 5" / pink, purplish to white, anther yellow/Caucasus		290

COLLETIA - Rhamnaceae

armata / HH Sh / 8-12' / fls waxy-white / s Chile	184 /	36
armata 'Rosea' - pale rose fls, deeper in bud	110 /	36

COLLINSIA - Scrophulariaceae

bicolor - see C. heterophylla	129	25
grandiflora / A / 15" /blue, violet, purple or wht. / B.C. to Calif.		25
heterophylla / A / 8-20" / white or violet fls / California		228
verna / A / to 2' / blue & white fls / N.Y., Wis., Ark.	224 /	25

COLLINSONIA - Labiatae
canadensis / P / 2-4' / light yellow fls / c North America 259

COLLOMIA - Polemoniaceae
biflora - see C. Cavanillesii
Cavanillesii / A / to 2'/ lvs linear; fls scarlet / Chile, Argentina 25
debilis v. larseni / R / sprawling / fls bluish-pinkish / Wash., Cal. 228 / 142
grandiflora / A / 4-40" / fls salmon to light yellow / w N. Am. 229 / 228

COLOBANTHUS
canaliculata / P / compact cushion / lvs stiff; fls colorless / N.Z. 15

COLUMNEA - Gesneriaceae
tulae 'Rubra' / Sh / corolla red 25

COLURIA - Rosaceae
geoides / R / to 10" / bright yellow fls / Siberia 7
laxmannii - see C. geoides 7

COLUTEA - Leguminosae
arborescens / Sh / to 12' / yellow fls / se Europe 211 / 36
X intermedia - see C. X media *
istria / Sh / 3-5' / fls coppery yellow / Asia Minor 36
X media / to 10' / fls brownish-red or coppery 36
X orientalis - see C. X media 25
persica / Sh / to 8' / fls yellow / Kurdistan 25

COLUTEOCARPUS - Crucuferae
reticulatus - see C. vescicaria 77
vescicaria / R / to 7" / white fls / Transcaucasia 77

COMARUM
palustre - see POTENTILLA palustris 209

COMBERA - Solanaceae
paradoxa / R / rosette / white & bluish fls / sub-alpine Argentina 64

COMMELINA - Commelinaceae
benghalensis / A / creeping / small pale blue fls / tropical Asia 200
caerulea - see C. virginica 121
coelestis / HH Tu / to 8" / fls blue / Mexico, in mts. 184 / 93
communis v. hortensis / A /creeping /lg showy blue fls/Jap,Sib, Kor 200
dianthifolia / HH Tu / 4-6" / clear blue fls / Mexico, Texas 229 / 185
sikkimensis / HH R / prostrate / sky-blue fls / Sikkim, Khasia Hills 121
tuberosa / HH Tu / to 18" / rich blue fls / Mexico, in mts. 25
virginica / P / creeping / petals blue / U.S. 99 / 121

COMPTONIA - Myricaceae
asplenifolia - see C. peregrina v. asplenifolia 25
peregrina v. asplenifolia / Sh / 3' / aromatic lvs / e United States 36

CONANDRON - Gesneriaceae
f. pilosa - see C. ramondioides f. pilosum
ramondioides / R / to 1' / fls purple / Japan 132 / 200
ramondioides f. pilosum/short pilose on scapes, cymes, calyx/Honshu 200

CONANTHERA - Tecophiliaceae
biflora / HH C / 3½-12" / fls purple / Chile 200
campanulata / HH C / to 1' / fls variable, white to purple 138 200
simsii - see C. campanulata 28

CONIMITELLA - Saxifragaceae
williamsii / P / 8-24" / white fls / Mont., Idaho, Wyo. 229 / 142

CONOPHOLIS - Orobanchiaceae
americana - parasitic plant on oak, scarcely ornamental 224 / 99

CONSOLIDA - Ranunculaceae
ambigua / A / 1-2' / blue, pink, white fls / Medit. region 210 / 25

CONVALLARIA - Liliaceae
Keiskei / P / to 6" / fls similar to C. majalis 25
Keiskei rosea - see CONCALLARIA majalis 'Rosea' 25
majalis / R / 6-8" / fls white / Europe, Asia 147 / 123
majalis 'Flore Pleno' - doubled fls 206
majalis 'Fortin' - fls ten days later than type 93
majalis fortunei - see C. majalis 'Fortin' *
majalis 'Rosea ' - pinkish fls 123

CONVOLVULUS - Convolvulaceae
althaeoides / R / trailing / fls pale red or lilac / Medit. region 268
calvertii / R / to 1' / fls white to pink / sw Asia, Crimea 288
cantabrica / R / 5-16" / fls pink / s & sc Europe 212 / 288
chinensis / Cl / 6' / purple fls / China 206
cneorum / HH Sh / 2-3' / fls white tinged pink / s Europe 129 / 36
demissus / R / dwarf / purple-pink fls / Andes of Chile 64
dissectus - see MERREMIA dissecta 115
lineatus / P / cushion forming / reddish-purple / s Europe 25
mauritanicus / P /prostrate / blue purple / N. Africa 25
sabatius / HH P / 5-20" / fls pink to blue / nw Italy 288
sabatius v. mauritanicus - see C. mauritanicus 25
sepium - see CALYSTEGIA sepium 288
soldanella - see CALYSTEGIA soldanella 288
tauricus - see C. calvertii 288
tricolor / A / to 1' / blue, yellow, white fls / s Europe 208 / 25
tricolor 'Royal Ensign' / 18" / deep blue fls, golden cen. 129

CONYZA - Compositae
bonariensis / A / to 8' / yellowish fls / tropical America 289

COOPERIA
drummondii - see ZEPHRANTHES brazoensis 25
pedunculata - see ZEPHRANTHES drummondii 25
Smallii - see ZEPHRANTHES Smallii 173 25

COPROSMA - Rubiaceae
acerosa / HH Sh / prostrate / fls white, small / New Zealand 36
areolata / HH Sh / to 15' / black frs / New Zealand 25
atropurpurea - see C. petriei v. atropurpurea *
brunnea / HH Sh / sprawling / translucent blue frs / New Zealand 238
cheesemannii / HH Sh / prostrate / orange globose frs / N. Zealand 238

depressa / HH Sh / prostrate / red globose frs / New Zealand 25
nitida / HH Sh / 3' / fls white; frs orange / alpine Tasmania 45
obconica / Sh / to 6'/ lvs in short branchlets; fls sm; fr wht /N.Z. 15
petriei / HH Sh / 2-3" / blue frs; dioecious / New Zealand 238 / 36
petriei v. atropurpurea - with port wine frs 36
propingua v. latiuseula / Sh / prostrate / many branches / N. Z. 15
pseudocuneata / HH Sh / variable hts. / orange-red frs / N. Zealand 238 / 36
pumila / HH Sh / prostrate / red globose frs / New Zealand 194 / 25
repens / Sh / 25' / lvs blotched yellow-green, fr red / N.Z., Calif. 25
rhamnoides / HH Sh / to 6' / dark red frs / New Zealand 25
robusta / Sh or T / to 15' / lvs leathery; fls orng-red / N.Z. 25
rotundifolia / Sh or T / to 15' / lvs dull gr; hairy; fr red/ N.Z. 25
serrulata / HH Sh / 3' / white flaking bark / New Zealand 238

COPTIS - Ranunculaceae
groenlandica / R / to 5" / fls white / e North America 22 / 99
japonica / R / 4-10" / white fls / Japan 165 / 200
japonica f. brachypetala - see C. japonica v. major 200
japonica major / leaves thrice ternate 200
laciniata / R / 4-6" / sepals yellowish / Wash., n Calif. 25
lanuginosus / P / to 4" / pale blue or purplish / Himalayas 25
quinquefolia / R / to 6" / white fls / Honshu 165 / 200
trifolia / R / to 4" / fls white / Europe, Asia 165 / 200
trifolia groenlandica - see C. groenlandica 99

CORALLODISCUS - Gesneriaceae
lanuginosus / P / to 4" / pale blue or purplish / Himalayas 25

CORALLORHIZA - Orchidaceae
maculata / P / 1½-2' / usually crimson-purple fls / United States 224 / 66
mertensiana / P / 8-15" / greenish or purplish fls / nw N. Am. 28
odontorhiza / P / to 16" / purplish fls / e N. Am. to Guatemala 224 / 25
trifida / R / 1' / yellowish-white to purple fls / n temp. zone 66

CORALLOSPARTIUM - Leguminosae
crassicaule / HH Sh / to 6' / leafless, green stems / New Zealand 238

CORDIA - Boraginaceae
decandra / Gh T / to 100' / orange fls / Guatemala 25

CORDYLINE - Liliaceae
australis / Gh T / 40' / used as juvenile seedlings / New Zealand 129 / 25
banksii / Gh Sh / narrow lvs to 5' long / New Zealand 93
indivisa - mostly referrable to C. australis 28
pumilo/Gh Sh/small/"grows well under Pinus radiata"/N. Aukland, N.Z. 193

COREOPSIS - Compositae
angustifolia / P / 1-3' / fls yellow / s United States 28
auriculata / P / 6-24" / fls yellow / Va. & s-ward 224 / 225
auriculata 'Nana' / 4-6" / orange-yellow fls 135
X 'Baby Sun' / 20" / bright yellow fls, uniform from seed 135
Bigelovii / P / 1-2' / single heads 1" across, yellow / c Calif. 228
'Goldfinch' - see C. grandiflora 'Goldfink' 129
grandiflora / P / 2' / yellow fls / Mo. & Kan. to Texas 308 / 25
grandiflora 'Badengold' / P/ 3'/single fl,buttercup-yel over ind-yel 268

grandiflora 'Goldfink' / 6-8" / bright yellow fls 300 / 129
grandiflora v. saxicola - wings of achenes comb-like 25
grandiflora 'Sunburst' / double yellow 184
lanceolata / P / 1-2' / yellow fls / e & c United States 229 / 227
lanceolata 'Grandiflora' - broader head 25
maritima / P gr. as A / 1-3' / yellow & gold fls / s California 227
palmata / P / 2-3' / yellow fls / Canada to Oklahoma 229
pubescens / P / 1½-3' / yellow fls / se United States 263
rosea / P / 2' / fls yellow & red / ne North America 25
saxicola - see C. grandiflora v. saxicola 25
Stillmanii / close to C. Bigelovii / c California 25
tenuifolia - see C. verticillata 28
tinctoria / A / 4' / dark red & purple, yellow & brown/ w U. S. 25
verticillata / P / 2½-2' / small yellow fls / e United States 269 / 129

CORETHROGYNE - Compositae
californica / HH Sh / prostrate / flesh-pink daisies / Calif. 228 / 132
leucophylla / HH Sh / low / fls lt. violet / San Francisco area 228

CORIARIA - Coriariaceae
japonica / Sh / 3-6" / fls green & red, fr bright red 25
myrtifolia / HH Sh /4-6'/greenish fls; black poisonous frs/Med. reg. 36
napalensis / HH Sh / large / black frs / n India 25
ruscifolia / HH T / 25' / frs purple-black / South America 25
terminalis v. xanthocarpa / Sh / 2-4' / yellow frs / Sikkim 138 / 36

CORIS - Primulaceae
monspeliensis / R / to 1' / fls pink, purple, blue / w & c Med. reg. 212 / 288

CORNUS - Cornaceae
alba v. Kesselringii /Sh/ to 10'/ bracts purp; frs wht/ Sib.,n China 25
alternifolia / Sh / to 20' / black frs, bloomy / e North America 36
amomum / Sh / to 10' / frs pale blue / e North America 36
canadensis / R / 6" / creamy-white bracts / North America 22 / 129
candidissimum - see C. racemosa 36
coreana / T / to 60' / purplish branches / Korea 221
florida / T / 10-20' / creamy-white bracts / e United States 110 / 36
florida f. rubra - bracts in some shade of pink 36
koreana - see C. coreana 221
kousa / T / to 10' / creamy-white bracts / Japan, China 110 / 36
mas / Sh / to 25' / early yellow fls / Europe 269 / 36
mas macrocarpa / T / 20' /fls yellow, fr dark red/C.,S. Am., w Asia 25
nuttallii / HH T / to 50' / white bracted / w North America 261 / 36
occidentalis / Sh / 6-18' / yellowish fls; white frs / B.C. to Cal. 36
paniculata - see C. racemosa 36
racemosa / Sh / 8-10' / white frs on red stalks / e & c U.S. 36
stolonifera / Sh / to 8' / red-barked; white frs / North America 36
stolonifera occidentalis - see C. occidentalis 294
suecica / R / to 10" / whitish bracts / n Europe 210 / 287

COROKIA - Cornaceae
cotoneaster / HH Sh / to 10' / fls yellow; frs red / New Zealand 36 / 25
X virgata / HH Sh / 10-15' / orange & yellow fls & frs / New Zealand 36 / 25

CORONILLA - Leguminosae
cappadocica / R / 3" / large yellow fls / Asia Minor (75) / 123
coronata / P / 12-28" / yellow fls / c Europe (133) / 287
emerus / Sh / 6' / bright yellow fls / c & s Europe 148 / 139
glauca - see C. valentina ssp. glauca 110 / 25
minima / Sh / 1' / fls yellow / c Europe (133) / 287
montana - see C. coronata 287
vaginalis / Sh / to 32" / fls yellow / Europe, in mts. 148 / 287
valentina / HH Sh / 3'+ / yellow fls / Portugal 212 / 287
valentina ssp. glauca - fewer pairs of leaflets / Medit. region 25
varia / P / to 2' / fls pink & white / Europe, native in U.S. 25

CORTADERIA - Gramineae
argentea pumila - see C. Selloana 'Pumila' 268
fulvida / HH Gr / 6'+ / drooping stems / New Zealand 279 / 268
richardii / HH Gr / 4-10' / panicles silvery-white / New Zealand 279 / 268
Selloana / Gr / to 10' / silvery-white to pink / S.Am. 25
Selloana 'Pumila' / 6' / silvery-white panicles 268
toe-toe - common name for C. richardii 268

CORTUSA - Primulaceae
matthioli / R / 6" / clean rosy-purple fls / Swiss Alps 148 / 107
matthioli 'Alba' - a white form 123
matthioli f. brotheri - orbicular lvs, larger fls (76)
matthioli 'Grandiflora' - larger fls 96
matthioli 'Hirsuta' - see C. matthioli f. villoso-hirsuta (76)
matthioli f. pekinensis / densely hairy / n Asia 164 / 93
matthioli f. protheri / large fls / Himalayas 96
matthioli f. villosa-hirsuta - lvs somewhat hairy (76)
matthioli v. yezoensis - the taller Japanese variety 200
protheri - see C. matthioli f. prothera
pekinensis - see C. matthioli f. pekinensis (76)
turkestanica / P / to 2' / violet fls / c Asia 10

CORYDALIS - Papaveraceae
alba / to 16" / cream or white, yellow at apex / Europe 286
altaica / P / 8-16" / few flw., purple umbels / Ural 288
ambigua/P/4-12"/many fls, blue-purple, rarely wht./Hokkaido, Honshu 200
angustifolia / R / 4-8" / pale yellow or white fls / Iran 286
aquae-gelidae / P / 3'+ / pinkish-lavender fls / Cascades, Oregon 142
aurea / A or B / 6-24" / fls golden-yellow / North America 224 / 25
aurea ssp. occidentalis / A or Bi / 6" / fls lg, erect, yel/ w N.A. 61
bracteata / Tu / 4-8" / yellow petals / Siberia 286
bulbosa / R / to 14" / fls purplish / Europe 148 / 286
bulbosa alba - see C. bulbosa f. albiflora *
bulbosa f. albiflora - attractive white form 268
bulbosa ssp. densiflora / P/ 8" / fls rose, white / n temp. zone 25
cashmeriana / R / 6" / clear blue fls / Kashmir 129 / 123
cava - see C. bulbosa 286
chaerophylla / P / to 30" / bright yellow fls / c Himalayas 25
cheilanthifolia / R / 8" / fls yellow / China 123
curvisiliqua / B / 4-16" / orange-yellow fls / e Texas 226
decumbens / P / 4-10" / few fls rose to blue-purple /Honshu to China 200
diphylla - see C. rutifolia 25
fabacea - see C. intermedia 286

flavula / R / 5-18" / fl pale yellow; diff. branched / ne N. America 268 / 99
glauca - see C. sempervirens 25
glauca rosea - see C. sempervirens 234
heterocarpa / A / 3' / pinkish-purple fls / sw Spain 286
incisa / B / 8-20" / rose-purple fls / Japan 165 / 200
intermedia / R / to 8" / purple fls / n & c Europe 286
lutea / P / to 28" / fls golden yellow / c & e Alps 148 / 286
lutea 'Alba' - cream-white fls, true from seed 96
micrantha / P / 8-20" / pale yellow fls / c United States 99
nobilis / P / 1-1½' / fls pale yellow / Siberia 75 / 268
ochroleuca / R / 1' / fls yellowish-white / Italy, Jugoslavia (78) / 93
ophiocarpa / P / to 32" / fls greenish-yellow / w Asia 165 / 200
pallida / R / 1-5" / golden yellow / China, Korea, Japan 25
popovii / R / 4" / violet-pink fls, darker lips / c Asia (12) / 5
rosea / P / to 2' / fls rose / China 25
rupestris / R / 8" / yellow fls / Iran 25
rutifolia / R / 8" / bright purple fls / w Asia 194 / 25
saxicola / HH R / 1' / decumbent / fls yellow / China 25
scouleri / P / 2-4' / fls pink or white / Oregon to B.C. 228
sempervirens / A or B / pink to purple fls / e & n North America 224 / 25
solida / R / 4-8" / purple fls / most of Europe 148 / 286
solida transylvanica - see C. transylvanica 185
speciosa / B / to 28" / fls bright yellow / n Asia 200
thalictrifolia - see C. saxicola (446) / 268
tomentella - glaucous, yellow fls 268 / 64
transylvanica / R / 4-6" / fls deep pinkish-terracotta / e Europe (344) / 185
wilsonii / HH R / 7" / fls bright yellow / China (79) / 107

CORYLOPSIS - Hamamelidaceae
glabrescens / Sh / to 20' / lvs to 4", teeth; sev. fls / Japan 25
pauciflora / Sh / 4-6' / primrose-yellow fls / Japan 269 / 36
sinensis / Sh / to 15' / pale yellow fls / c & w China 194 / 222
spicata / Sh / 6'+ / bright yellow fls / Japan 139
willmottiae / Sh / to 12' / pale yellow fls / w China 25

CORYLUS - Corylaceae
colurna / T / 60' / catkins in February / se Europe, w Asia 36 / 139
cornuta / Sh / to 10' / Beaked Hazel-nut / e North America 313 / 25

CORYNABUTILON - Malvaceae
vitifolium / Gh Sh / 10-25' / fls white to bluish / coastal Chile 194 / 25
vitifolium 'Album' / Sh / 10-25' / fls white / coastal Chile 25

CORYPHANTHA - Cactaceae
missouriensis / R / 2½" / fls yellow-green / N. Dak. to Texas (447) / 25
vivipara / R / 4" / showy pink fls / New Mexico, Arizona 229 / 227

COSMOS - Compositae
sulphurea / A / 7' / yellow-orange / Mexico 25

COTINUS - Anacardiaceae
coggyria / Sh / to 15' / purplish fr panicles / Eurasia 149 / 25
coggyria 'Purpureus' - lvs purplish 129 / 25

COTONEASTER - Rosaceae
acutifolius / Sh / to 10' / pink fls; black frs / China (448) / 25
adpressus / Sh / to 1½' / frs bright red / China 36 / 139
adpressus v. praecox / vigorous to 2' / large frs 36
apiculatus / Sh / to 6' / fls white or pinkish; frs red / Szechuan 309 / 36
bullatus / Sh / 10-12' / brilliant red frs / w China, Tibet 36
bullatus floribundus - larger fruit 36
buxifolius / Sh / to 6' / fls white; fr red / India 25
cochleatus - see C. microphylla v. cochleatus 36
congestus / Sh / to 2½' / frs bright red / Himalayas 36
conspicuus / Sh / low / white fls; purple-red frs / se Tibet 110 / 36
conspicuus decora - included in the sp. 36
cooperi / HH Sh / tall / frs dark purple / n India 36
X cornubia / HH Sh / to 20' / frs shiny scarlet 129
Dammeri / Sh / creeping / frs coral-red / c China 36
Dammeri 'Major' - vigorous cultivar with longer lvs 36
Dammeri minima - dwarf form of C. Dammeri 61
Dammeri v. radicans - with smaller lvs, w China (80)
dielsianus / Sh / to 7' / fls pink or white; frs bright red / China (449) / 25
dielsianus v. elegans - pendulous, orange-red fls 36
distichus / Sh / 4-8' / frs bright scarlet-red / Himalayas 36
divaricatus / Sh / to 6' / egg-shaped red frs / w China (450) / 36
foveolatus / Sh / 7' / fls pink; frs black / China (451) / 25
franchetii / Sh / to 10' / frs orange-scarlet / w China 129 / 36
franchetii v. sternianus - almost globose frs, orange-red 269 / 36
frigidus / Sh / 5-20' / fls white; frs bright red / Himalayas 269 / 36
frigidus 'Pendula' / HH Sh / pendulous branches / Darjeeling 36
frigidus 'Saint Monica' / Sh / to 15' / lvs 6"; fr in clusters 36
glaucophyllus / Sh / 10' / white fls; crimson frs / w China 36
harrovianus / HH Sh / 10-12' / dark red frs / w China 25
horizontalis / Sh / 2-3' / frs scarlet-red / China 129 / 36
X 'Hybridus Pendulus' - evergreen, red frs, prostrate branches 110
integerrimus / Sh / 4-7' / fls pinkish; fr red / Europe, N. America 25
lacteus / Sh / to 12' / lvs evergreen; fls white; frs red / China 25
lindleyi / Sh / to 10' / fls white; frs black / nw Himalayas 36
lucidus / Sh / 6-10' / fls rosy-white; frs black / Siberia 36
microphyllus / Sh / 2-3' / frs scarlet-red / sw China 218 / 36
microphyllus 'Cochleatus' - creeping, lvs spatulate 93 / 25
microphyllus 'Streib's Findling' /Sh/fl wht;fr red/Himalayas, w China 139
microphyllus v. thymifolius / nearly prostrate / lvs narrow 36
moupinensis / HH Sh / to 15' / pinkish fls; black frs / w China 222
multiflorus / Sh / 6-12' / fls white; frs red / e Asia 36 / 25
nashan - see C. adpressa 222
nitidifolius / Sh / 5-8' / white fls; crimson frs / Yunnan 36
nitidus / Sh / to 8' / fl white w pink centers; fr dark red / China 25
obscurus / Sh / to 10' / white fls, tinged pink / w Szechuan 36
pannosus / Sh / ? / fls white; fr red / China 25
radicans - see C. dammeri v. radicans 222
rotundifolius / HH Sh / 5-8' / scarlet frs / Himalayas 25
salicifolius / Sh / to 15' / bright red frs / China 138 / 36
simonsii / Sh / 10-12' / white fls; scarlet frs / Assam 36
sternianus - see C. franchetii v. sternianus 36
tomentosus /Sh/ 4-7' / fls pinkish, like C. integerrimus, lgr/ s Eur. 25
wardii / HH Sh / ? / leaf reverse white-tomentose / se Tibet 222
zabellii / Sh / 6-9' / rosy fls; red frs / w Hupeh, China (452) / 36

COTULA - Compositae
atrata / R / 6" / fl-heads blackish / South Island, New Zealand 193 / 268
atrata v. dendyi - flower heads yellow 25
atrata luteola - see C. atrata v. dendyi 15
coronopifolia / HH R / creeping / yellow button heads / N.Z. 212 / 238
dendyi / see C. atrata 25
pyrethrifolia / R / 5" / yellow daisies / New Zealand 64

COTYLEDON - Crassulaceae
oppositifolium - see CHIASTOPHYLLUM oppositifolium 93
simplicifolia - see CHIASTOPHYLLUM oppositifolium 129
spinosa - see OROSTACHYS spinosa
teretifolia / Sh / ? / fls yellow / Cape Prov. 25
undulata / Gh / to 2' / red fls / South Africa 93

COWANIA - Rosaceae
mexicana v. stansburyana / Sh / to 7' / fragrant creamy fls / w U.S. 25

CRAMBE - Cruciferae
cordifolia / P / to 7 ' / fls large panicle / Caucasus 25

CRASPEDIA - Compositae
alpina / HH R / to 1' / yellow fls / Tasmania (12) / 64
glauca / HH P / 18" / yellow fls / n Australia 206
incana / R / to 1' / yellow fls; snow-white lvs / high mts., N.Z. (81) / 15
lanata - see C. alpina 193 / 25
minor / HH R / 4-10" / yellow fls / New Zealand 15
richei / A / 1' / yellow fls / Australia 206
robusta / HH P / 2' / yellow fls / New Zealand 15
uniflora / HH P / 18" / fls yellow or white / N.Z., Australia 194 / 25
uniflora robusta - see C. robusta 15

CRASSULA - Crassulaceae
arborescens / Gh / 3' / white fls, aging pink / South Africa 129 / 128
corymbulosa / Gh / 1' / white fls / South Africa 25
intermedia / Gh / rosettes / fls white / se Cape Province 156
lanuginosa / mat forming, rarely 4" / few white fls / S. Africa 156
milfordae / HH R / prostrate / white fls / Basutoland 268
sarcocaulis / Gh / 8" / fls white to pink / S. Africa 194 / 25

CRATAEGUS - Rosaceae
dilitata / T / 18' / frs br. red, maturing early / Que & s 221
durobrivensis / Sh / 10-16' / lg white fls / N. Amer. 139
kansuensis / T / 24' / lobed lvs; red frs / n China 222
mollis / T / 30' / showy red frs / e North America 139
phaenopyrum / T / to 25' / lustrous bright red frs / se U.S. 313 / 25
punctata / T / 30'+ / white fls; frs dull crimson / e N. America 139
submollis / T / to 30' / bright red frs / e Canada, New York 25
tanacetifolia / T / to 40' / orange-yellow frs / w Asia 77 / 25
uniflora / Sh / to 5' / red to greenish-yellow frs / e & c U.S. 313 / 25

CRAWFURDIA - Gentianaceae
speciosa / HH P / twining / pale blue fls / Nepal 81

CREMANTHODIUM - Compositae

arnicoides / P / strong-growing / yellow fls / ? 64
delavayi / P / 2' / golden-yellow fls / G. Forrest coll. 64
nanum / R / 6" / orange-yellow fls / Himalayas 162
oblongatum / R / to 1' / fls yellow / Nepal, Tibet, Himalayas 268
reniforme / P / 18" / yellow fls / n India 123

CREPIS - Compositae

aurea / R / 2" / coppery-orange fls / European Alps 148 / 123
blattarioides - see C. pyrenaica 289
elegans / R / to 8" / fls yellow or orange / Alaska to Wyoming 20
hokkaidoensis / R / to 8" / fls yellow / alpine reg. Hokkaido 200
incana / R / 6" / fls magenta-pink / s Greece, in mts. 194 / 25
incana rosea - synonymous with incana 123
jacquinii / R / to 1' / bright golden fls / e Alps, Carpathians 148 / 219
nana / R / dwarf / crowded heads of yellow fls / n North America 209
purpurea / R / 4-16" / achenes dark brown / Anatolia 289
pygmaea / R / to 6" / fls pale to golden yellow / w Alps, Pyrenees 148
pygmaea 'Purpurea' - selection for reddish-purple coloring *
pyrenaica / P / to 28" / yellow fls / mts. of c Europe 314 / 289
rosea - see C. incana *
rubra / A / 10-16" / fls pink or white / e Europe 211 / 25
rubra 'Rosea' - pink selection 25
sibirica / P / 4'+ / bright golden fls / w Siberia 93
terglouensis / R / 1-2" / achenes yellow / c & e Alps 289

CRINODENDRON - Elaeocarpaceae

decipiens - see C. patagua 36
Hookerianum / Gh T / 10-30' / rich crimson fls / Chile 129 / 36
lanceolatum - see C. hookerianum 36
patagua / Gh T / 30' / fls white, bell-shaped / Chile 36

CRINUM - Amaryllidaceae

macowanii / Gh Bb / 2' / lilac-pink fls / Natal 111

CRISTARIA - Malvaceae

glaucophylla / HH R / prostrate / flesh-colored fls / Chile 64

CRITHMUM - Umbelliferae

maritimum / P / 1-2' / the edible Samphire / coastal Europe 149 / 25

CROCANTHEMUM

canadense - see HELIANTHEMUM canadense 99

CROCOSMIA - Iridaceae

aurea / HH C / 3' / golden-orange fls / Natal, Cape of Good Hope 267
citronella / Montbretia hybrid, to 3' / lg clear lemon yellow fls 111
masonorum / HH C / 3' / bright orange-red fls / South Africa 279 / 268

CROCUS - Iridaceae

adami - see C. biflorus v. Adamicus
aerius / C / 4" / fls usually pale lilac-blue / Turkey, Iran 208 / 185
aerius 'Grey Lady' - pale mauve-gray fls (55) / 267
alatavicus / Bb / ? / fls white inside, out variegated / c Asia 52

alativus / C / 4-12" / whitish & purplish fls / c Asia (453) / 4
albiflorus - see C. vernus 25
ancyrensis / C / 3" / brilliant orange fls / A. Minor, Turkey (345) / 267
angustifolius / C / 4-12" / bright yellow fls / sw Russia, in mts. (82) / 25
asturicus / C / 5" / purple fls in fall / Spain 267
asturicus 'Atropurpureus - fls dark mauve 25
asturicus purpureus - see C. A. 'Atropurpureus' *
aureau - see C. flavus 129 / 25
balansae / C / 4" / orange fls / w Asia Minor (454) / 267
banaticus / ? / white to deep purple fls / Balkans (132) / 185
biflorus / C / 4" / purple fls, striped white / Italy, A. Minor 185 / 267
biflorus v. Adamicus - fls pale purple, perianth faintly striped 25
biflorus v. alexandri / white fls, glossy purple outside/Levant 52 111
biflorus v. argenteus / earlier & smaller / Italy (55) / 25
biflorus crewei / wht with dk veins, black anthers / Greece, Turkey 185
biflorus v. parkinsonii - pale lilac, buff & purple outside 267
biflorus 'Weldenii Albus' - entirely white form (55) / 267
biflorus Weldenii / pure white v. of C. biflorus albus/ Dalmatia 52 / 25
boryi / Gh C / 3" / fls cream / Greece, Crete 175 / 185
byzantinus - see C. banaticus 185
cambessedesii / C / 3" / fls pale lilac-purple / Balearic Isls. 267
cancellatus / C / 3" / white to deep lilac fls in fall / Greece-Iran 267 / 25
cancellatus v. cilicicus / pale lilac-mauve fls / Asia Minor 267
cancellatus v. mazzianicus - included in the sp. 185
candidus / C / 4" / creamy-yellow fls / Levant (346) / 267
candidus 'Subflavus' / pale yellow fls / Asia Minor 267
carpentanus / C / 3" / fls pale violet / c & n Portugal, n Spain 212
cartwrightianus / C / 4" / purple-striped fls / Greece 25
cartwrightianus albus - see C. hadriaticus v. chrysobelonicus 185
caspius / C / to 12" / fls white to pale lilac / Iran, Russia 185
caspius v. lilacinus / C / 4-12" / fall-fl, fls rosy-lilac tinted 52
chrysanthus / C / 4" / yellow fls / Greece 267 / 129
chrysanthus 'Blue Bird' - outer petals purplish violet, white edge 268
chrysanthus 'Blue Pearl'-lt blue outside,paler inside, orange stigma 268
chrysanthus 'Canary Bird'-sm., bright orange-yel. fls, feathered brz 267
chrysanthus 'Cream Beauty' - pale crm-yellow w bronze gr. base 268
chrysanthus 'E. A. Bowles' - canary yellow, bronze vein toward base 268
chrysanthus 'Mariette' - petals straw-yellow, feathered royal purple 268
chrysanthus 'Moonlight' / C /?/ fls pale yel, crm at base /Asia Min. 25
chrysanthus 'Princess Beatrix'-clear lav, gold base;inside silver-bl 268
chrysanthus 'Snow Bunting' - fls wht, perianth spead, blue stripe 25
chrysanthus 'Warley White' - gray-white w purp. shading; inside white / 268
chrysanthus 'Zwaneberg Bronze'-dk brz outside, margin & inside yell. 268 / 25
clusii / C / 3" / mauve to purple fls in fall / s Spain, Portugal 267
corsicus / C / 4" / lilac & purple fls / Corsica 129
corsicus 'Albus' - white fls *
cretensis - see C. laevigatus 52
Crewci / C/ ? / white fls, dark veining / Greece, s Turkey 185
cyprius / Gh C / ?/ blue fls, yellow throat / Cyprus 185
dalmaticus / C / ? / fls rose-lilac to grayish-lavender / Jugoslavia 25
danfordiae / Gh C / mini / pale lemon-yellow fls / Turkey 25
etruscus / C / 4" / fls lilac, yellow throat / w Italy 267 / 186
flavus / C / 8-12" / yellow fls / Balkans 211 / 52
flavus aureus - a synonym, not a variety *
flavus 'Concolor' - pale yellow fls 17

flavus moesiacus - see C. flavus 25
fleischeri / C / 3" / white fls, outer purple stripes / Asia Minor 267 / 25
X fritschii - intermediate of C. napolitanus & albiflorus; sterile? 185
gargaricus / C / 2" / fls orange-yellow / w Turkey 185
graveolens / Bb / ? / pale yellow marked brown fls / Syria 52
goulimyi / C / 4-5½" / blue-mauve fls / Greece 149 / 268
hadriaticus / C / 4" / white fls, veined purple / Greece 267
hadriaticus v. chrysobelonicus / red-purple veins / w Greece 52
heuffelianus / C / 10-14" / fls white to violet / c Europe (78) / 93
heuffelianus 'Albus' - all white form of C. heuffelianus 185
heuffelianus scepusiensis - see C. scepusiensis 185
heuffelianus X vernus - see C. vernus ssp. vernus
imperati / C / 4" / bright mauve & buff-yellow fls / s Italy 267 / 129
imperati f. albus - white form found in wild 111
imperati ssp. suaveolens - included in the species 185
iridiflorus - see C. banaticus 185
karduchorum - see C. kotschyanus v. leucopharynx 17
'Kathleen Parlow' / pure white / conspicuous orange anthers 268
korolkowii / C / 4-8" / fls deep yellow / Afghanistan, Turkestan 185 / 267
kotschyanus / C / 4" / pale lilac fls in fall / Asia Minor, Lebanon 267 / 129
kotschyanus 'Albus' - all white fls 17
kotschyanus v. leucopharynx - lavender fls with white throat 267 / 17
laevigatus / C / 3" / fls white to lilac / Greece 129 / 185
laevigatus 'Fontenayi' - rosy-lilac & buff fls 25
longiflorus / C / 5" / pale to bright lilac fls / s Italy, Sicily 267 / 25
longiflorus v. melitensis / feathered & blotched purple / Malta 267 / 25
maesiacus - see C. flavus 52
Malyi / C / 1"+/ fls large, white / Croatia 121
medius / C / 4" / lilac & purplish fls in autumn / s France 184 / 267
minimus / C / 2" / fls in purple shades / Corsica, Sardinia 267 / 129
napolitanus - see C. vernus v. napolitanus 25
napolitanus 'Albus' - see C. vernus *
nevadensis / C / ? / fls whitish or pale lilac / s Spain, Algeria 212 / 185
niveus / C / 5" / fls white or pale lilac in fall / Greece 194 / 185
nudiflorus / C / 4" / fls pale to deep purple in fall / Pyrenees 212 / 185
nudiflorus banaticus - see C. heuffelianus 267
ochroleucus / C / 4" / creamy-white fls / Syria, Israel 267 / 185
olivieri / C / 4" / brilliant orange fls / Balkans, Greece 149 / 25
oreocreticus - see C. sativus 52
pallasii - see C. sativis pallasii 52
pestalozzae / Gh C / clear blue fls, blackish spot in throat / Turk. 185
'Pickwick' - striped garden crocus variety 17
pulchellus / C / 3" / pale lavender fls in fall / Turkey, Asia Minor 267 / 129
pulchellus 'Zephyr' - white fls shaded gray 17
purpureus - see C. napolitanus 185
X 'Purpureus Grandiflorus' - old hybrid, deep rich purple fls 129 / 267
reticulatus / C / ? / lilac & buff fls / n Italy to Russia (55) / 185
salzmannii / C / 3" / fls clear pale lilac in fall / s Spain 267 / 25
sativus / C / 4" / large deep purple fls / cult. clone for Saffron 267 / 185
sativus v. cartwrightianus - see C. cartwrightianus 25
sativus v. cartwrightianus f. albus - white fls, scarlet stigma 111
sativus cartwrightii - see C. sativus v. cartwrightianus 185
sativus v. elwesii / pale pinky-lilac fls / Turkey, Iran 184 / 185
sativus v. palasii / smaller lilac fls / Asia Minor 185
sativus v. thomasii / heavily veined lilac fls / Italy 185

scardicus / C / ? / fls yellow / Mt. Skardo, Bulgaria 52
scepusiensis / C / 12" / white to purple fls / Poland, Carpathians (347) / 185
scharojanii / C / 4-10" / fls bright yellow / Caucasus 4
serotinus / near C. clusii, but feathered form 52
serotinus ssp. clusii / C / fls lilac, throat pubesc./Port & w Spain 290
siculus - similar to C. albiflorus from Sicily 185
Sieberi / C / 5" / fls in lavender-purple shades / Greece, Crete 129 / 185
Sieberi f. atticus / deep mauve fls / Mt. Parnassus, Atticus 149 / 267
Sieberi 'Hubert Edelsten' - crimson marked cultivar of C. Sieberi 278
Sieberi v. tricolor / purple-white-yellow banded / n Peloponnense 149 / 185
Sieberi 'Violet Queen' - dark mauve fls 185
speciosus / C / 5" / pale blue-mauve fls in fall / e Europe, Iran 267 / 129
speciosus 'Aitchesonii' - large pale lavender-blue fls 185
speciosus 'Albus' - fls white 25
speciosus 'Artabir' / C / pale lav w dk feathering/e Eur, Asia Minor 267
speciosus 'Cassiope' - analine-blue fls 111
speciosus 'Globosus' - globose, nearly pure blue fls 25
speciosus 'Oxonian' - dark blue fls 129 / 111
speciosus 'Pollux' - pale blue fls 61
X stellaris - garden hybrid, fls pale yellow, sterile? (454) / 25
'Striped Beauty' / white w violet stripes, purple base 268
susianus - see C. angustifolius 25
suterianus / C / 3" / butter to deep golden-yellow fls / A. Minor 267
suaveolens / C / similar to C. Imperati / Rome, Italy 52
Suwarowianus - see C. vallicola Suwarowianus 52
tauri / C / ?/ short pale yellow pistil, wholly lilac fls / A. Minor 52
Tomasinianus / C / 4" / mauvish-blue or lilac fls / s Italy 186 / 129
Tomasinianus albus / C / ? / white form / c Jugoslavia 25
Tomasinianus 'Barrs Purple' - soft lilac-mauve fls (454) / 111
Tomasinianus 'Taplow Ruby' / dark rich red-purple, dark at tips 267
Tomasinianus 'Whitewell Purple' - less freely spreading 129 / 17
tournefortii / HH C / 4" / warm rosy-lilac fls in fall / Greek Isls 267 / 25
vallicola /C/ ? / creamy wht fls veined pale lilac/Caucasus, Armenia 52
vallicola v. Suwarowianus/petals veined lilac, both surfaces, rounded 268
veluchensis / C / 4" / lilac-purple fls / n Greece, in mts 149 / 267
vernus / C / 4" / fls white to lilac / c Europe 93 / 25
vernus albiflorus - included in the sp. 25
vernus caeruleus - see C. napolitanus 185
vernus 'Jeanne d'Arc' / C / large / white / Alps, Pyrenees 267
vernus f. leucostigma - creamy-white stigmas 52
vernus v. neapolitanus / fls purple / white throated / s Italy 25
vernus v. siculus - miniature variety from Sicily 52
vernus 'Vanguard' - pale silvery-lilac fls, early 128 / 267
vernus ssp. vernus /B/1-6"/fls purp or striped/Ital,e Med.,nat.Brit. 290
versicolor / Bb / ? / fls pale or dark purple / s France 28
versicolor 'Picturatus' / white fls, purple stripes / s France 185
vitellinus / Bb / ? / gold, yellow, orange fls / Asia Minor, Syria 267
weldenii - see C. biflorus Weldenii
zonatus - see C. kotschyanus 25

CROSSOSOMA - Crossosomataceae
bigelowii / HH Sh / 3'+ / fls white or purplish / Ariz to se Calif 25

CROTALARIA - Leguminosae
laburnifolia / Gh Sh / 5-6' / large yellow fls / Ceylon 81

CROTON - Europhbiaceae
alabamensis / Sh / 6-9' / lvs evergreen; fls green / Alabama 250

CRUCIANELLA
stylosa - see PHUOPSIS stylosa

CRUCKSHANKIA - Rubiaceae
glacialis / R / rosette / bright lemon-yellow fls / alpine Chile 64
[1] pumila / A / 3-6" / yellow bracted fls / Chile

CRYTOGRAMMA - Polypodiaceae
acrostichoides - see C. crispa v. acrostichoides 99
crispa / F / 3-6" / the Parsley Fern / England, Scotland 148 / 159
crispa v. acrostichoides / 8" / the North American Var. (7) / 99

CRYPTOMERIA - Taxodiaceae
fortunei - see C. japonica v. sinensis 127
japonica 'Cristata' / to 9' / conical & narrow 79
japonica 'Lobbii' - compact, conical tree 25
japonica 'Sekkwia (Sekkan)-sugi' - see C. japonica 'Cristata' 79
japonica v. sinensis / HH T / tall / looser habit than sp. 127

CRYPTOSTEGIA - Asclepiadaceae
madagascariensis / Gh Cl / ? / fls white or pinkish / Madagascar 28

CRYPTOTAENIA - Umbelliferae
canadensis / P / to 3' / white fls / e North America, in woods 224 / 99

CUCUBALUS - Caryophyllaceae
baccifer / P / 4', weak stems / greenish-white fls / s & c Europe 210 / 286
baccifer v. japonicus / P / 3'+ / fls white, black seeds / Far East 200

CUCURBITA
[2] palmata / A / trailing / fls yellow / sc California

CUMMINGIA
trimaculata - see CONANTHERA trimaculata 28

CUNILA - Labiatae
mariana - see C. origanoides 99
origanoides / R / 8-16" / fls pinkish-purple / e North America 224 / 25

CUPHEA - Lythraceae
cyanea / Gh Sh / 6'+ / fls yellow, pink, red, violet / Mexico 25
hookerana / Gh Sh / 2'+ / violet or red fls / Mexico, Central Am. 25
lanceolata / Gh Sh / to 4' / deep violet fls / Mexico 194 / 25

CUPRESSUS - Pinaceae
arizonica / HH T / 75' / evergreen with small needles / sw US 93 / 36

---

[1] T. Harper Goodspeed, Plant Hunters in the Andes, 1961.

[2] Liberty Hyde Bailey, Garden of Gourds, 1937.

bakeri / T / 35-50' / conspicuous resin glands / n California 127
funebris - see CHAMAECYPARIS funebris 25
goveniana / HH T / 20'+ / bright green needles / Monterey Co., Calif 25
macnabiana bakeri - see C. bakeri 25
macrocarpa / HH T / to 70' / large-coned / s California 139
sempervirens / HH T / to 150' / cedar-like / Medit. region 210 / 25
sempervirens v. horizontalis - Near-eastern form 36

CURTONUS - Iridaceae
paniculatus / HH Bb/ to 4' / orange-red fls / Transvaal, Natal 186 / 267

CUSSONIA - Araliaceae
paniculata / Gh Sh / 15' / white or yellowish fls / South Africa 25

CUTHBERTIA - Commelinaceae
graminea / HH R / to 16" / rosy fls / coastal, N.C. to Fla. 25

CYANANTHUS - Campunulaceae
inflatus / A / mats / small blue fls / Bhutan 64
integer - see C. microphyllus 46
lobatus / R / to 1' / deep violet-blue fls / Himalayas 30 / 129
lobatus 'Albus' - a white color form 123
lobatus v. insignis - with larger fls 107
lobatus 'Sheriff's Form' - see C. sheriffii 123
microphyllus / R / mat / fls violet-blue / Nepal, n India 129 / 25
sheriffii / R / procumbent / periwinkle-blue fls / s Tibet (498) / 123

CYANELLA - Tecophilaeaceae
capensis / Gh P / to 18" / fls bright blue / South Africa 184 / 25

CYANOPSIS - Compositae
muricata / A / to 20" / pink fls / s Spain 289

CYATHEA - Cyathaceae
smithii / F / to 24' / lowland tree-fern / New Zealand 15

CYATHODES - Epacridaceae
colensoi / HH Sh / 6-12" / white fls; rose-red frs / New Zealand 133 / 123
empetrifolia / HH Sh / prostrate / leathery lvs; reddish frs / N.Z. 25
fraseri - see LEUCOPOGON fraseri 25
juniperina / HH Sh / to 15' / minute lvs; fls white to purple / N.Z. 238
parviflora / HH Sh / to 6' / varii-colored fls / New Zealand 15
pumila / HH Sh / prostrate / minute fls; dark frs / New Zealand 15

CYCLAMEN - Primulaceae
africanum / HH Tu / 6" / dark to pale rose-pink fls / Algeria 267 / 25
africanum album / white form of species 184
alpinum / Tu / ? / rose-pink to deep carmine fls / sw Asia Minor 194 / 25
X atkinsii - blotched lvs; white fls (84) / 25
X atkinsii 'Albus' - white fls 267
X atkinsii 'Roseum' - fls deep pink; possible a form of C. coum 267
balearicum / HH Tu / ? / fragrant fls, white pink veins / s France 212 / 25
cilicium / HH Tu / 3-5" / pale pink fls / s Turkey 185 / 129
cilicium 'Album' - fls white 25
cilicium alpinum - see C. cilicium 'E. K. Balls - #669a' 239

cilicium 'E. K. Balls #669a' / 2" / fls wht or pale pink/A. Minor 239
cilicium v. intaminatum / without dark basal blotch 77
coum / Tu / 4" / fls white to pink & carmine / se Europe, A. Minor 185 / 25
coum f. abchasicum - see C. coum ssp. caucasicum 239
coum 'Album' - white fls with purple spot (86) / 25
coum atkinsii - see C. X atkinsii 25
coum atkinsii 'Album' - see C. X atkinsii 'Album' *
coum atkinsii 'Roseum' - see C. X atkinsii 'Roseum' *
coum ssp. caucasicum - corolla with darker rim / n Iran 185 / 25
coum ssp. coum/Tu/?/corolla wht-pnk-carm, rim wht, basal eye/Cauc. 25 / 239
coum 'Hiemale' - early-flowering, magenta-carmine fls / Turkey 267
coum ibericum - a synonym of the sp. 268
coum 'Roseum' - pink fls, purple spot 90 / 17
coum vernum - see C. coum ssp. caucasicum 25
creticum / HH Tu / 4" / fls mostly white / Crete 267 / 268
cyprium / HH Tu / 4" / fragrant white or shell-pink fls / Cyprus 267 / 25
europeum - see C. purpurascens 25
graecum / Tu / 6" / fls pale pink, streaked magenta / Asia Minor 267 / 25
graecum gaidurowryssi v. malingeri - see C. graecum 239
hederifolium / Tu / 6" / rose-pink to white fls / s Eu., w A. Minor 175 / 25
hederifolium 'Album' - beautiful albino form 239
hederifolium 'Fragrans' - "some forms are fragrant" 123
hiemale - see C. coum 'Hiemale' 129
ibericum - see C. coum 129
ibericum 'Tubergens Var.' - see C. pseudibericum 'Tubergens Var.' 239
indicum - see C. persicum 28
intaminatum - see C. cilicium intaminatum 239
libanoticum / HH Tu / ? / large salmon-pink fls / Lebanon 132 / 129
mirabile / Tu / 5" / pale pink fls / Turkey (348) / 239
neapolitanum - see C. hederifolium 25
neapolitanum 'Album' - see C. hederifolium 'Album' 239
orbiculatum - see C. coum 129
parviflorum / Tu / dwarf / dull green lvs; pale lav-pink fls/ Turkey 239
persicum / HH Tu / 6" / rose-pink to white fls / e Medit. reg. 267 / 25
persicum 'Album' - pure white fls 239
pseudibericum / HH Tu / 4-6" / near-purple fls / s Turkey 185 / 268
pseudibericum 'Tubergens Var.' - deeper in color 239
pseudograecum - Cretan form referred to C. graecum 239
'Puck' / Gh Tu / hybrid of C. purpurascens & C. persicum 203
purpurascens / Tu / 6" / rose-pink to slate-magenta fls / c & s Eu. 239 / 25
repandum / Tu / ? / fls white, pink, crimson / C & s Italy 194 / 129
repandum 'Album' / pure white form / Corsica 239
repandum X balearicum - cross made by donor, both species in text
repandum 'Pelops' / HH Tu / pink or carmine fls / Greece 239 / (31)
repandum rhodense / Tu / ? / carmine red, pink, white / c & e Medit. 25
Rohlfsianum / Tu / ? / rose-pink / Cyrenaica 25
trochopteranthum/Tu/?/corolla pink-magenta, dk basal blotch/Turkey 77
vernum - see C. coum ssp. caucasicum (455) / 25
vernum album - see C. coum 'Album' 267

CYDONIA - Rosaceae
japonica - see CHAENOMELES japonica 36
japonica v. maulei / Sh / to 3' / fls red / Japan 36
oblonga / T / to 20' / pale pink fls; edible frs / w Asia 210 / 25

CYMBALARIA - Scrophulariaceae
muralis / R / 2', trailing / fls lilac to violet / s Europe 224 / 288
muralis f. alba - from the rarely white wild strain 288
1 muralis 'Alba-Compacta' - dwarfed form
muralis 'Globosa' - neat little hummocks 46
muralis 'Globosa Alba' - with white fls *
muralis 'Globosa Rosea' - light pink fls 46
muralis 'Rosea' - light pink fls 46
pallida / R / 8" / fls pale lilac-blue / c Italy, in mts. 148 / 288

CYMOPHYLLUS - Cyperaceae
Fraseri / P / 8-24" / lvs green, leathery / Pa. to S. Car. 25

CYMOPTERUS - Umbelliferae
multinervatus / R / 3" / fls purple / w Texas to s California 226

CYNANCHUM - Asclepiadaceae
nigrum / P / erect or scrambling / fls purple-black / se mts. U.S. 25
purpurascens / P / 24-38" / fls purplish / Japan 200
vincetoxicum / P / 1-2' / fls greenish-white / Eu., escaped in N.Y. 99

CYNARA - Compositae
scolymus / HH P / 3-5' / the Garden Artichoke / cult. plant 28

CYNOGLOSSUM - Boraginaceae
amabile / B / 2' / bright blue fls / w China 75 / 93
amabile 'Firmament' / A / 18" / sky-blue fls 202
amabile f. roseum - pink fls 93
columnae / A / 10-18" / fls deep blue / c & e Medit. region 211 / 288
creticum / B / 1-2' / deep blue fls / s Europe 149 / 288
'Firmament' - see C. amabile 'Firmament' 202
grande / P / 1-3' / fls blue or purple / Wash., Calif., B.C. 228 / 25
magellense / R / 8-12" / fls reddish / c & s Apennines 288
nervosum / B or P / 3' / fls blue / Himalayas 129 / 25
officinale / B / to 2' / dull purple fls / Europe 148 / 25
zeylandicum /B/ 16-28" /densely flwd, bluish-whitish/ Korea, s Asia 200

CYPELLA - Iridaceae
coelestis / HH Bb / 2-3' / pale blue fls / s Brazil, Argentina 93
coelestis platensis - see C. plumbea v. platensis *
drummondii / HH C / 1' / purple & yellow fls / South America 206
herbertii / HH C / 12-20" / mustard-yellow & purple fls / S. America 171 / 185
peruviana / HH C / 8-16" / apricot-yellow fls, shaded purple / Peru 184 / 185
plumbea / HH C / 1½-3' / dull blue fls / South America 25
plumbea v. platensis - clear sky-blue fls 111

CYPERUS - Cyperaceae
alternifolius / Gh P / 1½-3' / Umbrella Plant / tropics 25
esculentus / P / to 2' / Nut Sedge, weedy / w Asia 220 / 25
vegetus / Gh P / 1' / grass-like / Chile 206

---

[1] Parker, Quarterly Bulletin, Alpine Garden Society, 37:3.

CYPHIA - Goodeniaceae
volubilis / A Cl / 1' / pale mauve fls / Cape of Good Hope 223 / 206

CYPRIPEDIUM - Orchidaceae
acaule / R / to 1' / fls rose-pink, pouched / ne & nc N. Am. 224 / 107
acaule f. albiflorum - lip white, sepals & petals pale 99
calceolus / R / 15" / yellow & brown fls / Europe 90 / 129
calceolus v. parviflorum / fragrant, deep yel. & brn-purp fls/ N.Am. 25
calceolus v. planipetalum / petals flat or undulate/ e Canada 99
calceolus v. pubescens / greenish-yellow fls / N. Am. 224 / 25
californicum / P / 10-18" / dull yellow fls / Ore., Calif. 228 / 66
candidum / R / to 1' / white fls / N.Y. to Missouri 224 / 25
cordigerum / R / 6" / white fls with markings / Himalayas 130 / 132
fasciculatum / P / 4-16" / fls purple or yellow / nw US 66
guttatum / R / to 1' / white fls marked purple / Siberia, China 13 / 85
macranthum / R / 10-16" / rose-purple fls / Far East 166 / 200
macranthum v. speciosum - included in the sp. 200
montanum / P / 10-40" / white-lipped fls / nw North America 229 / 66
parviflorum - see C. calceolus v. parviflorum (456) / 25
passerinum / R / to 16" / white & purple fls / Canada, Alaska (349) / 25
pubescens - see C. calceolus v. pubescens 25
reginae / P / to 1½' / white & pink fls / n US 224 / 25
spectabile - see C. reginae 25

CYRILLA - Cyrillaceae
racemiflora / HH Sh / 4'+ / fls white / S. Am., N. Am. to Virginia 36

CYRTANTHUS - Amaryllidaceae
angustifolius / Gh Bb / 18" / red fls / South Africa 223 / 25
falcatus / Gh Bb / 18" / pinkish-red fls / Natal 267
mackenii / Gh Bb / 1' / pure white fls / Natal 25
mackenii v. cooperi - fls cream to yellow 25
O'Brienii / B / 1'+ / fls bright scarlet / South Africa 25
parviflorus / Hh Bb / 1' / bright red fls / South Africa 25

CYRTOMIUM - Polypodiaceae
falcatum / F / 1-2' / shining, dark green fronds / se Asia 180 / 128
fortunei / F / to 2' / papery pinnae / Far East 200

CYSTOPTERIS - Polypodiaceae
bulbifera / F / 2½' / green bulblets beneath / e North America 73 / 25
dickiana / F / 2-3" / tight mass, deciduous / e Scotland 159
fragilis /F / 4-8" / gray-green fronds / n & s hemisphere 148 / 200
fragilis alpina / F / 1' / fragile fern / n hemisphere to Chile 25
fragilis v. Dickieana / F / 4-5" / dk green, finely downy / Scotland 61
regia - included in C. fragilis 286
sudetica /F /4-16"/lamina yel-gr/Carp.,Sudeten Mts., n Russia, Norw. 286

CYTISUS - Leguminosae
albus - see CHAMAECYTISUS albus 287
albus pallidus - see CHAMAECYTISUS banaticus 287
ardoinii / Sh / 4-5" / fls golden-yellow / Maritime Alps 134 / 36
australis / Sh / procumbent / yellow fls / Cz. to Caucasus 36
austriacus - see CHAMAECYTISUS austriacus 287
battandierii / HH Sh / to 15' / fls golden-yellow / Morocco 269 / 36

X beanii / Sh / 6-18" / deep golden-yellow fls	129 /	36
biflorus - see CHAMAECYTISUS ratisbonensis		287
X burkwoodii - cerise & maroon-red fls	110 /	36
cantabricus /Sh/3-7'/ solit. yel. fl, chgd to legume wht-blk/n Spain		287
'Cornish Cream' - see C. scoparius 'Cornish Cream'		36
decumbens / Sh / 4-6" / fls bright yellow / s Europe		36
diffusus / Sh / 3-10" / bright yellow fls / se Europe		36
emeriflorus / Sh / to 3' / fls yellow / s Alps		36
frivaldskyanus / Dw Sh / low mound, hairy lfy stems, fls bright yel.		139
glabrescens - see C. emeriflorus		287
hirsutus v. demissus - Sh / 3-4" / fls yellow clusters / Greece		36
X kewensis / Sh / to 1' / fls plae yellow	269 /	25
leucanthus - see CHAMAECYTISUS albus		287
monspessulanus - see TELINE monspessulana		287
multiflorus / HH Sh / to 10' / fls white / Spain, Portugal		36
multiflorus albus - is an old synonym		287
nigricans - see LEMBOTROPIS nigricans		287
'Peter Pan' - see C. scoparius 'Peter Pan'		184
X 'Porlock' / Gh Sh / fls butter-yellow, lg semi-evergreen plant		139
X praecox / Sh / to 10' / sulphur-yellow fls	269 /	36
procumbens / Sh / 8-16" / golden yellow fls / ec Europe, Balkans	28 /	287
pseudoprocumbens/ Sh/ 8-16"/ fls gold-yel, legume blk /Balkan Pen.		287
purgans / Sh / 3-4' / deep golden yellow fls / Spain, France		36
purpureus - see CHAMAECYTISUS purpureus	110 /	287
purpureus 'Roseus' - shade of pink		36
X racemosus / Sh / to 3' / fls bright yellow / Canary Is.		25
ratisbonensis / Sh / to 6' / fls bright yellow / Eur., w Asia		25
reverchonii / Sh / like C. scoparius, branches incurved / se Spain		287
Rochelii / Sh / 3-4' / pale yellow, brown spotted / c Europe		36
scoparius / Sh / 6' / golden-yellow fls / w, s & c Europe	210 /	287
scoparius 'Andreanus' / fls yellow & dark crimson / Normandy		25
scoparius 'Cornish Cream' - cream & yellow fls, bushy open habit		36
scoparius 'Firefly' / Sh / med. size/ fls yel, dp brnz stain / Eur.	36 /	139
scoparius ssp. maritimus / procumbent to 16" / coastal nw Europe		287
scoparius Pendulus - low prostrate habit, large showy fls		36
scoparius 'Peter Pan' / 1-1½' / crimson fls		184
sessilifolius / Sh / 5-6' / fls bright yellow / s Eu., n Africa		36
striatus / Sh / 3-10' / branches dry black, fls yellow /Port., Spain		287
supinus - see CHAMAECYTISUS supinus		287
supranubius / Gh Sh / 8-10'/ milky-wht fls tinged rose/ Canary Isls.		36

DABOECIA - Ericaceae

azorica / HH sh / 6-10" / rosy-crimson fls / Azores	75 /	123
azorica 'Bearsden' - seedling from D. azorica		291
azorica X cantabrica - see D. X 'William Buchanan		
azorica X polifolia - see D. azorica X cantabrica	36 /	132
cantabrica / Sh / to 2' / white to purple fls / w Europe	212 /	107
cantabrica 'Alba' / glistening white fls / Ireland	110 /	132
cantabrica 'Atropurpurea' - rich wine-colored fls	110 /	132
cantabrica 'Praegerae' / 1' / clear deep rose fls / w Ireland		132
'Jack Drake' / dwarf / fls garnet-red/'Seedling No. 3' of D. azorica	36 /	291
polifolia - see D. cantabrica		132
polifolia 'Alba' - see D. cantabrica 'Alba'		28

polifolia 'Atropurpurea - see D. cantabrica 'Atropurpurea' 28
polifolia praegeri - see D. Cantabrica 'Praegerae' 132
X scotica 'William Buchanan' - fls red-purple (147)
X 'William Buchanan' - fls garnet-red, D. azorica X cantabrica 36

DACRYDIUM - Podocarpaceae
bidwellii / Gh Sh / 2-15' / lvs very variable / New Zealand 127
cupressinum / HH T / to 18' / timber tree / N. Zealand 52
laxifolium / HH Sh / 2-3' / smallest of conifers / New Zealand 134 / 127

DACTYLIS - Gramineae
glomerata 'Variegata' / Orchard Grass / lvs striped green & white 268

DACTYLORCHIS
cruenta - see ORCHIS cruenta 25
incarnata - see ORCHIS incarnata 85
masculata - see ORCHIS maculata 85
sambucina - see ORCHIS sambucina

DACTYLORHIZA - Orchidaceae (referred to as Orchis by some authorities)
elata / P / 2½' / rosy-purple fls / Algeria (350) / 279
ericetorum - see ORCHIS ericetorum 85
foliosa / P / 18-24" / purple fls / Madeira (125) / 279
fuchsii - see ORCHIS fuchsii 85
incarnata - see ORCHIS incarnata 85
maculata / Tu / to 24" / fls yel-wht, pink, lilac, red, purp / Eur. 290
maculata ericetorum - see D. maculata ssp. maculata
maculta ssp. maculata / Tu / to 24" / fls pink, lilac, purp. / Eur. 290
majalis - see ORCHIS majalis 85
pratermissa - see ORCHIS praetermissa 85
purpurella - see ORCHIS purpurella 85
sambucina - see ORCHIS sambucina 85

DAHLIA - Compositae
1 'Bishop of Llandaff' / 3' / miniature red
merckii / P / 6' / disc fls yel.-purp. ray wht-lav-purp / Mex. 25

DALEA
foliosa - see PETALOSTEMON foliosus *
gattingeri - see PETALOSTEMON gattingeri
purpurea - see PETALOSTEMON purpureus 25
villosum - see PETALOSTEMON villosus 25

DALIBARDA - Rosaceae
repens / R / creeping / white fls / ne North America 28

DANAE - Liliaceae
racemosa / HH Sh / to 3' / small white fls; red frs / Syria to Iran 25

DAMNACANTHUS - Rubiaceae
indicus / Sh / to 5' / white / East Asia 25

---

1 Dahlias, Their History, Classification, Culture, Insects and Diseases. Agricultural Experiment Station, Michigan State College, 1935, Bull. 266

Name / description	Ref.
DAPHNE - Thymelaeaceae	
acutiloba / HH Sh / 3' / white fls / w China	102
alpina / Sh / decumbent / fragrant white fls / s & c Europe, in mts.	148 / 287
arbuscula / Sh / low / lilac-pink fragrant fls / Hungary	132
caucasica / Sh / to 5' / fragrant white fls / Caucasus	25
caucasica v. axilliflora / to 6' / fls white; frs red/ Transcaucasia	55
cneorum / Sh / 1' / fls pink,rose-red or white / Europe, in mts.	148 / 25
cneorum 'Eximia' - fls large, deep pink	55 / 25
collina - see D. sericea	129 / 287
Genkwa / Sh / to 3' / fls lilac / China	25
giraldii / Sh / 2' / golden-yellow fls / China	123
gnidium / HH Sh / 4' / fls creamy-white / Medit. region	149 / 25
X houtteana - purple lvs; deep purple to lilac fls, sterile ?	55
jasminea / HH Sh / to 1' / purple & whitish fls / se Greece	287
kamtschatica / Sh / ? / yellow fls / Korea, Russian Far East	36
Kosanina / Sh / 20"+ / deep pink fls, frs red / sw Bulgaria	287
laureola / Sh / 2-4' / fls yellowish-green / s & w Europe	149 / 36
laureola ssp. philippi / dwarf, dense Sh / smaller fls / Pyrenees	212 / 36
longilobata / Sh / to 6' / greenish fls; showy red frs / nw Yunnan	55
Mezereum / Sh / to 5' / deep rosy-purple fls / Europe, Asia Minor	148 / 129
Mezereum v. alba - white fls; yellow frs	113 / 287
Mezereum albiflora - see D. mazereum v. alba	139
Mezereum alpina/Sh/10'+/fls fragr pnk-purp, frs scarlet/Eur, A.Minor	139 / 287
Mezereum v. atropurpurea - see D. X houtteana	55
Mezereum 'Rubra' / Sh / 4-5' / fls reddish purple / Eur., w Asia	25
oleoides / HH Sh / 3' / white fls; orange-red frs / Medit. region	149 / 132
oleoides glandulosa - a synonym of the sp.	25
oleoides jasminea - see D. jasminea	36
papyracea / HH Sh / 10' / white fls; red frs / India e to Nepal	253 / 36
petraea / HH Sh / 6" / fragrant rose-pink fls / n Italy	55 / 25
pontica / Sh / to 3' / pale yellow fls / Bulgaria, Turkey	36 / 287
retusa / Sh / to 3' / wht fls tinged rose or violet / w China	269 / 129
retusa alba - white form of D. retusa	
sericea / HH Sh / 2'+ / fragrant pink fls / e & c Medit. region	149 / 287
striata / Sh / to 1½' / dark pink fragrant fls / Eu. Alps	53
tangutica / Sh / to 3' / rosy-purple fls / nw China	110 / 222
DAPHNIPHYLLUM - Daphniphyllaceae	
cuneatum / P / ? / minutely hairy / nc Ukraine	286
decorum / P / to 1' / petals white to bluish / California	25
depauperatum / P / to 1' / sepals dk blue / B.C. to Alta & Calif.	25
dictylocarpum / P / ? / heavily bearded / e Russia	286
macropodum / T / 10' / fls yellow-green / Japan, Korea	200
peregrinum / P / 12-30" / fls dirty violet / c & e Medit.	286
requienii / P / 12-40" / fls deep blue / s France, Corsica	286
rossicum /P/similar to D. simonkaianum, lvs diff./se Russia, Ukraine	286
simonkaianum / P / stems sparsely hairy / c Romania	286
tatsienense 'Album' / P / 1-3' / fls white / w China	25
variegatum v. superbum / P / 2' / sepals lt. blue to blue purp/Calif	25
DARLINGTONIA - Sarraceniaceae	
californica / Gh / to 30" / dark purple petals / n Calif., sw Ore.	228 / 25

DASYLIRION - Agavaceae
wheeleri / HH Sh / 3' / small white fls / Arizona & southwards 107

DATURA - Solanaceae
inoxia / A or P / 3' / fls white to pink / sw U.S., Mexico (458) / 25
inoxia ssp. quinquecuspidata / P / wht to pale lavender fls / sw US 25
Metel / A / 3'+ / fls variable in color / India, China 25
meteloides - see D. wrightii 226
stramonium / A / to 5' / white or violet-purple fls / N. Am., weedy 208 / 25
suaveolens / HH Sh / 10-15' / white trumpet fls / Mexico 28
Tatula - see D. stramonium; purple fld forms 25
wrightii - see D. inoxia ssp. quinquecuspidata 225 / 25

DAUCUS - Umbelliferae
carota ssp. maximus - Mediterranean wild form of the carrot 287
maximum - see D. carota ssp. maximus 287

DAVALLIA - Davalliaceae
mariesii / F / to 8" / epiphytic fern / Far East 200

DAVIESIA - Leguminosae
brevifolia / Sh/ small/ bl-grn lvs; red, yel, apricot fls/s Austral. 109
corymbosa / Gh Sh / 5' / red & yellow fls / New South Wales 45
latifolia / Sh/ 4-6'/ dull grn lvs; fls brn & yel / Tas., N.S.Wales 109
ulicifolia / Sh / lvs spiny; fls brown & yellow / Australia 109

DECAISNEA - Lardizabalaceae
fargesii / Sh / to 16' / large blue frs / w China 110 / 25

DECKERRA
aculeata - see PICRIS aculeata

DEGENIA - Cruciferae
velebitica / P / to 4" / silver gray, non-flowering / Jugoslavia 286

DEINANTHE - Hydrangaceae
bifida / P / to 2' / white fls, green bracted / Japan 165 / 200
caerulea / R / 10-12" / dull violet-blue fls / China 279 / 123

DELOSPERMA - Mesembryanthemaceae
brunnthaleri / HH P / 1' / fls violet-rose / S. Africa 9
cooperi / Gh / prostrate / purple fls / Orange Free State 25

DELPHINIUM - Ranunculaceae
ajacis - see CONSOLIDA ambigua 25
'Azure Fairy' - see D. grandiflorum 'Azure Fairy' 268
barbeyi / P / lower lvs 5-cut, upper 3-cut; fls wht, blue / Colo. 61
X belladonna / P / to 6' / rich blue fls 25
bicolor / P / to 18' / fls yellow & purple / nw North America 229 / 25
'Blue Butterfly' - see D. grandiflorum 'Blue Butterfly' 268
1 'Blue Fountain Series' / 24" / lg fls blue with dark & wht base
brownii / P / to 3' / blue or purplish fls / Mont. to Alaska 5

[1] Wayside Catalog, Spring, 1986.

brunonianum / R / 10-15" / fls violet shaded purple / Tibet (95) / 129
californicum / HH P / 7' / whitish or yellowish fls / Cal. 108
cardinale / HH P / to 6' / yellow & scarlet fls / California 26 / 25
carolinianum / P / 1-2' / blue fls / Ga. to Texas & Ark. 225 / 26
cashmerianum / P / 18" / blue & purple fls / Himalayas 93 / 25
caucasicum - of cult. is D. elatum 26
chinense - see D. grandiflorum 5
X coeruleum / P / 2' / blue fls / 1847 hybrid 206
'Connecticut Yankee' / P / 2½' / varii-colored fls 202
consolida - see CONSOLIDA ambigua 25
X cultorum - name for various garden plants of hybrid origin 25
cuneatum / P / ? / minutely hairy / nc Ukraine 280
decorum / P / to 1' / petals white to bluish / California 25
delavayi / P / lvs deep 5-cut; fls blue 61
denudatum / P / 2-3' / pale blue fls / Himalayas 61
depauperatum / P / to 1' / sepals dk blue / B.C. to Alta & Calif 25
dictylocarpum / P / ? / heavily bearded / e Russia 286
Duhmbergii / P / to 2' / brownish fls / Russia, Siberia 5
elatum / P / 16-80" / blue to violet fls / c Europe, n & s Russia 148 / 286
exaltatum / P / to 6' / fls blue or white / ec United States 105
fissum / P / 20-60" / blue, violet-blue, lilac fls / s Europe 138
flexuosum / P / 2-3' / dark blue fls / Caucasus 77
freynii - see D. schmalhausenii 34
geyeri / P / to 2½' / rich blue & yellow fls / Wyo., Neb., Utah 229 / 25
glareosum / R / 3-8" / purple & yellow fls / Olympic Mts., Wash. 228 / 25
glaucum / P / 3-8' / violet-purple fls / nw North America 228 / 25
grandiflorum / P / 1-3½' / violet or blue fls / Siberia, China 208 / 25
grandiflorum v. album - fls pure white 7
grandiflorum 'Azure Fairy' / 18" / Cambridge blue 268
grandiflorum 'Blue Butterfly' / 18" / rich deep blue fls, brown spot 268
grandiflorum 'Blue Mirror' - gentian-blue fls 135
grandiflorum v. chinense - synonymous 25
grandiflorum 'Tom Thumb' / 8" / gentian-blue fls 203
hansenii / P / to 3' fls dark purple, bluish, reddish, white/Calif. 227 / 25
hesperium / P / 3' / fls blue to white / California 26
likiangense / R / 4-8" / fls rich blue to lilac / Yunnan 26
menziesii / R / 6"+ / blue fls / California to Alaska 228 / 123
montanum / 6-25" / plant densely hairy; fls pale blue / Pyrenees 286
muscosum / R / 4-6" / large, deep blue-violet fls / Bhutan 268
nanum / A / dwarf / brownish-violet fls / Egypt 270
Nelsonii / P/ to 1½'/blue-purp to pale blue/S.Dak.& Ida-Ariz.& Nev. 25
nepalense / R / 4" / solitary large blue fls / Nepal (95) / 162
nudicaule / HH R / 1' / fls red & yellow / n California 147 / 123
nudicaule v. luteum / pale yellow fls / coast of c California 26
nuttallianum / P / 4-32" / fls bright blue to purplish / nw U.S. 229 / 228
occidentale / P / 40-120" / fls wht or pale blue / Ore & Mont 142
oreganum / P / 3' / narrowly segmented lvs / Oregon 64
[1] orfordii / P / 18" / leafy stem; big deep blue fls
oxysepalum / R / 4-20" / fls blue to blue-violet / Carpathians 286
parryi v. blockmannae / 1' / dark blue fls / s California 257
patens / P / 8-20" / fls white & blue / n California 228
peregrinum / P / 12-30" / fls dirty violet / c & e Mediterranean 286

---

1 Siskiyou Rare Plant Nursery Catalog, 1986.

przewalskii / R / 6-10" / brownish & white fls / w Mongolia 26
pylzowii / R / 6-10" / dark violet fls / w China 26
pyramidatum / P / 3'+ / pale blue fls / Caucasus 5
requienii / P / 12-40" / fls dp blue / s France, Corsica 286
rossicum / P /similar to D. simonkaianum, lvs different/se Rus., Ukr. 286
schmalhausenii / P / 8-40" / fls deep smoky blue / Transcaucasus 77
scopulorum v. glaucum - see D. glaucum 25
semibarbatum / P / 1-2' fls bright yellow / Iran 25
simonkaianum / P / stems sparsely hairy / c Romania 286
speciosum / P / to 2½' / black & violet fls / se Asia 25
staphisagria / 12-36"+ / lvs pubescent both surf; fls dp blue / Med. 286
tatsienense / R / 1' / prussian-blue fls / s China 93 / 24
tatsienense 'Album' / P / 1-3' / fls white / w China 25
Treleasei / P/ 1½-4'/fls blue-purp, rare wht/spotted yel-brn/sw Mo. 99
tricorne / P / 2' / deep blue to purplish fls / ec United States 224 / 225
trollifolium / P / 6' / fls deep blue / Oregon, California 228 / 26
uliginosum / P / 12-20" / lvs 3-cleft; fls wht & violet / n Calif. 228
variegatum / P / to 2' / yellow & blue-purple fls / c Calif. 227 / 25
variegatum v. superbum / P/ to 2'/ sepals blue to blue-purp /Cal. 25
vestitum / P / tall / blue fls / Himalayas 26
villosum / P / to 4' / blue fls / Upper Volga region 5
virescens / P / 1-3' / yellowish fls / c North America 224 / 26
xantholeucum / P / 12-32" / fls pale yellow / Wenatchee Mts., Wash 229
zalil - see D. semibarbatum 184 / 25

DENDROMECON - Papaveraceae
harfordii - see D. Rigida ssp. harfordii 25
rigida ssp. harfordii / HH Sh / to 18' / yellow fls / California 25

DENTARIA - Cruciferae
california / P / 4-16" / fls white to pale pink or lav / Baja, Cal. 228
digitata - see CARDAMINE pentaphyllow 184 / 286
enneaphylla - see CARDAMINE enneaphyllos 286
laciniata / R / to 1' / white fls / e & c North America 224
pinnata - see CARDAMINE heptaphylla 286
tenella - see CARDAMINE pulcherrima v. tenella 228 / 142

DESCHAMPSIE - Gramineae
caespitosa / P Gr / to 4' / spikelets pale grn. to purp. / n. hem. 166 / 25

DESFONTAINEA - Loganiaceae
spinosa / Sh / to 3' / fls scarlet & yellow / Pennsylvania, Ohio 25

DESMANTHUS - Leguminosae
illinoensis / P / to 3' / whitish fls / c & sc United States 257 / 25

DESMODIUM - Leguminosae
canadense / P / 4'+ / rose-purple fls / e & c North America 224 / 99
paniculatum / P / 2' / fls inconspicuous / ec U.S. 99

DEUTIZIA - Saxifragaceae
crenata - see D. scabra 167 / 222
X rosea / Sh / compact, pink fls 139
scabra / Sh / to 10'/ pure white fls, sometimes tinged pink/Far East 36

DIANELLA - Liliaceae
caerulea / Gh P / 1'+ / blue fls; blue frs / e Australia 28
intermedia / Gh P / 10-18" / white fls; purple-blue frs / N.Zealand 238 / 128
laevis / Gh P / to 3' / fls blue / e temperate Australia 45
nigra / B / short stems / fls grn-wht; frs blue or purp / N.Z. 183
tasmanica / Gh P / 4' / starry light blue fls / Tasmania 128 / 279

DIANTHUS - Caryophyllaceae
X allwoodii / P / 9-16" / fragrant fls, variable in color 300 / 27
X allwoodii 'Alpinus' / 6" / strain of above (499) / 282
alpestris - see D. furcatus 286
alpinus / R / 4" / pink or deep rose fls / Alps of Europe 148 / 129
alpinus 'Albus' - non-fragrant white fls 306 / 123
alpinus 'Allwoodii' - see D. X allwoodii 25
alpinus 'Joans Blood' - bronzy lvs; blood-red fls (103)
amurensis / R / 8" / mauve fls / Manchuria (100)
anatolicus / R / to 1' / white fls / Turkey 77 / 25
anatolicus - if yellowish-rose fls, as in Farrer, see D. arpadianus 77
anatolicus 'Albus' - white fls, correct for the sp. 77
andrzejowskianum - see D. capitatus ssp. andrzejowskianum 286
arboreus / HH Sh / 3-5' / rose fls, fragrant / Isles in Medit. Sea 286
arenarius / P / to 1½' / white fls / e Europe 27 / 286
1 arenarius glaucus /P/dense tuft 4-12"/fls wht, rare rose/Lap-Cauc.
X 'Ariel' / 4-6" / cherry-red fls 153
armeria / A or B / to 16" / reddish fls / c & s Europe to Iran 27 / 25
arpadianus / R / 2-6" / fls yellowish-rose / Greece, Turkey 77
arrosti / P / to 32" / pale pink fls / Sicily 286
X arvernensis / R / to 14" / purplish fls / c France 27
X arvernensis 'Alba' - white fls *
atrorubens - see D. carthusianorum 'Atrorubens' 25
banaticus - see D. giganteus 77
barbatus / B / to 2' / varii-colored fls / s.& e Europe 148 / 129
barbatus 'Albus' - fls white 25
barbatus 'Messenger' - an early-flowering strain, mixed-col. fls 268
barbatus 'Newport Pink' / 18" / salmon-pink fls 202
barbatus 'Puniceus Albus' - a white strain 27
barbatus 'Wee Willie' / 4-6" / crimson red and deep rose shades 268
bebius - see D. petraeus 286
bicolor - see D. marschallii 286
biflorus / P / 16" / color ? / mts. c & s Greece 25
'Blue Hills' / R / 3" / tufted / magenta-purple fls 26
brachyanthus - see D. subacaulis ssp. brachyanthus 286
'Bravo' - see D. chinensis 'Bravo' 268
brevicaulis / R / 2" / rose or carmine fls / Turkey 77 / 25
broteri - see D. malacitanus 286
caesius - see D. gratianopolitanus 27
calalpinus - X of D. alpinus or D. callizonus 61
callizonus / R / to 8" / carmine fls / Carpathians 27 / 286
calocephalus - see D. cruentus 286
campestris / P / 14" / pink or purplish fls / e & s Russia 286
capillifrons - see D. carthusianorum 286
capitatus / P / 12-16" / purplish fls / se Europe to Siberia 77 / 27

1 Akademiia naukSSR Botanicheskii Instituti, Flora of the USSR, vol. 6.

capitatus ssp. andrzejowskianus / petals not bearded / Black Sea reg. 286
carmelitarum / P / to 16" / carmine-red fls / Turkey 77
carthusianorum / P / to 2' / pink to purple fls / sw & c Europe 148 / 286
carthusianorum 'Atrorubens - fls dark red 25
carthusianorum v. minor /P/ 6-16"/ lvs linear; bract sm, brn /Russia 77
carthusianorum 'Nanus' / 4" / small purple fls 25
carthusianorum v. sanguineus /P/to 24"/lvs linear; fls pink-purp/Eur 286
caryophyllus / HH P / to 32" / fls variable in color / Medit. reg. 286
caryophyllus v. corsicus / HHP / to 32" / varii-colored fls 96
X 'Charles Musgrave' / R / 1' / pure white fls / green center 153
chinensis / A or B / 6-30" / fls rosy-lilac, purplish eye 208 / 25
chinensis v. amurensis - see D. repens
chinensis 'Bravo' / A / 6" / bright scarlet fls 268
chinensis 'Heddewigii Baby Doll' / 6" / variously colored fls 129
chinensis v. laciniatus - fringed petals 208
chinensis v. macrosepalus / P / to 1" / fls bright red 28
chinensis 'Rainbow' - see D. X 'Rainbow Loveliness' *
collinus / P / 8-30"/ pink to purplish fls / Austria, Poland 25
corsicus - see D. caryophyllus v. corsicus 96
corymbosus / A or P / to 14" / rosy fls / Balkans 286
crinitus / P / 7-25" / white to pale pink fls / Armenia 99
croaticus - see D. giganteus ssp. croaticus 286
'Crossways' / R / 6" / intense crimson-carmine fls 46
cruentus / P / to 40" / fls reddish-purple / Balkans 286
Cyri / A / to 16" / pink / Arabia, Syria, Turkey & Afghanistan 25
'Delight' - an annual race, varii-colored
deltoides / R / 6-9" / white, pink crimson fls / n Europe 147 / 129
deltoides 'Albus' - fls white 25
deltoides X 'Brilliancy' / R / 1' / vivid carmine fls (102) / 283
deltoides 'Brilliant' - bright deep rose fls 46
deltoides 'Erectus' / 6-8" / deep rose fls, upright 46
deltoides 'Flashing Light' / 4-6" / lvs dark green, fls red 49
deltoides 'Hansens Red' - fls bright crimson 268
deltoides 'Huntsman' - bright red fls 46
deltoides 'Splendens' / R / 6-9" / reddish / "The Maiden Pink" 25
deltoides 'Steriker' - see D. deltoides 'Hansens Red' 61
deltoides 'Wisley Variety' - dark red fls; purplish lvs 46
deltoides ' Zing' - has long flowering period 203
dentosus - see D. chinensis v. macrosepalus 28
erectus - see D. deltoides 'Erectus' 46
erinaceus / R / 2-6" / pink fls / Turkey, in mts. 77
erinaceus v. alpinus - not importantly different 77
Falconeri / P / 1-2' / lvs 3-6"; fls small, red / w Tibet 61
'Fanal' / 6" / deep red fls 48
ferrugineus liburnicus / P / woody stem 12-24"/ fls pink/Italy,Jugo 286 / 27
'Flashing Light' - see D. deltoides 'Flashing Light' *
fragrans / R / to 1' / white fls / Caucasus, n Africa 27
'F.(rederick) C. Stern' / 6" / rosy-red fls 184
freynii / R / 2" / fls pink / Balkan Peninsula (100) / 286
furcatus / R / 6-8" / purplish-rose fls / n Italy 314 / 27
furcatus geminiflorus / P / mound / lvs flat; fls green / Pyrenees 286
furcatus v. lereschei - fls light lilac with pink zone 27
furcatus ssp. tener / dwarf to 4" / n Italy 286
gallicus /P/ mound to 20"/lvs obtuse;fls pink/coast Fr.,Spain, Port. 286
gelidus - see D. glacialis ssp. gelidus 25

Name / description		
giganteus / P / 3'+ / purple fls / Balkans, Turkey	77 /	286
giganteus ssp. banaticus / 40" / purple fls / Balkans		286
giganteus ssp. croaticus - as above, but larger calyx		286
'Gingham Gown ' / Rock Pink / pink fls, bordered red		(418)
glacialis / R / 2" / fls purple-red / Alps, Carpathians	148 /	286
glacialis ssp. gelidus / petals ½" longer / Carpathians	306 /	25
'Gladys Cranfield' / 1' / fls clear pink		153
gracilis / R / 6-16" / petals pink above, yel. reverse / w Balkans		85
graniticus / R / 6-16" / reddish-purple fls / sc France		286
gratianopolitanus / R / 8" / fls in pink shades / c Europe	148 /	129
gratianopolitanus 'Icombe' - clear pink selection		153
gratianopolitanus 'Major' - robust form		153
gratianopolitanus montanus /dense mat/ fls solit.; lvs short, stiff		61
[1] gratianopolitanus 'Petite' / mounds / fr pink 3" fls		
gratianopolitanus 'Plenus' - double-flowered		189
gratianopolitanus 'Praecox' - early-flowering		*
grisebachii - see D. viscidus v. grisebachii		27
haematocalyx / R / 1' / purple above, yellow reverse / s Balkans	153 /	286
haematocalyx ssp. pindicola - similar in form, Greece, Albania		286
heuteri / T / tol' / reddish fls / s Carpathians		286
'Highland Fraser / 1'+ / fls velvety crimson		153
'Highland Queen' - crimson-pink fls		153
hispanicus / P / 6-20" / pink fls / Spain		212
hoeltzeri - see D. cruentus		286
hungaricus / P / 16" / white fls / Tatra Mts., se Europe		286
[2] 'Ian' / 15-18" / fragrant double fls, rich scarlet		
'Icombe' - see D. gratianopolitanus "Icombe"		153
inodorus - see D. sylvestris		286
ilgasensis / P / 10-18" / very slender calyx bright pink / Turkey		77
japonicus / P / 8-20" / rose or rose-purple fls / Japan	165 /	200
japonicus albiflorus - unrecorded white fls		*
'Joans Blood' - see D. alpinus 'Joans Blood'		(103)
kitaibellii - see D. petraeus		286
knappii / R / 14" / sulphur-yellow fls / w Jugoslavia	27 /	286
'La Bourbrille' / compact plant / bright pink fls	(103) /	153
'La Bourbrille Albus' - white sort		*
X latifolius 'Atropurpurea' / 6-18" / dark 'Button Pink'		27
X lemsii / 1' / hybrid of D. pavonius		27
lereschei - see D. furcatus v. lereschei		27
'Little Jock' / 3" / double light pink fls	153 /	132
luburnicus - see D. ferrugineus liburnicus	286 /	27
lumnitzeri - see D. plumarius v. lumnitzeri		25
lusitanus / HH P / 6-16" / pink fls / Spain, Portugal		286
malacitanus / HH P / to 20" / pink fls / s Spain, s Portugal		286
X 'Mars' / to 6" / vivid crimson fls		153
marschallii / R / to 1' / fls yellowish-white, pink reverse/ Crimea		286
masmenaeus / R / to 1' / pink fls / Taurus Mts.		77
microlepis / R / 4" / purple fls / Bulgaria, in mts.	(100) /	286
microlepis 'Albus' - fls white		25
microlepis v. musalae - longer stems		25

---

[1] Rocknoll Catalog, Spring 1986.

[2] Wayside Catalog, Spring 1986.

monspessulanus / P / to 10" / fls white or pink / s & c Europe	148 /	286
monspessulanus ssp. sternbergii / under 8" / c Alps	(100) /	286
'Mrs. Clarke' / 4" / deep rose fls		123
X 'Mrs. Sinkins' - fragrant white fls, fairly true from seed		153
multipunctatus - see D. strictus		286
musalae - see D. microlepis v. musalae		25
myrtinervius / R / procumbent / pink fls / Macedonia		286
nardiformis / R / 4" / pink fls / e Bulgaria		286
neglectus - see D. pavonius	129 /	286
neglectus roysii - see D. roysii		27
nitidus / R / 6-12" / showy fls, spotted rose / w Carpathians		286
noeanus - see D. petraeus v. noeanus		286
'Oakington Hybrid' - double pink fls		46
'Old Laced Pinks' / varii-colored, scented double fls / 1'		283
orientalis / P / to 16" / pink / Afghanistan to Iran, Turkey		25
pallens / P / to 18" / pink / Romania, Balkan Pen.		25
pancicii - see D. tristis		286
pavonius / R / to 4" / fls purplish-red / sw Alps	(100) /	286
pavonius 'Albus' - fls white		*
peristeri - see D. myrtinervius		27
petraeus / R / to 1' / fls white or pink / Balkans	(100) /	286
petraeus v. noeanus / to 1' / white fls / Bulgaria		286
'Pikes Pink' - gray mats, double soft pink fls		(103)
'Pink' - see D. latifolius 'Pink'		27
pindicola - see D. haematocalyx ssp. pindicola	(500) /	286
pinifolius / R / 6-10" / reddish fls / Balkans, Greece	77 /	27
plumarius / P / to 16" / fls white to bright pink / ec Europe	147 /	286
plumarius v. albiflorus / fls white / Carpathians		25
plumarius blandus /P/ to 16"/ lvs narrowed in upper;fls wht/ec Eur.		286
plumarius 'Highland Queen' - see D. 'Highland Queen'		153
plumarius v. lumnitzeri /calyx narrowly attenuate/Austria, Hungary		25
plumarius 'Mrs. Sinkins' - see D. X 'Mrs. Sinkins'		153
plumarius v. praecox / fls pale pink or white / Bulgaria		25
1 plumarius 'Spring Beauty'/ mix of clove scent dbl; fls white-salmon		
pontederae / P / 6-18" / red-purple fls, rose below / ec Europe	314 /	286
praecox - see D. plumarius ssp. praecox		25
preobraszhenskii / R / to 1' / fls pink, yellow below / Caucasus		(105)
X pulcherrimus - a 'Sweet William' of short stature		153
pungens / R / to 8" / pink fls / e Pyrenees		286
puniceus albus - see D. barbatus 'Puniceus Albus'		27
pyrenaicus / P / to 18" / pungent pink fls / Portugal to Pyrennes		25
X 'Rainbow Loveliness' / 1' / superbus type, varii-colored fls	300 /	268
2 repens / P / stems numerous 3-8" / fls pink / arctic Europe		
X roysii - large, deep rose fls		25
rupicola / R / ? / flat gray-grn lvs; fls clear pink / Italy, Sicily		(396)
ruprechtii / P / 8-20" / red fls / Caucasus		(105)
scaber / R / 6-16" / small, bearded fls / se France, ne Spain		286
scardicus / R / 4" / pink fls / s Jugoslavia, in mt. pastures		286
'Scaynes Hill' / 4-6" / mounded pl; fls carmine-purple		46
seguieri / P / to 20" / pink fls, yellowish below / c Europe	148 /	77

---

1 Wayside Catalog, Spring 1986.

1 Akademiia naukSSR Botanicheskii Instituti, Flora of the USSR, vol. 6.

seguieri collinus - see D. collinus 286
sequieri montanus - included in the sp. 77
serotinus / P / to 16" / cream-colored fls / n Balkans 292 / 286
serratus - see D. pungens 286
shinanensis / P / 8-16" / rose-purple fls / Honshu 200
simulans / R / tufted / deep pink fls / Bulgaria (501) / 132
sinensis 'Baby Doll' - see D. chinensis 'Baby Doll' 27
'Southmead Seedling' / to 5" / rosy pink fls, early *
X 'Spark' / to 6" / crimson-scarlet fls 153
X 'Spencer Bickham' / 4" / deep rose-pink fls 153
speciosus - see D. superbus ssp. speciosus 25
sphacioticus / P / ½-4" / lvs oblong; fls bearded / Greece? 286
spiculifolius / R / 5-7" / rose, pink, white fls / e Carpathians 27 / 286
'Spotty' / 5" / rose-red & white fls 300
squarrosus / R / to 1' / white or pink fringed fls / c Russia 153 / 107
squarrosus 'Nanus' - stems reduced to 6" 96
sternbergii - see D. monspessulanus ssp. sternbergii 286
Sternbergii v. Waldsteinii / 15"+/ 5-13 fld umbrel; lvs leathery 61
stribryni / P/ small/ bracts purplish; fls purp, smooth / Bulg, Alb 286
strictus / P / 24-32" / white fls / sw Asia 77 / 286
strictus v. bebius - narrow lvs, showy fls, purple calyx 27
subacaulis / R / 1-8" / pale pink fls / sw Europe, in mts. 148 / 286
subacaulis albus / R / to 8" / pale pink / mts. sw Europe 25
subacaulis ssp. brachyanthus / densely caespitose / Pyrenees 194 / 286
suendermannii - see D. petraeus 286
superbus / P / to 3' / pink or purplish fls / Europe 148 / 286
superbus 'Albus' / P / to 3' / white / Italy, Romania 25
superbus 'Alpester' - listed name 25
superbus v. amoenus - dwarfed in stature 273 / 153
superbus v. longicalycinus / longer calyx / Far East 165 / 286
superbus v. monticolus - listed name for species 25
superbus 'Rainbow Loveliness' - see D. X 'Rainbow Loveliness' 153
superbus ssp. speciosus - glaucous purplish calyx 273 / 286
superbus ssp. stenocalyx / calyx more greenish / c & s Russia 286
'Sweet Wivelsfield' / 1½' /multi-colored fls, 'Sweet-William' hybrid 153 / 129
sylvestris / R / 4" / pink fls / European Alps 148 / 286
sylvestris 'Purpurea' - purplish fl variation *
sylvestris ssp. tergestinus / neat dwarf / Jugoslavia 24
tatrae - see D. plumarius v. albiflorus 25
tenuifolius / P/ 20" / many heads, pink to purple / ce China 25
tergestinus - see D. sylvestris ssp. tergestinus 286
[1] tianschanicus / R /to 10" / fls pink or dark purple / c Asia
'Tiny Rubies' / 2" / masses of pink fls 300
tristis / P / 16" / pink fls / Balkans 286
velebiticus - included in D. carthusianorum 286
viscidus / R / 1' / fls purple, spotted / Greece 77 / 286
viscidus v. grisebachii - dwarfed form from Macedonia 27
viscidus v. parnassicus - included in the type 286
'Waithmans Beauty' / laced pink / 10" / deep maroon fls (100)
Waldensteinii v. Sternbergii - see D. Stanbergii v. Waldensteinii 61
'Windward Rose' / 6" / deep rose-pink fls; ash-gray lvs 153
'White Hills' / 4" / small pink fls, crimson zoned (100)

[1] Shishkin Flora U.S.S.R. vol. VI, 1970 (trans).

X winteri - compact hardy border, pink strain 27
'Zing' - see D. deltoides 'Zing' 202
zonatus / R / to 1' / deep pink fls, yellow below / Turkey 77

DIAPENSIA - Diapensiaceae
lapponica / R / 2" / glistening white fls / New England, mt. tops 148 / 107
lapponica ssp. obovata/R/cushion to 4"/fls wht,/Alaska,ne Asia,Japan 25

DIASCIA - Scrophylariaceae
barberae / A / 1' / rosy-pink fls / South Africa 129 / 25
cordata / HH R / 8" / fls rose-purple / Natal 24
X Ruby Field / A / to 1' / fls rosy-pink / S. Africa 25

DICENTRA - Papaveraceae
canadensis / R / 6-10"/ white-spurred fls / ne North America 224 / 107
chrysantha / HH P / to 5' / bright yellow fls / s California (145) / 25
cucullaria / R / to 10" / fls white, tipped gold / e & c N. Am. 224 / 107
cucullaria 'Ozark Form' - with pinkish fls 259
eximia / P / 1½' / rosy-purple fls / e United States 147 / 129
eximia 'Alba' - good white form 129
eximia 'Bountiful' / 18" / deep mauve-pink fls 280
eximia 'Snowdrift' - compact plant; pure white fls (64)
eximia 'Summer Beauty' - see D. X 'Summer Beauty' 135
formosa / P / 2' / rose to white fls / ns North America 93 / 268
formosa 'Alba' / 12" / all white fls 18
formosa ssp. oregana / 10" / cream & rose fls / Oregon & California 25
formosa ssp. oregana 'Rosea' - rose fls 25
glauca - see D. formosa ssp. oregana 25
iximia / P / 9-18" / fls reddish purple 61
ochroleuca / HH P / 5' / fls cream, tipped purple / s California 227
oregana - see D. formosa ssp. oregana 25
oregana 'Rosea' - see D. formosa ssp. oregana 'Rosea' 25
pauciflora / R / 3" / white to pink fls / California 228 / 25
peregrina / R / to 6" / fls white to purple / Siberia, Japan 165 / 25
peregrina alba - included in the sp 25
peregrina v. pusilla - included in the sp. 129 / 25
peregrina v. pusilla alba - included in the sp. 25
scandens / Cl / to 15' / fls yellow-white, pnk tipped / Nepal, China 25
spectabilis / P / 2' / rosy-pink & white fls / Siberia, Japan 147 / 129
spectabilis 'Alba' - white fls 25
uniflora / R / 3" / white or pink fls / nw North America 229 / 107

DICHELOSTEMMA - Liliaceae
capitatum - see D. pulchellum 25
congestum / Bb / 1-3' / fls blue-violet / Washington to California 25
ida-maia / Bb / 10" / bright crimson fls, green tipped / California 186 / 185
multiflorum / Bb / to 2½' / violet-lilac / Oregon to Baja, Calif. 25
pulchellum / Bb / 1-2' / lilac-blue fls / Oregon, California 185
volubile / HH Bb / to 5' / light to rose-pink fls / California 25

DICHROSTACHYS - Leguminosae
glomerata / T Sh / lilac, pink, white fls 25

DICOPOGON - Liliaceae
strictus / Gh / small / mauve fls / temperate Australia 45

DICRANOSTIGMA - Papaveraceae
franchetianum / A / 5-6' / clear yellow fls / w China 25
lactucoides / P / to 2' / yellow fls / Himalayas (502) / 25
leptopodum / A / low / small yellow fls; linear pod / China 25

DICTAMNUS - Rutaceae
albus / P / 2-3' / fls white / e Europe, Asia 229 / 129
albus v. caucasicus - large form, racemes longer, petals ovate 25
albus 'Purpureus' - purplish-pink fls 129
albus 'Rubra' - rosy-purple fls, veined 25
fraxinella - see D. Albus 28
rubra - see D. albus 'Rubra' 28

DIERAMA - Iridaceae
[1] igneum / HH C / 3½-4½' / rose-pnk bells, mauve-dk red fls. /S.Afr.
[1] medium / HH C / 2'+ / pinkish-mauve / South Africa
pendulum / HH C / to 3' white, pink, purple fls / South Africa 186 / 207
pulcherrimum / HH C / to 5' / fls dark violet-purple / South Africa 129 / 25
pulcherrimum 'Slieve Donard' - dwarfer, to 3' 111
pumilum / HH C / 2' / rosy-pink fls / South Africa 25

DIERVILLA - Caprifoliaceae
lonicera / Sh / 2-4' / fls yellow / Newfoundland to Sask., s to N.C. 95

DIETES - Iridaceae
grandiflora / HH P / to 4' / orange-yellow & brown fls/South Africa 25
iridioides - see D. vegeta 25
vegeta / HH P / 1-2' / white fls, brown flecked / South Africa 93

DIGITALIS - Scrophylariaceae
ambigua - see D. grandiflora 268
davisiana / P / 20" / fls pale yel, dark inner netting / Anatolia 268
dubia / HH P / 18" / soft purplish to pink fls / Balearic Isls. 212 / 25
eriostachys - see D. lutea 268
'Excelsior' - see D. purpurea 'Excelsior' 202
ferruginea / B or P / 6' / yellowish fls, marked rusty-red/Eurasia (106) / 25
'Foxy' - see D. purpurea 'Foxy' 202
gloxinaeflora - see D. purpurea v. gloxinaeflora 28
grandiflora / B or P / 2-3' / yellow fls / e & c Europe 148 / 288
grandiflora alba / P or B / to 3' / yellowish-white form / Eur, Asia 25
kishinskyii - see D. parviflora (398)
laevigata / P / 2-3' / fls yellow, purple marked / w & c Balkans 28
lamarckii / P / 20-32" / fawn fls, white lobed / Iran, Balkans 77
lanata / P or B / to 3' / nearly white fls, veined / Greece 149 / 25
lanata leucophaca / P or B / 1-3' /corolla wht w purple veins/Greece 288
lutea / P / to 28" / pale cream fls / w & nc Europe 207 / 93
lutea australis / P / to 3' / yellow to white / c & s Italy 25
mariana - see D. purpurea ssp. mariana 288
X mertonensis / P / 2' / rosy-mauve fls 279

---

[1] Eliovson, *Wild Flowers of Southern Africa*, 1980.

minor - see D. purpurea 288
obscura / HH Sh / 1-3' / fls orange-yellow & brown / Spain 212 / 288
orientalis - see D. lamarckii 77
parviflora / P / 1-2' / fls reddish-brown / n Spain 212 / 288
purpurea / B or P / to 4' / purple, pink, white fls / Medit. region 210 / 25
purpurea 'Alba' - fls white 25
purpurea 'Excelsior' / B / 5' / large-flowered 283
purpurea 'Foxy' / B / 3' / varii-colored fls 202
purpurea 'Gloxinaeflora' - robust, fls more flaring 28
purpurea ssp. heywoodii / B or P / wht fls, plant woolly / s Port. 288
purpurea ssp. mariana /white-tomentose lvs; purple fls /Spain, Port 288
sibirica - habit of D. grandiflora; fls of D. lanata / Siberia 28
X sibirica / P / 3' / yellowish fls, long middle-lobe lip 25
thapsi / P / 2½-3' / cream fls, suffused pink / Spain, Portugal 212 / 135
trojana / B or P/ 12-32" / many fls, red-brown; tube veined/Ia, Iowa 77
viridiflora / P / 20-32" / dull greenish-yellow fls / Balkans 288

DILLWYNIA - Leguminosae
hispida / Gh Sh / 4-8' / red, yel., crimson fls / Victoria, Austrl. 61

DIMORPHOTHECA - Compositae
1 aurantiaca - see D. sinuata
barberae - see OSTEOSPERMUM barberae 129 / 25
barberae v. compacta - see OSTEOSPERMUM v. 'Compacta' (503)
ecklonis / P gr. as A / 2' / white & blue daisies / South Africa 129
pluvialis / P / to 16" / disc. fls yellow, ray white-purp./ S. Afr. 25
sinuata / A / 1' / orange-yellow fls / South Africa 25

DIONAEA - Droseraceae
muscipula / Gh / 6" / white fls, insectiverous pl / N. & S. Carolina 128 / 25

DIONYSIA - Primulaceae
aretioides / R / loose cushion / white lvs; yellow fls / Iran 64
involucrata / Sh / compact / pale lilac fls / c Asia 122 / 10
teucroides / Sh / 6" / yellow fls / Turkey, on limestone 122 / 77

DIOSCOREA - Dioscoreacae
Battatus/Cl/3'/fls inconsp, edible yam, "Cinnamon Vine"/e Asia, trop 25

DIOSPHAERA - Ebenaceae
asperuloides - see TRACHELIUM asperuloides (504) / 132
dubia - see TRACHELIUM rumelianum 132

DIOSPYRUS - Ebenaceae
virginiana / T / 50'+ / pale orange edible frs / c United States 28

DIOTIS
candidissima - see OTANTHUS maritima 211

DIPCADI - Liliaceae
serotinum / HH Bb / 4-12" / fls greenish-brown / s Europe, n Africa 148/ 28

---

[1] Eliovson, Wild Flowers of Southern Africa, 1980.

DIPHYLLEIA - Berberidaceae
cymosa / P / to 3' / white fls; blue frs / Va. to Ga., streamsides 225 / 99
cymosa ssp. grayi - see D. grayii 200
grayii / P / 1-2' / white fls; blue frs / Japan, in high mts. 165 / 200

DIPIDAX - Liliaceae
triquetrum / Gh / 12-18" / whitish fls, veined brown / S. Africa (399) / 28

DIPLARRHENA - Iridaceae
latifolia / Gh / 3' / fls lilac and yellow / Tasmania 184
moraea / Gh / 2'-3' / white fls, flushed lilac / Tasmania, Australia 25

DIPLOPELTIS - Asclepiadaceae
hugelii / Gh Sh / 1' / rose & white fls / Australia 206

DIPSACUS - Dipsacaceae
fullonum / B / 6'+ / fls lilac & white / s, w & c Europe 194 / 289
sativus / B / 6' / fls pale lilac - 'Fullers Teasel' / Europe 25
sylvestris - see D. fullonum 220 / 289

DIPTERONIA - Aceraceae
sinensis / HH T / 30' / pinnate lvs; orbicular samaras / c China 28

DIRCA - Thymelaeaceae
palustris / Sh / to 6' / fls yellow; fr red or green / e U.S. 25

DISA - Orchidaceae
crassicornis / Gh / 1'+ / reflexed petals / South Africa 28
uniflora / HH P / 2' / red fls / South Africa (94) / 25

DISCARIA - Rhamnaceae
serratifolia / HH Sh / 10'+ / fragrant fls / Chile, Patagonia 139
toumatou / HH Sh / to 15' / white fls / New Zealand 238

DISPHYMA - Aizoaceae
australe / Gh / prostrate / fleshy lvs; pink to white fls / N.Z. 15

DISPORUM - Liliaceae
flavum / P / 3' / yellow fls / China (107)
hookeri / P / to 2½' / fls creamy to greenish-white / Ore., Calif. 229 / 25
hookeri v. oreganum / 2½' / creamy-white fls / nw North America 25
lanuginosum / P / to 2½' / fls yellowish-green / e United States 224 / 25
maculatum / P / 2' / yellow fls, spotted purple / ec United States 107
oreganum - see D. smithii
pullum / P / to 1½' / fls wht-dp purp, bell shaped / India, China 61
sessile 'Variegatum' / P/ 2' / creamy-wht fls; lvs striped wht/Japan 279
smithii / P / 1-3' / white fls / w North America (263) / 228
trachycarpum / P / 2½' / fls creamy-white / w & nw North America 229 / 25

DODECATHEON - Primulaceae
alpinum / R / 8-10" / fls bright purplish-crimson / Calif., Nevada 228 / 129
amethystinum / R / to 1' / fls deep red-purple / W. Va. to Minn 224 / 99
clevelandii / R / 8-16" / fls magenta to white / s California 227 / 228

[1] clevelandii 'Alba' - the white form
clevelandii ssp. insulare - yellow pollen-sacs, coastal (108)
clevelandii ssp. patulum / dark pollen sacs / in mts. (108)
clevelandii ssp. sanctarum/R/16"/fls wht, lvs, stamen diff.from sp. 227
conjugens / R / 3-10" / fls magenta to white / s B.C. to Ore., Mont. 25
cusickii / R / 5-10" / magenta or lavender to white fls / nw N. Am. 25
dentatum / R / 8-10" / fls white, purple at base / nw U.S. 229 / 107
dentatum ssp. ellisiae / with yellow filaments / se Ariz., N. Mex. (108)
ellisiae - see D. dentatum ssp. ellisiae (108)
frenchii - see D. meadia ssp. frenchii 99
frigidum / R / 3½-10"/ magenta to lavender / N. Am., Alaska, Yukon 25
glastifolium - see D. conjugens 142
hendersonii / R / 5-18" / light orchid to darker fls / nw N. Am. 227 / 228
hendersonii 'Alba' - white fls *
hendersonii 'Sooke' - very dwarf, cult. as D. littorale (109)
integrifolium / R / 10" / fls lilac-purple to pink / B.C. 132
integrifolium ssp. insulare / P / to 10" / lvs 5";fls 3-10 purp/B.C. 61
Jeffreyi / R / 6-24" / fls purple to white / nw North America 229 / 107
Jeffreyi redolens/ P / to 20"/ magenta-lav. or wht/ Alaska to Calif. 25
lancifolium - see D. jeffreyi 160
latifolium - see D. hendersonii 25
X lemoinei / R / 1' / pale rose fls 106
littorale - see D. hendersonii 'Sooke' (109)
macrocarpum - believed to be D. pulchellum *
Meadia / P / 1-1½' / fls rosy / c & sc United States 224 / 129
Meadia 'Alba' - graceful white form 129
Meadia v. frenchii / lvs abruptly contracted to petiole / Ill., Wis. 99
Meadia X pulchellum - crossed by donor
Meadia 'Rubrum' - deep colored selection *
Meadia 'Splendidum' - fls crimson 25
patulum - see D. clevelandii ssp. patulum 25
pauciflorum - see D. pulchellum 25
pauciflorum 'Red Wings' - see D. pulchellum 'Red Wings' *
poeticum / R / 5-15" / magenta to lavender fls / Oregon, Washington 25
[1] poeticum 'Red Wings' - see D. pulchellum 'Red Wings'
puberulatum - see D. cusickii 142
pulchellum / R / 2-20" / magenta or lavender fls / c U.S. to Mexico 229 / 227
pulchellum album / white form of D. pulchellum 25
pulchellum ssp. macrocarpum - dwarf form from Vancouver Island 276
pulchellum v. pulchellum / P / to 3" scapes sev. fls/Alaska to Mex. 142
pulchellum 'Red Wings' - brilliant rosy fls (108)
pulchellum ssp. watsonii / 2½' / Vancouver Is. to Montana 95
radicatum - see D. pulchellum 25
radicatum watsonii - see D. pulchellum ssp. watsonii 25
redolens / P / 10-24" / lavender to magenta fls / s Calif. to Utah 228
salinum / alkaline soil segregate of D. pulchellum / Idaho 64
tetandrum - see D. jeffreyi 25

DODONEA - Sapindaceae
viscosa / HH Sh / 15' / greenish fls / pan-tropical 238 / 25

---

[1] Ingram, Baileya, 11:3.

DONATIA - Donatiaceae
novae-zelandiae / HH R / to 4" / fleshy white fls / New Zealand 238 / 15

DOODIA - Polypodiaceae
aspera / Gh F / 18" / pinnatifid fronds / temperate Australia 28

DORONICUM - Compositae
austriacum / P / to 2' / yellow fls / Austria, in mts. 148 / 208
calcareum - see D. glaciale ssp. calcareum 219
carpaticum / R / 6" / yellow fls / e & s Carpathians 64
cataractarum / P / 3' / yellow fls / n Alps 93
caucasicum - see D. orientale 289
1 caucasicum 'Magnificum' / 2' / br yellow daisy like 3" fls
caucasicum 'Spring Beauty' - see D. X 'Spring Beauty' 279
clusii / R / to 14" / fls yellow / mts. of Europe 148 / 289
columnae / P / to 2' / golden-yellow fls / se Europe, Asia 148 / 93
columnae v. cordifolia - included in the sp. *
columnae 'Magnificum' - heads larger *
cordatum - see D. columnae 129 / 93
cordatum 'Magnificent' - see D. columnae 'Magnificum' *
corsicum / P / to 4' / fls yel with pappus / mts of Corsica 289
glaciale / R / to 8" / fls yellow / Swiss Alps 219 / 289
glaciale ssp. calcareum / without silky hairs / Austrian Alps 219
grandiflorum / R / to 16" / yellow fls / mts. of s Europe 148 / 289
hirsutum - see D. clusii 289
X 'Miss Mason' / 18" / bright yellow fls, early 279
orientale / P / 1-1½' / deep yellow fls / se Europe 289
pardalianches / P / 20-32" / yellow fls / e & n Europe 182 / 93
plantagineum / P / 3-5' / yellow daisies / w Europe 135
X 'Spring Beauty' / P / 18" / double yellow fls 129 / 279

DOROTHEANTUS - Aizoaceae
bellidiformis / A / ? / rose pink, varied / Cape Prov., S. Africa 25

DORYCNIUM - Leguminosae
germanicum - see D. pentaphyllum ssp. germanicum 287
herbaceum - see D. pentaphyllum ssp. herbaceum 287
hirsutum / P / 8-32" / fls white or pink / Medit. region 210 / 287
pentaphyllum / P /to 2½' / fls white / c Europe, Balkan Pen. 25
pentaphyllum ssp. germanicum / 8-32" / white fls / c Eu., Balkans 287
pentaphyllum ssp. herbaceum / 8-38" / white fls / c & se Europe 287
rectum / HH P or Sh / 1-5' / white or pink fls / Medit. region 287
suffruticosum - see D. pentaphyllum 25

DOUGLASIA - Primulaceae
dentata - see D. nivalis 25
laevigata / R / compact / carmine to rose-pink fls / Cascade Mts. 228 / 25
laevigata v. ciliolata - thin lvs, margins ciliate 25
montana / R / tight tuft / variable pink fls / Cascade Mts. 229 / 107
nivalis / R / loose tufts / pale pink fls / Washington 229 / 123
nivalis dentata - a synonym 25
Vitaliana / R / prostrate mat / clear yellow fls / Europe 129 / 123

---

1 Park Catalog, 1986.

Vitaliana 'Praetutiana' - lvs lime-encrusted 25
Vitaliana primuliflora / P / tufted / fls yel, ½" across/mts Spain 61

DOWNINGIA - Campanulaceae
elegans / A / 4-20" / blue or violet & white fls / Calif., Oregon 228

DOXANTHA
unguis-cati - see MACFADYENA unguis-cati 180 / 25

DRABA - Cruciferae
acaulis / R / cushion / golden-yellow fls / Cicilian Taurus Mts. 132
aizoides / R / 2-4" / fls yellow / c & s Europe, in mts. 148 / 286
aizoides bertonlonii - see D. aspera 25
aizoides eriocarpa - see D. aspera v. eriocarpa *
aizoon - see D. lasiocarpa 286
alpina / R / to 8" / bright yellow fls / arctic & subarctic Europe 148 / 286
alpina ssp. demissorum / 4" / lvs lanceolate (111)
alpina glacialis - see D. glacialis 286
altaica - see D. subcapitata 209
andina / R / 2" / pale yellow fls / Andes Mts., South America 132
androsacea - see D. lactea 209
arabisans / P / 8-20" / white fls / Newfoundland to New York (505) / 25
arabisans v. canadensis - lower, pods narrower 99
arbuscula / R / half-shrubby tuft / yellow fls / alpine Venezuela 64
arctica - see D. cinerea 209
aretioeides / minute dense tuffet / Columbian Andes 96
argyrea / R / silvery tufts / yellow fls / Idaho 64
armata - see D. aspera
aspera / R / 2-4" / yellow fls / Europe, in mts. 286
aspera v. eriocarpa - hairy frs 286
aspera v. erioscapa - hispid scape 286
athoa / R / to 5" / large yellow fls / Greece, w Balkans 286
athoa f. laicaita - smaller, tighter tuft 96
athoa f. leiocarpa - smooth fls *
aurea / B / 1'+ / deep yellow fls / arctic Canada, Greenland 312 / 209
austriaca - see D. stellata 286
barbata - see D. alpina 209
belli / R / 4" / fls dull yellow / Spitzbergen 286
bernensis - see D. incana 286
bertolonii - see D. aspera 25
borealis / R / to 8" / fls white / Hokkaido, Kuriles 165 / 200
brachystemon / near D. aizoides / e Pyrenees 286
brewerii / P / 1-5" / fls white / high slopes Wash. s to Calif. 225
bruniifolia / R / 4" / orange fls / Medit. region, in mts. 77 / 25
bruniifolia v. diversifolia - included in the sp. 77
brunifolia ssp. heterocomis - inflorescence many-flowered 77
brunifolia spp. olympica / villous scape / Turkey 77
bryoides - see D. rigida v. bryoides 77
bryoides v. imbricata / 1-2" / bright yellow fls / Caucasus 129
calycosa - see D. cappadocica 77
cana - see D. lanceolata 99
cappadocica / R / rounded tufts / yellow fls / Turkey, Kurdistan 77
carinthiaca / R / to 6" / white fls / c & c Europe, in mts. 148 / 286
chamaejasme - questionable name, possibly D. aizoides *
cinerea / R / 1'+ / fls white / fully circumpolar 209

compacta / R / very dwarf / deep yellow fls / n Balkans 286
corsica - see D. loiseleurii 61
cossonii / lg wht or pink flushed fls; ivy shaped lvs / n Africa 64
crassifolia / R / 2"+ / fls pale yellow / N. Am., Scandinavia 312 / 286
cretica / R / dwarf / yellow fls / Crete, in mts. 286
cuspidata / R / dwarf / large yellow fls / Crimea 286
daurica / R / to 10"/ cream fls / arctic Europe, Scandinavia mts. 25 / 286
daurica X dovrensis - cross by donor
dedeana / R / 3" / fls white / n & e Spain, in mts. 212 / 286
dedeana 'Mawii" - fewer blossoms, green sepals 96
dedeana v. zapateri - 1½" / bright white fls, purple calyx 96
demissorum - see D. alpine ssp. demissorum (111)
densifolia / R / 2½'" / yellow fls / nw North America 229 / 142
doerfleri - see SCHIEVERECKIA doerfleri 286
dorneri/similar to D. fladnizensis, longer, lf-pubescence/se Carp. 286
dovrensis / robust plant, hairy siliculae /Norway 286
dubia / R / to 8"/ white or cream fls / c & s Europe, in mts., 97 / 286
elegans / R / small mats / golden-orange fls / Cilician Taurus 96
elongata / similar to D. lasiocarpa, narrower obovate petals 286
eriophila verna ssp. praecox - see ERIOPHILA verna ssp. praecox 286
fladnizensis / R / 3" / fls white / arctic Europe 314 / 286
frigida - see D. fladnizensis 96
gigas - see ARABIS corduchorum 25
glabella / R / to 16" / fls white / circumpolar 209 / 99
glabella v. megasperma / 1'+ / white fls / ne North America 99
glacialis / R / to 8" / bright yellow fls / arctic Norway, Russia 286
globosa v. sphaerica / 1" / yellow fls, pilose plant / Oregon 29
gracilis / 1½" / yellow fls / Switzerland 206
haynaldii / R / very dwarf / yellow fls / Carpathians 286
helleriana / R / to 16" / yellow fls; lvs grayish / Col., N. Mex. 227 / 235
heterocoma / R / to 4" / fls yellow / Asia Minor 286
hirta - see D. daurica 286
hispanica / R / dwarf / yellow fls / e & s Spain 286
hispanica lebrunii / R / fls yellow (dwarf) / Spain 286
hispida / R / to 8" / fls bright golden yellow / Caucasia 77
hoppeana / R / very dwarf / yellow fls / European Alps 148 / 286
imbricata - see D. bryoides v. imbricata 132
incana / B / 1'+ / fls white / n Europe 210 / 286
incerta / R / 4" / large clear yellow fls / Wash., Mont., & n-ward 64 / 107
japonica / R / 2-4" / petals yellow, clawed below / Hokkaido, Honshu 200
kitadakensis / similar to D. borealis / 3" / Japan 165
kitadakensis koizumi - G. Koidzumi is botanical author 165
korabensis - see D. tomentosa or perhaps D. incana 286
kotschyii / R / 8" / white or creamy fls / s Carpathians 286
kurilensis / R / small / white fls / Kuril Islands 64
lacaithae - see D. athoa f. lacaita 96
lactea / R / 2-3" / fls greenish-white / arctic regions 25
lanceolata /R/1'+/resembling D. incana, hairy frs/n Russia,Can, U.S. 286
lanceolato - Canada & n United States 99
lasiocarpa / R / to 8"/ deep yellow fls / Carp., Balkans, Austria 286
lasiocarpa X hispanica - cross by donor
laxa - see D. norvegica 25
lemmonii / R / 1-5" / yellow fls / s Sierra Nevada, U.S. 228
linearis - see ERIOPHILA verna ssp. praecox 77
loiseleurii - Corsican form of D. hispanica 286
lonchocarpa / R / to 5" / white fls / nw North America 229 / 142

longipes / R / 2-9" / white fls / Alaska & nw Canada 209
longirostra - see D. Aspera 286
longirostra v. armata - see D. aspera 25
longirostra eriocarpa - see D. aspera v. eriocarpa 286
longirostra erioscapa - see D. aspera v. erioscapa 286
1 longisiligua / R / 4" / fls golden yellow
lutescens / A / 5½" / small pale yellow fls / s Spain, s France 286
macrocarpa - see D. oblongata 209
magellanica - see D. daurica 286
megasperma - see D. glabella v. megasperma 99
mollissima / R / 1½" / fls yellow / Caucasus 129 / 268
muralis / A / to 1' / white fls / Europe 314 / 286
nemerosa / A / 1' / pale yellow fls / across N. Am. & Eurasia 165
nivalis / R / 2" / white fls / arctic Europe 312 / 286
norica / R / to 9" / fls cream / Austria 286
norvegica / R / to 8" / white fls / Scotland, sub-arctic Europe 286
oblongata / R / to 4" / fls pale yel. to creamy / high-arctic N.Am. (325)
oligosperma / R / ½-4" / whitish fls / w N. Am., in mts. 229 / 228
oligosperma saximontana - cited as a synonym 64
olympica - see D. bruniifolia ssp. olympica 77
oreadum / HH R / ? / white fls / Morocco 64
oxycarpa / HH R / dense / yellow fls / Lebanon 96
pacheri / may be D. daurica / stellate hairs, branched rays 286
parnassica / like D. hispanica, longer villous hairs / Balkan Pen. 286
paysonii / R / ½-2½" / whitish fls / Calif. to B.C., e to Rockies 229 / 228
paysonii v. treleasii - yellowish fls 312
polytricha / R / 1½" / fls bright yellow / Transcaucasica 132 / 77
pyrenaica / see PETROCALLIS pyrenaica 25
ramosissima/ P / fl stem to 1'/ fls white, twining/Va., Ky., c Tenn. 224
repens - see D. sibirica 99
reuteri - see D. polytricha 77
rigida / R / 2-4" / yellow fls; lvs spreading / Armenia (59) / 77
rigida v. bryoides - lvs inflexed (245) / 77
rosularis / R / to 4" / yellow fls / Lazistan to Kurdistan 77
rupestris - see D. norvegica 286
sachalinensis / R / 4-10" / fls white / Hokkaido, Honshu 200
sakuraii / R / to 8" / fls white / Honshu, in high mts. 200
X salamonii / R / 2" / white fls 18
sauteri / R / 1½" / yellow fls / Alps 148 / 286
sauteri hoppe - Hoppe is the author of the sp.
saximontana - see D. oligiosperma 64
scardica / R / dwarf / small yellow fls / Balkans, in mts. 149 / 286
shiroumana / R / 2-4" / few fls white / Honshu 200
sibirica / R / creeping / yellow fls / e Russia, Asia, Greenland 300 / 286
siliguosa - see D. carinthiaca 286
simonkaiana / R / 3" / fls white or cream / se Carpathians, Roumania 286
sphaeroides / ½-6" / yellow / Nevada & se Oregon 229
stellata / R / to 4" / fls cream or white / e Alps, on limestone 314 / 286
stenocarpa - see ERIOPHORA verna ssp. praecox 286
streptocarpa / P / 1'+ / fls yellow, lvs scanty / Colo. & N. Mex. 227
stylaris - perennial, otherwise close to D. incana 286
subalpina / R / ? / large white fls / c Utah 229
subcapitata / R / 1½" / fls yellow / probably fully circumpolar 209

---

1 Siskayou Rare Plant Nursery Cat, 1985.

subnivalis / similar to D. tomentosa, more slender / e Pyrenees 286
X thomasii - see D. stylaris 286
tomasiniana - see D. stylaris 286
tomentosa / R / to 8" / white fls / Pyrenees, Alps, on limestone 148
ussuriensis / R / to 8" / fls white / ne Asia 25
vestita - see D. globosa v. sphaerula 29

DRACOCEPHALUM - Labiatae
altaiense - see D. grandiflorum 61
argunense / R / 6-14" / blue-purple fls / Far East 164 / 200
austriacum / P / to 2' / fls blue-violet / se France to Caucasus 148 / 25
botryoides / R / 6" / blue-purple fls / Caucasus 29
bullatum / R / ? / bluish-violet fls / Li Chang Range 67
calophyllum v. smithianum / woolly inflorescence / China, Tibet 25
forrestii - see D. calophyllum v. smithianum 67 / 25
grandiflorum - cult. material referred to D. rupestre 25
hemsleyanum / P / to 20" / fls purplish-blue / Tibet 29
imberbe / P / to 6" / branches erect; fls lilac-blue / Siberia 61
isabellae / P / 12-20" / dark lilac-blue fls / c China 93
mairei - see D. renatii [1]
moldavica / A / to 2' / fls violet or white / Siberia, c Asia 288
nutans v. alpinum / R / 12" / bright blue fls / c Asia 29
parviflorum / A or B / 2' / lt. blue fls / across N. Am., from Asia 224
purdomii / R / 8" / purple fls / c Asia 123
renatii / R / 10" / cream-white fls / Morocco 25
rupestre / P / 8-24" / dark bluish-violet fls / w China 25
ruyschiana / R / to 1' / blue-violet fls / Pyrenees, Alps 314 / 148
setigerum / R / 6" / whitish fls / Armenia 64
tanguticum - suggested to be D. argunense or D. calophyllum v. sm. 25
wendelboii / R / 1'+ / cobalt blue fls / Afghanistan *

DRACOPHYLLUM - Epacridaceae
muscoides / HH R / to 2" / white fls / New Zealand 193 / 15
pronum / R / hard wiry mats / waxy white fls / South Island, N.Z. 15

DRACUNCULUS - Araceae
canariensis /spathe pale grn to 13";spadix yel/Canary & Madiera Ils. 25
vulgaris / Tu / to 3' / reddish-purple fls / Medit. reg., A. Minor 129 / 25

DRAPETES - Thymelaeaceae
dieffenbachii / Gh Sh / prostrate / imbricate lvs / New Zealand (79) / 15

DREGEA - Asclepiadaceae
corrugata / C / to 10' / bark wrty / fls wht, red dots / China 61 / 95

DRIMYS - Winteraceae
aromatica - see D. lanceolata 93
lanceolata / HH T / 10-30' / pale brownish fls / se Australia 109 / 25
winteri / HH T / to 25" / ivory-white fls / South America 110 / 139
winteri v. andina - dwarfed to 3' 36
winteri v. chilensis - hardier form 36
winteri v. nana - see D. winteri v. andina *
winteri latifolia - see D. winteri v. chilensis 36

---

1 Furse, Quarterly Bulletin, Alpine Garden Society, 37.3.

DROSERA - Droseraceae
adelae / Gh / compact / lvs to 10" / Queensland, Australia (113)
anglica / R / to 8" / fls white or pinkish / w N. Am., Eurasia 194 / 99
arcturi / Gh P / 6" / white fls / New Zealand, Tasmania 15
auriculata / Gh / to 1' / pink fls / Australia, New Zealand 193 / 25
binata / stem short, scape to 20"+ / petals white /Australia, N.Z. 25
capensis / Gh / 6" / purple fls / South Africa 128 / 25
cistiflora / Gh / to 20" / white, scarlet, magenta fls / Cape coast 223
filiformis / R / 3-9" / purple fls / e Coast of United States 224 / 25
hilaris / Gh / short / rose-purple fls / Cape of Good Hope 223
indica / A / 3-6" /fls wht or rose /Honshu, China, Austrl, Ind, Afr 200
intermedia / R / to 10" / white fls / ne North America 224 / 99
pulchella / Gh / 1½" / pink fls / sw Australia 25
pygmaea / Gh / 1" / white fls / se Australia, New Zealand 109 / 25
rotundifolia / R / to 1' / white fls, rarely pink / N. Am., Eurasia 22 / 99
spathulata / Gh / short / fls white or rose / New Zealand, Japan 165 / 15
whittakeri / Gh / 3" / white fls / Australia 206

DROSOPHYLLUM - Droseraceae
lusitanicum / P / 1½" / fls yellow / s Spain, Portugal 286

DRYANDRA - Proteaceae
formosa / HH Sh / 10' / orange-yellow fls / w Australia 45
polycephala / Gh Sh / 6-8' / deep yellow fls / Australia 12
praemorsa / HH Sh / to 8' / yellow fls / w Australia 45

DRYAS - Rosaceae
Drummondii / R / 4" / creamy-yellow fls / n North America 13 / 107
Hookerana - see D. octopetala ssp. Hookerana 25
integrifolia / R / dense mats / creamy-white fls / arctic regions 229 / 209
integrifolia f. intermedia - lvs distinctly dentate 245
intermedia - see D. integrifolia f. intermedia 245
lanata - see D. octopetala v. argentea 25
octopetala / R / mat / fls white & egg-yellow / Alps, Scotland 148 / 129
octopetala v. argentea / lvs tomentose on both sides / e Alps 25
octopetala v. asiatica - slightly differing Far Eastern segregate 273 / 200
octopetala ssp. Hookerana / lvs broadest at middle / B.C. to Colo. 25
octopetala integrifolia - see D. integrifolia 209
octopetala lanata - see D. octopetala v. argentea 25
octopetala 'Minor' - very dwarf form (507) / 25
octopetala punctata /Sh/dwarf / ptls wht; lower surf lf wht/mts Eur. 287
octopetala tenella - see D. integrifolia 25
octopetala v. vestita - see D. octopetala v. argentea 25
punctata / Sh/dwarf / lg glands upper surface of lvs / arctic Russia 287
X suendermannii - yellow buds opening to white fls (458) / 107
tenella - see D. integrifolia 25

DRYMIS
winteri - error for DRIMYS winteri

DRYMOCALLIS
fissa - see POTENTILLA fissa 25

DRYMOPHILA - Liliaceae
cyanocarpa / HH Bb / 2' / fls white, frs blue / Australia, Tasmania 29

DRYOPTERIS - Polypodiaceae
abbreviata / F / small / blade dk gr above, pale reverse /Br. Isles 159
aemula / F / to 2' / stem purple, blade green / England, Scotland 159
arguta / F / 14" / evergreen fronds / w North America 249
atrata / F / 8-18" / evergreen fronds / Far East 272 / 200
austriaca / F / to 2' / fronds herbaceious / northern hemisphere 99
X bootii / F / 2½' / finely-toothed pinnules / ne N. Am. 25
borreri / F / 1'+ / reddish scales on petioles / Europe 286
borreri 'Cristata' / 2-3' / crested form true from spores / England 159
borreri grandiceps / F / 2' / heavy terminal crest / British Isles 159
borreri grandiceps 'Askew' / F/ 2' / raised by Mr. Askew / Br. Isles 159
carthusiana / F / to 2' / deciduous fronds / England 159
Clintoniana - see D. cristata
cristata / F / to 20" / Crested Wood-Fern / n hemisphere 292 / 99
erythrosora / F / 18" / young fronds coppery-pink / Far East (250) / 159
felix-mas / F / to 5' / dark green fronds / N. Am., Europe 72 / 159
felix-mas 'Angustata Cristata' - fancy Male Fern form 199
fragrans / F / to 1' / aromatic / cirumboreal 25
goldiana / F / to 4' / large deciduous frond / ne & c N. Am. 191 / 159
hirtipes - see D. atrata 200
marginalis / F / 18"+ / evergreen fronds / North America 191 / 159
phegopteris - see THELYPTERIS phegopteris 25
pseudo-mas angustata cristata - see D. felix-mas 'Angustata Cristata'
spinulosa / F / to 20" / Woods or Florist Fern / North America 25
thelypteris - see D. austriaca 99
wallichiana / F / 1'+ / golden-green fronds / Kashmir 286

DRYPIS - Caryophyllaceae
spinosa / R / to 1' / fls white or pink / Italy, Greece 25

DUCHESNIA - Rosaceae
indica / P / to 20" / yellow fls / s & e Asia, adventive elsewhere 224 / 287

DUDLEYA - Crassulaceae
candida / Gh / to 2½' / fls pale yellow / Mexico, Baja California 156
cymosa / Gh / 8" / yellow or red fls / s California 227
farinosa / Gh / 1' / lemon-yellow fls / sea cliffs, Ore. to Calif. 227
lanceolata / P / 1-3' / petals red to yellow / s Calif., Baja, Cal. 25
pulverulenta / Gh / 8-16" / red fls / s & S. Calif., Arizona 227

DUGALDIA - Compositae
[1] hoopesii / "Orange Sneezeweed"; poisonous/ Rocky Mts, w of Divide

DYCKIA - Bromeliaceae
brevifolia / P/ to 1½' / fls sulphur-orange / Brazil 25

DYSSODIA - Compositae
tenuiloba / P gr. as A / to 1' / orange-yellow fls / Texas, Mexico 226 / 25

---

1 Weber, Rocky Mountain Flora, 1976.

EBENUS - Leguminosae
cretica / HH Sh / to 20" / fls bright pink / Crete 149 / 287

ECBALLIUM - Cucurbitaceae
elaterium / A / trailing / yellow fls; frs green / Medit. region 194 268

ECCREMOCARPUS - Bignoniaceae
scaber / A cl Gh / 10-15' / vivid orange fls / Chile 194 / 128
scaber 'Aureus' - bright golden yellow fls 25
scaber f. lutea - see E. scaber 'Aureus' *
scaber 'Ruber' - reddish fls 93

ECHEVERIA - Crassulaceae
agavoides / P / 8-20" / fls red w yellow tips / Texas, Mexico & s. 25
cristata - crested form of E. secunda or E. agavoides 93
derenbergii / Gh / 4" / yellow & red fls/ s Mexico 156 25
elegans / Gh / short / fls rose, yellow tipped / Mexico (327) / 25
'Hoveyi' - pale blue-green lvs; fls white or pink-striped 156 / 25
Pulidonis / Gh / 7" / corolla yellow / e Mexico 25
pulverulenta - see DUDLEYA pulverulenta 28

ECHINACEA - Compositae
angustifolia / P / 18" / fls purplish to white / s North America 28
X 'Bright Star' / 2½-3' / bright rose-red fls 300 / 135
pallida / P / to 3' / fls pale rose-purple / c United States 226 / 99
pallida f. albida - fls white 259
paradoxa / P / to 3' / orange-yellow fls / Missouri, Ark. 259
purpurea / P / 3-5' / purplish fls / e & c United States 224 / 135
purpurea 'Alba' / white form of E. purpurea 25
purpurea nana / dwarf form 25
secunda / P / 4-12" / fls red & yellow / Mexico 25
tennesseensis / R / to 1' / purplish fls / Tenn., Ark. 250
'The King' / 3' / coral-crimson fls 135
'White Lustre' / P / 3½' / large white fls / center deep orange 300 / 268

ECHINOCACTUS - Cactaceae
texensis / R / to 1' / fls pink, reddish center / N. Mex., Texas 298 / 25

ECHINOCEREUS - Cactaceae
baileyi / R / 4" / deep rose fls / Oklahoma 93
fendleri / P / 18" / purple fls / s Utah to N. Mexico 25
pectinatus v. neomexicanus / R / 1' / yellow fls / e Arizona, Texas 298 / 25
pectinatus v. reichenbachii / 6" / deep rose fls / Texas 93
reichenbachii / R / to 8" / fls pink to purple / Kansas to New Mex. 298 / 25
reichenbachii v. perbellus/R/8"/fls straw-pink/w Tex., N.M., se Col. 25
viridiflorus / R / 1-8" / brownish to yellowish-green fls / c U.S. 25

ECHINOCHLOA - Gramineae
crus-galli / A Gr / to 5' / Barnyard Millet, weedy / e hemisphere 292 / 25

ECHINOCYSTIS - Cucurbitaceae
lobata / A Cl / 20' / greenish-white fls / e & c North America 224 / 25

ECHINOPANAX
japonicus - see OPLOPANAX japonicus 221

ECHINOPHORA - Umbelliferae
spinosa / P / to 20" / petals white, rarely pink / Medit. 287

ECHINOPS - Compositae
bannaticus / P / to 3' / grayish-blue fls / se Europe 289
chantavicus / P / to 3' / fls blue / c Asia 35
exaltatus / B / to 5' / fls white or grayish / ec Europe, Balkans 289
humilus / P / 6-12" / steel blue / w Asia 25
[1] humilis 'Taplow Blue' / 30" / lg fl heads, metallic blue
ritro / P / 2' / steely-blue fls / s, se, ec Europe 147 / 289
ritro 'Alba' - white form 279
sphaerocephalus / P / 6' / gray-white fls / Eurasia 77 / 279
spinosissimus / P / 20-32" / gray-green-blue fls / e Medit. reg. 289
'Veitch's Dwarf Blue' / 4' / vigorous, grey lvs, blue fl heads 268

ECHIOIDES - Boraginaceae
longiflorum / R / 1' / yellow fls, purple spotted / n Iran, Caucasus 25

ECHIUM - Boranginaceae
albicans / P / 8-30" / corolla pink-red to blue-purple / s Spain 288
fastuosum / P / to 6" / fls purple or dark blue / Canary Is. 25
rubrum - see E. russicum 288
russicum / B / to 20" / fls dark red / ec & se Europe 288
vulgare / P / 1-3' / brilliant blue fls / Eu., adventive in U.S. 148 / 99
wildpretii / Gh Sh / to 10' / fls pale red / Canary Isls. 194 / 25

EDRAIANTHUS - Campanulaceae
bosniacus - not given botanical recognition 25
caudatus - see E. dalmaticus 30
caudatus 'Albus' - see E. dalmaticus 'Albus' *
croaticus - see E. graminifolius 25
dalmaticus / R / low / fls violet-blue / Dinarian Alps 240 / 25
dalmaticus 'Albus' - white fls *
dinaricus / R / low / fls violet-blue / Dinarian Alps 30
graminifolius / R / 3" / narrow, purple bells / Dalmatia 148 / 123
graminifolius 'Albus' - see E. graminifolius ssp. niveus 289
graminifolius ssp. niveus / white fls / Jugoslavia 289
intermedius - see E. graminifolius 25
kitabelii - see E. graminifolius 25
niveus - see E. graminifolius ssp. niveus (508) / 289
parnassi / R / 1' / small purple bells / Balkans 96
parnassicus / R / 4"+ / violet fls / Macedonia, c Greece 25
pumilio / R / cushion / strong violet-blue fls / Balkans 30 / 129
pumilio ssp. stenocalyx - smaller fls in greater profusion 64
serbicus / R / tufted / blue-violet fls / Bulgaria, Jugoslavia 30
serpyllifolius / R / 4" / intense blue-violet fls / Dalmatia 24 / 129
serpyllifolius 'Albus' - fls white 25
serpyllifolius dinaricus - see E. dinaricus 30
serpyllifolius 'Major' - fls slightly larger 129

---

[1] Wayside Catalog, Spring 1986.

tasmanicus - see WAHLENBERGIA saxicola 25
tenuifolius / R / 4-6" / fls violet-blue / Jugoslavia 30
tenuifolius 'Albus' - white fls *

EHRETIA - Ehretiaceae
tinifolia / Gh T/ 18-40'/ wht or crm fls, fragrant / Mex., W. Indies 25

ELAEAGNUS - Elaeagnaceae
angustifolia / Sh / to 20' / lvs & frs silvery / Eurasia 210 / 25
umbellata / Sh / 12-18' / silvery-white fls; reddish frs / Far East 167 / 36

ELEORCHIS - Orchidaceae
japonica / R / to 12" / rose-purple fls / Hokkaido,Honshu,Kuriles 200

ELEPHANTOPUS - Compositae
P / 6-20" / fls blue or violet; lvs velvety / Fla. to Texas & n 226

ELMERA - Saxifragaceae
racemosa / R / 16" / fls white / Olympic & Cascade Mts., w U.S. 229 / 228
racemosa v. puberulenta - different glandular pattern 142
racemosa v. racemosa / P / 4-10" / lvs uniform; fls wht/ Cascades 142

ELSHOLTZIA - Labiatae
stauntonii / P / 4' / mauve fls / n China 93 / 279

ELYMUS - Gramineae
arenarius / Gr / to 4' / European Dune-Grass / Eurasia 210 / 25
dahuricus / P Gr / erect / foliage basal, good ground cover/ Siberia 303
interruptus / P Gr / 28-52" / of open rich soil / Wyo. to Mexico 141
sibiricus / P Gr / 1-3' / attractive spikes for drying / Eurasia 268

EMBOTHYRIUM - Protaeceae
coccineum / HH T / to 20' / bright crimson-scarlet fls / Chile 129 / 36
coccineum v. lanceolatum - in cult. is the Norquinco form 110 / 129
coccineum v. lanceolatum 'Norquinco' / hardier form / Argentina 129 / 36
lanceolatum - see E. coccineum v. lanceolatum 150
lanceolatum Norquincoform - see E. coccineum v. lanceolatum "N."

EMILIA - Compositae
flammea - see E. sagittata 93
sagittata / A / 1-2' / scarlet fls / India, China 93

EMPETRUM - Empetraceae
atropurpureum / Sh / trailing / fls inconspicuous; fr drupe red-purp 99
eamesii / Sh / prostrate / pink or red fls / Newfoundland 25
eamesii ssp. hermaphroditum / glabrous branchlets / N. Am., Eurasia 25
hermaphroditum - see E. eamesii ssp. hermaphroditum 25
nigrum / Sh / dwarf / purplish fls; black frs / N. A., Eur. 148 / 107
nigrum ssp. hermaphroditum - tetraploid form, hermaphrodite fls 209

ENCELIA - Compositae
virinensis Actonii/Sh/low/fls yel, lvs wht/se Utah to Calif, Ariz. 25

ENDYMION - Liliaceae
campanulatus ' Alba' - see E. hispanicus 'Alba' 25

hispanicus / Bb / 9" / varii-colored fls / Spain, Portugal 212 / 267
hispanicus 'Alba' - white fls 25
hispanicus 'Excelsior' - deep blue fls 129
hispanicus 'Pink Giant' - large-flowered pink form 279
hispanicus 'Rosea' - pink fls 25
hispanicus 'Rose Queen' - pink selection *
italicus / Bb / to 1' / fragrant lilac-blue fls / Italy, s France 25
non-scriptus / Bb / to 18" / fls in several colors / w Europe 210 / 129
non-scriptus alba / white fls 25

ENKIANTHUS - Ericaceae
campanulatus / Sh / to 8' / white or pink fls / Japan 129
campanulatus v. palibinii / Sh / 8' / red fls / mts. of Honshu 110 / 200
cernuus v. rubens / Sh / 15' / deep red fls / Japan 129 / 25
perulatus / Sh / 3-6' / white fls / Japan 36

ENTELEA - Tiliaceae
arborescens / HH Sh / to 18" / fls white / New Zealand 238 / 25

EOMECON - Papaveraceae
chionantum / P / to 1'+ / fls white / e China 25

EPACRIS - Epacridaceae
impressa / HH Sh / 3' / fls red or white / Australia 28
petrophylla / Sh / dwarf/ fls white / Alpine Tasmania 109

EPHEDRA - Ephedraceae
campylopoda - see E. fragilis ssp. campylopoda 286
distachya / Sh/ to 3'/gr. cover, source of ephedrine/ s Eur., n Asia 25
fragilis ssp. campylopoda /HH Sh/climbing & pendant/red frs/e Medit. 286
gerardiana / Sh / 2"-2' / yellow fls; red frs / Himalayas, Tibet 36
major ssp. procera /HH Sh/to 6'/ frs red or yellow /s Greece in mts. 286
minima / HH Sh / to 5" / red frs / China 25
procera - see E. major ssp. procera 286
viridis / Sh / dioecious; fr grn; catkins stalked / sw U.S. 61

EPIDENDRUM - Orchidaceae
schomburgkii / Gh / 2-3' / fls vermilion / Guiana, Brazil 28

EPIGAEA - Ericaceae
asiatica / Sh / prostrate / fls white to rose / Japan 238 / 200
gaultherioides / Sh / semi-prostrate / pink fls / Asia Minor (153) / (328)
repens / Sh / trailing / pink & white fls / e & c North America 224 / 107

EPILOBIUM - Onagraceae
alpinum / R / 2-12" / pink or lilac fls / w North America 228
angustifolium / P / 3-5' / fls violet or purplish / world-wide 148 / 287
angustisfolium albiflorum / P / 3-5' / white fls / N. Am., Eurasia 25
angustifolium 'Album' - the white form *
chloraefolium v. kaikourense / HH R / 6"+ / rosy-pink fls / N. Z. 15
colchicum / P / 20" / pink fls / w Caucasus 77
crassum / HH R / trailing / pink or white fls / New Zealand 96
dodonaei / P / 8-32" / violet or purplish fls / Europe 287
fleischeri / R / decumbent / purple to violet fls / Alps 148 / 287
foucandionum / P / stems 2-9" / fls rose / Hokkaido, Honshu 200

glabellum / HH R / to 1' / white to pink fls / New Zealand 238 / 25
glabellum 'Album' - whiter selection 46 / 96
glaberrimum / R / to 1' / fls rosy-purple to pink / nw N. Am. 227 / 142
hectori / HH R / 6" / fls white / New Zealand, in mts. 15
hornemannii / see E. fourcaudianum 200
kai-koense - see E. chloraefolium v. kaikourense 24
latifolium / R / decumbent / violet or purplish fls / n hemisphere 228 / 287
luteum / R / 6-15" / yellow fls / n California to Alaska 228 / 25
microphyllum / P / slender 4-6" / fls wht or pink;lvs opp./N.Z. 15
nerterioides / R / procumbent / white fls / N.A., escape in Eng. 238 / 287
niveum / R / 4-8" / red-violet fls / California in Coast Ranges 228
nummularifolium / R / ground-hugging / bronze-green lvs / N. Zealand 46
obcordatum / R / 4" / deep rose-pink fls / Nevada, California 229 / 123
obcordatum v. laxum / somewhat larger lvs / Siskiyou Co., Calif. 196
palustre / 4-16" / fls rose to white 200
paniculatum / A / tall / light pink to rose fls / w U.S. 229 / 142
purpuratum / P / prostrate / fls few, stms purplish-blue / N. Z. 25
pycnostachyum / R / screeplant / white fls / New Zealand 64
rigidum / R / close mounds / bright pink fls / Siskiyou Mts. 169 / 107
rosmarinifolium - see E. dodonaei 287

EPIMEDIUM - Berberidaceae
diphyllum / R / to 1' / fls white / Japan 200
grandiflorum / R / to 16" / fls white to pale purple / Japan 200
koreanum / R / ? / ? / Japan, Korea 200
macranthum v. violaceum - see E. grandiflorum 200
pinnatum / R / to 1' / yellow / n Iran, Caucasus 25
X sakakii - error for X sasakii
X sasakii - hybrid of diphyllum & grandiflorum 165
sempervirens / P / 1 1/4-2" /fls white or purplish / Honshu 200
X Warleyense / P / 8-22" / fls coppery red / Warley Gardens, England 61

EPIPACTIS - Orchidaceae
atrorubens / P / 8-32" / red & purple fls / Eurasia 314 / 85
gigantea / P / to 5' / pink, red, purple fls / w & wc N. Am. 229 / 228
helleborine / P / to 4' / fls of many colors / Eurasia, adv. in U.S. 224 / 123
latifolia - see E. helleborine 25
palustris / R / 10-12" / varii-colored fls / Europe, Siberia 182 / 123
papillosa / P / 1-2' / fls greenish / n Hokkaido 166
rubignosa / R / short / wine-red fls / Europe to Iran 25
thunbergii / P / 1-2' / greenish to purplish fls / Japan, Korea 200

ERAGROSTIS - Graminae
elegans / Gr / 1'+ / Brazil 28

ERANTHIS - Ranunculaceae
cilicica / Tu / 3" / finely-divided fl bracts / Cilicica 129
hyemalis / Tu / 8" / fls bright golden-yellow / Europe 148 / 129
hyemalis v. cilicica - see E. cilicica 93
pinnatifida / Tu / 4-6" / fls white / Japan 25
X tubergenii / Tu / 2" / fls shining golden-yellow 129 / 111
X tubergenii 'Guinea Gold' - larger fls, longer stems (444) / 185

EREMOSTACHYS - Labiatae
speciosa / HH R / 1'+ / yellow fls; woolly white calyces / Himalayas 64

EREMURUS - Liliaceae
aitchisonii / P / 5'+ / pink fls / Afghanistan 268
altaicus / P / to 4' / fls yellow / c Asia 25
altaicus v. fuscus / P / to 4' / fls yellow or brown / c Asia 4
bungei - see E. stenophyllus 129 / 268
elwesii / P / 6-9' / pink fls / habitat unknown 25
fuscus - see E. altaicus v. fuscus 4
himalaicus / P / to 5' / white fls / Himalayas 268
X isabellinus / P / to 8' / varii-colored fls 25
lactiflorus / P / 3' / milk-white fls, yellow in bud / Turkestan 25
olgae / P / 28-40" / fls pink or white / Iran to Tadjikistan 194 / 268
regelii / P / to 9' / rose fls / ? 4
robustus / P / to 10' / pink fls / Tien-Shan, Pamir Altai 129 / 268
X 'Shelford' - see E. X isabellinus 184 / 25
spectabilis / P / to 5' / yellowish fls / Lebanon, Iran 268
stenophyllus / P / to 4' / clear yellow fls / Iran 268
tauricus / P / to 6' / white fls, green banded / Crimea 4
X tubergenii - E. himalaicus X E. bungei 29
turkestanicus /B/1½-3'/ greenish yel and wht, milk grn band/c Asia 4

ERIANTHUS - Graminae
coarctatus / Gr / 3-6' / long awns / to Georgia & Virginia 99
ravennae / G / to 14" / panicle silvery to 2' / Europe 25

ERICA - Ericaceae
arborea / HH Sh / to 20' / fls almost white / Eurasia, Asia Minor 212 / 36
arborea v. alpina / HH Sh / 10'+ / fls white / Spain 129
australis / Sh / to 8' / fls purple to red / Spain, Portugal 25
baccans / Gh Sh / sturdy / rose-pink fls / Cape of Good Hope in mts 223
carnea - see E. herbacia & its cultivars 36
carnea 'King George' - see E. herbacea 'King George' 36
carnea 'Praecox Rubra' - see E. herbacea 'Praecox Rubra' 36
cerinthoides / Gh Sh / to 6' / scarlet fls / South Africa 223
ciliaris 'Maweana' / stiff, sturdy; fl ½" long; lvs dk grn/ Port. 36
cinerea / Sh / 6-24" / rosy-purple fls / w Europe 170 / 291
cinerea 'Coccinea' / 4" / dark scarlet fls 139
cinerea 'Mrs. Dill' - compact shrub with pink fls 134
cinerea 'Velvet Knight' - blackish-purple fls 139
cinerea 'Victoria' / 6" / large purple fls 291
X darleyensis 'Furzey' / 18" / deep rosy-pink fls 291
elegans / Gh Sh / 6" / green fls / Cape of Good Hope 206
formosa / Gh Sh / 1-2' / white fls / South Africa 28
glandulosa / Sh / to 2' / fls white / South Africa 25
glauca v. elegans / Gh Sh / pinkish fls / South Africa 223
glomiflora / Gh Sh / 1' / white fls / Cape of Good Hope 206
herbacea / Sh / 6-10" / fls deep rosy-red / Alps, mts. of c Europe 210 / 36
herbacea 'King George' / 6-10" / rosy-crimson fls; dark green lvs 36
herbacea 'March Seedling' / 6-9" / rosy-purple fls 291
herbacea 'Pink Pearl' / 6-9" / shell-pink fls 291
herbacea 'Praecox Rubra' / 10" / early deep red fls 36
herbacea 'Violacea' - fls purple-blue 25
herbacea 'Winter Beauty' / 6-9" / fls bright pink 291
hirtiflora / Sh / to 2' / fls pale purple or pink / South Africa 25
holosericea / Gh Sh / ? / pink & red fls / Cape of Good Hope 223

hybrida 'Furzey' - see E. X darleyensis 'Furzey' 291
inflata / Gh Sh / 18" / white & red fls / Cape of Good Hope 206
mammosa 'Coccinea' / robust greenhouse shrub / Cape of Good Hope 223
Monsoniana / 1½' / white / Cape of Good Hope 206
multiflora / Sh / to 1½' / fls pink / South Africa 25
peziza / Gh Sh / 18" / white fls / South Africa 25
plukinetii / Gh Sh / 6" / red fls / Cape of Good Hope 223 / 206
regia / Gh Sh / 2-3' / crimson fls / South Africa 61
sicula / HH Sh / 8-20" / white or pale pink fls / Sicily, Malta 288
speciosa / Sh / 2-4' / lvs in 3's; fls br red / South Africa 61
subdivaricata / Sh / 1-3' / lvs in 4's; fls frag. wht / S. Africa 61
taxifolia / Sh / to 1' / lvs in 3's; fls pink / South Africa 61
terminalis / Sh / to 4' / rosy purple / s Europe 25
tetralix / Sh / 18" / rose fls / n & w Europe 148 / 129
tetralix 'Alba' - fls white 129 / 25
tetralix 'Mollis' - light gray lvs; white fls 110 / 129
tetralix 'Rubra' / 6" / very compact / fine red fls 291
triflora / 1½' / white / Cape of Good Hope 206
umbellata / Sh / to 3' / fls flesh to pink / w Medit. 25
vagans / Sh / to 1' / fls purplish-pink / w Europe 170 / 25
vagans 'Mrs. D. F. Maxwell' / 15" / dark pink fls 129 / 208
vagans 'St. Keverne' / 15" / salmon-pink fls 208
walkeria / Gh Sh / 1-1½' / rosy to red fls / South Africa 61

ERIGENIA - Umbelliferae
bulbosa / R / to 8" / white fls / e North America 224 / 99

ERIGERON - Compositae
allocatus / P / 4-6" / Washington 229
alpicola / see E. Thunbergii glabratus 200
alpinus / R / 6-8" / mauve fls / alpine & arctic Europe 148 / 123
aphanactis / P / to 1' / fls yellow / sw Colo., s Calif. 25
atticus / P / to 2' / intense purple fls / Alps, Carpathians 219
aurantiacus / R / 1' / orange fls / Turkestan 50 / 123
aureus / R / 3" / fls bright golden-yellow / w North America 312 / 129
1 aureus 'Canary Bird / compact / much pler color / cuttings
2 'Azure Fairy' / 2½' / doubles
Bloomeri/P/ to 6"/ray fls absent, lvs satiny/ Ore., n Calif., Wash. 25
borealis / R / to 10" / white to pink-purple fls / n hemisphere 209
caespitosus / R / to 1' / fls blue, pink, white / n & wc N. Am. 229 / 25
caucasicus - syv. of E. pulchellum 61
chrysopsidis / R / dwarf / yellow fls / Wash., Calif., Nevada 142
chrysopsidis brevifolius / P /to 4" / tiny yel fls / Cal, Nev 142
compactus / P / to 2' / fls white or pink / Inyo Mts., to nw Utah 196
compositus / R / 3-4" / white or pale mauve fls / w North America 229 / 123
compositus v. discoideus - caespitose plant 25
compositus v. glabratus - high mt. form 142
coulteri / P / to 2' / white fls / California, Utah, New Mexico 228 / 25
X 'Dimity' / R / 10" / bright pink fls 207

---

[1] Alpines of the Americas. Report of the First Interim Inter. Rock Garden Plant Conference, 1976.

[2] Park Catalog, 1986.

'Darkest of All' / 2½' / deep blue fls		268
eatonii / R / 2-12" / fls white, rarely blue, pink / Wyo. & w-ward	228 /	229
elegantulus / R / dwarf / blue or pink fls / Oregon, California	228 /	142
filifolius / P / to 20" / fls white, pink, blue / w North America	228 /	25
flagellaris / carpeting / fls rose or white / w North America	229 /	93
flagellaris minor / P / trailing leafy stolons / B.C.		142
flavum xantham - error for ERIOGONUM flavum		*
flettii / R / 2-6" / white fls / Olympic Mts., Washington		228
'Foerster's Liebling' / P / 1½' / deep rose-pink fls	300 /	268
foliosus / P / ? / fls blue 15-65 rays / Calif. n to Oregon		142
foliosus v. confinis / R / 4-1-" / blue fls / nw United States		142
'Foster's Darling' - see E. 'Foerster's Liebling'		
frigidus / P / mat to 3" / lvs lg. dense hairs; fls violet / s Spain		289
glabellus / P / 12-16" / violet, purple, white fls / Wisc. to Utah	229 /	93
glabellus f. albus - a white form		28
glaucus / HH R / 6-10" / purple daisies / N. Am. Pacific Coast	227 /	123
glaucus 'Albus' - white rays		25
glaucus 'Elstead Pink' - clear lilac-pink fls		279
glaucus 'Roseus' - pink rays		25
gracilis / R / 2-10" / blue or rose fls / Montana, Idaho, Wyoming		229
[1] grandiflorus / R / to 6" / pale lavender fls / Alp. & arc. nw N. Am.		
howellii / P / 8-20" / white fls / Columbia River Gorge		228
humilis / R / 6" / white to purplish fls / Alaska & s-ward	229 /	209
hungaricus - see E. nanus		289
hyperboreus - see E. caespitosus		25
hyssopifolius/P/ ? /fls white-pink, heads solitary/Nfld. s to N.Y.		99
karvinskianus / HH R / 8" / white & purple fls / Mexico, S. America	212 /	93
lanatus / R / 2" / fls white, blue, pink / nw Montana, Colorado		229
latus / P / 1-3" / smooth basal lvs; fls white / Idaho		229
leiomerus / R / 4" / deep blue to white fls / Rocky Mts.	229 /	25
lemmonii / P / trailing to 2' / lvs hairy; fls wht-purp / Ariz, rare		227
linearis / R / 2-12" / yellow fls / Yosemite Park & s-wards	229 /	228
linifolius - see CONYZA bonariensis		289
macranthus - see E. speciosus		
mairei / HH R / 3" / lilac fls / n Africa, in mts.		25
melanocephalus / R / 2-6" / violet fls / Colorado, Wyoming		96
montanensis / R / 4-8" / white or pink fls / Wyoming	64 /	96
mucronatus - see E. karvinskianus	129 /	93
multiradiatus / P / to 2' / fls pinkish or purplish / Himalayas		25
nanus / R / 6" / lilac fls / Carpathians		289
ochroleucus v. scribneri /R /4" / fls blue, purple, wht / nw N. Am.		25
oreganus / R / 2-6" / blue, pink, white fls / Colorado River gorge		228
pappochromus /HH R/small tuft/ white, pink, violet fls /Australasia		64
peregrinus / P / 2½' / white to purple fls / Alaska, B.C.	13 /	25
peregrinus ssp. callianthemus - rose-purple fls		25
philadelphicus / P / 1-3' / fls pinkish to roseate / e N. America		99
'Pink Jewel' / 18-24" / lavender-pink fls		135
pinnatisectus / R / 4-6" / mauve daisies / Rocky Mts.	312 /	123
poliospermus / R / to 6" / fls pink or violet / Wash., w Oregon		229
polymorphous / R / to 1' / reddish-lilac fls / Alps, Balkans	314 /	219
'Prosperity' / P / to 1½' / light violet fls		268
pulchellus / P / 1½-2' / fls blue-purple to white / e&c N. Amer.	224 /	99

[1] A. E. Porsild, Illustrated Flora of the Canadian Archipelago, 1957.

purpuratus / R / ? / ? / Alaska, Yukon 213
radicatus / R / 2" / mauve fls / c Rocky Mts. 123
rosea - see E. speciosus 'Roseus' 28
roylei / R / 6" / bluish-purple fls / Himalayas 61
salsuginosus - see E. peregrinus ssp. callianthemus (509) / 25
scribneri - see E. ochroleucus v. scribneri 25
simplex / R / 8" / blue or pink fls / w United States, in mts. 227 / 25
speciosus / R / 12-16" / dark violet to lavender-blue fls / w Ore. 229 / 93
Thunbergii / R / 4-15" / fls blue-purple / Honshu 200
Thunbergii angustifolius / P/ 15"/ fls blue-purp., often wht /Japan 200 / 25
Thunbergii v. angustifolius / 2' / blue-purple fls / Japan 200
Thunbergii glabratus / less densely pubescent / sorbid white 200
trifidus - see E. compositus v. discoideus 25
Tweedyi / R / 8"/ fls blue to purple or wht/N.M., Wyo., Mont., Idaho 25
unalaschensis - see E. humilis 209
uniflorus / R / 6" / white fls, aging purple-blue / Alps, Pyrenees 148 / 25
ursinus / R / 2-10" / lavender-pink fls / Ariz., Mont., Idaho 227 / 107
villarsii - see E. atticus 289

ERINACEA - Leguminosae
anthyllis / HH Sh / spiny hummock / blue-violet fls / s & se Spain 148 / 212
pungens - see E. anthyllis 184 / 287

ERINUS - Scrophulariaceae
alpinus / R / 3-4" / rosy fls / w & c Europe, in mts. 148 / 107
alpinus 'Abbotswood Pink' - pink selection 282
alpinus 'Albus' - the white form is true seeding 107
alpinus 'Carmineus' - carmine-red fls 46
alpinus 'Dr. Hanele' - deep crimson fls 123
alpinus 'Mrs. Charles Boyle' - clear pink form 123
alpinus 'Purpureus' - red-purple fls *
hispanicus - softer, laxer, fls pink 96

ERIOCAULON - Eriocaulaceae
compressum / Aq / 6"+ / involucre lead-colored / sw United States 99
nudicuspe / R / 8-16" / dense white-pilose heads / Honshu 166 / 200
septangulare / A / 16-28" / fls lead color / e N. Am. & Asia 99 / 200

ERIOCEPHALUS - Compositae
africanus / Gh Sh / 5' / white-rayed fls, purple center / S. Africa 28 / 223

ERIODICTYON - Hydrophyllaceae
californicum / Sh / to 8' / fls lavender to white / Calif., Ore. 25

ERIOGONUM - Polygonaceae
allenii / P / 1-1½' / yellow fls / Virginia & W. Virginia, in mts. 313
caespitosum / R / mat / white lvs; fls yellow, orange / California 227
campanulatum / R / to 1' / yellow fls / Neb., Col., Wash. 236
compositum / P / to 16" / cream or yellow fls / n coast, California 229 / 228
compositum v. leianthrum / R/ 6-20"/ fls wht-yel;lvs not hairy/WA-OR 142
crocatum / HH R / 8-10" / fls yellow / coastal s California 227 / 25
douglassii / R / 8" / fls lemon, cream, pink / Wash., Calif. 229 / 25
fasciculatum v. flavorride / HH Sh / 1'+ / yellowish fls / sw U.S. 294
flavum / R / 4-10" / fls bright yellow / wc North America 229 / 78
flavum linguifolium - see E. flavum ssp. piperi 142

flavum ssp. piperi / greener lvs / w Rocky Mts. 142
[1] gracilipes - hardy species
grande f. rubescens / 12" / fls rose-red / Santa Cruz Island, Calif. 175
jamesii / R / 4-12" / fls whitish to cream / wc United States 229 / 78
kennedyi / HH R / mat / white fls / s California 227
lanatum - probably ERIOPHYLLUM lanatum *
latifolium / P / 2' / fls white, rose, yellow / s Calif. & n-ward 227
nudum / P / to 3' / fls white or tinged yellow, rose / Wash., Nev. 229 / 78
ovalifolium / R / 3-8" / fls white or yellow / w North America 312 / 123
ovalifolium v. nivale - the white form 312
proliferum / R / 6-12" / fls white turning purplish / e Wash., Idaho 78
saxatile / HH P / 3-24" / fls white to pale yellow / s California 227 / 61
sphaerocephalum / R / to 1' / yellow fls / nw United States 229 / 142
strictum / P / 18" / fls cream, yellow, white / Wash., Mont. 229
subalpinum / R / 4-12" / fls light yellow / B.C. to Nevada 78
thymoides / Sh / 1' / fls white, pink, very hairy / e Wash., e Ore. 25
torreyanum - see E. umbellatum v. glaberrimum 142
umbellatum / R / variable to 1' / fls yellow / nw North America 229 / 78
umbellatum v. glaberrimum - glabrous lvs 142
umbellatum v. subalpinum - fls cream 25
umbellatum v. umbellatum / R / to 4" / fls crm to yel/s Can. & Mont. 142
xanthum / R / dwarf mat / yellow fls / Colorado, above timberline 198

ERIOPHORUM - Cyperaceae
angustifolium / Aq / to 2' / white to creamy bristles / n hemis. (343) 99
polystachium / synonym of E. angustifolium 290
scheuchzeri / P / 1'+ / bristles bright white / N. Am., Europe 243 / 99
spissum /Gr/6"-4'/bristles bright white/n Canada, Alaska, Orient 99 / 200
vaginatum - see E. spissum 99 / 200

ERIOPHYLLUM - Compositae
caespitosum - see E. lanatum 25
lanatum / R / 8"+ / golden-yellow daisies / w N. Am. 229 / 123
lanatum v. integrifolium / to 8" / lvs entire, tomentose 25
lanatum v. monoensis - see E. lanatum v. integrifolium 25
staechadifolium / P / 1-4' / fls yellow/ coastal s Ore., to s Calif. 25

ERIOSYCE - Cactaceae
cerastites / Gh / 3' / fls yellow or reddish / Chile 25

ERITRICHIUM - Boraginaceae
aretiodes / R /mats/ corolla violet, turns bright blue/Russia, N.A. 288
argenteum - see E. elongatum v. argenteum 25
caucasicum - from subalpine belt, n slope, Great Caucasus 265
elongatum / R / mats / fl stems to 2½"/ blue w yellow crest / Colo. 25
elongatum v. argentium /P/tufted/dense hr; toothed nutlets/Mont,Wyo 25 / 229
howardii / R / tufts and mats / bluest of all blue fls / Col., Mont. 25
jankae - see E. nanum ssp. jankae 288
nanum / R / dense cushion / soft blue fls / Alps 129 / 123
nanum ssp. jankae / lvs densely white-villous / e Alps 288
nanum ssp. nanum / R / villous stems to 2" / Eur., Alps, Carpathians 288

---

[1] Alpines of the Americas. Report of the First Interim International Rock Garden Plant Conference, 1976.

rupestre / R / 4-8" / fls sky-blue / Kashmir 93
rupestre v. pectinatum / 6" / blue fls / Himalayas 123
strictum - see E. rupestre v. pectinatum (510) / 24
terglovense - see E. nanum ssp. nanum 288

ERODIUM - Geraniaceae
acaule / R / 6" / lilac fls / Portugal, Medit. region 287
atlanticum / R / to 1' / violet-pink fls / Atlas Mts 96
boissieri/P/stemless/petals lilac w purp veins/s Spain, Sierra Nev. 287
carvifolium / R / 6" / purple fls, black at base / nc & wc Spain (119) / 287
cazorianum / R / 7" / pale lilac to purplish fls / se Spain 287
chamaedryoides flora pleno - see E. reichardii flora pleno 287
chamaedryoides 'Roseum' - may be E. corsicum (290) / 25
cheilanthifolium - see E. petraeum ssp. crispum 287
chrysanthum / HH R / to 5" / yellow fls / Greece 149 / 25
cicutarium / A / to 18" / fls purple or pink / Medit. reg., weedy 308 / 25
corsicum / HH R / 3" / pale magenta-pink fls / Corsica 184 / 129
corsicum 'Album' - white fls *
corsicum 'Rubrum' - fls deeper pink 129
daucoides / R / 5" / fls pale pink or purplish / Spain 212
daucoides cazoriense - see E. cazorianum 287
gruinum / A or B / to 20" / violet fls / Sicily, Aegean reg. 194 / 287
Guicciardii / R / to 8" / fls rose pink / Greece 25
guttatum / HH R / to 1' / deep violet fls / s Spain 212 / 287
hymenoides / R / 1' / lilac-white & crimson fls / nw Africa (522) / 46
X kolbianum - soft shell-pink fls; E. supracanum X E. macradenum 93
macradenum - see E. petraeum ssp. glandulosum 287
macradenum 'Roseum' - See E. petraeum ssp. glandulosum 'Roseum' *
Manescavii / R / to 1' / purple fls / w & c Pyrenees 148 / 287
pelargoniiflorum / R / 6-12" / white fls, violet veins / Turkey 77
petraeum / R / stemless / finely cut lvs / Spain, sw France 148 / 212
petraeum ssp. crispum - white-woolly lvs; apple pink or lilac fls 212
petraeum ssp. glandulosum - acaulescent, viol-purp fls, Pyrenees 194 / 212
petraeum ssp. glandulosum 'Roseum' - pink selection *
petraeum ssp. petraeum / lvs +/- hairy, fetid; petals bright pink 287
reichardii / HH R / acaulescent / white fls / Balearic Isles 287
reichardii flora pleno - multiple petals
reichardii 'Roseum' / HH R / 2" / deep pink fls / Majorca 132
rodiei / R / 8" / bright pink fls / se France 287
romanum / R / 8" / purplish fls / Medit. region 25
rupestre / R / acaulescent / fls pale pink / ne Spain 287
stephanianum / R / 6" / blue fls / Dahuria 206

EROPHILA - Cruciferae
verna macrocarpa / A / stems to 8" / fls wht or reddish / e Medit. 286
verna ssp. praecox / A / 4" / fls white to reddish / Medit. region 286

ERYNGIUM - Umbelliferae
agavifolium / HH P / 5' / spiny-toothed lvs / Argentina 279 / 25
alpinum / P / 1½-2' / fls metallic-blue / Europe 148 / 129
amethystinum / P / 8-20" / fls bluish / Balkans, Italy 147 / 287
aquaticum / P / to 4' / whitish fls / New Jersey to Florida 224
bourgatii / P / 18" / blue fls / Pyrenees, Spain 148 / 212
bromeliifolium - see E. agavifolium 25
caeruleum / P / 3' / blue fls / Caucasus 28

campestre / P / to 28" / fls pale greenish / Eur., s England 210 / 287
coeruleum / P / to 3' / blue fls / Caucasus to c Asia 25
delaroux / HH P / 30" / gunmetal-blue fls / Mexico 287
ebracteatum / P/lvs sword shaped; fls small, bluish / Para. to Argen. 61
giganteum / B / 3-4' / silvery-blue fls / Caucasus (418) / 129
giganteum 'Miss Willmott's Ghost' - associational name for above 108
glaciale / R / 3-4" / bluish fls / s Spain (511) / 212
maritimum / P / 6-24" / bluish fls / seacoasts of Europe 93 / 287
X oliverianum / P / 3' / blue fls 28
pandanifolium / Gh P / to 8" / dark purple fls / s Brazil, Argentina 28
planum / P / 1-3' / bluish fls / c & se Europe 147 / 287
proteiflorum / HH P / 3' / steely-white fls / Mexico 279 / (245)
proteiflorum delaroux - botanical authority for above *
spinalba / P / 8-24" / bluish fls / sw Alps 148 / 287
tricuspidatum / HH P / 6"-2½' / greenish fls / sw Spain 287
tricuspidatum variifolium - see E. variifolium *
X tripartitum / P / 2½' / blue-gray fls 49 / 279
tripartitum variifolium - see E. variifolium 268
variifolium / HH P / 2' / silver-blue fls / Atlas Mts., Morocco 49 / 268
X 'Violetta' / 2-2½' / large violet-blue fls 135
yuccifolium / P / to 6' / greenish fls / e United States 224 / 226
X Zabelii / P / 18" / blue or amethystine fls 28

ERYSIMUM - Cruciferae
X allionii - P / 1½' / orange fls 93
alpinum / R / 6" / sulphur-yellow fls / Scandinavia 123
alpinum 'Moonlight' - luminous primrose-yellow fls 123
arkansanum - see E. asperum 25
arkansanum 'Golden Gem' - see E. capitatum 'Golden Gem' 259
asperum / B / 6-10" / coppery-yellow fls / Calif. to Texas 229 / 123
caespitosum / R / 4-6" / lemon-gold fls / n Iran 96
canescens - see E. diffusum 286
capitatum / B / 8-10" / pale yellow fls / c & w United States 229 / 259
capitatum 'Golden Gem' - fls deeper color 61
concinnum - see E. suffrutescens 25
crepidifolium / B or P /6-24"/ yellow, petal-limb oblong-ovate/Eur. 286
cuspidatum / A or B / 3'+ / yellow fls / se Europe 286
decumbens / P / to 16" / yellow fls / Pyrenees, sw Alps 148 / 286
diffusum / B / to 3' / yellow fls / c & se Europe 286
dubium - see E. decumbens 286
helveticum / P / 2' / fls yellow / Balkans, Alps, Pyrenees 148 / 286
helveticum v. dubium - see E. decumbens 93
hieraciifolium / B or P / to 3½' / fls orange or red /nc & e Eur. 25
Kotschyanum / R / to 6" / fls br yellow / high mts. Asia Minor 25
linifolium / R / 12" / fls soft purple / Spain, Portugal 212 / 24
moesiacum - see E. canescens australe (E. diffusum) 262
myriophyllum / P / 6-16" / lower lvs gray or white / s Spain 286
1 nivale / P / low mound / fls clear yellow / Rocky Mts.
ochroleucum - see E. decumbens 286
odoratum / B / petals bright yellow / se Eur. 286
pachycarpum / P / to 2' / fls orange-yellow / Tibet, Sikkim 286
perofskianum / A/ 1-2' / fls bright orange / Caucasus 194 / 96

---

1 William A. Weber, Rocky Mountain Flora, 1976.

pulchellum / R / to 1' / fls deep orange / Greece, Asia Minor 25
pumilum - see E. helveticum 286
pumilum 'Golden Gem' - see E. capitatum 'Golden Gem' *
rupestre - see E. pulchellum 184 / 25
suffrutescens / HH P / 2' / yellow fls / coastal California 227 / 25
torulosum / B / to 8" / fls yellow / w North America 228 / 25

ERYTHRAEA
centaurium - see CENTAURIUM erythraea 288
chloodes - see CENTAURIUM chloodes 93
conferta - see CENTAURIUM chloodes 288
scilloides - see CENTAURIUM scilloides 93

ERYTHRINA - Leguminosae
crista-galli / HH Sh / to 15' / crimson fls / Brazil, Argentina 180 / 177
herbacea / HH P / 4'+ / fls deep scarlet / Fla., Texas to N.C. 308 / 224

ERYTHRONIUM - Liliaceae
albidum / Tu / 4-8" / fls white / e North America 229 / 224
albidum v. mesochorum / shades of blue-lavender fls / c U.S. 259
americanum / Tu / to 1' / solitary yellow fls / e North America 224 / 25
californicum / Tu / to 1' / fls creamy-white / California 228 / 267
californicum v. helenae - see E. helenae 25
californicum v. praecox - slightly earlier flowering 267
citrinum / Tu / 6" / fls creamy-white / Oregon, California 267
Dens-Canis / Tu / 6" / white or pink fls / Europe 148 / 129
Dens-Canis 'Purple King' - fls rich cyclamen-mauve, edged white 111
Dens-Canis v. sibiricum - large fls with yellow base 28
giganteum - either E. grandiflorus or E. oreganum 185
grandiflorum / Tu / 1-2' / golden yellow fls / Calif., Wash. 228 / 267
grandiflorum ssp. candidum - fls greenish or creamy-white 25
grandiflorum v. pallidum - included in the sp. 142
helenae / Tu / to 1' / white fls / Mt. St. Helena, California 267 / 227
hendersonii / Tu / 1' / pale lilac fls / Oregon 228 / 267
howellii / Tu / 6" / fls creamy / sw Oregon 267 / 185
japonicum / Tu / to 1' / fls rose-purple / Japan, Korea 166 / 200
klamanthense / Tu / 6" / white fls / nw California, sw Oregon 228
X 'Kondo' - yellow fls, vigorous hybrid from E. tuolumense 185
mesochorum - see E. albidum v. mesochorum 259
montanum / Tu / 1'+ / white fls / Oregon, Washington, B.C. 228 / 25
multiscapoideum / R / to 8" / fls wht., pale gr-yellow base/n Calif. 25
obtusatum / Tu / 1' / fls white or cream / Wyoming to Montana 29
oregonum / Tu / to 1' / white, pink, yellow fls / Ore. to B. C. 228 / 25
oregonum 'Giganteum' - synonym of the sp. 25
oregonum ssp. leucandrum - included in the sp. 142
X 'Pagoda' - deep canary-yellow fls (122) / 17
X 'Pink Beauty' - see E. revolutum 'Pink Beauty' 267
purdyi - see E. multiscapoideum
revolutum / Tu / to 16" / rose-pink fls / Calif. to B.C. 228 / 25
revolutum album - white form
revolutum v. johnsonii - included in the sp. 142
revolutum Pink Beauty' - clear pink fls, extremely reflexed (433) / 111
revolutum 'Rose Beauty' - deep rose selection 249
revolutum v. smithii - included in the sp. 25
revolutum 'White Beauty' - see E. X 'White Beauty' 17

sibiricum - see E. dens-canis v. sibiricum 28
tuolumnense / Tu / 1' / golden yellow fls / s Calif., Nevada 267 / 228
tuolumnense 'Pagoda' - see E. X 'Pagoda' 185
X 'White Beauty' / 2' / large fls, good grower 267 / 17

ESCALLONIA - Escalloniaceae
illinita / HH Sh / to 10' / fls white; evergreen lvs / Chile 36
pulverulenta / Gh Sh / to 12' / white fls; lvs evergreen / Chile 36

ESCHSCHOLZIA - Papaveraceae
caespitosa / R / 1' / fls bright yellow / c California & s-ward 228 / 227
californica / A / 9-19" / orange, yellow, cream fls / Calif. 228 / 129
lobbii / A / 4-12" / yellow fls / Sierra Nevada Mts., Calif. 228
stauntonii - error for ELSHOLTZIA stauntonii *

EUCALYPTUS - Myrtaceae
alpina / Gh Sh / ? / large frs. / Victoria, Australia 25
archeri - see E. gunnii 36
citriodora / Gh / juvenile / lemon-scented lvs / Australia 28
cloeziana / HH T / 30-45' / dark brown bark / Queensland, Australia 25
coccifera / HH T / large / pruinose shoots / Tasmania 268
cordata / T / to 50' / bark sometimes smooth white / Tasmania 25
globulus / HH T / large / bluish bark; aromatic lvs / Tasmania 212 / 25
gunnii / HH T / to 100' / flaking bark / Tasmania 36
johnstonii/T/to 200'/deciduous orange-red to br-green bark/Tasmania 36
leucoxylon / T / 50-100' / adult lvs 3-6"; wht-pink aux fls/s Austrl 61
macrandra / Sh / to 25' / bark smooth / w Australia 25
macrocarpa / Gh Sh / to 13' / red fls; lvs mealy / Australia 194
Nicholii / T / to 40' / fls white / New South Wales 25
niphophila / HH T / 20' / good foliage & bark / Victoria, Australia 129 / 268
parvifolia / HH T / to 30' / small green lvs / Australia 268
pauciflora / T / to 100 / adult lvs, 3-8"; fls white / Austrl, Tas. 61
Perriniana / T / to 20' / bark smooth / N.S. Wales to Tasmania 25
platypus / HH T / to 30' / smooth gray bark / w Australia 25
rhodantha / Sh / to 8' / fls red / w Australia 25
rubida /HH T/to 100'/pale salmon bark, peeling/s Austrl., Tasmania 25
[1] saligna / HH T / tall / lanceolate lvs / Australia
stellulata / HH T / med. size / leathery lvs / N. S. Wales, Victoria 25
urnigera / T / to 50' / bark smooth, deciduous mottled / Tasmania 25

EUCHARDIUM
breweri - see CLARKIA brewerii 25

EUCOMIS - Liliaceae
autumnalis / HH Bb / 1' / green to whitish fls / South Africa 267
bicolor / HH Bb / 1-1½' / greenish-yellow fls / South Africa 194 / 129
comosa / HH Bb / 2' / green fls / South Africa 128 / 25
pole-evansii / Gh Bb / 5-6' / greenish-yellow fls / Transvaal 111
punctata - see E. comosa 25
zambesiaca / HH Bb / 2' / fls green, unspotted / Africa 61

---

[1] Warda H. Birchner, Gardens of the Hesperides, 1960.

EUCRYPHIA - Eucryphiaceae
cordifolia / HH T / to 100'/ large white fls / Chile 25

EUGENIA - Myrtaceae
uniflora / HH Sh / to 7½' / white fls; edible red frs / Brazil 25

EUONYMUS - Celastraceae
alata / Sh / 8' / corky branches, fine fall color / Far East 129
alata v. aptera - branches, not or slightly, winged 222
alata 'Compacta' - subglobose habit 127 / 222
alata v. subtriflorus - see E. alata v. aptera 222
americanus / Sh / 8' / pink frs / New York to Texas 28
atropurpurea / Sh / to 25' / purplish fls; crimson frs / e N. Am. 257 / 25
Bungeana / T / to 30' / yellowish fls; orange & pink frs / n China 36
Bungeana 'Pendula' / Sh or T / to 15' / branches drooping / China 25
europea / Sh / to 18' / scarlet capsules / Europe 110 / 139
europea 'Alba' - white frs 222
europea 'Aldenhamensis' - bright pink frs on long stems 127 / 25
europea v. atropurpurea - narrow purple lvs 300 / 222
europea 'Red Cascade' / 15' / brilliant scarlet frs 129 / 268
fortunei v. vegeta / Sh / 5' / evergreen lvs, orange capsules 25
hamiltoniana v. maackii - 12' / pink frs / ne Asia 25 / 268
hamiltoniana v. yedoensis 'Calocarpa' - bright crimson frs 222
hians - see E. sieboldianus 200
latifolia / Sh / 20' / frs bright red / sc & se Europe 36 / 287
maackii - see E. hamiltonianus v. maackii 268
nana / Sh / 1-3' / brown-purple lvs; pink, orange frs / Caucasus 36
nana turkestanica / Sh /3'/ fls pink & orange / Caucasus to w China 25
occidentalis / Sh / 18' / frs red; fls brown-purple / nw N. Am. 25
oxyphylla / T / 25' / rich carmine frs / Japan, Korea 61 / 36
phellomana / Sh / 6-10' / corky twigs; pink frs / n & w China 36
planipes - see E. sachalinensis 25
radicans 'Vegetus' - see E. fortunei 'Vegetus' 139
'Red Cascade' - see E. europea 'Red Cascade' 268
sachalinensis / Sh / 12' / carmine-red & orange frs / ne Asia 194 / 25
sieboldianus / Sh / 10'+ / pale green fls / Far East 167 / 200
yedoersis v. calocarpa - see E. hamiltoniana v. yedoensis 'C.' 25

EUPATORIUM - Compositae
album / P / 8-36" / fls white / Florida to La. to mts of S.Car. 99
cannabinum / P / to 4' / fls reddish-mauve to whitish / Eurasia 210 / 25
chinensis v. sachalinense / P /6'/ 5 fl., glandular dots/Sakhalin 200
coelestinum / P / to 3' / fls bluish-violet / e & c U.S. 224 / 99
glechonophyllum / Gh Sh / low / white fls / Chile 28
glehnii - see E. Glehnii v. hakonense 200
Glehnii v. hakonense / P / to 2'/ pappus white, infl. loose / Japan 200
ligustrinum / Gh Sh / to 9' / creamy-white fls / e Mexico 36
macrophyllum / Gh Sh / to 6' / whitish fls / Mexico & s-ward (365)
maculatum / P / to 5' / purple fls / North America 224 / 99
micranthum - see E. ligustrinum 93
occidentale / P / to 2½' / fls red, purple or white / Wash., Calif. 25
perfoliatum / P / to 5' / white fls / e North America 224 / 25
purpureaum / P / 3-6' / fls purplish-pink / North America 147 / 129
rugosum / P / to 4' / fls bright white / North America 224 / 99
sordidum / Gh Sh / 6'+ / fragrant violet fls / Mexico 184 / 25

EUPHORBIA - Euphorbiaceae
acanthothamnos / Gh / low cushions / yellowish fls / Greece 149 / 268
amygdaloides / P / 1'+ / yellowish fls / n Europe 210 / 279
biglandulosa - see E. rigida 287
characias / HH P / to 32" / pale green & purple fls / Medit. reg. 129 / 211
characias ssp. wulfenii / to 6' / fls yellowish / e Medit. reg. 210 / 147
corollata / P / to 3' / fls white / Ont. to Fla. & Texas 224 / 25
coralloides / HH P / to 2' / purplish fls / c & s Italy, Sicily 268
epithymoides / P / 8-16" / fls yellow or purple-tinged / c & se Eu. 147 / 287
griffithii / P / 2½' / fls bright pinkish-red / e Himalayas (418) / 268
hyberna / P / 1-2' / warty seed capsules / w Europe, Ireland 314 / 268
lathyris / A / 2-3' / yellow-green fls / Europe (366) / 268
marginata / A / 3'+ / green & white fls / North America 268
mellifera / HH Sh / 6'+ / yellow fl heads / Madiera (125) / 268
myrsinites / R / 6" / greenish-yellow fls / s Europe 148 / 279
nicaeensis / P / to 32" / yellowish fls / s, e & ec Europe 287
palustris / P / to 4½' / yellow fls / most of Europe 129 / 268
paralias / P / 2'+ / succulent lvs / Europe, Asia Minor, coastal 210 / 268
polychroma - see E. epithymoides
portlandica / P / to 16" / yellowish fls / w Europe 268
rigida / P / 12-20" / yellow fls / Sicily, Greece, n Africa 149 / 268
robbiae / R / 9" / flat green lvs / Asia Minor 194 / 279
[1] robusta / P / lvs broad; woody stem / Rocky Mts.
seguieriana ssp. niciciana / P / 2' / lvs spreading / Balkans (445) / 287
serrulata / A / to 32" / greenish-yellow fls / sw & w Europe 287
sikkimensis / P / 2'+ / fls yellow / e Himalayas (607) / 268
stricta - see E. serrulata 287
villosa / P / 1-4' / yellowish fls / s & e Europe 287
wulfenii - see E. characias ssp. wulfenii 287

EUPHRASIA - Scrophulariaceae
brevipila / see E. stricta 288
glacialis / HH R / very dwarf / fine flower clusters / Tasmania 64
picta / R / 1' / fls lilac or lilac & white / c Europe, in mts. 314 / 288
stricta/A/14"/plant strongly tinged purple, corolla lilac, wht/Eur. 288

EUROTIA - Chenopodiaceae
lanata / Sh / 2'+ / grayish mold-like vesture / w North America 229 / 142

EURYOPS - Compositae
acraeus / HH Sh / 2' / fls canary-yellow / Basutoland 129 / 268
evansii - see E. acreus 184 / 268
pectinatus / Sh / to 3' / all parts soft wht, pubescent / S. Africa 25

EUSCAPHIS - Staphylaceae
japonica / HH T / 10' / reddish, leathery pods / e Asia 25

EUSTOMA
grandiflorum / P / to 3' / pale purple / Colo., Nebr. to Texas 25

---

[1] William A. Weber, Rocky Mountain Flora, 1976. Illus.

EWARTIA - Compositae
nubigena / R / prostrate / "Australian Edelweiss" / New South Wales 109 (123)

EXOCARPUS - Thymelaeaceae
cupressiformis / T/ small/ fls minute grn; lvs tassel-like/ Austrl 109
humifusus / Gh / trailing / apetalous / Australia 206

EXOCHORDA - Rosaceae
'Korolkowii' / Sh / 12' / white fls / Turkestan 28
X macrantha 'The Bride' / 4' / white fls 187
racemosa / Sh / 10-12' / pure white fls / China 75 / 25

FABIANA - Solanaceae
violacea / HH Sh / fls violet / Chile 25

FAGOPYRUM - Polygonaceae
esculentum / A / to 3' / fls white / c & n Asia 25

FAGUS - Fagaceae
sylvatica 'Atropunicea' / T / 80' / copper lvs 25

FALLUGIA - Rosaceae
paradoxa / Sh / to 7' / white fls / Calif., Utah & s-ward 36 / 25

FARSETIA
clypeata - see FIBIGIA clypeata 286
eriocarpa - see FIBIGIA eriocarpa 286

FATSIA - Araliaceae
japonica / HH Sh / to 12' / creamy-white fls / Japan 61 / 129

FAURIA - Gentaianaceae
crista-palli / P / to 1' / fls white / Japan & Alaska to Washington 25

FELICIA - Compositae
aethiopica - see F. amelloides 268
amelloides / Gh Sh / 1-2' / sky-blue fls / South Africa 208 / 25
amoena / P gr. as A / 1½' / bright blue fls / South Africa 25
bergeriana / A / 4-6" / deep azure-blue fls / South Africa 75 / 123
coelestis - see F. amelloides 28
fragilis - see F. tenella 268
fruticosa / Gh Sh /2-4' / purple fls / South Africa 25
pappei - see F. amoena 25
rosulata / HH R / 1' / blue fls / South Africa 25
tenella / A / 12-15" / pale blue fls / South Africa 208

FENDLERA - Saxifragaceae
rupicola / Sh / to 6' / fls white or rose-tinged / Texas, N. Mex. 222

FERRARIA - Iridaceae
crispa / HH C / 18" / fls brownish to purplish / South Africa 25
undulata - see F. crispa (459) / 25

FERULA - Umbelliferae
communis / P / to 12' / fls small yellow-green / Medit. 25
galbaniflua - see FERULAGO campestris 268
tingitana / HH P / 6'+ / yellow fls / Spain, Portugal, n Africa 287

FERULAGO - Umbelliferae
campestris / P / 6'+ / yellow fls / se Europe 268

FESTUCA - Gramineae
alpina / Gr / 2-8" / bright green lvs / Alps, Apennines 268
amethystina / Gr / 20" / dark violet fls / Alps 93
caesia - see F. longifolia
capillata / Gr / 1' / slender leaf-blade / Eu., introduced to U.S. 99
duriuscula glauca - see F. ovina v. glauca 25
eskia / Gr / mat, stems to 1½'/ green, gold, purple fls / Pyrenees 268
Gautieri / Gr /8-20"/ panicle dense yellow-green/sw France, ne Spain 290
gigantea / Gr / tufts, 1½-5' / bright green lvs, purple nodes / Eu. 268
glacialis / R / 6" / dense tuft / Pyrenees, Alps (177) / 268
glauca - see F. ovina v. glauca
longifolia / Gr / 4-15" /lvs pale green, awns short/n France, e Eng. 290
ovina / Gr / tufted to 2' / useful lawn grass / n temperate zone 268
ovina capillata - see F. capillata 141
ovina v. duriuscula / tufted to 2' / wide & smooth blade / Europe 25
ovina v. glauca / R / bluish lvs / Europe 129 / 25
ovina tenuifolia - see F. tenuifolia 268
pseudeskia/Gr/dense mat to 15"/lvs, spikes pungent gr, viol./s Spain 290
punctoria / Gr / to 5" / sheaths blue, prickly / Greece & Turkey 124
rubra / Gr / 3' / meadow & forage grass / N. Am., Eurasia 124 25
rubra duriuscula - see F. ovina v. duriuscula *
rubra v. pruinosa - Sand Fescue, useful for sand dunes 268
rubricaprina / Gr / 3-9" / soft bright gr spikes / c Eur Alps 290
scoparia - see F. Gautieri
tenuifolia / Gr/ to 1" / acid soil turf grass / Europe 268
valesiaca / Gr / 10'15" / florets pruinose, awns sh./ c Ger.-nc Rus. 290

FIBIGIA - Cruciferae
clypeata / P / 12-30" / fls yellow / Italy, Balkans 210 / 286
eriocarpa / differs but slightly from above / Greece 286
macrocarpa / R / to 1' / yellow fls / Turkey, Syria, Iran 77

FILIPENDULA - Rosaceae
hexapetala - see F. vulgaris 287
hexapetala 'Flore Plena' / 1½' / double fls 129
multijuga / P / to 2' / fls rose to white / Japan 25
palmata / P / 3' / fls pale pink / Siberia 184 / 25
palmata 'Nana' - to 1' 184
rubra / P / 6-8' / pink fls / e U.S. 224 / 279
Ulmaria / P / to 6' / fls cream to white / Eur., Asia, not in U.S. 25
Ulmaria 'Aurea-Varigata' / P / to 6' / lvs yellow-variegated 210 / 25
vulgaris / P / to 32" / fls pale creamy & purplish / Europe 210 / 287

FIRMIANA - Sterculiaceae
simplex / HH T / to 60' / lemon yellow fls / e Asia (460) / 25

FOENICULUM - Umbelliferae
vulgare / P / 4-6' / yellow fls / Europe 129

FORSTERA - Stylidiaceae
tenella / R / mats / white fls dotted red / New Zealand 238

FORSYTHIA - Oleaceae
X intermedia / Sh / 6-8' / F. suspensa X viridissima (418) / 36
X intermedia 'Beatrix Farrand' / Sh / large canary-yellow fls 129

FOTHERGILLA - Rosaceae
gardenii / Sh / to 3' / white fls / Virginia to Georgia 300 / 25
major / Sh / 9' / white & yellow stamens / Georgia 28
parvifolia - see F. gardenii 216

FRAGARIA - Rosaceae
indica - see DUCHESNEA indica
glauga virginiana/P/lvs blue-gr above, petals varied/B.C, Alta, N.M. 142
moschata / P / to 16" / edible European Strawberry / c Europe 287
vesca / R / to 1' / frs red / Eurasia, North America 314 / 25
vesca f. alba - white frs 25
vesca 'Alpine' / R / 10" / red or yellow frs 230
vesca 'Alpine Yellow' - the yellow form 230
vesca leucocarpa - see F. vesca f. alba *
virginiana / P / low / white fls; red frs / e North America 224 / 25

FRANCOA - Saxifragaceae
appendiculata / HH P / 2½' / fls pale rose / Chile 182 / 25
ramosa / HH P / 3' / fls white / Chile 25
sonchifolia / HH P / 2-3' / fls white or deep pink / Chile 184 / 279

FRANGULA - Rhamaceae
alnus / Sh or T / 13-17' / fls axillary drupe red turning black/Eur. 287 / 36

FRANKENIA - Frankeniaceae
laevis/P/to 15"/lvs hairy beneath, fls purp or wht/w Eur., s Eng. 287
thymifolia / R / 1' / purplish fls / c, e & s Spain 287

FRANKLINIA - Theaceae
alatamaha / T / 15-20' / white fls / along Altamaha River, Ga. 300 / 36

FRASERA - Gentianaceae
albicaulis v. columbiana / P / 2'+ / pale to dk. blue fls / w N. Am. 142
caroliniensis - see SWERTIA caroliniensis 99
speciosa / P / to 6' / fls greenish-white, purple spotted / nw U.S. (352) / 25

FRAXINUS - Oleaceae
americana / T / to 120' / gray, furrowed bark / e North America (342) / 36
excelsior / T / 100'+ / black winter buds / Europe 314 / 36
latifolia / T / to 80' / with stalkless leaflets / w North America 36
pennsylvanica / T / 60' / Red Ash / e & c North America 28
quadrangulata / T / to 70' / square, winged branchlets / se & c U.S. 313 / 36
tomentosa / T / 50' / Pumpkin Ash / e & c United States 99

FREESIA - Iridaceae
- Fergusoniae / C / stem 4" / 5-6 fld, yel splashed orange / S. Africa 121
- Muirii / to 24" / frag fls, clear white, rarely yellow / S. Africa 121
- refracta v. xanthospila / Gh C / 18" / yellow fls / South Africa 61
- xanthospila - see F. refracta v. xanthospila 61

FREMONTIA
- californicum - see FREMONTODENDRON californicum (418) / 25
- mexicana - see FREMONTODENDRON mexicanum 36

FREMONTODENDRON - Bombacaceae
- californicum / Gh Sh / 15' / yellow fls / California, Arizona 187 / 25
- mexicanum / Gh T / 10-20' / orange-yellow fls / Baja California 36

FRITILLARIA - Liliaceae
- acmopetala / Bb / 1½' / brown & green fls / Asia Minor 211 / 129
- agrestis / Bb / 1-2' / greenish-white & brown fls / California (378) / 228
- alfredae / Bb / 12" / 1-2 sm. green-yellow fls, brown tip / Lebannon 268
- 1 alfredae ssp. glaucoviridis - see F. glaucoviridis
- 'Antemis' - see F. meleagris 'Antemis' 267
- armena / Bb / 4" / brown-purple fls / Armenia 25
- assyrica / Bb / 9" / maroon fls, gold tipped / Iraq 40
- atropurpurea / Bb / 6-24" / brown, green, purple fls / Oregon 229 / 228
- aurea / Bb / 4" / bright yellow fls / Turkey, Cilicia (128) / 185
- biflora / HH Bb / 6-18" / greenish to dark purple fls / s Calif. 227 / 28
- bithynica / Bb / 6" / yellow fls / nw & w Asia Minor 149 / 267
- bucharica - see RHINOPETALUM bucharica 4
- camtschatcensis / Bb / 4-20" / fls dark brown-purple / Alaska, Japan 166 / 200
- canaliculata / Bb / to 8" / green & chocolate-purple fls / w Iran 185
- carduchorum / Bb / ? to 6" / brick red fls; lvs shiny green/e Turkey 185
- 1 carica / 2-6" / lvs grey-grn; fls yel or grn-yel /Turkey, e Aeg. Is.
- caucasica / Bb / 12" / mahogany fls / Caucasus 40 / 111
- cirrhosa / Bb / 1½' / yellowish fls, purplish checks / Himalayas 25
- citrina / Bb / 8-10" / clear yellow fls / Greece, Taurus 194 / 123
- conica / HH Bb / 5" / greenish yellow fls / Greece 185
- crassifolia / Bb / 6" / green & chocolate-maroon fls / Iran, Lebanon 267 / 185
- crassifolia kurdica - see F. kurdica (444) / 185
- davisii / Bb / 6" / deep chocolate fls / Mani Pen., Greece (353) / 185
- degeniana - see F. orientalis 290
- delphinensis - see F. tubiformis 40
- drenovskii / Bb / 8" / purple-crown & yellow fls / Bulgaria 267 / 132
- eduardii / Bb / to 2' / bright red fls / c Asia 4
- ehrhartii / Bb / 6-7" / dk red-br, inside golden green / Europe 40
- elwesii / Bb / 18" / green fls flushed purple / Asia Minor 25
- falcata / Bb / 4" / white fls, reddish speckling / California 185
- forbesii / considered doubtful 40
- gentneri / Bb / 1'+ / fls bluish-red / Oregon (129)
- gibbosa / Bb / 6-8" / light pink and darker fls / c Asia (354) / 185
- glauca / Bb / 4" / yellow or brownish-yellow fls / Calif., Ore 228 / 185
- glaucoviridis / Bb / 6" / bright green fls / s Turkey 185
- gracilis / Bb / 18" / purple fls, checkered / Dalmatia 26 / 25
- graeca / Bb / 6" / green fls margined red / Greece 267 / 132

---

1 Brian Mathew and Turhan Baytop, The Bulbous Plants of Turkey, 1984. Illus.

graeca v. gussichae / 8-16" / green fls striped brown	19	/	268
graeca v. thessalica - vigorous form from n Greece	267	/	185
gussichiae / Bb / 7-12" / fls green, brown tinged /Macedonia			290
hermonis ssp. amana / Gh Bb / 6" / green fls / s Turkey, Syria	(378)	/	(315)
hispanica / Bb / 6-15" / green fls striped purple or brown / Spain	212	/	268
imperialis / Bb / to 4' / fls yellow to red / India	93	/	129
imperialis 'Aurora' - deep reddish-orange fls, vigorous			267
imperialis 'Lutea' - deep lemon-yellow fls	(461)	/	267
imperialis 'Lutea Maxima' - deep lemon-yellow fls, large	129	/	267
imperialis v. raddeana - see F. raddeana			268
imperialis 'Rubra' - fls bright red			111
involucrata / Bb / 12-15" / fls pale green, brown markings / Italy	148	/	185
ionica / Bb / to 1' / green and purple fls / nw Greece, Corfu			185
japonica / Bb / diminutive / white / Japan			40
karadaghensis / Bb / 6" / yellow-green fls, marbled / Iran	40	/	132
karelinii - see RHINOPETALUM karelinii			4
kurdica / Bb / to 8" / purple fls, yellow banded / Caucasus			4
lanceolata / Bb / 3'+ / purple-brown fls, mottled / nw N. Am.	227	/	25
lanceolata v. gracilis - smaller black-purple fls			28
lanceolata tristulis / Bb / 6" / 1-2 lg bells/w Canada, s to Calif.			185
[1] latakiensis / 4-10" / fls purp outside, gr-yel inside/Turk, n Syria			
latifolia / Bb / to 1' / deep chocolate fls / ne Turkey, Caucasus	267	/	185
latifolia v. lutea - Caucasian yellow-flowered form			185
latifolia v. nobilis / accaulescent, chocolate fls / ne Turkey	(512)	/	185
libanotica - see F. persica			185
liliacea / Bb / to 1' / white fls streaked green / coastal Calif.			25
lusitanica / Bb / 1' / red-maroon & yellow fls / Portugal	40	/	185
lutea / Bb / small / yellow fls / c Caucasus			40
macandra - see F. rhodokanakis			
meleagris / Bb / 1' / fls reddish-purple & white / Europe			129
meleagris 'Alba' - white fls	129	/	25
meleagris 'Aphrodite' - strong-growing, white fls	(357)	/	129
meleagris 'Artemis' - fls dusky-purple			267
meleagris 'Charon' - dark purple fls			129
meleagris 'Emperor' / 15' / gray-mauve fls, marked purple			111
meleagris 'Orion' - deep claret-red blooms, finely mottled			112
meleagris 'Pomona' - tall, white fls marked violet-mauve			111
meleagris 'Poseidon' - soft purplish-rose fls			267
meleagris 'Saturnus' - fls light reddish-purple			129
messanensis / Bb / 10" / reddish-blue to brown fls / Crete	149	/	132
messanensis gracilis-upper lvs alternate, perianth rarely tesselated			290
micrantha / P / to 3' / purplish to greenish white / Sierra Nevadas			228
michalovskyi / Bb / 4-8" / purple-brown & gold fls / Turkey	(112)	/	268
montana / Bb / 8-18" / dark red fls / Balkans, e Europe			4
neglecta - see F. nigra			
nigra / Bb / 6" / deep purple-red fls / Medit. region			185
obliqua / Bb / to 1' / dark mahoghany fls / Greece	194	/	132
olivieri / Bb / 15" / yellow fls flushed red-brown / Iran	149	/	185
orananensis / Bb / 12" / purple fls / nw Africa	(513)	/	132
orientalis / Bb / 6-16" /greenish, tesselated purplish-brown/se Eur.	40	/	111
pallidiflora / Bb / 2' / creamy or greenish-yellow fls / Iran			290
persica / Bb / 6-40" / fls blackish-plum to straw / Iran	194	/	129
	(353)	/	185

[1] Brian Mathew and Turhan Baytop, The Bulbous Plants of Turkey, 1984. Illus.

persica 'Adiyaman' - a promising selection, plum-colored fls		279
pinardii / HH Bb / 9" / deep purple & olive-green fls / n Turkey	267	111
pineticola / Bb / 5-9" / fls yel-grn; bot lvs broad / Samos		268
pluriflora / Bb / 18" / pink-purple fls / n California	19	25
pontica / Bb / 6" / brown fls, green tipped / sw Europe	267	17
pontica ionica / smooth seed head / n Greece & Corfu	267	185
pudica / Bb / to 1' / yellow to orange fls, aging red / Mont., B.C.	267	25
purdyi / Bb / 6-12" / fls white tinged purple / California		28
pyrenaica / Bb / 1½' / dark maroon-purple fls / Pyrenees	148	129
pyrenaica 'Alba' - fls white		*
pyrenaica 'Lutea' - yellow fls	(336)	185
raddeana / Bb / 2½' / pale creamy-yellow fls / Iran, Turkey	185	268
recurva / Bb / 1' / scarlet & orange fls / California, Oregon	267	107
reuteri / Bb / 8-12" / purple-brown & gold fls / Iran		268
rhodokanakis / Bb / 6" / purplish-maroon & yellow fls / Greece	40	185
roderickii / Bb / ? / fls creamy-green / nw North America	(64)	(129)
roylei / Bb / 6-40" / pale green to purplish fls / w Himalayas	194	185
ruthenica / Bb / 1-2' / fls dark purple / Caucasus, se Europe		25
'Saturnus' - see F. meleagris 'Saturnus'		267
schliemanii - see F. citrina		185
sewerzowii - see KOROLKOWIA sewerzowii		4
sibthorpiana / Bb / 4" / clear golden-yellow, nectary dk. gr/Eur.		40
stenanthera - see RHINOPETALUM stenanthum		4
stribrnyi -/ near to F. sibthorpiana / Bulgaria, Macedonia, Taurus		40
striata / Bb / 8-15" / pale pink fls, brown veined / California		268
1 syriaca - see F. pinardii		
tenella - see F. nigra		185
thunbergii - see F. verticillata v. thunbergii		28
tubiformis / Bb / 2-3" / reddish-purple fls / se France, N Italy	148	123
tubiformis delphinensis - a synonym		28
tubiformis v. moggridgei - yellow fls		185
tuntasia / Bb / 12" / white & green fls / Altai Mts.	267	123
1 uva-vulpis / 4-8" / fls green; upper lvs whorl of 3 / Turkey		
verticillata / Bb / to 2' / cream fls, veined green / Far East		185
verticillata v. thunbergii / 1½' / fls light yellow / China		200
Walujewii/ distinct bell transparent wht; inside brn/ Kashmir,Afghan		64
whitallii - similar to F. meleagris, w end of Taurus Mts.		40

FUCHSIA - Onagraceae

excorticata / HH T / 36' / dark purple fls / New Zealand		194
magellanica / HH Sh / 12' / crimson-red & purple fls / Peru	208	129
magellanica 'Discolor' - dwarf habit		268
magellanica 'Pumila' - dwarf form		25
magellanica 'Tom Thumb' / 8-16" / red fls		93
procumbens / HH R / prostrate / dark red fls / New Zealand	194	206
pumila - see F. magellanica 'Pumila'		25
thymifolia / Sh / tp 3' / fls white to pink / Mexico		25

FUMANA - Cistaceae

nudifolia - see F. procumbens		
procumbens / Sh / to 16" / yellow fls / wc & s Europe	149	287
thymifolia / HH R / to 8" / yellow fls / Medit. region, Portugal		287

---

1 Brian Mathew and Turhan Baytop, The Bulbous Plants of Turkey, 1984. Illus.

FUMARIA
africana - see RUPICAPNOS africana 286

GAGEA - Liliaceae
alexeenkoana / R / to 6" / yellow & green fls / Caucasus 4
arvensis / HH R / 6" / yellow fls / Spain, n Africa 211
fistulosa / R /to 6" / yellow & green or reddish fls/ c Eu., Crimea 149 / 4
graeca / HH R / to 6" / fls white, streaked purple / Greece 210 / 149
lutea / R / to 10" / bright yellow starry fls / Europe 149
villosa / R / 3" / yellow fls / Caucasus 206

GAILLARDIA - Compositae
aristata / P / 2' / fls yellow & red / w North America 229 / 129
Burgundy / P / to 2½' / narrow red ray fls / N., S. Am. 25
X grandiflora 'Baby Cole' /P/ to 2½'/ yel red banded fls/N., S. Am. 25
X grandiflora 'Goblin' - yellow fls, red banded 147 / 25
X grandiflora 'Nana Nieski' / 15" / yellow & red fls 208
X grandiflora 'Sun Dance' / 8" / red & yellow fls 135
pinnatifida / P / 6-24" / yellow fls / Utah, Colorado, Oklahoma 229
suavis / P / ? / sometimes lacking ray fls / Texas 226

GALACTITES - Compositae
tomentosa / A or B / 2½' / purple fls / Medit. region 149 / 25

GALANTHUS - Amaryllidaceae
autumnalis - see LEUCOJUM nicaeense 256
byzantinus / Bb / 6" / white & green fls / w Turkey (355) / 111
caucasicus / Bb / 4-6" / glaucous lvs / Iran, Caucasus (356) / 17
corcyrensis - see G. nivalis ssp. reginae-olgae 149
elwesii / Bb / 6"+ / large white fls, green tipped / Jugoslavia 269 / 17
elwesii 'Cassaba' - form distinguished in 1894 256
elwesii 'Whittallii - larger leaf form 268
fosteri / HH Bb / 12" / white fls marked green / Lebanon 111
graecus / Bb / 6" / white & green fls / n Greece (338) / 111
ikariae / Bb / 6" / white & green fls / Nikaria, island off A. Minor 111
ikariae ssp. ikariae / Bb / 5½" / white w gr. markings/Is. Nikaria 25
ikariae ssp. latifolius - with wider segments 25
imperati 'Magnet' - see G. X 'Magnet' 256
X 'Ketton' - strong grower with large fls 256
X 'Magnet' - tall, with long arching pedicels 256 / 185
X 'Melvillei' - medium height, globose fl 256
nivalis / Bb / 6-9" / fls white / Europe 267 / 25
nivalis 'Atkinsii' - early-flowering, larger fls (514) / 25
nivalis caucasus - see G. caucasus 256
nivalis melvillei - see G. X 'Melvillei' 256
nivalis olgae - see G. nivalis ssp. reginae-olgae 267
nivalis ssp. reginae-olgae / autumn fls / mts. of Greece 149 / 267
nivalis 'Sam Arnott' - robust cultivar (61) / 185
nivalis scharlockii / Bb / 4½" / fls white w gr. markings / Europe 25
nivalis 'Straffan' / 4-6" / late flowering 17
nivalis 'Viridapicis' / 3" / white fls, spotted green 17
platyphyllus - see G. ikariae ssp. latifolius 185

plicatus / Bb / 6" / white & green fls, fragrant / s U.S.S.R. 267 / 111
plicatus X nivalis - crossed by donor? 17
plicatus 'Warham' / 6" / large white fls / Crimea 17
rizenensis / Bb / lvs dull green; fls white / Asia Minor 268
viridapices - see G. nivalis 'Viridapicis' 17
woronowii / Bb / 8-10" / lvs pale green; fls white / Caucasus 4

GALAX - Diapensaceae
aphylla - see G. urceolata 224 / 25
urceolata / P / to 2½' / small white fls / Virginia to Georgia 25

GALEGA - Leguminosae
officinalis / P / to 5' / fls white to lilac / c Eu., to Iran 210 / 25
orientalis / P / 2½-4' / fls purplish-blue / Caucasus 28

GALEOPSIS - Labiatae
angustifolia / P / to 16" / fls reddish-pink & yellow / wc & e Eu. 288

GALIUM - Rubiaceae
anisophyllum / R / dwarf / pale yellow fls / Europe 64
californicum / R / 2-8" / fls yellowish / coastal California 228
odoratum / R / to 1' / fls white / Eurasia, n Africa 210 / 25
purpureum / R / 10" / brownish-red fls / near Lake Como, Italy 96
verum / P / 1-3' / Yellow Bedstraw / Europe, weedy in U.S. 28

GALTONIA - Liliaceae
candicans / HH Bb / to 4' / white fls, bell-shaped / South Africa 129
clavata - see PSEUDO CLAVATA
princeps / HH Bb / 3' / pale green bells / South Africa 186 / 279
viridiflora / HH Bb / ? / fls greenish-white / South Africa 267

GAMOLEPIS - Compositae
Tagetes / A / to 1' / ray fls fr. yellow or orange / South Africa 25

GARRYA - Garryaceae
buxifolia - see G. flavescens v. buxifolia 294
flavescens v. buxifolia / HH Sh / 5' / catkin-bearer / California 294
fremontii / HH Sh / 10' / lvs yellowish-green / California, Oregon 28
Veatchii / HH Sh / Canyon Tassel-bush / coast & mts. of California 25 / 294

GASTERIA - Liliacaea
lingua (G. disticha) / P / to 10" / fls red or rose / S. Africa 25

X GAULNETTYA - Ericaceae
wisleyensis / Sh / dwarf / fls white; frs purplish 184 / 268
wisleyensis 'Wisley Pearl' / HH Sh / 3' / pearl-wht fls; purp-red frs 36

GAULTHERIA - Ericaceae
adenothrix / Sh / to 1' / fls white; frs red / Japan 273 / 200
antipoda / HH Sh / prostrate to 3' / frs red or white / N. Zealand 238 / 139
crassa / HH Sh / 1'+ / leathery lvs / New Zealand 36
cuneata / Sh / 1' / fls & frs white / w China 36 / 107
depressa / Sh / 8-10" / pink & white fls; scarlet frs / N.Z. (221) / 123
hispida / HH Sh / 6' / fls & frs white / Australia, Tasmania 45
hispidula / R / creeping / white fls & frs / ne North America 22 / 99

Hookeri / Sh / to 6' / fls & fr white / Himalayas 25
humifusa / Sh / 4" / fls white-pinkish; frs scarlet / Br. Columbia 229 / 132
itoana / HH Sh / creeping / bright red frs / Formosa 139
merrilliana - see G. itoana 36
microphylla / Sh / low / white fls; pink frs / Antarctic region 222
miqueliana / Sh / 1' / white, pink tinged fls & frs / Japan 273 / 107
nummulerioides / Sh / 1' / white fls; blue-black frs / Himalayas 129
nummularioides elliptica/Sh/fls sol. petals pink-wht/Himal., w China 25
ovatifolia / Sh / procumbent / fls pink & white / B.C., Oregon 229 / 107
procumbens / Sh / creeping / fls white; frs scarlet / e N. Am. 129 / 25
pyroloides / HH Sh / to 1' / white fls; blue-black frs / w China 25
rupestris / Sh / 1' / white fls; frs not fleshy / New Zealand (132) / 93
semi-infera / Sh / to 6' / fls white, fr blue / Himalayas, w China 25
shallon / Sh / 5-6' / pink fls; frs dark purple / w North America 93 / 129
sinensis / Sh / 6" / wht fls; blue frs / Burma (515) / 123
thymifolia / Sh / 3-5" / pink fls; blue frs / w China 123
trichophylla / Sh / 3-6" / pink fls; blue frs / Himalayas, w China 75 / 36

GAURA - Onagraceae
coccinea / P / to 2' / fls white & pink to red / Alta. to Mexico 25 / 229
lindheimeri / P / 2-3' / fls white & rosy / Texas, Louisiana 226

GAYLUSSACIA - Ericaceae
baccata / Sh / to 3' / fls reddish; edible black frs / ne N. Am. 25
brachycera / Sh / 6-12" / fls white; frs blue / e United States 36

GAZANIA - Compositae
linearis / P gr. as A / short / fls yellow or orange / South Africa 138 / 25
longiscapa - see G. linearis 25
rigens / P gr. as A / decumbent / yel. to orange fls / South Africa 194 / 289
X splendens - see G. rigens 129 / 289

GEISSORRHIZA - Iridaceae
corrugata / 2" stems / solitary bright yellow fls / South Africa 121
[1] inflexa v. erosa /C/10'/fls cerise, a/k/a 'Red Sequines'/S. Africa
rochensis / HH C / 4-8" / intense purple-blue fls / sw Cape Province 194 / 185
secunda / HH C. / 9" / purple-blue fls / Cape Province 111
splendissima / Gh / 6-9" / brilliant dark purple fls / S. Africa 25

GELASINE - Iridaceae
azurea / HH P / 18" / blue fls / Brazil 206
pilosa / to 1' / tangle mat, semi procumbent stems / Eur. & Brit. 154

GENISTA - Leguminosae
aetnensis / HH Sh / 15' / yellow fls / Sardinia, Sicily 269 / 287
anglica / Sh / to 3' / yellow fls / w Europe 287
cinerea / HH Sh / to 3' / bright yellow fls / s Europe, n Africa 110 / 25
dalmatica - see G. sylvestris v. pungens 287
decumbens - see CYTISUS decumbens 222
delphinensis - see CHAMAESPARTIUM sagittale ssp. delphinensis 287
depressa - see G. tinctoria ssp. depressa *
depressa v. moesiaca/Sh/4"(not sep. from G. tinctoria in Fl. Eur.) 262

---

[1] Sima Eliovson, Wild Flowers of Southern Africa, 1980.

fragrans - see G. tinctoria		262
germanica / Sh / 2' / small yellow fls / Europe	148 /	287
hirsuta - see CYTISUS hirsutus		
hispanica / HH Sh / to 3' / golden-yellow fls / sw Europe, Spain	148 /	129
hispanica compacta / Sh/2'/fls yel, dense habit/ w Pyrenees, n Spain		25
hispanica 'Nana' - dwarf shrub		25
horrida / HH Sh / 18" / spiny branchlets; yellow fls / France, Spain		25
humifusa - see CYTISUS diffusus		222
lydia / HH Sh / 3' / deep yellow fls / s & se Europe	(187) /	129
micrantha / Sh / fls yellow, in-turned raceme / ne Spain, n Portugal		287
pilosa / Sh / to 3' / yellow fls / w & c Europe	(441) /	287
pilosa procumbens - see CYTISUS procumbens		107
pulchella / Sh / spreading / yellow fls / se France to Albania		287
radiata / Sh / 3' / dark yellow fls / s Europe	148 /	208
sagittalis - see CHAMAESPARTIUM sagittale		287
sagittalis ssp. delphinensis - see CHAMAESPARTIUM s. ssp. d.		287
scorpius / HH Sh / ? / yellow fls / Spain, s France		287
sericea / Sh /? / fls yellow, born singly / Balkans, n Italy		287
subcapitata / Sh / to 1' / yellow fls, sessile in heads / c Balkans		287
sylvestris / Sh / decumbent / yellow fls / Albania, Jugoslavia		287
sylvestris v. pungens / Sh / 4-6" / golden-yellow fls / Dalmatia		36
tenera / HH Sh / to 12' / bright yellow fls / Madeira		36
tinctoria / Sh / 3' / fls deep yellow / s Europe, Britain	148 /	129
tinctoria ssp. depressa / to 8" / decumbent / few-flowering		287
tinctoria ssp. depressa f. moesiaca - see G. depressa f. moesiaca		262
tinctoria v. humifusa / prostrate / to 4"		123
tinctoria ssp. littoralis / 8" / procumbent / fls few		287
villarsii - see G. pulchella	184 /	287
virgata - see G. tenera		36
GENTIANA - Gentianaceae		
acaulis / R / 2-3" / fls bright blue / Alps, Pyrenees	212 /	129
acaulis 'Alba' - fls white		25
acaulis alpine - see G. alpina		288
acaulis v. angustifolia - see G. angustifolia		288
acaulis clusii - see G. clusii		28
acaulis 'Coelestina' - large, vivid sky-blue fls		168
acaulis dinarica - see G. dinarica		288
acaulis 'Undulatifolia' - medium-blue fls; wavy-edged lvs		168
affinis / 6-18" / fls deep blue, streaked green / Calif. to B.C.	229 /	228
alba / P / to 8' / fls light yellowish-green / ne N. Am		25
algida / R / 4-10" / pale yellow fls, green spots / e Asia, w N.Am.	312 /	200
algida f. igarashii / 4" / large fls / Hokkaido in high mts		200
alpina / R / 2" / bright blue fls / European ranges	148 /	123
amarella - see Gentianella amarella	224 /	25
andrewsii / P / 2' / blue-purple fls, white tips / e N. Am.	224 /	123
andrewsii 'Alba' - see G. andrewsii f. albiflora	(462) /	*
andrewsii f. albiflora - fls white		25
andrewsii v. dakotaea /P/ 1-3'/fls bl-viol, purp-white/ N.Dak., Kans.		229
angulosa - see G. verna		25
angustifolia / R / 3" / blue fls; narrowed lvs / Alps	148 /	25
aquatica / A / 4" / fls pale blue / e Asia		77
asclepiadea / P / 1½-2' / deep blue fls / Europe	148 /	129
asclepiadea 'Alba' - white fls		*
asclepiadea 'Nana' - dwarf form		123

asclepiadea 'Nana' alba / white form in cultivation 25 / 288
[1] asclepiadea 'Phyllis' - light blue fls, reasonably true from seed
asclepiadea 'Rosea' - roseate fls *
[2] 'Asiatic Hybrids' / R / very similar to G. farrari
aspera / B / to 16" / violet, pink or white / S. Germany, Czech. 288
aurea - see GENTIANELLA aurea 288
austriaca - see G. austriaca 288
autumnalis / P / to 2' / fls indigo-blue / New Jersey to S. Car. 224 / 99
autumnalis f. porphyrio - fls brown-spotted within 99
axillaris - see G. amarella 288
axilliflora - see G. triflora v. japonica 200
baltica - see GENTIANELLA campestris ssp. baltica 288
bavarica / R / to 8" / two-toned blue fls / Alps 208 / 288
bavarica v. subacaulis / 2"+ / deep blue fls / Alps 93
bellidifolia / HH R / 4-6" / fls white / New Zealand 35 / 123
bellidifolia v. australe / cream flowers / s Alps, New Zealand 15
biflora - akin to G. dahurica 64
bigelovii / P / 12-16" / fls violet / New Mexico, Arizona, Colorado 227
bisetea / R / 5" / fls blue / inner coastal ranges of Oregon 228
brachyphylla / R / 2"+ / deep blue fls / mts. of c Europe 148 / 288
brachyphylla ssp. favratii - corolla lobes as wide as long 288
bracteosa / 12" / funnel shaped, deep purple-blue fls/ mid w. U.S. 168
X brevidens (G. Makinoi X G. scabra) -hybrid of G. s. f. stenophylla 200
burseri / P / to 2' / clustered yellow fls, brown spots / sw Alps 148 / 288
cachemirica / R / to 8" / striped fls, azure-blue / Kashmir 35 / 25
calycosa / R / 6-12" / deep or pale blue fls / nw North America 312 / 228
campestris - see GENTIANELLA campestris 25
cashmerica - preferred spelling is cachemirica *
ciliata / R / 5-10" / fringed pale blue fls / Alps, Apennines 208
clausa / P / 2' / blue to blue-violet fls / ne & ec North America 224 / 259
clusii / R / 3" / deep blue unspotted fls / c & s European mts 148 / 288
clusii 'Alba' - white form
clusii ssp. costei - less warty leaf margins 35
concinna / A / 4" / white fls, red streaked / New Zealand 64
cordifolia - see G. septemfida 168
corymbifera / R / 1'+ / white fls, dark veined / New Zealand 238 / 168
crassicaulis / P / rank leafy plant / whitish fls / ? 64
crinita - see GENTIANOPSIS crinita 224 / 25
cruciata / R / 10" / bluish to greenish fls / sc & e Europe 148 / 288
cruciata ssp. phlogifolia / shorter corolla / Carpathians 288
dahurica / R / 6-8" / deep blue fls, paler throat / A. Minor, China 61 / 25
decora / R / 8-10" / white fls striped blue or violet / Ga. to Va. (352) / 25
decumbens / R / to 1' / blue fls / c & ne Asia (463) / 25
dendrologii / R / 6-15" / fls white to mauve / w China 35 / 25
depressa / R / to 2" / greenish-blue fls / Nepal, Sikkim 194 / 93
X 'Devon Hall' / R / 4"+ / medium blue fls 168
diemensis / see G. diemensis (123)
dinarica / R / 3" / bright blue fls / Balkans, limey mts. 35
divisa / HH R / 6" / fls white / New Zealand 15
'Drakes Strain' - compact mat, Cambridge-blue fls 35

---

[1] Hodgkin, Jour. Royal Hort. Soc., April 1973.

[2] Siskiyou Rare Plant Nursery Catalog, 1984.

elegans - see GENTIANOPSIS thermalis			25
excisa - see G. acaulis			288
X Farorna - soft sky-blue fls, banded outside			168
farreri / R / 3-5" / bright sky-blue fls / se Tibet, w China	132	/	129
fetisowii / P / to 2' / intense dark blue fls / w Siberia			10
fetissowii 'Alba' - unrecorded white form			*
fischeri / P / 8-30" / blackish-blue fls, green spots / w Siberia			10
flavida / P / 1-2' / yellowish fls / c Ontario, c United States	224	/	168
Freyniana / R / to 1' / blue fls in autumn / Asia Minor	182	/	25
frigida / R / 2-4" / fls yellowish-white, blue spots / Carpathians	148	/	288
froelichii / R / to 4" / clear blue fls / se Alps	148	/	288
gebleri - included in G. decumbens			10
gelida / R / 6-16" / yellow fls / Caucasus	35	/	10
germanica - see GENTIANELLA germanica			93
germanica austriaca alba - see G. austriaca			10
glauca / R / to 6" / blue fls / nw North America, Asia	13	/	25
gracilipes / R / 6-9" / clear rich blue fls / China	(133)	/	93
gracilipes 'Alba' - fls white			184
grayi - see C. linearis ssp. rubricaulis			114
grombezewskii / R / to 16" / fls yellow / Turkestan			25
Grossheimii / R / 4" / yellow-blue & dark blue fls / Caucasus			10
X hascombensis - good blue fls, seedlings nearly true	(134)	/	168
hexaphylla / R / 4-6" / lt. blue, green spotted fls / e Tibet			123
holopetala - see GENTIANOPSIS holopetala	227	/	25
hookeri - see G. saposa			15
humilis - see G. aquatica			77
'Inshriach' - cambridge-blue trumpets			249
'Inverleith' - deep sky-blue fls on low plants			168
jesoana - see G. triflora v. japonica			200
Kaufmanniana / P / 6-8" / fls 1-3, terminal, stalked, blue / Asia			61
kesselringii - see G. walujewii			10
kirishima rindo - see G. scabra saxatilis			168
kochiana - see G. acaulis			288
Kurroo / R / to 10" / blue fls, blue & white spots / Himalayas	(100)	/	25
Kurroo v. brevidens - see G. dahurica			61
lagodechiana / R / prostrate / deep blue fls / Caucasus	35	/	25
linearis / P / to 2' / blue fls / ne North America	22	/	25
linearis ssp. rubricaulis / 8-20" / cluster of blue fls/ Gr. Lakes			114
ligustica / corolla has gr. spots in throat / Calicole			288
loderi - see G. cachemerica			64
lutea / P / to 4' / fls yellow / c & s Europe, in mts.	148	/	288
X macauleyi - large deep blue trumpets			123
macrophylla / P / 16-28" / blue-violet fls / Siberia to China			10
makinoi / P / 1-2' / dark blue fls / Honshu, in high mts.	164	/	200
makinoi f. albiflora - white fls			35
montana / HH P / 2' / white fls / South Island, New Zealand			96
moorcroftiana - see GENTIANELLA moorcroftiana			25
nesophila / A or B / 1-10"/ blue-violet to creamy fls /Newfoundland			99
newberryi / R / 2-4" / white inside, blue outside / s Ore., Calif.	228	/	25
nipponica / R / to 4" / blue fls / alpine Hokkaido & Honshu	164	/	200
nivalis / A / 6" / deep blue fls / circumboreal, s in Eu. mts.			148
nummularifolia / Andean alpine / woody, dull white fls			64
occidentalis / corolla w few or no green spots in throat/Calcicole			288
ochroleuca - see G. villosa			28
olivieri / R / to 1' / blue or blue-violet fls / c Asia	64	/	10

orbicularis / R / tufts / sessile deep blue fls / c Eu., in mts.		148
oregana - see G. affinis		142
ornata / R / to 4" / fls pale blue, spotted / Nepal	(463) /	25
pannonica / P / 2' / fls purple, reddish-black spots / c Europe	148 /	288
pannonica 'Alba' - white fls		*
paradoxa / R / / fls solitary bells / Caucasus		64
parryi / differing from G. affinis in succulent lvs / Wyo., Col.	227 /	25
phlogifolia - see G. cruciata ssp. phlogifolia		25
platypetala / R / 8-10" / brilliant blue fls / Alaska		168
pneumonanthe / R / 6-12" / dark blue fls, striped green / Eurasia	148 /	208
pneumonanthe 'Alba' - white form		168
pneumonanthe v. latifolia / 6" / deep blue fls		35
porphyrio - see G. autumnalis f. porphyrio		99
praecox - see GENTIANELLA praecox		10
procera - see GENTIANOPSIS procera	224 /	25
procumbens / A / tufted / stems 1-2" long / Nepal		81
prolata / R / to 6" / blue fls, purple striated / w Himalayas	(516) /	25
propinqua - see GENTIANELLA propinqua	229 /	25
prostrata / A or B / to 6" / fls blue / arctic Asia & North America	148 /	209
przewalskii / R / 8" / fls blue & white / nw China, ne Tibet		25
puberula - see G. puberulenta		25
puberulenta / P / to 2' / blue fls / c North America	229 /	25
pumila / R / 3" rosette / fls blue / s, w & e Alps	148 /	25
punctata / P / to 2' / fls yellow, maroon spotted / Alps	(401) /	148
purdomii - of cult. may be G. decumbens, gracilipes or dahurica		25
purpurea / P / to 2' / reddish purple, coppery fls / c Eu., in mts.	210 /	148
purpurea nana - small form		168
pyrenaica / R / 3" / violet or green fls / Pyrenees, Caucasus	148 /	25
quinquefolia - see GENTIANELLA quinquefolia	224 /	25
robusta / R / 12" / white or whitish fls / s Tibet		61
rochelii / R / 2-3" / blue fls / Hungary		25
rostani / R / 3" / light azure-blue fls / Cottian Alps, Pyrenees	148 /	96
rubricaulis / P / to 28" / fls pale violet or white / c & ne N. Am.		25
saponaria / P / 16-32" / purplish fls / nc United States	224 /	313
saponaria 'Alba' - white fls		*
saxatile - see G. scabra v. saxatilis		*
saxosa / HH R / 2-4" / fls white / New Zealand	(418) /	129
scabra / R / 1'+ / uniformly dark blue fls / e Siberia, Korea	70 /	10
scabra v. buergeri - the Japanese form., later, handsomer	164 /	96
scabra v. orientalis - see G. scabra v. buergeri		200
scabra 'Kirishima-Rindo' - see G. scabra saxitalis		168
scabra v. saxatilis - procumbent, fls deep blue	(100) /	168
scabra f. stenophylla - narrow leaved		200
sceptrum / P / 1-4' / blue fls, marked green / Calif. to B.C.	35 /	228
schistocalyx - var. G. asclepiadea, calyx tube split halfway 1 side		61
septemfida / R / 6" / deep blue fls, dotted inside / Asia Minor	147 /	123
septemfida 'Doeringiana' - dark blue fls		35
septemfida freyniana - see G. freyniana		93
septemfida hascombensis - see G. X hascombensis		64
septemfida v. lagodechiana - see G. lagodechiana		25
septemfida v. latifolia - wider leaved		96
septemfida olivieri - see G. olivieri		28
septemfida procumbens - see G. grossheimii		10
serotina / B / 6" / white fls / South Island, New Zealand		15
setigera / R / 1' / blue fls / n California, coastal		25

sikkimensis / R / 6" / one inch blue trumpets / alps of Sikkim (517) / 96
sikkimensis alba / R / white form of G. sikkimensis / Sikkum 168
sikokiana / R / to 8" / fls blue, green spots / Japan 200
simplex / R / 2-8" / deep blue fls / s California to c Oregon 228
sino-ornata / R / mats / gentian-blue fls / se Tibet, w China 129
siphonantha / R / 1' / blue fls / Himalayas 194 / 168
speciosa - see CRAWFURDIA speciosa
squarrosa / B / 1-4" /corolla long, pale /Japan, Korea, India, China 200
X 'Stevenagensis' - prostrate / fls deep purple & yellowish (518) / 35
stragulata / R / to 3" / bright purplish-blue fls / Yunnan 184 / 67
stylophora / P / 3'+ / greenish-yellow fls / Asia (519) / 64
X 'Susan Jane' - trailing, ultramarine fls, white throated 35
tenuifolia / R / 12-16" / white fls / New Zealand 15
terglouensis / R / 3" / sky-blue fls / s & e Alps 148
thermalis - see GENTIANOPSIS thermalis 227 / 25
thunbergii / B / to 6" / blue fls / Japan, Korea, China, Manchuria 200
thunbergii minor / R / to 6" / pale blue fls, smaller / Japan 25
tianshanica / R / 8-16" / dark blue fls / c Asia 10
tibetica / P / 2' / greenish to yellowish fls / Tibet 284 / 168
townsonii / HH R / 1'+ / white fls, lobed corolla / New Zealand 15
triflora / P / 2'+ / dark blue fls / e Siberia, Korea 164 / 10
triflora alba - white form
triflora v. horomuiensis - with narrow lvs 200
triflora v. japonica - from Japan, Kuriles & Sakhaline 200
triflora v. montana - alpine form, dwarf 200
triflora v. montana 'Alba' - white form 200
trinervis - see TRIPTEROSPERMUM japonicum 200
tubiflora / R / 1' / blue fls / alps of Kashmir & Kumaon 96
undulatifolia - see G. acaulis 'Undulatifolia' *
utriculosa / A / 4-8" / intense blue fls / c Europe, Balkans 148 / 288
vietchiorum / R / 4" / fls deep royal blue / w China 132
verna / R / tufts / variable blue fls / Alps, Asia Minor, Pyrenees 148 / 129
verna alba - white form of G. verna
verna v. angulosa - synonym of the sp. 25
verna ssp. pontica / corolla lobes obtuse / ce Balkan Pen. 288
verna ssp. tergestina/corolla lobes acute/Balkan Pen, Bulg., Italy 288
villosa / P / 4-24" / greenish-white to purplish fls / ec U.S. 224 / 168
waltonii / R / to 16" / deep blue or purplish fls / Tibet (135) / 25
walujewii/ P / 12-16" / pale yellow fls, blue lined / c Asia 10
wilsonii / R / 4" / sea-blue fls / w China 64
wutaiensis / R / to 8" / whitish fls, dotted pale blue / China 25
yakushimensis / P / 3-12" / few stems / solitary blue-purple / Japan 200

GENTIANELLA - Gentianaceae
amarella / A / to 1' / fls pale lilac / North America, Eurasia 148 / 25
amarella ssp. acuta/A or B/3-15"/fls lilac/Lab., Alask., Nfld., etc. 99
aurea / A or B / 2-6" / fls pale yellow, rare blue / Arctic Eur. 288
austriaca / B / 6"+ / purplish or whitish fls / ec Europe 288
campestris / A or B / to 14" /bright purple to white fls /n & c Eu. 148 / 25
campestris ssp. baltica / A / blue-lilac to wht fls / Sweden, France 288
ciliata / B / to 1' / blue fls / Europe, except n 288
dentosa /A or B / 2-8" / dark blue fls / arctic & subarctic 288
diemensis / HH R / 2-18" / purple-veined white fls / New South Wales (123)
germanica / B / to 16" / fls violet, pink, whitish / w & c Europe 148 / 288
moorcroftiana / A / to 1' / fls pale blue / Himalayas 25

praecox / B / 6"+ / fls dark blue to white / c Europe 10
propinqua / A / to 14" / fls pale lilac or violet / nw N. Am. 25
quinquefolia / A or B / to 2' / fls pale lilac / e North America 25
ramosa / B / 1-5" / fls pale violet or white / csw Alps 288

GENTIANOPSIS - Gentianaceae
crinita / A / 3' / bright blue fls / e North America 25
holopetala / A / to 1' / blue fls / California, in mts. 25
procera / B / 3' / corolla fringe only on sides / ne & nc N. Am. 25
thermalis / A / to 1'/deep blue fls, streaked lighter/ Col. to Ariz. 25

GERANIUM - Geraniaceae
aconitifolium - see G. sylvaticum ssp. rivulare 160
'Alpenglow' - see G. sanguineum 'Alpenglow' 16
anemonifolium - see G. palmatum 25
argenteum / R / 3" / fls pink, darker veins / n & c Italy 287
armenum - see G. psilostemon 25
asphodeloides / P / 2'+ / pinkish-lilac fls / s Europe 77 / 287
X 'Ballerina' / R / 6" / pink fls veined darker 129 / 24
bohemicum / A or B / 10-30" / br-bl-violet / ec Europe 287
canariense - see G. palmatum
candidum / R / decumbent / white fls, rosy center / China 184
cataractarum / R / to 16" / fls bright pinkish-purple / s Spain 287
chinense / P / 1½' / fls blue to rosy-purple / China (136)
cinereum / R / 3-6" / fls in varying pink shades / Pyrenees 148 / 129
cinereum 'Album' - white fls of purity & brilliance 96
cinereum ssp. subcaulescens / deep red-purple fls / Balkans 129 / 287
cinereum ssp. subcaulescens 'Splendens' - Tyrian-rose fls 300
'Claridge Druce' / vigorous plant / lilac-pink fls (443) / 49
collinum / P / to 1½' / fls purple-violet w dark veins 25
columbinum / A / to 2' / purplish-pink fls / Europe, escaped in US 287
dahuricum / R / 3" / lobed rosy-mauve fls / China 200
dalmaticum / R / 6-8" / clear pink fls / Dalmatia 107 / 123
dalmaticum album / A / low / fls white / Damatia 268
delavayi - see G. sinense 279
dissectum / A / to 2' / purplish-pink fls / most of Europe 287
Donianum / P / from rhizome / fls purple / Himalayas 25
endressii / P / 1½' / clear pink fls / sw France 182 / 259
endressii 'Wargrave Pink' / 2' / bright salmon-pink fls 300 / 279
erianthum / R / 1'+ / bluish-purple fls / n hemisphere 165 / 200
eriostemon / P / 2' / violet-blue fls / Far East 165 / 25
eriostemon v. reinii f. ononei / 6" / purple fls / Japan 200
farreri / R / 6" / fls pale rose to lilac / s China 184 / 25
fremontii / P / 1-2' / rose-purple fls / Rocky Mts. 229 / 107
gracile / P / 16-28" / fls pink, deep red veins /n Iran, w Caucasia 77
grandiflorum - see G. himalayense (464) / 25
grandiflorum v. alpinum - see G. Regelii 25
grevillianum - see G. lambertii 25
himalayense / R / 1'+ / lilac fls, purple veins / India, Tibet 25
ibericum / P / 1-1½' / fls deep purple / Caucasus 269 / 287
incanum / Gh / 6" / fls mauve or purple / South Africa (138)
lambertii / R / prostrate / rose fls / Himalayas 279 / 25
lancastriense - see G. sanguineum v. prostratum 25
macrorrhizum / P / 1-1½' / fls bright red / s Europe 148 / 129
macrorrhizum album - white form of G. macrorrhizum

maculatum / P / 1½' / fls pale rose / North America 224 / 107
maculatum f. albiflorum - fls white 129 / 25
maculatum 'Album' - see G. maculatum f. albiflorum 99
manescavii - see ERODIUM manescavii *
microphyllum / HH R / 1' to prostrate / white fls / New Zealand 15
montanum / R / to 1' / violet fls, purple veins / Caucasus (402)
napaligerum - see G. farreri 25
napuligerum / R / 6" / pale rose fls / Yunnan 123
nepalense / P / to 18" / rose-purple to white fls / Asia in mts. 25
nepalense 'Album' - white form, red veins *
nepalense v. thunbergii - see G. thunbergii 200
nodosum / P / 18" / fls lilac, paler center / Europe 148 / 279
nodosum 'Album' - white fls *
palmatum / Gh / 1½' / pale purple fls / Canary Isls. 25
parryi / P / 2' / pinkish-purple fls / Wyo., Col., Ariz. 229 / 25
pelopenensiacum / P / ? / fls bluish-violet / Albania, s Greece 287
phaeum / P / 18-28" / fls blackish or brownish-purple / c Europe 148 / 287
phaeum v. lividum - fls slate-colored 279
platypetalum / P / 6-28" / blue fls / Caucasia, n & nw Iran 77
polyanthes / P / 2'+ / "more ordinary Cranesbill" / Himalayas 64
pratense / P / 12-32" / fls bright violet-blue / c Europe 148 / 287
pratense 'Album' / 2' / pure white fls / n Europe 129
pratense striatum - see G. striatum 61
pratense v. transbaicalicum / 18" / bluish-rose recurrent bloom *
procurrens / R / prostrate / rose-purple fls / Himalayas 154
psilostemon / P / 2' / fls dark red, black base / Armenia 251 / 25
pylzowianum / R / 3' / large clear pink fls / Kansu 123
pyrenaicum / P / 10-28" / fls purple to lilac / s & w Europe 148 / 287
oreganum / P / 16-32" / fls sm., red-purple / Ore., Wash., n Calif. 228
reflexum / P / 16-28" / fls dull lilac or purple / Balkans 287
Regelii / R / to 8" / blue fls / Turkestan to Pakistan 25
renardii / R / 9" / pale lavender fls, veins crimson / Caucasus (141) / 46
richardsonii / P / 1-3' / fls white or pink, red veins / w N. Am. 229 / 228
richardsonii 'Album' - selection of the white form *
Robertianum / A or B / to 20" / fls bright pink / Eu., N. Am. 148 / 287
Robertianum f. albiflorum/A or B/to 1½'/fls white / N.A., Eurasia 25 / 99
Robertianum 'Album' - white fls 25
Robertianum ssp. celticum - see G. Robertianum 'Album' 25
sanguineum / P / 18" / fls intense magenta / Europe 148 / 129
sanguineum 'Album' / 9" / white fls 129
sanguineum 'Alpenglow' - prostrate, dark red fls 16
sanguineum v. lancastriense - see G. sanguineum v. prostratum 93
sanguineum 'Nanum' - low compact habit 25
sanguineum v. prostratum / 6" / fls a deeper rose 300 / 93
[1] sanguineum 'Walney Form' / 6-9" / fls a deeper rose
sessiliflorum / R / 4" / white fls / Andes Mts. 25
sessiliflorum 'Nigrum' / bronzy-leaved form of G. glabrum / N. Z. 15
sibiricum / P / to 3' / fls lilac to pale rose / e Europe to Korea 25
sinense / P / 2' / maroon fls / w China 279
soboliferum / P / 24-32" / fls rose-purple / Japan, Korea, China 200
stapfianum / R / 4" / fls wine-purple, reddish lvs / ? (142)
stapfianum 'Roseum' - marbled lvs, crimson-purple fls (143)

---

[1] Carl R. Worth in correspondence. 8/7/1974.

striatum / P / to 18" / petals white, veined red / c Europe 25
subcaulescens - see G. cinereum ssp. subcaulescens 287
subcaulescens 'Splendens' - see G. cinereum ssp. subcaulescens 'S' 46
sylvaticum / P / to 2' / fls variable in color / most of Europe 148 / 287
sylvaticum 'Album' - white fls 279
sylvaticum ssp. rivulare / 1' / fls white, red-veined / Alps 287
thunbergii / P / to 20" / pale rose fls, deeper nerves / Japan 165 / 200
thunbergii 'Album' - white fls *
transbaicalicum - see G. pratense v. transbaicalicum *
traversii / HH P / 18" / rose or white fls / Chatham Isls, N.Z. 193 / 25
traversii 'Elegans' - mat, fls pink 61
tuberosum / HH R / 15" / rose-purple to violet fls / s Europe 149 / 25
tuberosum 'Charlesii' / 9" / fls rose 1-1½" across / Aftghanistan 61
versicolor / P / 12-32" / white or pale lilac fls / Balkans 287
viscosissimum/P/2'/sticky lvs, pinkish-purple fls/S. Dak. to Calif. 228 / 25
wallichianum / R / 1' / fls violet-blue / Himalyas 306 / 129
wallichianum 'Buxton's Blue' - fls a deeper shade 279 / 129
'Wargrave Pink' - see G. endressii 'Wargrave Pink' 279
wilfordii / P / 1-2' / pale pink fls / Manchuria, Korea, Hupeh (402)
wlassovianum / P / 2' / dark violet fls / Siberia, Manchuria 279
yesoense / P / 12-32" / fls rose-purp, wht base / Japan 200
yesoense v. nipponicum / P/ 12-32"/ rose-purple fls / Honshu in mts. 273 / 200

GERARDIA - Scrophulariaceae
flava / P / 3-6' / fls yellowish / Maine to Georgia 263
grandiflora / P / 3'+ / yellow fls / Wisconsin to Texas 259
pendicularia / A / 3' / fls yellow, purple tinged / Me. to W. Va. 263
tenuifolia / P / to 3' / roseate fls / e & c U.S. 99
virginica / P / to 4' / yellow fls / New Hampshire to Louisiana 263

GERBERA - Compositae
anandria - see LEIBNITZIA anandria 25
anandria nana - see LEIBNITZIA anandria nana 25
jamesonii / HH P / 8" / fls orange-flame / Transvaal 28
kunzeana - see LEIBNITZIA kunzeana 25
nana - see CREMANTHODIUM nanum 64
nivea / R / 6" / deep rose-purple fls / Asia 64

GEUM - Rosaceae
aleppicum / P / to 4½' / orange or deep yellow fls / Eurasia 165 / 99
aleppicum v. strictum - nearly or quite glabrous achenes 25
anemonoides - see G. pentapetalum 200
X borisii - (in hort. is) / R / 1' / fls orange-scarlet 49 / 129
bulgaricum / P / 12-20" / whitish to pale yellow fls / Balkans 287
calthifolium / R / 6" / yellow fls / nw North America 81
calthifolium v. nipponicum / to 1' / fls yellow / Japanese alps 200
campanulatum - see G. triflorum v. campanulatum 142
canescens - neater form of G. triflorum, locale not given 64
chiloense / P / 1-2' / fls scarlet / Chile 182 / 129
chiloense v. miniatum - lighter red fls 61
chiloense 'Mrs. Bradshaw' - see G. X 'Mrs. Bradshaw' 129 / 135
coccineum / P / 1-2' / red fls / Balkan Peninsula, in mts. 194 / 129
elatum / P / 2' / golden fls / Himalayas 96

1'Georgenburg' / 12" / deep yellow
glaciale / R / to 8" / pale yellow fls / Siberia, Alaska 209
* heldreichii - orange fls, referred to G. montanum 28
japonicum - see G. macrophyllum 287
X 'Lady Strathenden' - semi-double golden-yellow fls 184 / 135
leiospermum / HH R / 8" / fls white / New Zealand 15
macrophyllum / P / to 3' / bright yellow fls / N. America, Eurasia 99
magellanicum - see G. urbanum 15
montanum / R / 2-4" / fls golden yellow / c & s Europe, in mts. 148 / 129
X 'Mrs. Bradshaw' / 2½-3' / semi-double scarlet fls 300 / 135
parviflorum / HH R / to 1' / fls white / Chile, New Zealand 193 / 15
peckii / R / 6-16" / yellow fls / White Mts., New Hampshire 22 / 107
pentapetalum / R / 4-8" / fls white / Japan to Aleutians 165 / 200
pyrenaicum / R / 8" / golden yellow fls / Pyrenees 274
radiatum / P / to 2' / fls deep yellow / North Carolina, Tennessee 28
reptans / R / 6" / fls bright yellow / Alps, Carpathians 148 / 287
reptans glacialis - see G. glaciale 209
X 'Red Wings' / 2' / fls brilliant bright scarlet 268
X rhaeticum / 4-5" / golden yellow fls / Engadine 123
rhodopeum / P / 1-2' / yellow fls / s Bulgaria 287
rivale / R / 1' / reddish fls, nodding / Europe, North America 148 / 129
rivale 'Album' - creamy-white selection *
rivale 'Leonard' / 1½' / salmon or orange tinted fls 129
rossii / R / 1' / bright yellow fls / Asia to Canada, e arctic 312 / 209
'Sibiricum' / 10" / coppery-red fls / not determined botanically 25
triflorum / P / to 18" / fls purplish to straw / e & c N. Am. 224 / 25
triflorum v. campanulatum / neater form / Oregon in mts. 142
triflorum v. ciliatum / lvs more dissected / western 25
uniflorum / HH R / 3" / white fls / New Zealand 238 / 96
urbanum / P / to 2' / yellow fls / Europe, w Asia, n Africa 314 / 25
'Waights's Hybrid' / compact, / 6" / orange-red fls 16

GEVUINA - Proteaceae
avellana / HH T / 15' / white fls; red frs / s Chile 36 / 139

GILIA - Polemoniaceae
achilleaefolia / A / 4-24" / blue-violet, white fls / s & c Calif. 228
aggregata - see IPOMOPSIS aggregata 25
capitata / A / 8-32" / white to blue-violet fls / Calif. to B.C. 229 / 228
longiflora - see IPOMOPSIS lingiflora 25
rigidula / R / 3-10" / fls blue or purple / Col., Kansas, Oklahoma 229
rubra - see IPOMOPSIS rubra 25

GILLENIA - Rosaceae
trifoliata / 2-4' / fls white or pinkish / e North America 224 / 25

GINKGO - Ginkgoaceae
biloba / T / 100'+ / fan-shaped lvs / China 127 / 36

GLADIOLUS - Iridaceae
alatus / HH C / 10" / brick-red fls / Cape Province (465) / 111
* atroviolaceus / C / 2' / violet-purple fls / n Turkey 111

---

[1] International Growers Exchange, Inc. Catalog, Fall 1985.

aurantiacus / Gh C / 3' / bright orange-yellow fls / S. Africa 61
blandus - see G. carneus 25
blandus carneus - see G. carneus *
brevifolius / HH C / 2' / pink fls / South Africa 223 / 25
byzantinus / C / to 2½' / fls bright magenta-pink / e Medit. reg. 210 / 129
callianthus / HH C / 3' / white fls, dark blotches / Ethiopia 186 / 279
1 cardinalis / HH Bb / to 2' / bright crimson /Cape S. Africa Prov.
carinatus / Gh C / 2' / blue, mauve, yellowish fls / S. Africa 25
carmineus / HH C / 1' / fls deep rose / Cape of Good Hope 223
carneus / HH C / to 2' / fls pale pink or cream / South Africa 25
caucasicus / HH Bb / 1'+ / pink / Caucasus 296
X citrinus - yellow fls 61
communis / C / 16-32" / rosy-purple fls / Spain to Greece 149 / 211
cooperi / HH C / 3' / fls red striped yellow / Natal 61
crassifolius / HH C / 3' / bright red fls / South Africa 25
1 dalenii / HH Bb / 4" / fls red-orange / S. Africa
debilis / HH C / to 2' / ivory-white fls, red marked / S. Africa 223
floribundus / HH C / 1' / citron-yellow fls / Cape of Good Hope 206
Garnieri / large, bright red, hooded fls / c Madagascar 121
gracilis / HH C / 2' / pink-cream fls / Cape Province 111
grandis - see G. liliaceus 25
illyricus / C / 3' / fls crimson-purple / England to Balkans 212 / 25
imbricatus / C / to 28" / fls purplish-violet / c Eu., Caucasus 186 / 4
involutus / HH Bb / to 1½' Cape of Good Hope 206
italicus / C / 20-40" / purp.-red to light pink / s Europe, w France 25 / 290
kotschyanus / C / 1-2' / fls light violet / Afghanistan, Iran 61
liliaceus / HH C / 2' / cream to red-brown fls / South Africa 25
macowianus / HH C / 2' / cream fls / Cape Province 223 / 111
X nanus - yellow G. tristis x red G. cardinalis, tender 267
natalensis / HH C / 3-4' / dark crimson fls / South Africa 25
orchidiflorus / HH C / 18" / green-brown fls / Natal 184 / 111
palustris / C / 1-2' / purple-red fls / c Europe 29 / 93
papilio / HH C / 2' / cream fls / Cape Province 111
primulinus / 1½' / seg. yel., marg viol-purp, red spot / s Africa 61
psittacinus - see G. natalensis 25
punctatus - see G. recurvus 61
punctulatus / HH C / to 2½' / fls pink to mauve / South Africa 25
recurvus / HH C / to 2' / lilac & yellow fls / South Africa 25
1 scullyi / HH Bb / 12" / yell-purple tipped, scented / S. Africa
segetum / C / to 2' / fls bright purple / Medit. reg. to Turkestan 212 / 25
sericeo-villosus / HH C / to 2' / yellow fls / Cape of Good Hope 206
tenellus/HH Bb/1½'/yel or gr.-wht, tinged lilac, blk throat/S. Afr. 25
triphyllus / C / small / fls purple & cream-white / Cyprus 121
tristis / HH C / 18" / cream fls flushed purple / Natal (145) / 279
tristis v. concolor - fls pale yellow to white 186 / 25
undulatus / HH C / 3' / white or pale pink fls / South Africa 25
viperatus - see G. orchidiflorus 267
vittatus / HH C / 1' / white fls / South Africa 28

GLAUCIDIUM - Ranunculaceae
fimbrilligerum - error for CLAUCIUM fimbrilligerum *
flavum - see GLAUCIUM flavum *

---

1 Sima Eliovson Wild Flowers of Southern Africa, 1980

palmatum / R / 12"+ / pale lilac fls / Japan 132 / 129
palmatum 'Album' - see G. Palmatum 'Leucanthum" 279
palmatum 'Leucanthum' - fls white 279

GLAUCIUM - Papaveraceae
corniculatum / A / 12-16" / fls orange or reddish / s Europe 211 / 286
elegans / A / to 1' / fls yellow to red / c Asia, Iran 5
fimbrilligerum / B / 1-2' / fls yellow / c Asia 5
flavum / P gr. as A. /1-3' / yellow or orange fls / Eu., n Africa 194 / 129
flavum v. fulvum - found in wasteland in California 228
grandiflorum / P / 12-20" / fls dark orange to crimson/Greece, Iran 77
phoeniceum - see G. corniculatum 25
squamigera / A / to 1½' / fls yellow or orange / Russia to c Asia 25

GLEDITSIA - Leguminosae
japonica / T / 35' / lvs with up to 30 leaflets / Japan 139

GLEHNIA - Umbelliferae
leiocarpa / R / 4" / white fls / Calif. to Alaska on sandy coasts 228

GLEICHENIA - Gleicheniaceae
circinata / Gh F / 1'+ / coriaceous pinnules / New Zealand 93 / 15
circinata v. alpina / mt. form / Tasmania, New Zealand 15
dicarpa v. alpina - see G. circinata v. alpina 15

GLOBULARIA - Globulariaceae
alypum / Sh / to 3'+ / fls usually blue / Medit. 288
aphyllanthes - see G. punctata 288
arabica / HH Sh / dwarf / blue fls / w Medit. coast 270
bellidifolia - see G. meridionalis 288
cambessedesii / R / 1' / blue fls / ne Spain, in mts. 212
cordata 'Bellidifolia' - see G. meridionalis 288
cordifolia / Sh / to 4" / gray-blue fls / c & s Europe 148 / 288
cordifolia ssp. nana / 1" / lilac-blue fls / Pyrenees, Spain 24
cordifolia rosea / mat forming / fls rose / Europe, w Asia 61
elongata - see G. punctata 288
X fuxeensis (G. nudicalus X G. repens) / P /to 1'/fls blue / s Eur. 288
incanescens / R / to 4" / violet-blue fls / Italy 148 / 123
meridionalis / Sh / to 6" / gray-blue fls / se Alps 314 / 288
nana - see G. repens 25
nudicaulis / R / 6" / large medium-blue fls / s Europe 148 / 123
nudicaulis 'Alba' - white fls *
nuficaulis 'Nana' - dwarf form of G. nudicaulis 288
[1] orientalis / Sh / to 10" / lvs in basal rosettes /
punctata / R / to 1' / fls gray-blue / s & c Europe 314 / 288
punctata 'Alba' - white fls *
pygmaea - see G. meridionalis *
repens / R / 3" / gray-blue fls / sw Europe, in mts. 212 / 129
repens v. pygmaea - see G. meridionalis
spinosa / R / 8" / clear blue fls / se Spain (61) / 212
stygia / Sh / dwarf / ice-blue fls / Greece 134 / 288
tricosantha / R / 6-10" / steel-blue fls / e Balkans 50 / 107

---

[1] SCHWARZ in Quar. Bull. Alpine Garden Soc. 35:4.

vulgaris / R / to 1' / blue fls / s Sweden, s Europe 210 / 25
vulgaris 'Alba' - white fls *
wilkommii - see G. punctata 288
wilkommii 'Alba' - see G. punctata 'Alba' *

GLORIOSA - Liliaceae
rothschildiana / Gh Cl / tall / crimson fls / tropical Africa 28
simplex / Gh Cl / 4½' / yellow fls / tropical Africa 194 / 93
superba / Gh Cl / 5'+ / fls yellow aging red / Africa, Asia (327) / 25
virescens - see G. simplex 93

GLYCYRRHIZA - Leguminosae
lepidota v. glutinosa / P / to 4' / Licorice-root / w N. Am. 142

GLYPHOSPERMA - Liliaceae
Palmeri / P / 12-18" / fls white, starry / n Mexico 61

GNAPHALIUM - Compositae
dioicum - see ANTENNARIA dioica 160
hoppeanum /P/1-6"/fls reddish purple/c Europe, Italy, Balkan Pen. 289
japonicum / P / 3-10" / bracts red-brown / Korea, China, Japan 200
norvegicum / R / 4-8" / fls light brownish / Europe 93
obtusifolium / B / to 4½' / whitish fls / e North America 224 / 99
supinum / R / to 4" / brownish fls / Europe & e U.S. in mts 93
trinerve / HH Sh / few inches / pure white fls / New Zealand 184 / 61

GODETIA - Onagraceae
amoena - see CLARKIA amoena 25
quadrivulnera / A / ? / lilac or pale crimson fls / Oregon 28

GOMPHOCARPUS - Asclepiadacae
fruticosus / HH Sh / 3-6' / fls white / S. Africa 288

GOMPHOLOBIUM - Leguminosae
huegelii / HH Sh / 3' / yellow pea fls / New South Wales, Australia 45
tomentosum / Gh / 2½' / yellow fls / Australia 206

GONIOLIMON - Plumbaginaceae
caucasicum - densely pubescent variant of G. tataricum 10
speciosum / P / 4-20" / reddish purple fls / e Russia (466) / 288
tataricum / P / 18" / rose-pink fls / Eurasia 25

GOODIA - Leguminosae
lotifolia / Gh Sh / to 12' / fls yellow, purplish at base / Austrl. 109 / 45

GOODYERA - Orchidaceae
oblongifolia / P / to 18" / fls white, tinged green / n N. Am. 229 / 66
pubescens / P / to 16" / white fls, white-netted lvs/ ne N. Am. 224 / 25
repens / R / 4-8" / white fls; lvs reddish-veined / Europe 66
tesselata / R / to 1' / white fls / n United States & Canada 28

GORDONIA - Theaceae
lasianthus / HH T / to 60" / white fls / coastal N. C. to Fla. 28

GREENOVIA - Crassulaceae
aizoon / Gh / rosette / yellow fls / Canary Isls. 156

GREIGIA - Bromeliaceae
sphacelata / Gh / 3' / rose fls / Chile 28

GREVILLEA - Proteaceae
alpina / Sh / 2-4' / bushy, fls few, red base, yel upward/ Australia 61
lavandulacea / HH Sh / 5' / red fls / Swan River area, Australia 206
rosmarinifolia / HH Sh / 6-7' / fls deep rosy-red / New South Wales (418) / 36
sulphurea / HH Sh / to 6' / pale yellow fls / New South Wales 129 / 36
wickhamii / Gh Sh / to 6' / red fls / w Australia 45

GRINDELIA - Compositae
chiloensis / HH Sh / 3-5' / yellow fls; evergreen lvs / Argentina 171 / 139
hirsutula / HH P / 1-3' / yellow fls / coast ranges of c California 228
integrifolia / P / 6-30" / bracts hairy and narrow / Wash., Ore. 25 / 228
squarrosa / P / 8-40" / yellow fls / Washington to South Dakota 224 / 229
stricta ssp. venulosa /HH R/3'/broad fleshy lvs/coastal Ore., Calif. 25

GROSSHEIMIA
macrocephala - see CENTAUREA macrocephala 77

GUICHENOTIA - Sterculiaceae
macrantha / Gh Sh / ? / purplish fls / tropical Australia 28

GUNDELIA - Compositae
tournefortii / HH P / 15" / fls yellow-green / Asia Minor 270 / 206

GUNNERA - Halorgidaceae
chilensis / HH P / to 6' / red frs / Patagonia 182 / 25
flavida / HH R / 4" / deciduous ground cover / New Zealand 15
magellanica / R / 5" / stoloniferous / Patagonia 194 / 25
monoica / P / rosettes crowded; drupes white / New Zealand 15
prorepens / HH R / creeping / brownish lvs / New Zealand 238 / 25

GUTIERRHIZA - Compositae
sarothrae / P / 2' / yellow fls / w North America 229 / 25

GYMNADENIA - Orchidaceae
conopsea / P / 8-20" / bright pink fls / Eurasia 148 / 85
fragrans - see HABENARIA dilitata *
odoratissima / P / to 18" / pale pink fls / c Europe 85

GYMNASTER - Compositae
savatieri / P / 8-20" / fls blue or white / Japan, in mts. 164 / 200

GYMNOCARPIUM - Polypodiaceae
dryopteris / F / 9" / Oak Fern / Great Britain 159
Robertianum / F / lvs to 8" long, scented / n N. America 25

GYMNOCLADUS - Leguminose
dioica / T / 100' / greenish-white fls / New York to Oklahoma 28

GYNANDIRIS - Iridaceae
setifolia / Gh C / 6" / pale blue to lilac & yellow fls / Cape Prov. 185
sisyrinchium / C / 4-8" / bright blue or mauve fls / Medit. reg. 211 / 129

GYNURA - Compositae
sarmentosa / Gh / 18', twining / orange fls / India 129

GYPSOPHILA - Caryophyllaceae
aretioides / R / cushion / white fls; gray-green lvs / Iran 132
bicolor / P / to 5' / fls white to pale pink / Iran, Turkey 77
1 bungeana - see G. sericea
cerastioides / R / 2-3" / fls white, veined purple / Himalayas (133) / 123
'Dorothy Teacher' - see G. repens 'Dorothy Teacher' *
dubia - see G. repens 25
fastigiata / P / to 3' / fls white or pale purplish / ec Europe 286
'Franzii' - name of no Botanical standing
gmelini / P / to 18" / clustered white fls / Europe 64
libanotica / R / 4-16" / fls white to pink / e Medit. region 77
microphylla / R / mats / pink fls / alps of Alatau 96
muralis / A / to 8" / pink to white fls / Eurasia 25
nana / R / to 6" / pale purplish fls / s Greece, in mts. 286
paniculata 'Rosy Veil' - see G. X 'Rosy Veil' 135
papillosa / P / to 3' / white, pale purplish fls / n Italy 286
petraea / R / caespitose / fls white or pale purplish / Carpathians 286
repens / R / 3-6" / fls lilac-pink to white / Alps 148 / 129
repens 'Alba' - white fls 25
repens 'Dorothy Teacher' / 3" / pink fls 48
repens 'Fratensis' - blue-gray lvs; pink fls 46
repens 'Rosea' - showy pink form 147 / 107
rosea - see G. viscosa 25
X 'Rosy Veil' / 18" / doubled soft pink to white fls 135
sericea / R / to 1' / white & lilac fls / Siberia, Mongolia 25
silenoides / R / 1' / fls white or pinkish, purple veins / Caucasus 77
tenuicaulis - error for G. tenuifolia *
tenuifolia / R / to 8" / fls white to pink / Caucasus 107
viscosa / A / 4-22" / white to pale pink / Near East 77

HAASTIA - Compositae
pulvinaris / Sh / dense mound / minute fls / New Zealand 238 / 194
pulvinaris v. minor - smaller in all parts 15
recurva / Sh / loose flops / tawny-hair coated / New Zealand 64
sinclairii / Sh / fewer branches / shining hairs / New Zealand 64
sinclairii v. fulvida - lvs with buff tomentum 15

HABENARIA - Orchidaceae
bifolia - see PLATANTHERA bifolia 85
blephariglottis / P / to 3' / fls bright white to creamy / N. Am. 224 / 99
chlorantha - see PLATANTHERA chlorantha 85
ciliaris / P / to 3½' / fls bright yellow to deep orange / e U.S. 224 / 25
conopsea / P / to 16" / fls rose or purple-violet / Eurasia, Japan 25

---

[1] V. L. Komarov, ed., Flora of U.S.S.R., Vol. VI.

dilitata / P / to 4' / fls white / n North America in bogs — 22 / 25
dilitata v. leucostachys / long-spurred / w United States & Canada — 66
fimbriata / P / to 3' / pink-purple fls / ne & c North America — 224
hookeri / P / to 16" / fls yellowish-green / ne & c North America — 66
hyperborea / P / to 3' / small fls, greenish / n North America — 224 / 25
leucostachys - see H. dilitata v. leucostachys — 66
psycodes / P / to 3' / purple, lilac, white fls / Nfld. to Tenn. — 224 / 25
psycodes v. grandiflora/P/3'/fls purp., lilac, wht. 2X H. p./e N.A. — 25
radiata - see PECTEILIS radiata — 25
viridis - see COELOGLOSSUM viride — 224 / 85

HABERLEA - Gesneriaceae
ferdinandi-coburgii / R / 4-6" / pale lilac fls / Balkans — 132 / 123
rhodopensis / R / 6-8" / fls pale lilac / Greece — 149 / 129
rhodopensis 'Virginalis' - white fls, less vigorous — (446) / 129

HABRANTHUS - Amaryllidaceae
advenus - see HIPPEASTRUM advenum — 267
andersonii - see H. tubispathus — 14
andersonii 'Aureaus' - see H. tubispathus 'Aureus' — (148)
andersonii 'Cupreus' - see H. tubispathus 'Cupreus' — (148)
andersonii 'Roseus' - see H. tubispathus f. roseus — (148)
andersonii texanus - see H. texanus — 25
bifidus - see RHODOPHIALA bifida — (139)
brachyandrus / HH Bb / to 1' / fls pale rose-pink / South America — 267
citrinus - see ZEPHYRANTHES citrina — *
cupreus - see H. tubispathus 'Cupreus' — (148)
gracilfoius / B / to 1½' / fls tube green, lobes purple/Uruguay — 25
robustus / HH Bb / 2' / lime-green, whitish & pink fls / S. Am. — (149) / (148)
texanus / HH Bb / 8" / fls yellow, copper outside / Texas — 25
tubispathus / HH Bb /to 6"/sulphur-yellow fls, purplish out./S. Am. — 14
tubispathus 'Aureus' - Herbert's golden variety — (148)
tubispathus 'Cupreus' - Herbert's coppery variety — (148)
tubispathus f. roseus - rose-colored fls — (148)

HACKELIA - Boraginaceae
brachytuba / P / ? / fls pale to dark blue / Sikkim to Yunnan — (150)
diffusa / P / fls white, larger lvs / Wash. & B.C. — 142
jessicae / P / 12-40" / blue fls / Cascades to Canada — 226
mundula / P / 1½-3' / sticky seeds; fls white / California, Oregon — 228

HACQUETIA - Umbelliferae
dondia - see H. epipactis (Dondia, a generic synonym) — 81
epipactis / R / 3-4", ground cover / yellow fls / Europe — 148 / 129

HAEMANTHUS - Amaryllidaceae
albiflos / Gh / 1' / white fls / South Africa — 194 / 25
katherinae / Gh / 1' / bright red fls / South Africa — 128 / 25
natalensis / Gh / 1' / fls flesh or pale green / South Africa — 25

HAKEA - Proteaceae
laurina / HH T / to 30' / crimson fls, yellow styles / w Austrl. — (327) / 25

HALACSYA - Boraginaceae
sendtneri / R / to 10" / fls yellow / Albania, Jugoslavia — 288

HALENIA - Gentianaceae
deflexa / A or B / 1-9' / greenish or bronze fls / North America 224 / 99

HALESIA - Styracaceae
carolina / T / 30'+ / white bells / se United States (135) / 36
monticola / T / 50'+ / white fls / North Carolina to Georgia 36
monticola v. vestita - young lvs white-woolly 36

HALIMIUM - Cistaceae
alyssoides / HH Sh / 2' / bright yellow fls / sw Europe 212 / 25
atriplicifolium / HH Sh / 3'+ / fls yellow / sw Europe 212 / 287
halimifolium /Sh /3-4' / lvs gray-gr.; fls yellow / S. Eur., n Africa 25
lasianthum ssp. formosum - 3' / spotted yellow fls / s Portugal 287
ocymoides / HH Sh / 3' / yellow fls / Spain, Portugal 212 / 25
striplicifolium / HH Sh / 3'+ / fls yellow / sw Europe 287
umbellatum / HH Sh / 10" / fls white / sw Europe 287

X HALIMOCISTUS - CISTUS X HALIMIUM - Cistaceae
sahucii / HH Sh / 15" / fls white / s France 221

HAMAMELIS - Hamamelidaceae
'Hiltingbury' - see H. X intermedia 'Hiltingbury' 139
X intermedia 'Arnold Promise' / Sh / to 17' / bright yellow fls 300 / (370)
X intermedia 'Hiltingbury' / Sh / large, coppery fls 139
japonica / Sh / to 30' / early yellow fls / Japan 167 / 200
japonica 'Arborea' - golden yellow fls, larger lvs 28
japonica flavo-purpurascens/Sh or T/30'/fls yellow w red base/Japan 25
japonica mollis 'Brevipetala' /Sh or T /30'/ fls deep orange/w China 25
japonica 'Zuccariniana' / 15' / pale lemon-yellow fls 36
mollis / Sh / to 30' / golden yellow fls / c China 28
mollis 'Arnold Promise' - see H. X intermedia 'Arnold Promise' 36
vernalis / Sh / 10' / fls pale yellow to red / c United States 194 / 139
virginiana / Sh / to 15' / golden-yellow fls / e North America 208 / 25

HAMELIA - Rubiaceae
patens / T/ to 25'/gray lvs; scarlet or orange fls / Fla., W. Indies 25

HAPLOPAPPUS - Compositae
acaulis / R / 2-6" / yellow fls / nw North America 229 / 25
clementis / R / 4-16" / golden-yellow fls / Wyo., Col., Utah 229
coronopifolius - see H. glutinosus 25
croceus / R / 1'+ / orange-yellow fls / Wyoming, Colorado, Utah 229
glutinosus / HH R / to 1' / yellow fls / Chile, Argentina 184 / 25
lanuginosa andersonii / P / few, smooth lvs; fls yel / nw U.S. 142
linearifolius / P / 1-5' / yellow fls / inner coastal mts., Calif. 228
lyallii / R / to 4" / yellow fls / Olympic Mts., Washington 312 / 228
nuttallii - see MACHERANTHERA grindelioides 227 / 142
rydbergii / P / 8-18" / twigs brittle; fls yel / ne Utah 229
spinulosus / P / 1-2' / yellow fls / Mexico, Texas & n-ward 229 / 227
stenophyllus / P / 2-4" / fls yellow / Washington s to Calif., Nev. 229

HAPLOPHYLLUM - Rutaceae
boisseranum / R / to 10" / yellow fls; simple vls / Jugo., Alb. 287

patavinum / R / to 1' / yellow fls, middle lvs 3-sect. / w Balkans		287
HARDENBERGIA - Leguminosae		
comptoniana / Gh Cl / to 7' / purplish-blue fls / w Australia	187 /	128
monophylla f. rosea - see H. violacea f. rosea		93
violacea / Gh Cl / 6' / fls purple, pink, white / Australia	(327) /	194
violacea 'Rosea' - fls pink		25
HARRIMANELLA - Ericaceae		
stellerana - see CASSIOPE stellerana		25
stelleriana / Sh / mats / white fls / ne Asia, nw N. Am.		268
HARTIA		
sinensis - see STEWARTIA pteropetiolata		139
HAWORTHIA - Liliaceae		
tesselata / P / stemless/ lvs dk green, marked white / S. Africa		25
HEBE - Scrophulariaceae		
albicans / HH Sh / dwarf / dense rounded glaucous lvs / N. Zealand	110 /	139
allanii / with rather larger lvs / New Zealand		64
X andersonii / bedding plant / fls bluish-purple fading white	184 /	36
armstrongii / HH Sh / to 3' / golden-green whipcord type / N.Z.		15
'Autumn Glory' - see H. elliptica 'Autumn Glory'		241
X balfouriana / dwarf Sh / purple fls, lvs opaque-green, purple edge	90 /	15
bidwillii - see PARAHEBE bidwillii		*
brockii / HH Sh / 8-12" / white fls / New Zealand		15
buchananii / HH Sh / 12" / fls white / South Island, N.Z.		36
buchananii 'Minor' / 6-8" / white fls		36
buxifolia / HH Sh / to 5' / fls white / New Zealand		25
X 'Carl Teschner' - dwarf, violet & white fls	90 /	139
carnulosa - see H. pinguifolia		15
catarractae - see PARAHEBE catarractae		15
catarractae 'Rosea' - see PARAHEBE catarractae 'Rosea'		*
chathamica / HH Sh / prostrate / dark purple fls / coastal N.A.		15
cheesemanii / Sh / to 12"/ fls white sessile		15
ciliolata / HH Sh / to 1' / imbricated lvs / New Zealand, in mts.		15
colensoi / HH Sh / 1-2' / fls white / North Island, New Zealand		36
decumbens /HH Sh / 1-3' / white fls / South Island, New Zealand	238 /	36
diffusa - see PARAHEBE catarractae		15
diosmaefolia / HH Sh / 2-5' / lav.-blue, white fls / N. Is., N.Z.		36
divaricata / Sh / to 10' / lvs slightly hairy, fls white/ N.Z.		15
elliptica / HH Sh / to 6' / white or bluish fls / N.Z., Falklands		15
elliptica 'Autumn Glory' / 2' / purple spikes, late		241
epacridea / HH Sh / decumbent / white fls / New Zealand		15
X 'Franciscana' / Sh / ? / earliest horticultured form / N. Z.		15
fruticeti / Sh / 3'+ / fls white / New Zealand		15
gibbsii / HH Sh / 12" / white fls / New Zealand		134
glaucophylla / HH Sh / to 3' / white fls / New Zealand		15
haastii / HH Sh / prostrate / fls small, white/ New Zealand		132
hookeriana - see PARAHEBE hookeriana		*
hulkeana / HH Sh / to 5' / pale blue-lilac fls / New Zealand	193 /	129
X kirkii / HH Sh / tall / white fls / New Zealand		36
lavaudiana / HH Sh / 6" / fls rosy-pink / South Island, N.Z.		132

lyallii - SEE PARAHEBE lyallii 15
lycopodiodes / HH Sh / 1-2' / white fls / South Island, N. Zealand 36
macrantha / HH Sh / 1-2' / fls white / South Island, N.Z. 36 / 268
X 'McEwanii' / dwarf, erect / HH Sh / fls tinged blue 36
nivea / HH Sh/tall /white, lilac, pale blue fls / Tasmania, Austrl. 64
odora / Sh / varies to 5' / fls sessile in opposite pairs/ N. Z. 15
pageana - see H. pinguifolia 'Pagei' 36
parviflora / HH Sh / to 25' / fls in small dense racemes / N.Z. 15
perfoliata / HH Sh / 1' / blue fls / New South Wales, Austrl. 206
pimeleoides / HH Sh / to 1' / fls blue to purple / New Zealand 133 / 15
pimeleoides glauca (v. glaucocaerulea) - normal sp. variation 36
pimeleoides v. minor / 4" / fls dark blue / New Zealand 15
pimeleoides nana - small form of H. pimeleoides 25
pinguifolia / HH Sh / decumbent / fls white / New Zealnd 222
pinguifolia 'Pagei' - 1½' / white fls 48 / 129
pubescens / HH Sh / to 6' / fls white to lavender / New Zealand 15
rakaiensis / HH Sh / dwarf / fls white / New Zealand 139
raoulii / HH Sh / procumbent / fls lavender to white / N. Zealand 15
raoulii v. maccaskillii - from limestone, fls white 15
raoulii v. pentasepala / Sh / 8"/erect; wht, lav./Hodder River, N.Z. 15
recurva / Sh / to 3' / fls white / New Zealand 25
rigidula / Sh / 6-12" / fls white / New Zealand 15
salicifolia / HH Sh / to 12' / fls white, tinged lilac / N. Z. 238 / 25
speciosa / HH Sh / to 6' / reddish magenta fls / New Zealand 238 / 15
subalpina - of hort. see H. rakaiensis 139
treadwellii / HH Sh / dwarf / white fls / New Zealand 64
vernacosa / Sh / to 10" / lust. green lvs; fls white / New Zealand 222
venustula / HH Sh / to 5' / white fls / New Zealand 238 / 25
X youngii / HH Sh / low / purple fls 249

HEBENSTREITIA - Scrophulariaceae
dentata / Gh Sh / 1-2' / white fls, flecked yellowish / S. Africa 93
integrifolia / Gh Sh / 1' / white fls / Cape of Good Hope 206

HEDEOMA - Labiatae
pulegioides / A / low / American Pennyroyal / e & c N. Am. 224 / 99

HEDYCHIUM - Zingiberaceae
gardnerianum / Gh P / 6' / light yellow fls / n India 93 / 128

HEDYOTIS
caerulea - see HOUSTONIA caerulea 25
caerulea Faxonorum - see HOUSTONIA caerulea Faxonorum 99
longifolia - see HOUSTONIA longifolia 25
purpurea - see HOUSTONIA purpurea 61
tenuifolia - see HOUSTONIA tenuifolia

HEDYSARUM - Leguminosae
alpinum ssp. americanum / P / to 3' / pink or magenta fls / n N. Am. 99
americanum - see H. Alpinum ssp. americanum 99
boutignyanum / P / to 2' / fls cream or white / sw Alps 314 / 287
boreale mackenzii - see H. mackenzii 209
coronarium / P or B / deep red, fragrant fls / Europe 212 / 25
hedysaroides / P / 16" / red-violet fls / sc Europe, in mts 148 / 287
mackenzii / P / to 20" / violet-purple fls / arctic regions 137 / 209

obscurum - see H. hedysaroides 287
occidentale / P / to 2' / reddish purple fls / nw United States 229 / 25
semenowii / P / to 4' / fls yellow / fodder pl / c Asia 9
sibiricum 'Albiflorum' / P / to 4' / fls white / Siberia 61
spinosissimum / A / 1' / fls white to pinkish-purple / Medit. reg. 287
ussuriense / P / to 20" / fls pale yellow / Far East 9
vicioides / P / 10-32" / fls pale yellow / Japan, Korea, Siberia 200

HELENIUM - Compositae
autumnale / P / 4-6' / yellow fls / e Canada & United States 147 / 129
autumnale v. pumilum / 1-2' / free bloomer 28
Bigelovii / P / to 3' / fls yellow & brown / sw Oregon to s Calif. 25
Hoopesii / P / to 3½' / fls orange-yellow / Wyo., Ore., Calif. 229 / 25

HELIANTHELLA - Compositae
guinguenervis / P / 2-3' / fls br.-yellow / Rocky Mts. 25

HELIANTHEMUM - Cistaceae
alpestre - see H. oelandicum ssp. alpestre 287
'Amabile' / P / orange fls tinged pink / cultivar 25
apenninum / Sh / to 20" / fls white & yellow / s & w Europe 148 / 287
appeninum roseum / Sh / to 18" / fls reddish / Eur., Asia Minor 25
articum - see H. nummularium 287
'Ben Hope' - see H. nummularium 'Ben Hope' 139
broussonetii / HH Sh / tufted / orange fls / Canary Isls. 61
canadense / P / 8-16" / yellow fls / e & c North America 224 / 99
canum / R / to 8" / yellow fls / c & s Europe 148 / 287
cinereum / Sh / dwarf / yellow fls / Medit. region (523) / 287
croceum / Sh / to 12" / fls yellow & white / w Medit. 25
globulariaefolium - see TUBERARIA globularifolia 287
grandiflorum - see H. nummularium ssp. grandiflorum 287
guttatum - see TUBERARIA guttata 287
hirsutum - see H. nummularium ssp. obscurum 287
hirtum / Sh / to 1' / fls yellow or white / sw Europe 25
hymettium / Sh / to 4" / fls small yellow / Greece 287
italicum alpestre - see H. oelandicum ssp. alpestre 287
ledifolium / A / to 2' / yellow fls / s Europe 287
lippii / HH Sh / dwarf / yellow fls / n Africa, Asia Minor 270
lunulatum / R / to 8" / yellow fls, orange base / Maritime Alps 148 / 287
majus - see H. ledifolium 206 / 287
marifolium / Sh / dwarf / lvs grayish; fls yellow/ s Port., se Spain 287
nitidum - see H. nummularium ssp. glabrum 287
nummularium / Sh / to 20" / fls varii-colored / Europe 148 / 287
nummularium 'Ben Hope' - carmine fls; light gray lvs 129 / 139
nummularium 'Ben Nevis' / 6" / yellow fls, orange center 129
nummularium ssp. glabrum / fls yellow / c, s & sw Europe, in mts. 287
nummularium ssp. grandiflorum - c, s & sw Europe, in mts 314 / 287
nummularium 'Jubilee' - double yellow fls 46
nummularium 'Mutabile - fls changing lilac to white 25
nummularium ssp. obscurum / sepals & lvs pubescent / c Europe 287
nummularium ssp. pyrenaicum / Sh / dwarf / fls vari-colored/Pyrenees 287

nummularium 'Red Orient' - glowing deep red fls 268
nummularium 'Wisley Pink' - soft clear pink fls 48 / 268
oelandicum / R / to 8" / yellow fls / most of Europe, except n 148 / 287
oelandicum ssp. alpestre / 8" / yellow fls / c & s Europe 287
oelandicum ssp. alpestre 'Serpyllifolium' - smaller, prostrate to 2" 123
ovatum - see H. nummularium ssp. obscurum 287
pilosum / R / to 1' / white fls, yellow clawed / w Medit. region 287
praecox - see TUBERARIA praecox 61
pyrenaicum - see H. nummularium pyrenaicum 287
rubellum - see H. cinereum 287
salicifolium / A / 1' / yellow fls / s Europe 287
serpyllifolium - see H. nummularium ssp. glabrum 287
tuberaria - see TUBERARIA lignosa 287
umbellatum - see HALIMIUM umbellatum 287
violaceum - see H. pilosum 287
viscidulum / Sh / dwarf / fls simple or branched yellow / s Spain 287
vulgare - see H. nummularium 287
vulgare grandiflorum - see H. nummularium ssp. grandiflorum 287
'Wisley Pink' - see H. nummularium 'Wisley Pink' 268

HELIANTHUS - Compositae

angustifolius / P / to 6' / fls deep yellow / New York to Texas 226 / 25
annuus / A / 1-14' / Yellow Sunflower / Great Plains, U.S. 226
cordata - see H. mollis v. cordatus 28
debilis / A / to 6½' / fls red-purple / coast of Fla. 25
decapetalus / P / to 5' / fls yellow / Me. to S.C. coast 25
mollis / P / 2-5' / yellow to orange-yellow fls / c U.S. 259
mollis v. cordatus - lvs deeply cordate & clasping 99

HELICHRYSUM - Compositae

acuminatum / Sh / 9" / white fls / Australia 109 / 154
alveolatum - see H. splendidum 36
ambiguum / Sh / dwarf / 2-24" / fls yellow / Balearic Isles 289
amorginum / 4-12" / fls yellow, 1-2" across / Greece 289
angustifolium - see H. italicum 129 / 289
apiculatum / HH P / to 2' / golden-yellow fls / Australia 25
arenarium / R / to 1' / citron to golden-yellow fls / e & n Europe 208 / 93
backhousii / HH Sh / small / white fls / Tasmania (152)
baxteri / P / to 1' / lvs narrow; fls white / s Australia 109
bellidioides / HH R / 3" / fls white / New Zealand 46 / 129
bellidioides v. prostratum - congested form 123
bracteatum / P gr, as A / to 3' / varii-colored fls / Australia 194 / 25
[1] 'Bright Bikinis' / 1' / double, bright red 2" fls
coralloides / HH Sh / 10" / fls cream / New Zealand 90 / 132
dealbatum / HH P / 18" / white fls / Tasmania 206
depressum / HH Sh / to 3' / yellow fls; hairy lvs / New Zealand 25
diosmaefolium / HH P / tall / white fls / Australia 28
frigidum / HH Sh / 2-3' / white fls / Corsica 129
fulgidum / HH R / ? / golden everlasting / ? 64
graveolans / P / 3-12" / sweet smelling, fls yellow / sw Asia 289
hookeri / HH Sh / to 5' / strawy bracts / Victoria, Tasmania 25
humile / Gh P / low, spreading / rosy fls / South Africa 28

---

1 Park Catalog, 1986.

italicum / P / to 20" / shining yellow fls / s Europe 289
italicum ssp. microphyllum / HH R / to 1' / Medit. islands 289
italicum ssp. seotinum / P / 8-20" / aromatic, fls yellow/ sw Eur. 289
lanatum / Gh Sh / 1½-2' / fls bright lemon-yellow / South Africa 129
ledifolium - see OZOTHAMNUS ledifolius 139
litoreum / HH R / ? / yellow fls / coastal s Italy 289
marginatum / HH R / to 10" / glossy snow-white fls / South Africa 25
microphyllum / HH Sh / to 32" / yellow fls / New Zealand 238
milfordiae / HH R / 4" / silvery-white & yellow fls / Basutoland 48 / 129
milliganii / A / 6" / bracts colored as H. bracteatum / Australia 64
nanum / Gh / 3-6" / fls golden-yellow / South Africa 61
orientale / HH Sh / to 2' / bright yellow fls / Medit. & Aegean regs 182 / 77
parvifolium - see H. microphyllum 193
petiolare - see H. petiolatum *
petiolatum / HH Sh / to 4" / ivory fls / South Africa 25
plicatum / R / 6" / deep, vivid yellow fls / Medit. reg. 77 / 64
plumeum / HH Sh / 2' / woody plant / New Zealand 15
retortum / Gh / ? / silver lvs / ? 154
rosmarinifolium - see OZYTHAMNUS thrysoides 139
scorpioides / HH P / 2' / fls yellow / s Australia 45
selaginoides / Sh / lvs tiny, blunt; fls white / Tasmania 109
selago / HH Sh / 8" / fls creamy-white, white stems / N. Zealand 238 / 25
selago v. tumidum - tomentum, loose, sometimes scanty 15
semipapposum / A / to 3' / small golden fls / Australia 109 / 64
sibthorpii / R / to 4" / white fls / ne Greece on cliffs 133 / 289
splendidum / Sh / 3' / white-woolly lvs; yellow fls / Africa, in mts 36
stoechas / HH R / 1' / yellow fls / Spain, Italy, n Africa 211
thianschanieum / P / to 2' / woolly yellow-orange fls / Turkestan 25
tumidum - see H. selago v. tumidum 15
virgineum - see H. sibthorpii 184 / 289

HELICOTRICHON - Gramineae
sempervirens / P Gr / 1-4' / yellow & purple fls / sw Europe 147 / 268

HELIOPHILA - Cruciferae
[1]coronopifolia / A / 1½-2' / many stems; fls blue / S. Africa
longifolia / A / 1-1½' / blue & white or yellow fls / S. Africa 129

HELIOPSIS - Compositae
helianthoides / P / to 4½' / yellow fls / e & c North America 99
helianthoides scabra / stems pubescent/ ray fls orng-yel / sc U.S. 25
scabra - see H. helianthoides scabra 25

HELIOSPERMA
albanicum - see SILENE pusilla 286
alpestre - see SILENE alpestris 286
alpina - see SILENE vulgaris ssp. prostrata 286
pusillum - see SILENE pusilla 286
quadridentata - see SILENE pusilla 286
quadrifidum - see SILENE pusilla 286

---

[1] Sima Eliovson, Wild Flowers of Southern Africa, 1980.

HELIOTROPIUM - Boraginaceae

amplexicaule / P / stems to 24" /fls purple /S. Am., native in Italy	289
anchusifolium - see H. amplexicaule	288
arborescens / P / to 4' / fls violet or purple / Peru	25

HELIPTERUM - Compositae

albicans / HH R / to 9" / white & yellow fls / alps of s Australia	25 / 45
albicans v. alpinum / 4-6" / higher mt. form	(524)
alpinum - see H. incanum v. alpinum	*
anthemoides / HH P / to 18" / white bracts / s Australia, Tasmania	45
incanum - see H. albicans	25
[1] incanum v. alpinum / HH R / 4-6" / white fls / N.S. Wales, Australia	

HELLEBORUS - Ranunculaceae

abschasicus / P / 18" / maroon to red-purple fls / Caucasus	25
abschasicus 'Coccineus' - wine-crimson fls	207
antiquorum / R / 1' / reddish purple fls / Turkey	25
[2] 'Appleblossom' - pale pink flowered Lenten Rose	
argutifolius - see H. Lividus ssp. corsicus	*
atrorubens - see H. dumetorum ssp. atrorubens	286
X 'Ballard's Black' - a clone with dark purple fls	279
bocconei / HH P / 2' /green fls, few-segmented lvs / Italy, Sicily	286
caucasicus / P / 2' / fls pale-greenish to yellow-brown / Caucasus	5
corsicus - see H. lividus ssp. corsicus	129 / 286
cyclophyllus / P / 2' / fls light glaucous green / Balkans	211 / 286
cyclophyllus odorus - see H. odorus	286
dumetorum / P / to 18" / violet or green fls / ec Europe	286
dumetorum ssp. atrorubens / 2½' / fls violet / Jugoslavia	194 / 286
foetidus / P / 8-32" / fls green, purple margins / sw Europe	49 / 286
guttatus / P / 2' / fls white, red spots & purple margins / Caucasus	5
kochii - yellow fls, origin uncertain	279
lividus / HH P / to 32" / fls pale green / Balearic Isls.	212 / 286
lividus ssp. corsicus / with spinier lvs / Corsica, Sardinia	(418) / 286
lividus ssp. corsicus 'Nanus' - dwarf strain	*
macranthus - see H. niger ssp. macranthus	286
multifidus / P / to 2' / fls green / Albania, Jugoslavia	286
multifidus ssp. istriacus / to 2½' / fls green / Jugoslavia	286
niger / R / to 1' / fls white or pink tinged / e Alps	148 / 286
niger v. altifolius - see H. niger ssp. macranthus	279
niger 'Ballards Black - see H. X 'Ballards Black'	*
niger ssp. macranthus / lvs spinulose-serrate; white fls / Italy	286
niger 'Potter's Wheel' / 1-1½' / pure white fls with broad petals	90 / 129
odorus / P / 2' / scented green fls / e, c & s Europe	286
olympicus - see H. orientalis	77
orientalis / P / 2' / varii-colored fls / Greece, Asia Minor	269/ 129
orientalis 'Atropurpureus' - dark purple fls	25

---

[1] Valder, Jour. Royal Hort. Soc., Sept. 1964.

[2] Fish, Bull. Hardy Plant Soc. 1:7.

orientalis atrorubens - see H. dumetorum ssp. atrorubens *
[1] orientalis 'Peach Blossom' - cream-pink fls
orientalis 'Slate Hybrids' - wide range of colors *
'Prince Rupert' - pale green fls, heavily spotted maroon (154)
purpurascens / P / 14" / fls purplish-violet / ec Europe, Ukraine 286
[1] 'Rupert' - see H. 'Prince Rupert'
'St. Brigid' - old variety of H. niger 277
X sternii - variable strain of the Corsican clan 255 / 268
X torquatus / 1' / purplish fls / Jugoslavia 19
viridis / P / 9-18" / fls yellowish-green / c Europe, Maritime Alps 148 / 286
viridis ssp. occidentalis / lvs glabrous / w Europe 286

HELONIAS - Liliaceae
bullata / P / 12-18" / fls clear pink / coastal bogs, N.J., s-ward 224 / 107

HELONIOPSIS - Liliaceae
breviscapa - see H. orientalis v. breviscapa 200
japonica - see H. orientalis 200
japonica alba - see H. orentalis v. flavida 200
orientalis / P / to 2' / fls rose-purple / Far East 166 / 200
orientalis alba - see H. orientalis v. flavida 200
orientalis v. breviscapa / 1' / fls white to pale rose / Japan 200
orientalis v. flavida / P / 4-24" / fls white; lvs thin / Japan 200

HELWINGIA - Cornaceae
japonica / Sh / 2-6' / fls pale green; fr. blade / Japan 95

HEMEROCALLIS - Liliaceae
aurantiaca / P / to 3' / orange fls / China 25
citrina / Tu / to 4' / fls fragrant, lemon-yellow / China 25
Dumortieri / P / to 20" / orange-yellow fls / Japan, Siberia 49 / 200
esculenta - see H. middendorffii v. esculenta 200
exaltata / P / 4' / soft orange fls / Tobishima, Japan (467) / 279
flava - see H. lilioasphodelus 25
'Francis Fay' / dormant type / 32" / yellow & gold fls (457) / (156)
fulva / P/ to 6'/fls not frag., rusty orange/Asia, Eur., not in U.S. 25
'Hyperion' - greenish-yellow cultivar 49
'Jake Russell' / Tu / to 36" / fls broad, clear gold 111
lilioasphodelus / P / to 3' / fragrant yellow fls / e Sib. to Japan 210 / 25
middendorffii / P / to 28" / orange-yellow fls / Far East (457) / 200
middendorffii v. esculenta / longer-petalled fls / Honshu 200
minor / P / 18" / fls clear yellow, brown tinted / e Asia 279
nana / P / to 1½' / fls fragrant orange / China 25
'Pink Damask' / 34" / old rose fls, yellow throated 208
thunbergii - see H. vespertina (457) / 200
vespertina / P / 3'+ / fragrant lemon-yellow fls / Korea, Japan 200
yezoensis / P / 17-34" / fls greenish-yellow / Hokkaido 200

HEMIGRAPHIS - Acanthaceae
alternata / Gh / pot plant / purplish lvs; white fls / tropics 93

---

[1] Fish, Bull. Hardy Plant Soc. 1:7.

HEMIMERIS - Scrophulariaceae
montana / Gh / 1' / scarlet fls / Cape of Good Hope 206

HEMIPHRAGMA - Scrophulariaceae
heterophyllum / HH R / prostrate / small pink fls / Himalayas 28

HEMITELIA
smithii - see CYATHEA smithii 15

HEPATICA - Ranunculaceae
acutiloba / R / 6" / white, pink, blue fls / e United States 224 / 107
americana / R / 4" / deep blue to white fls / e North America 224 / 107
americana f. rhodantha - the rosy-pink variant 99
angulosa - see H. transsylvanica 286
angulosa alba - see H. transsylvanica 'Alba' *
angulosa rosea - see H. transsylvanica 'Rosea' *
japonica - see H. nobilis v. japonica *
X media 'Ballard's Var.' / 6" / mauve fls 123
nobilis / F / 6" / white, blue, pink fls / most of Europe 210 / 286
nobilis 'Alba' - a cultivated white *
nobilis 'Bicolor' - two-colored form *
nobilis v. glabrata - glabrous forms not botanically significant *
nobilis v. japonica - from Far East 165 / 200
nobilis v. japonica f. magna - with variegated lvs 165
nobilis 'Rosea' - rosy fls *
nobilis 'Rubra' - reddish fls *
nobilis 'Stellata' - said to be red-flowered *
transsylvanica / R / 5" / fls mauve-blue / c Roumania 147 / 286
transsylvanica 'Alba' - white fls *
transsylvanica 'Rosea' - pink fls *
triloba - of Europe see H. nobilis; of Am. see H. americana 129
triloba alba - white fls on H. nobilis or H. americana *
triloba asiatica - presumably see H. nobilis v. japonica *
triloba rosea - if H. americana see H. americana f. rhodantha *
triloba 'Rubra' - reddish form of either Eu. or Am. round-lobed spp. *

HERACLEUM - Umbelliferae
dissectum - see H. sphondylium ssp. montanum 287
lanatum - see H. sphondylium ssp. montanum 224 / 25
mantegazzianum / B or P / 6-16' / white or pink fls / w Asia 93 / 287
maximum - see H. sphondylium ssp. montanum 25
minimum 'Roseum' / R / 1' / pink fls / se France, in mts. 287
sphondylium / B or P / 4'+ / white umbels / c Europe 287
sphondylium ssp. montanum / to 9' / white fls / N. Am., Eurasia 25
sphondylium ssp. sibericum/B, P/cauline lvs seg.; fls gr/ ne,nc Eur. 287
stevenii / B or P / 3'+ / lvs white beneath; white fls / Caucasus 287
villosum - see H. stevenii 287

HERBERTIA
amoena - see ALOPHIA amoena *
drummondii - see TRIFURCIA lahue ssp. caerulea (23)
platensis - see ALOPHIA platensis *
pulchella - see ALOPHIA pulchella 184 / 25

HERMINIUM - Orchidaceae
alpinum - see CHAMORCHIS alpina 85
monorchis / R / to 10" / green or yellow fls / Eurasia 210 / 85

HERMODACTYLIS - Iridaceae
tuberosus / Tu / 1' / purple-black fls / e Medit. region 211 / 129

HERNIARIA - Caryophyllaceae
glabra / A or P / matted / green granular fls / Eurasia 286

HERPOLIRION - Liliaceae
novazealandica / R / mats / fls blue, lilac / New Zealand 96

HERTIA - Compositae
cheirifolia / Gh P / trailing / yellow fls / Algeria, Tunisia 93

HESPERALOE - Liliaceae
parviflora / HH P / to 4' / fls dark to light red / Texas 226 / 25
parviflora v. engelmannii /HH P / 3-4' /rosy bell-shaped fls / Tex. 28

HESPERANTHA - Iridaceae
baurii / HH C / 6-8" / bright pink fls / Natal, Lesotho 185
buchrii / HH C / 9" / white & pale pink fls / Natal 111
falcata / HH C / to 1' / claret-red fls / South Africa 28
inflexa v. stanfordiae / HH C / to 1' / yellow fls / Cape Province 25
montana / HH C / 4' / white fls, purple flushed / South Africa 121
radiata / R / to 6" / fls white, striped red-brown / S. Africa 61
standfordiae - see H. inflexa v. stanfordiae 25
vaginata / HH C / 4-20" / yellow fls / sw Cape, S. Africa 185
vaginata v. metelerkampiae - fls yellow, brown center 185

HESPERIS - Cruciferae
lutea - see SISYMBRIUM luteum 200
matronalis / B / 1-4' / fls purple to white / c & s Europe 210 / 286
matronalis 'Nana Candidissima' - pure white selection 28
tristis / B or P/ to 20"/ fls yellow-green, rose-violet veins/ Eur. 25

HESPEROCALLIS - Liliaceae
undulata / Gh Bb / to 2' / white fls, green striped / s Calif, Ariz. 227 / 25

HESPEROCHIRON - Hydrophyllaceae
californicus / R / 3-4" / white fls, veined / California 229 / 123
pumilus / R / 2-3" / fls white, blue veins / Oregon 229 / 107

HESPEROYUCCA
whipplei - see YUCCA whippleyi 302
whipplei ssp. intermedia - see YUCCA whippleyi v. intermedia 302

HETEROPAPPUS - Compositae
altaicus / P / 18" / fls pale blue to mauve / c Asia 25
hispidus / B / to 3' / blue-purple to white fls / Far East 200
hispidus v. insularis / R/ 1'+/fls blue-purp. to wht/coastal Shikoku 200
insularis - see H. hispidus v. insularis 200

HETEROTHECA - Compositae
1 fulcrata / P / lvs green; fls orange, strong stems / Rocky Mts.
graminifolia- see PITYOPSIS graminifolia
villosa - see CHRYSOPSIS villosa 25

HEUCHERA - Saxifragaceae
americana / P / to 3' / fls purplish / ne & c North America 225 / 99
bracteata / R / 4-6" / petals very narrow / s Wyoming, Colorado 229 / 25
2 X 'Bridget Bloom' / R / luxuriant lvs; fls pink, white
chlorantha / P / to 3' / greenish fls / B.C. to Oregon 25
cylindrica / P / 2½' / yellowish-green to cream fls / nw N. Am. 229 / 25
cylindrica v. alpina / lvs glandular-pubescent / nw United States 142
cylindrica v. glabella - petioles & lvs glabrous 25
2 cylindrica "Greenfinch' / R / tufted fls, green-sulphur / Alan Bloom
cylindrica v. ovalifolia - lvs dark green to red-brown 25
glabella - var. of H. cylindrica 61
glabra / P / 20" / white fls / Oregon & n-ward 228 / 25
'Greenfinch' - see H. cylindrica 'Greenfinch' 49
grossularifolia / P / 18" / white fls / along Columbia River 228
grossularifolia v.grossularifolia/P/18"/goosebry lvs;fls grn/Ore,Ida 142
grossularifolia v. tenuifolia / 2½'+ / Idaho, Oregon, Washington 142
hallii / R / to 1' / greenish-white fls, suffused red / Colorado 229 / 25
hispida / P / to 4' / stems white, hairy / mts. Va., W. Va. 25
longiflora / P / 2'+ / plant mostly glabrous / W. Virginia to Ala. 99
longiflora v. aceroides / pink fls / N.C., Tenn., Ky. 250
maxima - see H. sangunea 'Maxima' 28
merriamii / P / 8-24" / fls pale pink / n California 228
micrantha / P / 3' / blush-white fls, lvs gray-marked / w N. A. 229 / 279
micrantha 'Erubescens' -of gardens belongs here 28
ovalifolia - see H. cylindrica v. ovalifolia 25
ovalifolia alpina - not separated in latest botanical treatment 25
parviflora / P / to 18" / pl & fls villous / Va., N.C., Ky. 99
parvifolia /P/ to 15"/ fls yel. & wht / Rocky Mts., Alta. to Ariz. 25
pilosissima / P / 8-24" / pale pink fls / n California 228
'Pluie de Feu' - see H. sanguinaria 'Pluie de Feu' 189
Pringlei / R / to 15" / lvs narrow; fls hemispheriod / California 25
pubescens / P / to 3' / purplish fls; bronzed lvs / Pa., N.C. (433) / 224
pubescens 'Alba' - white fls *
racemosa - see ELMERA racemosa 107
richardsonii / R / 1'+ / greenish fls / c & nw North America 224 / 99
richardsonii v. hispidior / more hispid / Wisconsin to Colorado 99
rubescens / R / 4-12" / fls pink / n California 25
rubescens v. alpicola / pink fls / s Cascades 312
sanguinea / P / 2½-2' / fls deep bright red / Mexico, Arizona 147 / 129
sanguinea 'Bressingham Hyb.' / to 2' / white, pink, coral-red fls 147 / 129
sanguinea 'Maxima' - dark crimson fls 28
sanguinea 'Pluie de Feu' / 18" / cherry-red fls 135

---

1 William A. Weber, Rocky Mountain Flora, 1976.

2 Fritz Kohlein, Saxifrages and Related Genera, 1985.

sanguinea 'Snowflake' - large white fls 135
villosa / P / 2' / fls deep creamy / Virginia to Tennessee 224 / 279

X HEUCHERELLA - Saxifragaceae
tiarelloides / R / to 1' / light pink fls 49

HEXAGLOTTIS - Iridaceae
flexuosa / HH C / 15" / yellow fls / Cape Province 111
virgata / HH C / 12" / yellow fls / Cape of Good Hope 206

HIBBERTIA - Dillenaceae
dentata / HH Sh / trailing / yellow fls / Victoria, New South Wales 109 / 25
procumbens /P/ spreading mat/lvs soft, blunt; fls yel/Tas, se Austrl 109

HIBISCUS - Malvaceae
'Bluebird' - see H. syriacus 'Bluebird' 268
cameronii / HH P / tall / pink fls, dark basal blotch / Madagascar 25
coccineus / HH P / to 6' / deep red fls / Ga., Fla., in swamps 225 / 25
diversifolius / Sh gr. as A / yellow or purplish fls / Am. tropics 25
huegelii / HH Sh / 3-6' / fls lilac, purplish-red / s & w Australia 194
lasiocarpus / P / 3-7' / white or pinkish fls / c United States 25
lasiocarpus v. californicus - the California plant not distinct 25
manihot - see ABELMOSCHUS manihot 95
militaris / P / 4-6' / fls white to pale rose / Pa.-Minn., & s-ward 28
Moscheutos / P / to 8' / white, pink, rose fls / e United States 224 / 25
Moscheutos ssp. palustris - the northern race / Mass. to N.Car. 25
Moscheutos roseus / rose fls of H. Moscheutos 25
palustris - see H. Moscheutos ssp. palustris 25
syriacus / Sh / 6-12' / fls rose, purple, white / Asia 28
syriacus 'Bluebird' / Sh / to 8' / single mauve-blue fls 110 / 268
trionum / A / 1½' / fls creamy-white & purple-black / tropics 224 / 129

HIERACIUM - Compositae
alpinum / R / to 6" / fls bright yellow / Eurasia, Greenland 148 / 25
amplexicaule / P / 18" / yellow fls / Pyrenees 206
aurantiacum / R / 6-8" / orange-red fls / Eu., weedy escape in U.S. 148 / 123
aurantiacum croceum - see H. brunnecroceum *
bombycinum / R / 12" / silvery lvs; fls yellow / c Spain 25
brunneocroceum - brownish-orange fls, weedy 194 / 237
canadense / R / 1'+ / yellow fls / North America, in woods 99
cynoglossoides / P / 1-4' / fls yellow / nw North America 229
X fuscatum - mixed red & yellow fls 289
glabratum / R / 8-16" / fls yellow / mts. of c Europe 289
gronovii / P / to 3'+ / yellow fls / e & c United States 99
intybaceum / R / to 10" / light yellow fls / e & c Alps 97 / 93
jankae / R / 8-16" / fls yellow / Balkans 289
japonicum / R / to 1' / deep yellow fls / Honshu 200
lanatum / P / to 20" / bright golden fls / w Alps 47 / 93
longiberbe / P / 1-2' / fls yel / Colorado Riv. gorge, Ore. to Wash. 228
maculatum / P / 1½' / yellow fls / Bulgaria 61
murorum / P / to 2½' / fls yellow / Eur., adventive in ne N.Am. 25
X pamphilii / R / to 16" / fls yellow; white-hairy lvs 25
paniculatum / P / to 4' / yellow fls / open woods, e N. Am. 224 / 99
pilosella / R / to 1' / fls pale yellow / Eurasia, weedy in U.S. 210 / 25

pilosella niveum - see H. tardans 289
pratense / P / to 2' / yellow fls; aggressive King Devil / Europe 99
rubrum / P / 1½' / densely woolly plant / ec Europe 29
scabrum / P / to 5'/yellow fls; stems white-tomentose/ e & c N. Am. 224 / 99
scouleri / P / 1-1½' / orange fls / California to Canada 228
tardans / R / rosettes, weak short stolons / yellow fls / w Alps 289
umbellatum / P / 3'+ / dark brown infloresence / N. Am., Eu., Asia 99
venosum / P / 1-3' / bright yellow fls / Maine to Nebraska, weedy 224 / 25
villosum / R / 1' / bright yellow fls / c Europe 219 / 129
waldsteinii / P / 1½' / fls yellow / Balkan Pen. 25
'Welwitschii' / 1' / gray felty lvs; yellow fls / unidentified name 25

HIMANTOGLOSSUM - Orchidaceae
hircinum / P / 2'+ / green & red fls / Europe 212 / 85
longibracteata - see BARLIA robertiana 149

HIPPEASTRUM - Amaryllidaceae
advenum / Gh Bb / 1' / fls crimson-scarlet / South America 267
andicola - RHODOPHIALA andicola 14
aulicum / Gr Bb / to 2' /fls red w. green throat/ c Brazil, Paraquay 25
Bagnoldii / Gr Bb / to 1' / fls yellow tinged w red / Chile 25
bicolor / Gh Bb / 12-18" / fls bright red, yellowish at base / Chile 61
bifidum / HH Bb / to 1' / fls bright red / Argentina, Uruguay 25
blumenavia / Gh Bb / to 8" / fls wht, mauve-crims. bands / se Brazil 61
elwesii - see RHODOPHIALA mendocina 14
pratense / HH Bb / 1' / bright red or violet-purple fls / Chile 25
rhodolirion - see RHODOPHIALA rhodolirion 14
roseum / Gr Bb / to 6" / bright red, perianth tube green / Chile 25

HIPPOCREPIS - Leguminosae
ciliata / A / to 2½' / fls yellow / s Europe 287
comosa / R / mat / yellow fls / c & s Europe 107

HIPPOPHAE - Elaeagnaceae
rhamnoides / Sh or T / to 30' / orange frs / Eurasia 93 / 36

HOHERIA - Malvaceae
glabrata / HH T / 30' / white fls / New Zealand 238 / 93
lyallii / HH T / to 30' / inflores. white, densely pubescent/ N.Z. 194 / 25
lyallii glabrata - see H. glabrata 93
populnea / HH T / to 30' / white fls / New Zealand 193 / 15

HOLCUS - Graminae
mollis / Gr / 8-12" / variable panicle, white to purple / Europe 290

HOLODISCUS - Rosaceae
discolor / Sh / to 20' / fls cream-white / B.C., Calif., Mont. 269 / 25

HOMALOCEPHALA
texensis - see ECHINOCACTUS texensis 25

HOMERIA - Iridaceae
Breyniana / HH C / to 1½' / salmon-pink or yellow fls / S. Africa 25
Breyniana 'Aurantiaca' / P / to 1½' / fls w. yellow claw 25
collina - see H. Breyniana 184 25

collina v. ochroleuca - see H. ochroleuca 25
comptonii / HH C / 30" / flame-red & yellow fls / South Africa 121
lilacina / HH C / to 9" / lilac fls veined darker / South Africa 25
miniata / HH C / 24" / salmon fls / Cape Province 111
ochroleuca / HH C / to 2½' / gold-yel, salmon-pink fls / S. Africa 25
ochroleuca v. aurantiaca - see H. Breynianca c. v. 'Aurantiaca' 25

HOMOGLOSSUM - Iridaceae
huttonii / HH C / 1-2' / fls crimson, yellowish streaks / S. Africa 61
merianella / HH C / 1-2' / fls bright red / South Africa 61
watsonium / Gh C / to 2' / red fls / Cape of Good Hope 223

HOMOGYNE - Compositae
alpina / R / to 6" / pale violet fls / c Europe, Balkans 148 / 25
discolor / P / to 11"/fls brt purp.; pappas dirty wht /e Alps., Jug. 289

HONKENYA - Tiliaceae
peploides / R / trailing / greenish-white fls / n & w Eur., coastal 286

HORDEUM - Gramineae
jubatum / A Gr / 1-1½' / silvery-gray seed heads / N. & S. Am. 90 / 129
murinum / Gr / 2-24" / spikelets winged / s & w Europe, England 290

HORKELIA - Rosaceae
californica / HH P / 8-40" / fls rust to purplish / coastal Calif. 228
sericata / R / 6-12" / white or pink fls / nw Calif., sw Oregon 228

HORMATOPHYLLA
pyrenaica - see PTILOTRICHUM pyrenaicum *
reverchonii - see PTILOTRICHUM reverchonii 286

HORMINIUM - Labiatae
pyrenaicum / R / 6-8" / bluish-purple fls / Pyrenees, Tyrols 148 / 123
pyrenaicum 'Roseum' - roseate fls *

HOSTA - Liliaceae
albo-marginata - see H. sieboldii 25
albo-marginata v. alba - see H. sieboldii 'Alba' 25
1 'Allan P. McConnell' / tiny narrow green lvs, edged white
atropurpurea / P / 18-24" / fls deep purple / mts. of Hokkaido 200
capitata / P / to 24" / fls purple; lvs dull green / Japan 200
clavata / P/ 18" / fls white to pale purplish / mts. of Honshu 200
coerulea - see H. ventricosa 268
crispula / P / 1½-2½' / fls violet; lvs white-margined / Japan 49 / 129
decorata / P / to 22" / lt purple fls; white-margined lvs / Honshu (432) / 268
decorata minor alba - this name not associated with H. decorata *
decorata f. normalis / lvs green / Honshu native 200
decorata 'Thomas Hogg' - an obsolete synonym of the sp. 151
elata / P / to 3' / fls light bluish-violet / Japan 300 / 268
erromena - see H. undulata 'Erromena' 25
fortunei / P / 2'+ / fls pale lilac to violet / Japan 147 / 25

---

[1] Rocknoll Catalog, Spring 1986.

fortunei 'Albomarginata' / lf to 11", broad white margin / Japan 25
fortunei v. albo-picta - see H. fortunei 'Aureomaculata' 129 / 25
fortunei 'Aureomaculata' - lvs yellow, narrow green margin 300 / 25
fortunei 'Aureomarginata' - lvs narrowly yellow-margined 49
fortunei v. gigantea - see H. elata 268
fortunei v. marginato-albo / blades 5-8"/fls pale blue;lvs wht marg. 200
fortunei robusta - see H. sieboldiana v. elegans 151
fortunei v. variegata - green & white lvs 208
'Frances Williams' - see H. sieboldiana 'Frances Williams' 135
[1] 'Ginko Craig' / dk green lvs, white margin; fls lav, 15" stems
glauca - see H. sieboldiana 268
glauca alba - see H. sieboldiana 200
gracillima / R / 7-10" / pale purple fls / Shikoku 200
kikuti v. caput-avis / R / 6-10" / fls pale pink / Japan 166
kikuti polyneuron - see H. kikuti v. yakusimensis 200
kikuti v. yakusimensis / white fls / Kyushu 200
kiyosumiensis / R / 1'+ / white fls / Honshu 200
lancifolia / P / 12-16" / dark violet fls / Honshu 192 / 268
lancifolia alba - see H. albomarginata v. alba 151
lancifolia albomarginata - see H. albomarginata 268
lancifolia fortis - see H. undulata v. erromena 151
lancifolia 'Kabitan' - lvs yellow-green, bordered olive-green (161)
lancifolia tardiflora - see H. tardiflora 268
lancifolia 'Wogan Gold' - yellow-leaved form (162)
longipes / R / 12" / fls pale purple / Honshu, Hyushu 166 / 200
longissima / P / 20" / fls pale rose-purple / Honshu 166 / 200
longissima v. brevifolia / purplish fls / Honshu 200
'Louisa' / 14" / pure white fls; lvs edged white (161)
marginata 'Thomas Hogg' - see H. decorata 151
minor / R / short / campanulate-funnelform fls / Japan, Korea (432) / 25
minor alba - see H. albomarginata v. alba 151
montana - see H. elata 268
nakaiana / P / 1-1½' / purple fls / Honshu, Kyushu (161) / 268
nakaiana minor - included in the sp. (161)
plantaginea / P / 16-26" / waxy white fls / China (145) / 200
plantaginea 'Grandiflora' / 1½-2' / long white fls 300 / 129
rectifolia / P / to 32" / fls whitish / Japan, Kuriles 166 / 200
rupigraga / P / lvs leathery; fls pale purple / Japan 200
[1] 'Satin Beauty' - 9" lvs; satin sheen near-white fls
sieboldiana / P / 1½-2' / fls pale lilac / Japan 129
sieboldiana 'Aurea Marginata' - yellow-margined lvs 25
sieboldiana v. elegans - fls white, shorter 49 / 268
sieboldiana fortunei - see H. fortunei *
sieboldiana 'Frances Williams' - lvs bordered cream & yellow (161) / 135
sieboldiana v. hypophylla - larger lvs, shorter scapes, fls darker 200
sieboldii / P / 2½' / violet fls / Japan 25
sieboldii 'Alba' / lvs grn; fls white 25
[2] 'Snowflake' - seedling of H. albomarginata v. alba
takudama / P / 1-1½' / fls white to pale purple / Honshu 200
tardiflora / R / 6-12" / late pale purple fls / Japan 268
[2] 'Tinker Bell' - seedling of H. albomarginata v. alba

---

[1] Rocknoll Catalog, Spring 1986.

[2] John Wister, Am. Horticultural Magazine, 50:3.

'Thomas Hogg' - see H. decorata 151
undulata 'Erromena' / 3' / fls light violet / Japanese clone 25
ventricosa / P / 3' / fls dark violet / e Asia 93 / 268
ventricosa 'Aureo Maculata' / 2½' / lvs splashed yellowish-white 279
ventricosa 'Variegata' - dark green lvs, edged cream 279
venusta / R / 8-12" / pale lilac fls / Korea, Japan 268
'Wogan's Gold' - see H. lancifolia 'Wogan Gold' (162)

HOUSTONIA - Rubiaceae
caerulea / R / 3-6" / tufts / light blue fls / ne United States 224 / 107
caerulea v. Faxonorum / A / to 1" / fls pale lilac-blue / N.H. 99
longifolia / R / to 1' / purplish fls / nc North America 224 / 107
purpurea / P / 1-2' / purple fls / ec & sc United States 225 / 107
purpurea v. calycina / R / 6" / lilac to white fls / n & c U.S. 313
serpylifolia / A / prostrate / fls blue / Pa. & W. Va. to Ga. 99
tenuifolia / R / to 1' / purple fls / Pennsylvania to Mexico 225 / 107
wrightii / P / to 8" / fls cluster white to pink / Tex., Mexico 226

HOUTTUYNIA - Saururaceae
cordata / P / 15" / white bracts / Japan, temperate e Asia 165 / 25

HOVEA - Leguminosae
celsii / HH Sh / 3-6' / fls blue / w Australia 61
pungens / Gh Sh / 3' / purplish fls / w Australia 45

HOVENIA - Rhamnaceae
dulcis / T / 30'+ / yellow fls; edible fruit-stalks / Far East 61 / 36

HUDSONIA - Cistaceae
ericoides / Sh / 6-8" / bright yellow fls / e United States 224 / 36
tomentosa / Sh / 6" / frosty gray lvs; yellow fls / e United States 107

HULSEA - Compositae
algida / R / 6-14" / fls yellow / high Sierra Nevadas 228
nana / R / 2-6" / fls yellow / volcanic mts. Washingto to N. Car. 25

HUMEA - Compositae
elegans / Gh Sh / 6' / aromatic lvs; red fls / Australia 128

HUNNEMANNIA - Papaveraceae
fumariifolia / P gr. as A / to 2' / yellow fls / Mexico 208 / 25

HUTCHINSIA - Cruciferae
alpina / R / 4" / fls white / c & s Europe, in mts. 148 / 286
alpina ssp. auerswaldii / to 6" / stems flexuous / n Spain 286
alpina ssp. brevicaulis / 2" / petals narrower 286
auerswaldii - see H. alpina ssp. auerswaldii 286
brevicaulis - see H. alpina ssp. brevicaulis 286
petraea / A / 1½-4" / small white fls / Europe, North Africa 77
stylosa / B / to 3" / fls white in corymbs / Italy 61

HYACINTHELLA - Liliaceae
azurea - see PSEUDOMUSCARI azureum 185
azurea 'Alba' - see PSEUDOMUSCARI azurea f. album 185
azurea v. amphibolus - see PSEUDOMUSCARI azurea v. amphibolus 185

dalmatica / Bb / 4" / blue fls in dense spikes / Jugoslavia 185
leucophaea / B / to 10" / fls pale, rarely white / se Europe 290

HYACINTHOIDES - Liliaceae
hispanica / B / 9-20" / fls blue petals, not recurved/w Iberia Pen. 290
hispanica alba - white form of H. hispanica 290
nonscripta /B/ 8-20"/fls viol. bl. rarely pnk or wht/w Eur., n Scot. 290
vincentina /B/ 4"/fls bl-viol. w wht-yel anthers/se France, nw Italy 290

HYACINTHUS - Liliaceae
amethystinus - see BRIMEURA amethystina 129 / 268
amethystinus alba - see BRIMEURA amethystina 'Alba' 184 / 268
amethystinus gavarnie - see BRIMERA amethystina 290
azureus - see PSEUDOMUSCARI azureum 129 / 185
azureus albus - see PSEUDOMUSCARI azureus f. album 184 / 185
ciliatus - see BELLEVALIA ciliata 268
dalmaticus - see HYACINTHELLA dalmatica 25
dubius - see BELLEVALIA dubia 181
fastigiatus - see BRIMEURA fastigiata 185
festaus - see BRIMEURA fastigata 290
litvinovii / Bb / 4" / pale blue fls, dark striped / e Iran 185
orientalis / Bb / 1' / white to mauve-purple fls / c & e Medit. reg. 211 / 267
pouholtzii (correctly pouzolzii) - see BRIMERA fastigiata 268
romanus - see BELLEVALIA romana 271
spicatus - see STRANGWEIA spicata 185
tabrizianus / Bb / 2" / whitish fls shaded blue / Iran 185

HYDRANGEA - Hydrangeaceae
anomala ssp. petiolaris / Cl high / white sepals / Japan 25
aspera ssp. robusta / HH Sh / 6' / blue fls / e Asia 25
hirta / Sh / small / blue-purple fls / Japan 46 / 139
paniculata / Sh / 12-20' / fls white, aging purplish / Far East 269 / 36
quercifolia / Sh / 3-6' / fls white / se United States (206) / 268
robusta - see H. aspera ssp. robusta 25

HYDRASTIS - Ranunculaceae
canadensis / R / 1' / white fls; crimson frs / e North America 224 / 107

HYDROCHARIS
Morsus-ravae / Aq / several ft. / fls white, 1" across / Eur., Asia 25

HYDROCOTYLE - Umbelliferae
peduncularis / HH R / creeping mats / small white fls / Tasmania 25

HYDROPHYLLUM - Hydrophyllaceae
appendiculatum / P / 1-2' / fls pale violet / c & n N. Am. 224 / 257
canadense / P / 1-1½' / fls green, white, purple / e N. Am. 225 / 28
capitatum / P / 8-18" / fls purplish-blue to white / nw N. Am. 229 / 25
macrophyllum / P / 2-3' / violet or purple fls / Va., W. Va. 263
virginianum / P / to 3' / white to purplish fls / c N. Am. 224 / 25

HYDROTAENIA - Iridaceae
Van-Houttei / HH Bb / 2-3' / pale lilac, yellow fls / Mexico 28

HYLOMECON - Papaveraceae
japonicum / R / 1' / golden yellow fls / Japan 49 / 129
vernalis / R / 8-16" / yellow fls / Mancuria 5

HYMENANTHERA - Violaceae
alpina / HH Sh / 1-2' / coriaceous lvs / alps of Canterbury, N.Z. 238 / 15
crassifolia / HH Sh / to 5' / purplish frs / New Zealand 36 / 25

HYMENOCALLIS - Amaryllidaceae
Amancaes / Gh Bb / 2' / bright yellow fls / Peru (468) / 25
caroliniana / HH Bu / to 22" / fls white / Ga. w to La. & Indiana 25 / 99

HYMENOPAPPUS
newberryi - see LEUCAMPYX newberryi 64

HYMENOSPORUM - Pittosporaceae
flavum / HH T / to 40' / cream & yellow fls / Australia 45

HYMENOXYS - Compositae
acaulis / R / 1-15" / yellow fls / sc Canada to Texas 229 / 25
acaulis v. caespitosa - lvs woolly-villous 25
acaulis v. glabra / lvs thinly villous / glabrous / Ont., Ohio., Ill. 25
acaulis v. simplex - included in the sp. (164)
Cooperi / B / to 2' / fls yellow / Idaho, Arizona, California 25
grandiflora / R / 2-18" / yellow or orange fls / N. Mex. to Canada 229 / 227
richardsonii / R / 2-38" / yellow or orange fls / N. Mex. to Can. 227
scaposa / R / to 1' / yellow fls / c United States 25

HYOSCYAMUS - Solanaceae
albus / A or P / 1-3' / yellowish-white fls / s Europe 210 / 288
aureus / B or P / weak-stemmed / golden-yellow fls / sw Asia 210 / 288
niger / A or B / 1-2½' / fls greenish-yellow / Europe, Asia 28

HYPERICUM - Hypericaceae
adenotrichum / R / to 1' / yellow fls / Turkey 77
aegypticum / HH Sh / low / yellow fls / Mediterranean islands 287
anagalloides / R / mats / yellow fls / w coast U.S., e to Montana 229 / 142
androsaemum / Sh / 2-3' / light yellow fls / w Europe, Balkans 77 / 36
annulatum / P / to 2' / sulphur-yellow fls / Balkan Peninsula 287
archibaldii / see H. orientale
ascyron / P / to 32" / fls yellowish to reddish / n temperate zone 200
athoum / P / less than 1' / fls few w. black glands / n Aegean reg. 287
balearicum / HH Sh / to 4' / yellow fls / Balearic Isls. 212 / 287
bithynicum / P / 6-22" / black-dotted yellow fls / Turkey 77 / 287
buckleyi / Sh / 1' / yellow fls / N. C. & Ga., in high mts. 225 / 25
bupleuroides / P / 18-32" / yellow fls / Russian Armenia 77
calycinum / HH Sh / 10-15" / rich yellow fls / se Europe, Asia Minor 129 / 208
canariense / HH Sh / 2' / yellow fls / Canary Isls 81
cerastoides / R / 10" / yellow fls / se Europe 77 / 25
chinense / HH Sh / to 3' / golden fls / China, Formosa 139
coris / P / 18" / yellow fls / n & c Italy, Switzerland 148 / 287
cuneatum - see H. pallens (525) / 77
degenii - see H. annulatum 287

delphicum / P / 1' / yellow fls / Aegean region		287
densiflorum / Sh / 6' / yellow fls / se United States	225 /	25
drummondii / A / to 32" / small yellow fls / s & sc United States	227 /	99
elatum - see H. X inodorum		139
elatum 'Elstead' - see H. X inodorum 'Elstead'		139
elegans / P / 6-34" / yellow fls / c & e Europe		287
elodes / P / ? / red glands on sepals / w Europe to n Scotland		287
empetrifolium / HH Sh / 12-15" / pale golden fls / Greece	211 /	36
empetrifolium v. oliganthum / procumbent / Crete, in mts.		36
empetrifolium 'Prostratum' - see H. empetrifolium v. oliganthum	133 /	36
erectum / P / 8-24" / yellow fls / Japan		200
foliosum / HH Sh / to 3½' / fls yellow / Azores		287
'Forrestii' / Sh / to 4' / fls yellow / w China, Burma, Asia		25
formosum / P / 4-32" / yellow fls / nw North America		229
fragile - in cult. is H. olympicum f. minus		(166)
frondosum / Sh / 3'+ / bright yellow fls / s & se United States	25 /	139
gentianoides / A / to 2' / yel. fls; lvs scaly / Me. to Fla.		225
gracile - see H. lanuginosum		77
gramineum / HH R / 1' / grassy plant / New Caledonia		81
graveolens / Sh / 3' / yellow fls, faint brown stripes / N.C., Tenn.	225 /	250
haplophylloides / Sh / 10"/ yellow fls, marg. black glands / Albania		287
hircinum / Sh / to 3' / fls yellow / Medit. region		25
hirsutum / P / 1-3½' / yellow fls, yellow veined / Europe		287
hookerianum / HH Sh / 2-5' / deep to pale yellow fls / Himalayas		36
hypericoides / Sh / 2½' / yellow fls / e N. Am. to Central America		25
hyssopifolium / Sh / to 2' / fls yellow / sw Europe		25
X inodorum / HH Sh / to 4½' / small pale yellow fls		139
X inodorum 'Elstead' - selection for salmon-red frs		139
japonicum / A or P / 8-20" / yellow fls / New Zealand, Far East	193 /	200
kalmianum / Sh / 2-3' / small yellow fls / ne North America		28
kamtschaticum / R / 4-12" / yellow fls / n Japan & Kuriles		200
kamtschaticum v. senanense / R / 1' / yellow fls / alps of Honshu		200
kelleri / HH R / procumbent / yellow fls / w Crete		287
kouytchense / HH Sh / 2-4' / rich yellow fls / w China		36
lanuginosum / HH P / to 2½' / yellow fls; whitish lvs / Near East		77
lanuginosum v. gracile / 6-12"/ woolly-coated plant/ coast of Syria		96
linarifolium / P / to 2' / yellow fls, black str. / w Eu.		287
linarioides / R / 7" / yellow fls, red tinged / c Balkans		287
lloydii / HH Sh / 18" / bright yellow fls / S. Carolina to Florida		(165)
lydium / P / 4-30" / yellow fls, rarely red-veined / n Iraq		77
maculatum / P / 3-40" / yellow fls, black-dotted / most of Europe		287
montanum / P / 10-28" / yellow fls / w & c Europe	148 /	287
X moserianum - half-hardy shrub / 2' / fls golden yellow		28
mutilum / A or P / 1-3' / yellow fls / e & ec United States		99
nummularium / R / to 1' / yellow fls, red veined / Pyr., Alps		287
oblongifolium / Sh / to 1' / fls yellow / w Himalayas		25
olympicum / P / 4-20" / large yellow fls / Balkan Peninsula	149 /	287
olympicum 'Citrinum' - see H. olympicum f. uniflorum 'Citrinum'		(166)
olympicum f. minor - see H. olympicum f. minus		(166)
olympicum f. minus - lvs narrowly elliptic		(166)
olympicum f. minus 'Sulphureum' - pale yellow fls		(166)
olympicum f. uniflorum 'Citrinum' - pale yellow fls		(166)
orientale / P / 3-18" / yellow fls / Turkey		77
pallens / HH Sh / to 10" / yellow fls / e Medit. region		77
patulum / HH Sh / 3-4' / evergreen lvs; large yellow fls / China	208 /	25

patulum 'Hidcote' / HH Sh / to 3' / large yellow fls 129
penduliflorum - see H. kouytchense 139
perforatum / P / 4-40" / fls yellow / all of Europe 229 / 287
polifolium / P / near prostr./ fls golden with scarlet /sw A. Minor 61
polyphyllum - see H. olympicum f. minus (166)
polyphyllum 'Citrinum' - see H. olympicum f. minus 'Sulphureum' (166)
polyphyllum 'Sulphureum' - paler fls 123
prolificum / Sh / 5' / yellow fls / N.J., Iowa, Ga. 99
pseudopetiolatum v. yakusimense / 2" / tiny yellow fls / Japan 165
ptarmicaefolium - see H. orientale 77
pulchrum / P / 4-36" / fls yellow, red tinged / nw Europe (469) / 287
repens - SEE H. Linariodes 287
reptans - in cult. may be H. olympicum f. minus (166)
rhodopeum - see H. cerastoides 129 / 25
richeri / P / 4-20" / black-dotted yellow fls / s & sc Europe 148 / 287
rumelicum / P / 2-15" / petals covered w black dots / Balkan Pen. 287
samiense / Sh / to 2' / fls yellow, blades spotted / China, Japan 61
thasium / R / 6-16" / yellow fls / se Balkan Peninsula 77 / 287
tomentosum / P / decumbent to 3' / yellow fls / sw Europe.
trichocaulon / HH R / 10" / yellow fls / w & s Crete 287
undulatum / P / to 40" / yellow fls. red tinged / sw Europe 287
virginicum / P / to 32" / flesh to mauve fls / e North America 99
yakusimense - see H. pseudopetiolatum v. yakusimense 165
yezoense / P / 4-12" / small yellow fls / Japan, Siberia 200

HYPOCHOERIS - Compositae
glacialis / R / rosette / white fls / Andes Mts. 64
lanata / R / ? / silvery lvs / Andes Mts. 64
radicata / P / 1-3' / yel. fls / much of U.S., introduced from Eur. 229
robertia / R / to 1' / light yellow fls / Italy, Isls. of Medit. 289
uniflora / P / 6-20" / yellow fls / Europe, in mts. 210 / 148

HYPOXIS - Amaryllidaceae
hirsuta / P / 4-24" / starry yellow fls / e & c United States 224 / 107
hygrometrica / R / 6" / fls yellow / Australia (526) / 25
micrantha / HH R / 12" / yellow fls / Florida to Texas 225

HYPSELA - Lobeliaceae
longiflora - see H. reniformis 25
reniformis / R / 2' mats / white or rose fls / South America 194 / 25

HYSSOPUS - Labiatae
officinalis / P / 8-24" / brilliant blue-violet fls / Eurasia 210 / 148
officinalis 'Alba' - fls white 25
officinalis aristatus - synonym of the sp. 25
officinalis 'Rosea' - fls roseate 25
officinalis 'Rubra' - fls red 25

HYSTRIX - Gramineae
patula / P Gr / 2-4' / decorative spikes / North America 57 / 268

IBERIS - Cruciferae
amara / A / to 1' / fls white, fragrant 25

amara v. coronia / A / 1' / large white fls 28
amara 'Little Gem' - presumably I. sempervirens 'Little Gem' *
attica / A / to 1' / fls rose or white / e Medit region 77
aurosica / R / to 6" / fls purple-lilac / w Alps 148 / 286
candolleana - see I. pruitii 286
conferta - see TEESDALIOPSIS conferta 286
gibraltarica / HH R / 1' / reddish-lilac to white fls / Gibralter 212 / 286
jordanii - see I. pruitii 286
jucunda - see AETHIONEMA coridifolium 28
'Little Gem' - see I. sempervirens 'Little Gem' 46
pinnata / A / to 1'/ fls white to lilac / s Europe, Asia Minor 25
procumbens / HH R / 4-12" / lilac fls, rarely white / Portugal 286
pruitii / A or P / 6" / white to lilac fls / Medit. reg., in mts. 212 / 286
pygmae - name of no botanical standing 25
saxatilis / R / 4-6" / fls white / s Europe, Pyrenees 129
saxatilia cinerea / Sh/ sm. to 16"/ gray-gr lvs; fls wht / c,s Spain 286
semperflorens / HH Sh / to 32" / fls white, evergreen lvs / w Italy 286
sempervirens / Sh / 10" / white fls / high mts of Medit. region 148 / 286
sempervirens 'Compacta' - dwarfer 25
sempervirens v. correifolia 'Little Gem' - free-flowering 46
sempervirens v. correifolia 'Snowflake' / 9" / large white heads 129
sempervirens 'Little Gem' - see I. s. v. correifolia 'Little Gem' 300 / 25
sempervirens 'Snowflake' - see I. s. v. correifolia 'Snowflake' 300 / 25
simplex / A or P / to 1' / fls white to pale purple / s Russia 286
spathulata / R / to 4" / fls purplish to white / Pyrenees 286
taurica / A or B / 8" / fls lilac to white / Asia Minor 25
tenoreana - see I. pruitii 286
umbellata / A / 9-15" / fls in purple shades / s Europe 129

IDAHOA - Cruciferae
scapigera / A / to 5" / white fls; red or purple sepals/e Cascades 142

IDESIA - Flacourtiaceae
polycarpa / HH T / to 40' / frs deep red / Far East 138 / 36

ILEX - Aquifoliaceae
ambigua v. monticola - see I. ambigua v. montana 25
ambigua v. montana / T or Sh / to 35' / red frs / e U.S. 25
aquifolium 'Fructo-Lutea' - yellow frs; evergreen lvs 25
aquifolium shepherdii / T/ to 80'/ lvs 4"; fr red /cultivar Far East 61
bioritensis / HH T / to 30' / thick leathery lvs / China, Taiwan 25
'Blue Princess' - clone of I. X meserva (I. rugosa X aquifolium) 300 / (405)
cornuta / HH Sh / 8-10' / spiny-leaved evergreen / China, Korea 36
crenata / Sh / to 15' / fls white, fr black / Japan 25
crenata 'convexa' / Sh or T/ 15'/bullate convex lvs; black fr/ Japan 25
crenata mariesii / Sh or T / 15' / lvs dwarf / Japan 25
crenata 'Repandens' /Sh or T/15'/spreading lvs, closely spaced/Japan 25
fargesii / HH T / to 20' / frs red, globose / w China 36
integra / Sh or T / 10-25' / fls yel-green; frs red / Japan 25
X Meservaea/Sh/7'/fr red-cuttings incl. 'Bl. Boy', 'Bl. Girl'/H.Kong 25
montana - see I. ambigua v. montana 25
opaca / T / to 40' / red frs; evergreen lvs / e & c United States (470) / 36
paraguariensis / Gh Sh / to 20' / frs reddish-brown / Brazil 28
pedunculosa / Sh / to 15' / fr red sometimes yellow / China, Japan 25
pernyi v. veitchii - see I. bioritensis (471) / 25

rugosa / Sh / creeping / white fls; red frs / n Japan, Kuriles 200
serrata / Sh / 4-10' / globose red frs / Japan 167 / 25
verticillata / Sh / to 15' / globose red frs / n & c N. Am. (113) / 25
verticillata v. chrysocarpa /Sh / 15'/ fr yellow / Nfld, s to Ga. 25

ILLIAMNA - Malvaceae
corei / P / 3' / rose fls / Giles County, Virginia 263
remota / P / 3-6' / pale rose-mauve fls / Illinois 25
rivularis / P / 3' / rosy fls / nw North America 229 / 25

ILLECEBRUM - Caryophyllaceae
verticillatum / A / to 1' / small white fls / w & c Europe 210 / 286

ILLICIUM - Magnoliaceae
floridanum / HH Sh / 6-10' / dark crimson to purple fls / Fla., La. 28

IMPATIENS - Balsamaceae
Balsamina / A / 2½'/ fls wht-yel. or red /India, China, Malay. Pen. 25
biflora - see I. capensis 99
capensis / A / 2-3' / orange fls, spotted / North America 224 / 99
cristata / A / 2' / yellow fls / China 206
glandulifera / A / 3-7' / purple to wine-red fls / Himalayas 210 / 93
Noli-tangere / P / 1½' / br. yel-white spots / Gr. Britain, Eur. 25
pallida / A / 3-6' / fls lemon-yellow / ne & nc North America 224
parviflora / A / 3' / lemon-yellow fls / Eurasia 210 / 99
roylei - see I. glandulifera 184 / 25
scabrida / A / 2-3' / stems purple; fls yellow / India 61
textori / A / branched 16-32" / fls purple-red / Japan 200

INCARVILLEA - Bignoniaceae
'Bees Pink' - see I. mairei 'Bees Pink' *
1 bonvalottii - see I. compacta
brevipes - see I. mairei 25
compacta / R / 6" / purple fls / nw China 268
Delavayi / P / 2' / fls rosy-red / w China, Tibet 208 / 279
Delavayi 'Alba' - white form 61
Delavayi 'Bees Pink' - see I. mairei 'Bees Pink' 279
Farreri - see I. sinensis ssp. variabiles 95
grandiflora - see I mairei v. grandiflora 129 / 268
grandiflora brevipes - see I. mairei 268
grandiflora 'Frank Ludlow' see I. mairei 'Frank Ludlow' 268
mairei / R / to 1' / fls crimson-purple / Tibet, China 279 / 268
mairei 'Bees Pink' - soft pink fls 279
mairei 'Frank Ludlow' / 6" / rich deep pink fls / Bhutan 268
mairei v. grandiflora - solitary crimson fl 268
marei v. grandiflora brevipes - synonym of the sp. 25
mairei 'Nyoto Sama' / 6" / light pink fls / se Tibet 268
olgae / P / 3' / rose fls / c Asia 208 / 25
sinensis / A / to 2' / fls reddish / China 25
sinensis ssp. variabilis / A or P / 2'/ fls rose-pnk/China, Manchu. 25
sinensis ssp. variabilis f. przewalskii / P / fls yellow 25
variabilis / P / 1-2' / rose fls / e Tibet 93

1 Grierson, Notes from the Royal Bot. Gard., Edinburgh, 23:3.

variabilis przewalskii - see I. sinensis ssp variabilis. f. p. 25
younghusbandii / R / acaulescent / purplish-pink fls / Tibet, Nepal 194 / 25

INDIGOFERA - Leguminosae
australis / HH Sh / to 4' / red fls / Australia 45 / 25
decora / HH Sh / 18" / reddish-purple fls / Japan, China 200
decora f. 'Alba' - white form 25
gerardiana / Sh / to 6' / rosy-purple fls / India 180 / 25
heterantha / HH Sh / 2-4' / rosy-purple fls / nw Himalayas 36
incarnata - see I. decora 200
kirolowii / Sh / to 4' / fls bright rose / n China, Korea 25
pweudotinctoria / Sh / 12-20" / fls pale red-wht; sd. gr-yel / Japan 200

INULA - Compositae
acaulis / R / 1½" / stemless golden daisies / Asia Minor 123
brittanica / P / to 3' / yellow fls / Eurasia 210 / 258
brittanica v. chinensis - included in the sp. 231
candida / P / to 12" / fls yellow / Greece 289
ensifolia / R / 12" / yellow daisies / Europe, n Asia 147 / 107
ensifolia 'Compacta' / 9" / golden daisies 46
ensifolia nana / P/ ? / fls yel, dwarf form of I. ensifolia /e&c Eur 289
glandulosa - see I. orientalis 77
grandis / tall / big lvs in rosetts / semi-arid zone, c Asia 266
helenium / P / 3-6' / yellow fls / Eurasia, escape in U.S. 224 / 77
hirta / P/ 18" / yellow fls / Eurasia 25
hookeri / P / 2½' / greenish-yellow fls / Himalayas 129 / 279
magnifica / P / 6' / golden yellow fls / Caucasus 25
montana / R / to 14" / fls yellow / Spain, Italy, Switzerland 25
oculis-christi / P / 2' / bright yellow fls / e Europe 292 / 25
orientalis / P / to 2' / orange-yellow fls / Caucasus 49 / 77
rhizocephala / R / low / yellow fls / Iran, Afghanistan 93
royleana / P / 20" / orange-yellow fls / Himalayas 28
spiraefolia / P / 12-30"/fls yel; lvs prominent/w,c France, Bulgaria 289
squarrosa - see I. spiraefolia 289
viscosa / HH P / 3-6' / yellow fls; sticky lvs / sw Europe 149 / 77

IONOPSIDIUM - Cruciferae
acaule / A / 2-3" / fls white, purple, lilac / s Europe, n Africa 182 / 25

IPHEION - Liliaceae
[1] brevipes / Bb / near ground / fls white, green keeled / Chile
uniflorum / Bb / 6" / white or pale lilac-mauve fls / South America 75 / 129
uniflorum 'Album' - the white form *
uniflorum 'Violaceum' - light blue fls 17
uniflorum 'Wisley Blue' / 2" / fls deepest blue 111

IPOMOEA - Convolvulaceae
coerulea - see I. nil 206
dissecta - see MERREMIA dissecta 25
hederifolia / A / climber / fls rose to pale purple/ West Indies 25
lacunosa / A Cl / to 10' / white fls / c & sc United States 229 / 99
leptophylla / P / 2-4' / pink or purplish fls / Kansas 229 / 257

---

[1] Traub and Moldenke, Herbertia 11, 1955.

Nil / P gr. as A / high / varii-colored fls / circum-tropical 25
purpurea / A Cl / 6' / purple fls / American Tropics 206 / 116
quamoclit / A Cl / to 20' / scarlet fls / American tropics 25

IPOMOPSIS - Polemoniaceae
aggregata / P / 1-3' / bright red fls / nw coast ranges & e-ward 229 / 228
longiflora / P / 4-16" / fls white or pink / Nebraska to Utah 227
multiflora / R / 4-12" / violet-blue fls / New Mexico, Nevada 229 / 227
rubra / B / to 3' / fls scarlet to yellow / se United States 308 / 226
tenuituba / P / 1-3' / pink fls / California to Colorado 227 / 229

IRIS - Iridaceae
acutiloba / R / to 8" / creamy-white fls, brown veins / Caucasus 129 / 25
afghanica / HH R / 8-12" / yellow & purple fls / Afghanistan 183
albomarginata / Bb / 15" / fls bright lilac / Turkestan 268
anglica - see I. xiphoides
aphylla / R / 6-16" / red-purple fls / Balkans & e-ward 217
arenaria / R / 2-10" / golden fls / Balkans 306 / 96
attica - see I. pumila ssp. attica 183
atropurpurea / C / to 8" / fls red & purp; falls grn, black / Syria 217 / 25
aucheri / Tu / 6-22" / pale lilac-blue fls / Mesopotamia 129
aurea - see I. crocea 268
aurea bastardii - see I. pseudacorus v. bastardii *
babadagica / R / 6-10" / purple-violet fls / ? 183
bakeriana / Bb / 6" / pale blue fls / se Iran, e Turkey 129
baldshuanica / R / to 4" / yellowish fls, purple veins / Afghanistan (527) / 185
biflora - see I. subbiflora 212
biglumis - see I. lactea 183
bloudowii / R / 3" / bright yellow fls / Altai Mts., e Europe 132
bracteata / R / 4-12" / yellow fls, brown-purple veins / Ore., Calif 229 / 25
brandzae / R / 8-10" / blue-purple fls / se Eu., A. Minor (170) / 183
brevicaulis / P / 18" / deep blue to blue-purple fls / c U.S. 212 / 25
brevituba - see I. ruthenica 86
bucharica / Tu / 1½' / fls creamy-white / Turkestan 129 / 267
bucharica orchioides - synonym of the sp. 4
bulleyana P / 1½' / cream & blue-purple fls / w China 86 / 25
californica - see I. macrosiphon 25
carthaliniae / P / to 3' / sky-blue fls, dark veins / Caucasus 25
caucasica / R / 4-8" / yellow fls / n Iran, Asia Minor 185 / 4
Cengialtii / R / to 1' / blue-purple & white fls / n Italy, s Tyrol 25
chamaeiris / R / 1-6" / blue & yellow fls / Maritime Alps 194 / 132
chamaeiris v. campbellii / 8" / deep purple fls 29
chamaeiris 'Lutea' - yellow fls *
cretica / HH R / stalkless fls / blue-lilac fls / e Medit. region 211
X chrysofor - I. chrysographes X forrestii 132
chrysographes / R / 12" / deep blue fls, gold markings / w China 86 / 206
chrysographes X Inforrestii 217
chrysographes 'Inshriach' - deep blue-black fls 83
chrysographes 'Nigra' - nearly black fls 279
chrysographes 'Rubella' / maroon with dark falls / China (170) / 24
chrysophylla / R / 9" / creamy yellow fls / Oregon 306 / 268
clarkei / P / 2' / blue to purple fls / Himalayas 86 / 217
coerulea - either I. albo-marginata or pumila 25
confusa / HH P / 3' / lilac-white fls / w China (170) / 279
cristata / R / 6" / pale mauve fls / e United States 224 / 123

cristata 'Alba' - fls white 25
crocea / P / 3-3½' / fls bright yellow / Himalayas 268
cycloglossa / 24-36" / pale blue, white crest 183
cypriana / HH P / to 3' / bright blue fls / Cyprus 25
danfordiae / Bb / 3" / yellow fls / Caucasus 132 / 111
darwasica / R / 2'+ / green-brown fls / Uzbek, Turkestan 25
decora / HH Tu / 20" / fls bright lavender / sw China, Himalayas 93
delavayi / P / 3-4' / violet-purple fls / Yunnan 129 / 25
demetrii / P / 3' / deep blue fls, yellow striped / Armenia 25
dichotoma / P/ to 3'/green-white fls, dark spots / n China, Siberia 25
'Dorothea K. Williamson' - I. fulva x I. brevicaulis 217
douglasiana / R / 6-16" / purplish fls / nw United States 175 / 129
[1] douglasiana 'Agnes James' - collected albino
douglasiana 'Alba' - fls white 25
douglasiana 'Amiguita' - fls white blotched purple (173)
douglasiana 'Pegasus' - selection for white fls (174)
douglasiana X tenax - natural hybrid 217
douglasiana watsoniana - in synonymy with the sp. 25
dykesii / P / 20" / fls dark blue-purple / China 61
ensata / P / to 16" / white to dark red-purple fls / Far East 306 / 25
ensata f. spontanea / more robust / red-purple fls / Japan 25
farreri / R / 8" / white & purple fls / w China 25
fernaldii / P / 18" / cream-yellow fls / c California 228
filifolia / Bb / 1½' / rich red-purple fls / s Spain, n Africa 212 / 267
flavissima / R / 3-6" / golden fls / Hungary, ne Asia 292 / 107
flavissima arenaria - see I. arenaria 217
foetidissima / P / 2-3' / gray-purple or yellow fls; red frs / Eu. 129 / 25
foetidissima citrina/Bb/2'/lg. fls citron & pale mauve/w Eur., Brit. 279
foetidissima 'Lutea' - yellow-flowered form *
foliosa - see I. brevicaulis 268
fontanesii / HH Bb / 2' / dark blue fls / Morocco 217 / 183
forrestii / P / 12-18" / yellow fls / Asia 86 / 107
fosterana / R / to 1' / pale yellow & purple fls / Afghanistan 185 / 25
fulva / P / to 4' / reddish-brown fls / c United States 225 / 25
furcata / P / 6-18" / purple fls / ? 183
X germanica / P / 2½' / deep & light violet fls / origin unknown 210 / 25
Gatesii / Tu / 1½' / falls pale greenish, beard gray-gr/Asia Minor 25
giganticaerulea / P / 5' / violet-blue fls / s Louisiana 25
'Gold Wing' 217
gormanii - see I. tenax (472) / 217
gracilipes / R / to 9" / pinkish-mauve fls / Japan 166 / 132
gracilipes alba / Tu / to 1'/ fls white / Japan 25
graeberiana / Tu / 4-8" / fls mauve, variable markings / Turkestan (175) / 268
graminea / P / to 3' / fls yellowish-white, purple veins / c & s Eu. 149 / 25
gueldenstaedtiana - an undetermined form of I. spuria 217
halophila - see I. spuria v. halophila 25
Hartwegii / Tu / to 1' / falls creamy or pale yellow / California 25
hexagona / P / to 4' / lilac fls / s United States 225 / 25
hispanica - see I. xiphium
histrio v. aintabensis / 4" / pale blue fls / n Syria 267
histrioides / Bb / 9" / bright blue-purple fls / n Asia Minor 267 / 25
histrioides 'Major' - early-flowering, bright blue fls 267 / 25

[1] Roy Davidson, Bull. Am. Rock Garden Soc. 23:3.

hoogiana / P / to 2' / lilac-violet or white fls / c Asia		4
hookerana / R / 5" / lilac-blue fls / Kashmir		183
hookeri - see I. setosa v. canadensis		25
humilis / R / to 1' / fls blue-purple/ Trans. to Caucasus		25
iberica / R / 3-6" / whitish fls overlaid brown & purple / Caucasus	(473) /	25
iberia elegantissima - similar to I. iberia		183
imbricata / P / 18" / greenish-yellow & brown fls / sw Asia		25
innominata / R / 8" / fls yellow, purple or white / Oregon	90 /	129
innominata 'lutea' / R / to 8" / yellow fls / sw Oregon to Calif.	25 /	228
japonica / P / 3½' / lavender-blue fls / Japan, China	208 /	25
'Jemmy O'Goblin' - cv. of I. innominata x I. douglasiana		217
juncea / HH Bb / 12" / yellow fls / s Spain		111
kaempferi / P / 2' / red-purple fls / Japan		25
kaempferi 'Rose Queen' - see I. laevigata 'Rose Queen'		*
kamaonensis / R / 4-6" / mottled lilac fls / w Himalayas	218 /	183
Kernerana / R / to 1' / yellow fls / Asia Minor	279 /	25
klattii / P / 1-3' / dark blue-violet fls; wide lvs / Caucasus		183
koreana / R / 10" / yellow fls / Korea		29
kumaonensis - see I. kamaonensis		268
kuschakewiczii / Gh Bb / 3" / pale violet fls / Turkestan		183
lactea / R / 4-12" / white or blue fls / Far East		183
lactea chinensis - see I. ensata		86
lacustris / R / 1-3" / blue fls / Gt. Lakes' shores, Mich., Wisc.	129 /	25
laevigata / P / 2'+ / fls blue-purple or white / Far East	166 /	200
laevigata alba / P / to 2' / fls white / e Asia, China, Japan		25
laevigata 'Albo-Purpurea' - falls streaked, standards white		25
laevigata 'Rose Queen' / ribbed lvs, soft old rose fls / Ireland		279
laevigata 'Variegata' - leaf variegation		*
latifolia / P / 10-20" / violet blue, yellow center / Pyrenees	183 /	290
longipetala / P / to 2' / fls wht, veined lilac / c Calif., coastal	228 /	25
lusitanica - see I. xiphium		25
lutescens / R / 3-9" / yellow fls / ?		183
lutescens v. italica - see I. chamaeiris		
macrosiphon / P / 16" / yellow, purple, lavender fls / c & n Calif.	86 /	228
magnifica / Tu / 1½-2' / pale violet fls / c Asia	(176) /	268
magnifica 'Alba' - cultivated white form		268
mellita / R / 3-4" / pale smoky brown fls / Malta, se Europe	(177) /	123
mellita 'Aurea' - smoky yellow fls		217
mellita v. rubromarginata - lvs edged red		183
microglossa / Bb / 6" / pale mauve & white fls / c Afghanistan	185 /	268
milesii / P / 3' / fls light mauve / Himalayas	(178) /	279
minuta - see I. minutaurea		132
minutaurea / R / 4" / golden yellow & brown fls / Japan		132
missouriensis / P / 8-20" / pale or deep blue fls / nw U.S.	229 /	228
missouriensis 'Alba' - fls white		25
missouriensis v. arizonica / 18-20" / white fls veined purple / sw US		183
missouriensis latifolia / B / 8-20" / blue / s California		228
missouriensis v. pelogonus - shorter, thicker lvs		25
monnieri / HH P / 3' / lemon-yellow fls / Crete		279
montana - see I. missouriensis v. pelogonus		25
'Mountain Lake' / 2½' / st. hyacinth-blue; falls lobelia blue	217 /	268
munzii/ HH P / 2½' / pale lavender & darker fls / California		268
mussulmanica / P / 2' / fls bluish or yellowish-white / Azerbaijan		25
nepalensis - see I. decora		183

Entry	Pages
nertschinksia - see I. sanguinea	183
nusairiensis / 4-6" / blue, violey or gray	183
X ochrgaceae / I. aurea X I. ochroleuca	217
ochroleuca / HH P / to 6' / fls white or pale yellow / Crete, Aegaen	129 / 149
ochroleuca sulphurea - see I. orientalis	
orchioides / Bb / 1' / golden yellow fls / Turkestan	185 / 267
orientalis - see I. ochroleuca	(418) / 149
Pacific Coast' hybrids - see I. innominata X I. douglasiana	129
palestina / Gh / 6" / yellow to blue-green fls in winter / Syria	61
pallida / P / 2-3' / mauve-purple fls / Tyrols	208 / 86
pallida 'Variegata' - variegated lvs	129 / 25
paradoxa / R / 6" / fls in purple shades or white / n Iran	61
persica / R / 2" / grayish-blue fls, but variable / Near East	182 / 185
persica v. tauri - falls deep black-purple, streaked white	183
planifolia / HH Bb / 8-12" / fls bright lilac-blue/ n Africa, Sicily	212 / 185
planifolia 'Alba' - recorded white form	185
polakii / R / 3" / maroon & purple fls / n Iran	25
polakii urmiensis - see I. urmiensis	268
prismatica / P / 1-2' / fls pale lilac, veined / e North America	183 / 28
prismatica alba - white form of above	
pseudacorus / P / 2-3' / fls deep golden-yellow / n Europe	147 / 129
pseudacorus v. bastardii - lemon-yellow fls	129
pseudacorus 'Variegata' - lvs striped creamy-yellow	28
pseudocaucasica / Bb / 4" / fls azure blue / Caucasus	4
pseudocyparus - see I. graminea	86
pseudopumila / R / 7" / varii-colored fls / Italy	25
pumila / R / 4-5" / stemless varii-colored fls / Europe, Asia Minor	147 / 123
pumila alba / C / short / fls cream / c Europe to Asia Minor	25
pumila ssp. attica - fls various colors; lvs falcate	149 / 183
pumila 'Lutea' - fls pale yellow	28
purdyi / R / 6-12" / fls white or pale lavender / n California	86 / 228
X purdyi (x innominata) - natural hybrid	217
reginae / P / 12-20" / pale viol-wht, beard yel & wht / c Macedonia	25
reichenbachii / R / 3-12" / fls yel or chocolate / Hungary, Balkans	25
reticulata / Bb / 6-8" / deep bluish-mauve fls / Asia Minor, Iran	267 / 129
reticulata 'Cantab' / 6" / fls brilliant Cambridge blue	129 / 113
[1] reticulata v. 'Clairette' / skyblue and purp with tigered falls	
reticulata 'J. S. Dijit' - reddish-purple fls	267
reticulata 'Joyce' / Tu / deep sky blue, gold markings / Caucasus	267
reticulata 'Springtime' - blue & white fls	267
rosenbachiana/4"/pale mauve or wht w bl.-purp. fall, orng ridge/USSR	183
rubromarginata - like I. mellita, red-edged lvs	217
rudski / R / 12" / lavender & yellow fls / ?	183
ruthenica / R / 6" / deep blue fls / Siberia, China	86 / 123
sambucina / P / 2' / fls violet to bluish / probably of hybrid orig.	61
sanguinea / P / 18" / white & blue-purple fls / Manchuria, Japan	166 / 25
sanguinea v. alba / 36-40"/ wht form of I. sanguinea / Manch., Japan	183
sanguinea 'Kamayama' - see I. sanguinea v. violacea / Manch., Japan	200
sanguinea v. violacea/fls dk viol., lvs narrow, deeper gr/Kor.,Japan	200
sari / R / 6" / yellow & reddish fls / Asia Minor	25
schachtii / R / 9" / fls greenish-yellow / Turkey	268
scorpioides - see I. planifolia	217

---

[1] McClure and Zimmerman Catalog, 1985.

setosa / P / 2'+ / fls blue-purple / Far East 166 / 200
setosa arctica / 12" / purple, white blotched 183
setosa v. canadensis / smaller / coastal Maine to Labrador 25
setosa hookeri - see I. setosa v. canadensis 25
shrevei / P / 18-24" / purple to white fls / c United States 192 / 183
sibirica / P / to 3½' / blue-purple to grayish fls / c Eu., Russia 147 / 25
sibirica 'Alba' - fls white 25
1 siberica 'Carrie Lee' / 28" / med. rose pink, white signals / 4" fls
sibirica 'Ceasar' -3' / rich violet-purple fls 207
sibirica 'Caesar's Brother' / 4' / dark velvety-violet fls 129 / 268
sibirica 'Eric the Red' / heavily veined wine red fls 184
sibirica v. flexuosa - synonym, not a var. 86
sibirica 'Helen Astor' / 2½' / rosy-red fls 268
1 sibirica 'Marlya' - deep blue, lighter falls; purple infl.
sibirica 'Mt. Lake' / 2½' / st. hyacinth-blue, fl. lobelia blue 268
sibirica 'Nottingham Lace' / 36' /A & St. pale wine-purp., laced wht 268
1 sibirica 'Orville Fay' - viol-blue darker, extra wide fls
sibirica 'Perry's Blue' / 3½' / sky blue fls 268
1 sibirica 'Polly Dodge' / 26" / ruffled viol, text red, gold blaze
sibirica 'Skeena' - sibirica-orientalis hyb.; blue-violet fls (179)
sibirica 'Snow Queen' / 2' / white fls, yellow blotch 268
1 sibirica 'Swank' - clear blue, round and flaring
sibirica 'Tycoon' / 3½' / large rich violet fls 268
sibirica 'White Swirl' / 3' / pure white fls 129 / 268
sikkimensis / R / 6" / fls in lilac shades / Sikkim 86 / 25
sintenisii / R / 9" / blue fls, white veins / s Europe 86 / 46
sisyrinchium - see GYNANDIRIS sisyrinchium 217
songarica / P / 12-18" / fls bluish but variable / Iran, w China 268
sogdiana - see I. spuria v. halophila 25
spathacea - see MORAEA spathacea 160
spuria / P / 2-4' / cream, yellow, blue fls / c & s Europe 212 / 129
spuria cathaliniae / vigorous; pale lilac fls / Caucasus 121
spuria v. halophila / short / fls dull yel. wht, gray-purp. / Asia 25
spuria v. maritima / 10-12"/red-purp., viol.-blue fls/coastal France 86
spuria musulmanica - variable, diff. height, robustness, color 183
spuria ochroleuca - undetermined I. spuria ssp. 217
spuria sogdiana - see I. spuria v. halophila 25
spuria v. subbarbata / vigorous, visible pubescence on falls / c Eu. 86
stolonifera / P / to 2' / fls blue-purple / Uzbek, Turkestan 86 / 25
subbiflora / HH P / 8-16" / deep purple fls / s Portugal, s Spain 86 / 212
sulphurea - color form of I. pumila 28
suoveolens / P / lvs 8" x 1/4"; fls viol-brn / Balkan P. to Romania 290
taschia / P / 12-24" / yellow or violet fls / ? 183
tauri - see I. persica v. tauri 183
taurica - see I. pumila 121
tectorum / R / 1' / fls light blue, streaked darker / Japan 129
tectorum 'Album' - frosty white fls 107
tenax / R / 6-12" / fls varii-colored / sw Oregon, Washington 86 / 228
tenax alba / P /6-12" / fls white / Oregon, Washington 228
tenax v. gormanii / no longer separated from the sp. / yellow fls 217
tenax ssp. thompsonii - fls purple 25
tenuifolia / P / to 16" / blue purple / c Asia to China 25

---

1 Rocknoll Catalog, Spring 1986.

tenuis / R / 1' / fls white or cream, veined purple / n Oregon 228
tenuissima / P / ? / fls white or cream, purple veins / California 228
tenuissima ssp. purdyiformis /P/ 16"/crm fls inflated stigmas/n Cal. 25
X thompsonii - see I. tenax ssp. thompsonii 25
timofejewii / R / 5" / red-purple fls / Daghestan
tingitana / HH Bb / 2' / pale purplish-blue fls / nw Africa 129 / 267
tingitana fontanesii - see I. fontanesii 183
Tubergeniana / R / to 4" / yellow fls/Turkestan 25
unguicularis / R / 6-8" / lilac-mauve fls / s & e Medit. region 129
unguicularis 'Alba' - white fls, streaked yellow 129
unguicularis ssp. cretensis - violet-blue fls; narrow lvs 149 / 183
unguicularis lazica - violet with wide leaves; stemless 183
unguicularis 'Mary Barnard' - dark purple fls 279
unguicularis 'Walter Butt'/ free-growing, fls palest lilac /Algeria 279
urmiensis / R / to 1' / lemon-yellow fls / Iran 268
variegata / P / 15" / yellow fls, dark veined / Austria, Balkans 292 / 86
verna / R / 6" / violet fls / Maryland, Virginia & s to Florida 224
verna'Alba' / R / 6" / white form / Maryland, Va., s Florida, Miss. 224
versicolor / P / to 3' / lav. to red or blue-violet fls / e N. Am. 224 / 25
versicolor alba / P / to 3' / falls (rarely) white / e Canada to Pa. 25
versicolor 'Kermesina' / 2' / fls wine-magenta 25
vicaria / Bb / 8-16" / fls violet, darker veins / c Asia (475) / 4
violacea / P / to 3' / sky-blue fls, veined dark blue / Caucasus 4
virginica / P / 2-3' / lavender, lilac, viol., wht fls / Va. to Tex. 224 / 25
virginica v. shrevei / lvs firmer, flower-stem erect / mainly c U.S. 99
warleyensis / Bb / 1½' / pale violet fls / Bokhara 267
watsoniana - see I. douglasiana 93
wilsonii / P / to 2' / fls pale yellow, veined brown / w China 86 / 279
winogradowii / Bb / 4" / clear pale yellow fls / w Caucasus 267 / 132
xanthochlora / 6" / yellow-green / Juno form 183
xiphioides / Bb / 2' / fls dark purple-violet / n Spain 148 / 279
xiphium / HH Bb / 2' / purple fls / s Spain 212 / 111
xiphium 'King Mauve' / 24" / pale mauve / s France, Port., Spain 268
xiphium 'L'Innocence' - white variety, late blooming 267 / 25
xiphium lusitanica / 12-15" / yellow or bronze / Portugal 183

ISATIS - Cruciferae
glauca / P / 3' / yellow fls / Turkey, Iran 77
tinctoria / B / 4' / yellow fls / n Europe 210 / 77

ISODON
umbrosus - see PLECTRANTHUS umbrosus 200

ISOPLEXIS - Scrophulariaceae
canariensis / Gh / 4' / yellow-brown fls / Canary Isls. 194 / 25
sceptrum / Sh / 4-6' / lvs oval; fls yellow / Madeira 61

ISOPOGON - Proteaceae
cuneatus / HH Sh / 4-8' / fls pale blue-purple / w Australia 61

ISOPYRUM - Ranunculaceae
biternatum / R / to 1' / fls white / e & c North America 224 / 257

Hallii / P / 1-3' / fls white / Washington, Oregon 25 / 228
nipponicum / R / 10" / pale greenish-yellow fls / Honshu 200
stipitatum / R / 2-3" / white or pinkish fls / n Calif., Oregon 107
thalictroides / R / 9"+ / nodding white fls / Europe 149 / 25
thalictroides 'Flore Plena' / 9-15" / double white fls / Europe *
trachyspermum / P / smooth 2-8" / fls pale yellow-green / Japan 200

ISOTOMA - Lobeliaceae
axillaris / P gr. as A / 6-12" / bluish purple fls / Australia 109 / 93
longiflora / HH P / 2' / fls pure white / Carib. & S. America 93

ISOTRIA - Orchidaceae
verticillata / R / to 1' / purple & yellow-green fls / e U.S. 99

ITEA - Iteaceae
illicifolia / HH Sh / 6' / greenish-white fls, fragrant / c China 139
virginica / Sh / 3-5' / creamy-white fls / e United States 138 / 36

IVESIA - Rosaceae
baileyi / P / to 10" / fls cream or white / se Oregon, sw Idaho 229
gordonii / R / 2-6" / fls yellowish / Washington to Montana 228

IXERIS - Compositae
chinensis / P / to 20" / fls yellow, violet, whitish / e Asia 258
dentata v. alpicola / low / yellow fls / Japanese alps 200

IXIA - Iridaceae
capillaris / Br Tu / 18" / fls lilac / South Africa 61
dubia / HH C / 9" / red fls / Cape of Good Hope 206
maculata / HH C / 15" / yellow fls / Cape Province 111
monodelpha / HH C / 12" / lilac fls / South Africa 25
paniculata / HH C / 3' / creamy-white fls / South Africa (434) / 25
polystachya / HH C / 2' / white fls / South Africa 25
scariosa / HH C / 8-12" / maroon & purple fls / Cape Province 25
speciosa / HH C / 6-12" / crimson fls / South Africa 194 / 25
viridiflora / HH C / 12"/ grn fls shaded Prussian-blue / Cape Prov. 194 / 111

IXIOLIRION - Amaryllidaceae
ledebourii - see I. montanum 267
montanum - / Bb / 1-1½' / fls blue-viol; lvs slender / c Asia 267
montanum v. pallasii - fls rose-purple / Caspian 267
pallasii - see I. montanum v. pallasii 267
tataricum - see I. montanum 182
tataricum v. ledebourii - see I. montanum 267
tataricum v. montanum - see I. montanum 267
tataricum v. pallasii - see I. montanum v. pallasii 267

IXODA - Compositae
achilleoides / HH Sh / 1-3' / fls yellow, bracts white / s Australia 61

JAMESII - Saxifragaceae
Americana /Sh /6'/flaking bark; fls wht, pnk/mts. Calif., Nev., N.M. 25

JACARANDA - Bignoniaceae
mimosifolia / Gh Sh / 10'+ / blue fls / Brazil 128
ovalifolia - see J. mimosifolia 128

JANKAEA - Gesnariaceae
Heldreichii / R / 1-3' / fls lavender / Greece 25

JASIONE - Campanulaceae
crispa / R / to 4" / blue fls / e Pyrenees, ne Spain 314 / 289
crispa amethystina / P / to 15" / bracts purplish / s Spain 289
heldreichii / P or B / 8-16" / bluish-lilac / Balkans 289
humilis - see J. crispa 289
jankae - see J. heldreichii 289
laevis / P / 8-16" / bright blue fls / w & wc Europe, in mts 289
montana / A or B / 2-20" / blue fls, varying to pink or white / Eu. 210 / 289
parkeri - referrable to JASMINUM
perennis - see J. laevis (476) / 289
perennis ssp. carpetana / presumably more dwarf / Spain 64
supina / P / 4" / fls violet / Turkey 77
supina supina / bracts glabrous inside / nw Turkey 77
supina ssp. tmolea - outer bracts narrow; few short hairs inside 77
tmolea - see J. supina ssp. tmolea 77
tuberosa - see JASONIA tuberosa *

JASMINUM - Oleaceae
beesianum / HH Cl / 3'+ / fls red or rose / w China 25
fruiticans / Sh / 9' / fls yellow / s Europe, n Africa, sw Asia 25
humile 'Revolutum ' / HH Sh / 3-6" / fragrant yellow fls 93 / 268
parkeri / HH Sh / 1' / clear yellow fls / nw India 132 / 123
revolutum - see J. humile 'Revolutum' 268
X stephanense / Climbing / ? / fls pink, fragrant / w China 25

JASONIA - Compositae
tuberosa / P / 4-18" / yellow fls / sw Europe 289

JEFFERSONIA - Berberidaceae
diphylla / R / 10" / white fls / ne United States 224 / 107
dubia / R / 6" / pale lavender-blue fls / Manchuria 90 / 129
manschuriensis - see J. dubia 5

JOVELLANA - Scrophulariaceae
sinclairii / Gh / 3' / white fls, purple spots / New Zealand (181) / 15

JOVIBARBA - Crassulaceae
allionii / R / to 6" / yellowish fls / s Alps 148 / 268
arenaria / R / to 5" / lvs bright green, red toward apex / e Alps 286
heuffelii / R / 4-8" / pale yellow fls / se Europe 149 / 268
heuffelii 'Glabra' - the glabrous phase *
hirta / R / 4-6" / pale to yellow-brown fls / e Alps, Hungary 148 / 268
kapaonikense / R / 6" / gray-green lvs / Servia, Bulgaria 93
simonkiana / near J. hirta / Carpathians 64
sobolifera / R / 4-8" / greenish-yellow fls / n Europe, Asia 148 / 268

JUNCUS - Juncaceae
alpinus / R / 2-20" / black or brownish infloresence / n Europe 209

castaneus / P / to 1' / brownish fls / circumpolar 3
effusus / P / to 6' / straw to brown fls / all temperate regions 210 / 99
effusus spiralis/P/1-6'/fls yel-gr.,brown, stem twisted/Eur,N.A,N.Z. 25
ensifolius / Gr / creeping to 3'/many fl heads, sides dark / Finland 290
mertensianus / P / 10" / blackish brown fls / N. Am., Hokkaido 200
trifidus/Gr/mat-forming/fls dk brown/Arctic, sub-Arctic, Europe & s. 290

JUNIPERUS - Pinaceae
chinensis /T / to 60' / adult & juvenile lvs on same plant / Orient 167 / 127
chinensis v. procumbens / low Sh / needles with white bands / Japan 200
communis / Sh / 6-12' / black frs with blue bloom / Eu., N. Am. 148 / 36
communis ssp. alpina / may be either J. c. saxitalis or horizontalis 25
communis v. depressa / 3' / erect branchlets / n North America 25
communis v. hemisphaerica / 3' / mts of s Europe (182)
communis v. montana / mats / mts. of n Europe (182)
communis v. montana hemisphaerica - see J. communis v. hemisphaerica *
communis nana - see J. communis v. montana 25
communis v. saxatilis - see J. communis v. hemisphaerica (182)
deppeana v. pachyphlaea / HH T / 35' / glandular lvs / Tex., Ariz. 25
horizontalis / Sh / procumbent / bluish needles / ne North America 79
horizontalis 'Blue Rug' - see J. horizontalis 'Wiltonii' 300 / 139
horizontalis 'Wiltonii' - br. long, prostrate, glaucous-blue carpet 139
[1] nana - see J. communis v. montana
occidentalis / HH T / 35' / grayish-green needles / w United States 79
osteospenna /T/to 20'/bark peels,lvs scale,cones long /Mont. to N.M. 25
oxycedrus / HH t / 30' / red or brown frs / Medit. region 36 / 25
pachyphlaea - see J. deppeana v. pachyphlaea 25
phoenicea / Sh / to 10' / needles blue-grn / Med., Port., n Africa 25 / 36
procumbens - J. chinensis v. procumbens 200
sabina / Sh / to 10'/low,spreads/needles dk grn, aromatic/s & c Eur. 25
[1] saxatilis - see J. Communis v. hemisphaerica
virginiana / T / 35'+ / brownish-violet frs / N. Am., e of Rockies 79
wallichiana / HH T / to 60' / blue frs / Himalayas 127

JURINEA - Compositae
alata / P / 12-20" / fls purplish-blue / Caucasus 28
arachnoidea - see J. Mollis spp. arachnoidea 262
humilis / R / stemless / pale purple fls / s Europe 212
macrocephala / R / 6" / fls pinkish blue / Himalayas 182
mollis / P / to 2½' / fls rose-purple / s Europe 314 / 25
mollis arachnoidea / P / to 2½' / fls rose purple / s. Europe 262 / 25
pinnata / R / to 4" / purple fls / s & c Spain 212
spectabilis / P / 1' / fls purple / Caucasus 61

KADSURA - Schisandraceae
japonica / HH C. / to 12' / showy red frs / Far East 167 / 25

KALANCHOE - Crassulaceae
flammea / Gh / 1½' / red & yellow fls / Somali Republic (327) / 25

---

[1] D. R. Hunt and H. J. Welch, Taxon, 17:5.

KALMIA - Ericaceae
angustifolia / Sh / 3' / fls purple to crimson / e North America 157 / 25
angustifolia 'Candida' - fls white 25
angustifolia pumila - a dwarf form 61
angustifolia 'Rubra' - fls dark purple 25
cuneata / HH Sh / to 5' / fls creamy-white, red-banded / N. & S. Car. 157
latifolia / Sh / 6'+ / fls pale to deep rose-pink / e N. Am. 269 / 129
latifolia v. fuscata /Sh or T/to 10'/fls chocolate band inside/eN.A. 25
latifolia 'Myrtifolia' / Sh or T/ dwarf form/fls rose-wht/ e N.A. 36 / 25
latifolia v. rubra - fls deep pink 28
microphylla / Sh / to 2' / rose-purple to pink fls / w N. Am. 157
polifolia / Sh / to 2' / fls rose-purple / nc North America 75 / 25
polifolia v. microphylla - see K. microphylla (477) / 157
polifolia 'Nana' - a dwarf form 29

KALMIOPSIS - Ericaceae
leachiana / Sh / 1' / rose-purple fls / Oregon 36 / 123
leachiana 'Curry Co. Form' - rose-purple fls (185)
leachiana 'M. LePiniec' - designating an early plant collection (186) / (185)
leachiana 'Umpqua Form' - trailer, with good bloom (478) / 249

KALOPANAX - Araliaceae
pictus / T / to 80' / white fls / Far East 167 / 25

KECKIELLA
corymbosa - see PENTSTEMON corymbosa

KELSEYA -Rosaceae
uniflora / R / cushion / small white fls / Mont., Ida., Wyo. 107

KENNEDIA - Leguminosae
1 Beckxiana / Cl / woody? / fls red / w Australia
eximia / Gh Sh / prostrate / scarlet fls / w Australia 61
coccinea / Gh / trailing / scarlet fls / Australia 28
1 glabrata / P / prostrate / fls violet? / w Australia
macrophylla - see HARDENBERGIA comptoniana 28
nigricans / Gh / twining / dark purple fls / w Australia 184 / 93
prostrata / Gh / prostrate / scarlet fls / Australia (333) / 25
rubicunda / Gh / twining / purplish fls / Australia 45

KENTRANTHUS - see CENTRANTHUS 25

KERNERA - Cruciferae
alpina - see RHIZOBOTRYA alpina
saxatilis / R / to 1' / white fls / s & c Europe, in mts. 219 / 286

KICKXIA - Scrophulariaceae
spuria / A/ decumbent, to 20" / yellow & purple fls / s, w, c Eu. 288

KIRENGESHOMA - Saxifragaceae
palmata / P / 3' / fls bright yellow / Japan 165 / 208

---

[1] William E. Blackall and Brian J. Grieve, How to Know Westerm Australia's Wild Flowers, Parts, I, II, III, 1974.

KLEINIA
  repens - see SENECIO repens 93

KNAUTIA - Dipsaceae
  arvensis / P / 12-32" / lilac-blue fls / n Europe, Siberia 224 / 93
  drymeia / P / 3½' / fls reddish-violet to purple / Balkans 29
  macedonica / P / 2' / fls dark crimson / Macedonia 279
  sylvatica / P / 3' / violet fls / Europe 29

KNIPHOFIA - Liliaceae
  caulescens / HH P / 4½' / pink fls, suffused yel.-grn / S. Africa 208
[1] ensifolia / 2-3' / fls grn-wht, coral bds; or wht & grn bd/Transvaal
  galpinii - see K. triangularis 129 / 279
  macowanii / P / to 2' / fls yellowish to orange-red / S. Africa 25
  nelsonii / HH P / 1½-2' / bright scarlet fls / Kalahari region 28
  X pfitzeri / to 7' / fls red 25
  'Pfitzer's Hybrid' - see K. X. pfitzeri 25
  pumila / P / 2' / fls orange-red / South Africa 25
  rooperi / HH P / 4' / pale red or yellow fls / South Africa 28
  sarmentosa / HH R / 6-12" / yellow & red fls / South Africa 28
  snowdenii / HH P / 5' / fls orange-yellow / Mt. Eglon, Uganda 61
  triangularis / HH P / 2-3' / orange-flame fls / South Africa 279
  tuckii / P / 4-5' / yellow & scarlet fls / South Africa 251
  uvaria / P / to 5' / red fls / Cape Peninsula 93 / 279

KOCHIA - Chenopodiaceae
  prostrata / Sh / 2'+ / green fls / s Europe to sc Russia 286
  scoparia 'Trichophylla' / A / lvs red in fall / China, Japan 129

KOELERIA - Gramineae
  cristata / P Gr / 1-1½' / open, dry ground plant / prairies, N. Am. 28
  glauca / P Gr / 20" / blue-green lvs / Europe 93
  gracilis - see K. nitida 73
  nitida / P Gr / to 2' / native forage grass / c North America 73
  splendens - see K. vallesiana 268
  vallesiana / P Gr / 4-16" / silver green or purplish spikes / sw Eu. 268

KOELREUTERIA - Sapindaceae
  paniculata / T / 30' / fls yellow / Korea, China 200

KOHLRAUSCHIA
  prolifera - see PETRORHAGIA prolifera 286

KOLKWITZIA - Caprifoliaceae
  amabilis / Sh / to 15' / pink fls, yellow throat / c China 208 / 25

KOROLKOWIA - Liliaceae
  sewerzowii / Bb / 8-20" / brownish-violet fls / c Asia (416) / 4

KRIGIA - Compositae
  biflora / P / 8-28" / yellow fls / e & c North America 99

---

[1] Sima Eliovson, Wild Flowers of Southern Africa, 1980.

montana / R / 9-12" / yellow dandelions / Carolinas & Georgia 28
virginica / A / 12" / yellow dandelions / Maine to Florida, Texas 224 / 225

KUNZEA - Myrtaceae
ambigua / HH Sh to small T / ? / white stamens / Australia 45
baxteri / Gh Sh / 4'+ / red-bottlebrush fls / w Australia 45 / 25
ericifolia / HH Sh / 8-12' / fls yellow / Queensland, Australia 61
parvifolia / Gh Sh / 4-8' / fls purplish-crimson / Australia 109 / 61
peduncularis / HH Sh / tall / fls white / Australia 29

LABURNUM - Leguminosae
alpinum / T / 15'+ / fls yellow / sc Europe, in mts. 287
alpinum v. watereri - see L. X watereri
anagyroides / T / 15'+ / yellow fls / c & s Europe 148 / 139
X watereri - small tree with glossy lvs 139

LACHENALIA - Liliaceae
aloides 'Quadricolor' / 1' / four-colored fls 25
bachmannii / HH Bb / 10" / fls white, ridged red / South Africa 25
bulbifera / HH Bb / 1' / varii-colored fls / South Africa 267
contaminata / HH Bb / to 8" / white fls, flushed red / South Africa 25
glaucina / HH Bb / 1' / blue-lilac fls / Cape Province 189 / 111
glaucina v. pallida - light yellow fls 223
1 juncifolia / HH Bb / 6" / fls white tinged red / South Africa
liliflora / HH Bb / 1' / white fls / South Africa 93
mediana / HH Bb / to 16" / fls greenish wht. irid. / South Africa 25
mutabilis / HH Bb / to 1' / fls var. blue aging red-brn / S. Africa 25
orchioides / HH Bb / to 14" / fls in several colors / South Africa 25
pallida / HH Bb / to 10" / fls white / South Africa 25
purpureo-coerulea / HH Bb / 6-8" / reddish-purple fls / South Africa 93
pustulata / Bb / to 12" / lvs 6-9"; fls wht, red tinged / S. Africa 61
reflexa / HH Bb / 6" / fls yellowish-green / South Africa 25
roodeae - fls 4-6" tall, rosy purp-blue purp; lvs to 1' 95
2 rosea / HH Bb / 5-6" / fls pink / sw Cape, South Africa
2 rubida / HH Bb / 6" / fls deep crimson / Cape Province, South Africa
2 unicolor / HH Bb / to 15" / fls violet / western Cape, South Africa
2 violacea / HH Bb / to 7" / blue-grn, lined purple / Namaqualand

LACTUCA - Compositae
alpina / P / to 6' / fls pale blue / arctic & alpine Europe 25
canadensis / B / 1½-8' / fls yellow / ne United States & Canada 99
floridana / B / 2-5' / fls bluish / Fla. to Texas & n-ward 99
floridana v. villosa /B / 2-5'/ fls blue; lvs uncleft/Fla.-Tex.& n 99
hirsuta / B / tall / fls yel, dry purplish / Pennsylvania & Virginia 99
perennis / P / 1-2' / lilac or blue fls / s, c & e Spain, s France 148 / 212
plumieri (cicerbita)/ P / 2-3'/ fls blue,violet/ Pyrenees of France 289
quercina / A or B / 1-3' / fls yel., dense panicle / c & e Europe 289
saligna / A or B /slender willow-lvs; fls yel./ Pa.-Mich, s to Mo. 99
tenerrima / P / 8-20" / lilac fls / sw Europe 289

---

[1] J. Ingram, Baileya, 14:3.

[2] Sima Eliovson, Wild Flowers of Southern Africa, 1980.

LAGENOPHORA - Compositae
pinnatifida / HH R / to 10" / white to purple fls / New Zealand 15
pumila / HH R / 6"+ / white to purple fls / New Zealand 184 / 15
stipitata / HH R / low / purple or dark red fls / s hemisphere 109 / 184

LAGERSTROEMIA - Lythraceae
indica / HH T / 10-35' / fls usually bright pink / China 28
speciosa / Gh T / to 80' / purple or white fls / tropics (327) / 25

LAGOTIS - Scrophulariaceae
glauca / R / 4-10" / pale blue-purple fls / n Japan 273 / 200
stelleri v. yesoensis /R/ 4-6"/fls dense, blue-purp/ Hokkaido, Japan 200

LAGURUS - Gramineae
ovatus / A Gr / to 1' / panicle pale & downy / Medit. region 129 / 25

LALLEMANTIA - Labiatae
canescens / P gr. as A / to 18" / bluish-lavender fls / sw Asia 25

LAMARCKIA - Gramineae
aurea / A Gr / 6-12" / golden yellow panicles / Medit. reg; Afghan. 28
aurea 'Golden Top' / Gr/ to 16"/ panicle yel. to purp / sw U.S. 25

LAMIUM - Labiatae
Galeobdolen / P / to 18" / fls yellow / Europe, w Asia 28
Galeobdolen 'Variegatum' / 6" / fls yellow; lvs silver & green 129
maculatum / P / to 18" / pink to purplish fls / Eurasia 210 / 25
[1] maculatum 'Beacon Silver' / 6-8" / silver lvs, grn edge; pink fls
maculatum 'Chequers' - lvs silvery white, margin green 95
maculatum 'Roseum' - shell-pink fls 129
orvala / P / 20-40" / pink to dark purple fls / n Italy, Jugoslavia 129

LAPAGERIA - Liliaceae
rosea / Gh / twining / rose to rose-crimson fls / Chile (327) / 129

LAPEIROUSIA - Iridaceae
anceps / HH C / 6-10" / dark purple to magenta fls / South Africa 25
corymbosa / P / 8-12" / fls pale to deep violet / South Africa 25
cruenta - see L. laxa 25
cruenta 'Alba' - see L. laxa 'Alba' 25
divarietae / C / 6-12" / perianth white / South Africa 25
erythrantha / HH C / to 16" / bluish, white, reddish fls / Africa 185
erythrantha rhodesiaca - included in the sp. 185
jacquinii / HH C / 6" / blue fls / Cape of Good Hope 223 / 185
laxa / HH C / 4-10" / bright red fls / South Africa 25
laxa 'Alba' - fls white 25
laxa ssp. grandiflora - larger fls 185
rhodesiana - see L. erythrantha 185
sandersonii - see L. erythrantha 185
schimperi / HH C / to 1' / white fls / Ethiopia, w & sw Africa 185
viridis - see TRITONIA viridis 121

---

[1] Wayside Catalog, Spring, 1986.

LAPIEDRA - Amaryllidaceae
martinezii / HH Bb / 6" / white fls / s Spain 212 / (116)

LAPPULA - Boraginaceae
echinata / A / 6-20" / blue fls; barbed frs / United States 227

LARDIZABALA - Lardizabalaceae
biternata / Gh Cl / to 40' / fls white; edible frs / Chile (479) / 25

LARIX - Pinaceae
X 'Dunkeld' - see L. X eurolepis *
X eurolepis - intermediate of parents: European & Japanese Larch 25
Kaempferi / T / to 100' / needles bluish / Japan 25
laricina / T / to 60' / light bluish green needles / ec N. Am. 28
laricina f. depressa - prostrate branched American Larch 99

LASERPITIUM - Umbelliferae
gallicum / P / 4' / white or pink fls / s Europe, in mts. 148 / 287
latifolium / P / 2-5' / white fls / much of Europe 148 / 287
peucedanoides / P / 1-3' / rays smooth, petals white / Jugoslavia 287
Siler / P / 1-5' / lvs midrib whitish; petals wht/ sc Europe 287

LASTHENIA - Compositae
chrysotoma / R / to 16" / golden fls / Calif., Oregon, Arizona 228
macrantha / R / 4-16" / yellow fls / coastal California, Oregon 228

LATHRAEA - Orobanchaceae
clandistina / pale gray-purple fls; parasitic plant on willow / Eu. (187) / 28

LATHYRUS - Leguminosae
albus - see L. pannonicus 287
aphaca v. affinis /A/trailing to 20"/fls crm or sulphur/e Medit reg. 77
articulatus / A / to 3' / crimson & white or pink fls / Medit reg. 287
aurantius - see VICIA crocea 9
aureus / P / to 2' / fls brown or orange-yellow / Black Sea region 287
cirrhosus / P / 1-4' / pink fls / Pyrenees, Cebennes 314 / 287
cyaneus - see L. digitatus 287
davidii / P / 3-5' / fls yellow changing to brown / Japan, China 200
digitatus / R / 4-16" / fls bright reddish-purple / se Europe 287
drummondii - see L. rotundifolius 61
filiformis / P / 6-20" / fls red-purple / n Italy to e Spain 287
gmelinii / corolla yellow / c & s Ural Mountains 287
grandiflorus / P / 6' / rose-purple fls / s Europe 149 / 25
hirsutus / A / to 4' / red & pale blue fls / c & s Europe 225 / 287
inermis - see L. laxiflorus 287
japonicus / P / decumbent stem to 2' / purple fls / Eurasia 210 / 25
japonicus ssp. maritimus / 3' / fls purple / w Europe 287
laevigatus / P / 8-24" / fls yellow / ec & e Europe 212 / 287
latifolius / P / 2'+ / fls purple-pink / c & s Europe 257 / 287
latifolius 'Albus' - white fls 25
laxiflorus / P / 8-20" / fls blue-violet / se Europe 287
littoralis / P / climbing to 2' / purp, wht fls / Eur., nat. N.A. 25
luteus aureus - see P. laevigatus 93
magellanicus / P gr as A / 3-5'/ drk purp-blue fls /Straits of Mag. 28

maritimus - see L. japonicus ssp. maritimus 224 / 287
montanus / P / 6-20" / crimson fls aging bluish / s, w & c Europe 314 / 287
niger / P / to 3' / fls purple turning blue / Europe 194 / 287
nissolia / A / to 3' / fls crimson / w & s Europe, England 210 / 287
odoratus / A Cl / to 8' / fls in wide range of colors / Sicily 229 / 129
pannonicus / P / 6-20" / pale cream fls / ec Europe 314 / 287
pauciflorus / P / ? erect / fls 4-7, orchid or purp/ n Cal, s Ore. 228
pisiformis / P / 3' / reddish-purple fls / ec Europe, USSR 287
pratensis / P / 3' / yellow fls / Eurasia 148 / 25
pubescens / Gh Sh / 4' / purplish blue fls / Chile, Argentina 206
roseus / P / to 5' / pink fls / Anatolia, Caucasus 287
roseus 'Plenus' - doubled fls 268
rotundifolius / P Cl, by tendrils / fls deep rose / e Russia (189)
rotundifolius albus / P / 16-32" / corolla purp-pink / Russia 287
sativus / A / 3' / fls white, pink, blue / fodder pl in Europe 210 / 287
sibthorpii - see L. undulatus 28
sphaericus / A / 4-20" / fls orange-red / s Europe, n to Sweden 287
sylvestris / P Cl / 2-6' / fls purple-pink / most of Europe 138 / 287
tingitanus / A / 2-4' / bright purple fls / Iberian Peninsula 208 / 287
tomentosus / Cl / 3' / fls lilac / Buenos Ayres 206
tuberosus / P / 4" / rose fls, edible tubers / Eurasia 210 / 25
undulatus / P / 2'+ / purple-pink fls / European Turkey 184 / 287
variegatus - see L. venetus 287
venetus / P / 8-16" / fls reddish-purple / se & ec Europe 287
venosus / P / to 3' / fls purple / e & c North America 224 / 25
vernus / P / 8-16" / fls reddish-purple / Europe 210 / 287
vernus 'Albus' - fls white 25
vernus 'Albo-roseus' - light pink & white fls 147

LAURELIA - Monimiaceae
aromatica - see L. serrata 139
serrata / Gh Sh / 5'+ / evergreen aromatic lvs / Chile 139

LAURENTIA - Campanulaceae
axillaris / A / 1-2' / lav-blue, dk mid vein, sol. fl / Australia 95
gasparrini / A / 4" / tiny lilac fls / Portugal, w & n Spain 212
minuta / R / acaulescent / pale violet fls / Medit. region 25
tenella - see L. minuta 25

LAVANDULA - Labiatae
angustifolia / Sh / to 3' / purple fls / Medit. region 148 / 288
angustifolia ssp. angustifolia / Sh / 2-3' / fls purple / Europe 25
angustifolia ssp. angustifolia 'Hidcote'/slow-growing/fls purple 300 / 25
angustifolia ssp. angustifolia 'Hidcote Pink' - fls pinkish wht 36
angustifolia ssp. angustifolia 'Munstead'/fls lav-blue, early bloom 25
1 angustifolia 'Jean Davis' / 15"/ bl-grn lvs; frag wht fls, tinge pnk
angustifolia 'Munstead Dwarf' / 1' / mound of fl spikes 300 / (190)
angustifolia 'Nana' - dwarf 25
angustifolia 'Nana-Alba' / 1' / fls white 36
angustifolia nana atropurpurea - sim.to 'Hidcote', calyx less woolly 36
angustifolia ssp. pyrenaica - longer floral bracts 25
angustifolia rosea / Sh / 2-3' / fls rose pink / Europe 25
angustifolia 'Twickel Purple' - purple fls in fan-like clusters 25

---

1 Wayside Catalog, Spring, 1986.

dentata / Sh / 1-3' / fls purplish / s & e Spain 25
lanata / HH Sh / 1-2' / lilac fls; lvs gray-green / s Spain 212 / 288
latifolia / HH Sh / to 3' / purple fls / w Medit. reg., Portugal 184 / 288
multifida / Sh / to 2' / fls blue violet / w Mediterranean, Portugal 25
officinalis - see L. angustifolia 25
officinalis 'Hidcote' - see L, a. ssp. angustifolia 'Hidcote' 25
officinalis 'Hidcote Pink' - see L. a. ssp. a. 'Hidcote Pink' 36
officinalis nana 'Munstead' - see L. a. ssp. a. 'Mumstead' 25
pedunculata - see L. stoechas ssp. pedunculata 288
pedunculata 'Atlas' - see L. stoechas s. pedunculata
pinnata / Sh / to 3' / lavender / Madeira, Canary Islands 25
spica - see L. angustifolia 288
spica angustifolia - see L. angustifolia *
spica 'Hidcote' - see L. a. ssp. angustifolia 'Hidcote' 184 / 25
spica 'Hidcote Pink' - see angustifolia L. a. ssp. a. 'Hidcote Pink' 36
spica 'Munstead Dwarf' - see L. angustifolia 'Munstead Dwarf' 25
spica 'Nana' - see L. angustifolia 'Nana' 25
spica 'Nana-Alba' - see L. angustifolia 'Nana-Alba' 25
spica 'Twickel Purple' - see L. angustifolia 'Twickel Purple' 25
stoechas / HH Sh / 1' / fls dark purple / s Europe 194 / 129
stoechas ssp. lusitanica / short fl spike / c & s Portugal 288
stoechas ssp. pedunculata / longer peduncle / s Portugal 212 / 288
vera nana alba - see L. angustifolia ssp. angustifolia 25
vera 'Munstead' - see angustifolia 'Munstead Dwarf' 25
viridis / HH Sh / 2' / white fls / sw Spain, s Portugal 212 / 288

LAVATERA - Malvaceae
acerifolia / Sh / 5' / fls pink; evergreen / Teneriffe 206
arborea / HH P / to 10' / fls purple-pink, veined darker / s Eu. 129 / 208
assurgentifolia / HH Sh / 6-15' / fls purple / S. Calif. Isla. 28
cachemiriana / P / 5' / pale rose fls / Kashmir 279 / 282
cretica / A or B / 4-6' / fls lilac to purple / s Europe 25
'Loveliness' / 3' / compact / rich deep rose-pink fls 268
maritima / HH Sh / to 5' / pale lilac fls / s France 212 / 139
olbia / Gh Sh / 6' / pink to reddish-pink fls / s France 139
olbia 'Rosea' - see L. X 'Rosea' 129 / 279
plebeia / Gh / 2' / pale fls / Australia 206
X 'Rosea' / HH P / 5'+ / clear rose-pink fls / reverts from seed 279
thuringiaca / P / 2-5' / fls purplish-pink / c & se Europe 292 / 287
trimestris / A / to 4' / fls bright pink / Medit. region, Portugal 287

LAVAUXIA
mutica - see OENOTHERA mutica

LAWSONIA - Lythraceae
inermis / Gh Sh / to 20' / fls white to red; dye-plant / trop. Asia 138 / 25

LAYIA - Compositae
elegans - see L. platyglossa 129 / 25
platyglossa / A / to 1' / yellow fls tipped cream-white / Calif. 228 / 25

LEDEBOURIA - Liliaceae
socialis / Gh Bb / 4" / fls greenish; gray-blotched lvs / S. Africa 185

LEDUM - Ericaceae
columbianum / Sh / 3' / white fls / Washington, Oregon 28
glandulosum / Sh / to 6' / fls white / nw North America 25
glandulosum v. columbianum - see L. columbianum 28
groenlandicum / Sh / to 3' / white fls / n North America 22 / 25
groenlandicum 'Compacta' - dwarfed 93
groenlandicum nanum - see L. groenlandicum 'Compacta' 93
hypoleucum - see L. palustre v. diversipilosum 200
palustre / Sh / to 3' / white fls / n Eurasia 25
palustre v. decumbens / small linear lvs / n Asia, n North America 25
palustre v. diversipilosum - lvs glaucous beneath 25
palustre japonicum - see L. palustre v. diversipilosum 200
palustre nipponicum - see L. palustre v. diversipilosum 200
palustre yesoense - see L. palustre v. diversipilosum 200

LEGOUSIA - Campanulaceae
falcata / A / to 1' / bluish or pinkish fls, weedy / Eurasia 12
speculum-veneris / A / 8-18" / fls violet blue to white / Eurasia 212 93
speculum-veneris 'Grandiflora' / 9" / blue fls 283

LEIBNITZIA - Compositae
anandria / R / to 8" / white & red-purple fls / Far East 164 / 200
kunzeana / R / to 1' / pappus chestnut-brown / Himalayas 25

LEIOPHYLLUM - Ericaceae
buxifolium / Sh / to 3' / fls white to pink / e North America 36 / 25
buxifolium v. hugeri / pedicels glandular / Kentucky, Carolinas 25
buxifolium v. prostratum / prostrate / Tenn., Ga., N.C., in mts. 133 25
prostratum - see L. buxifolium v. prostratum 28

LEMBOTROPIS - Leguminosae
nigricans / Sh / to 3' / fls golden yellow / sc Europe, Italy 210 / 287

LEONOTIS - Labiatae
dubia / Gh / ? / deep orange-yellow fls / South Africa 61
leonurus / Gh Sh / 6-7' / orange fls / South Africa (327) / 25

LEONTICE - Berberidaceae
albertii / R / 5" / fls ocre-yellow / Turkestan 28
chrysogonum - see BONGARDIA chrysogonum 211
ewersmannii - see L. leontopetalum ssp. eversmannii 77
leontopetalum ssp. ewersmannii / to 8" / narrow lvs / c Asia 77

LEONTODON - Compositae
montanum ssp. pseudotaraxaci / R / 4-8" / deep yel. fls./Carpathians 312
pseudotaraxaci - see L. montanus ssp. pseudotaraxaci 312

LEONTOPODIUM - Compositae
alpinum / R / 4-6" / silvery-white bracts / European Alps 148 / 129
alpinum nivale - see L. nivale 93
calocephalum / R / to 16" / white or yellowish fls / Tibet 93
campestre - see L. leontopodioides 93
discolor / R / to 1' / bracts brown / Hokkaido, Sakhalin 200
fauriei / R / 6" / white or brown bracts / Honshu 164 / 200
haplophylloides / R / to 14" / lvs lemon-scented / c China 138 / 25

hayachinense / R / to 8" / white-woolly bracts / Honshu 164 / 200
himalayanum / R / 10" / silvery-woolly lvs / Himalayas 25
jacotianum / R / to 12" / like L. alpinum, stoloniferous / Himalayas 25
japonicum / P / 10-22" / yellowish bracts / Far East 164 / 200
japonicum f. orogenes /P/ 8-16"/many heads, inflor.br/Honshu, Hyushu 200
kamtschaticum - see L. kurilense 25
kurilense / R / 8" / lvs forming floccose star / Kurile Isls. 25
leontopodioides / R / to 1' / white bracts / Altai & Himalayas 93
X lindavicum - hybrid of L. alpinum X L. japonicum 29
monocephalum / R / to 5" / yellowish, woody / Himalayas 25
nivale / R / 4" / white woolly bracts / Bulgaria 148 / 93
ochroleucum / R / to 4" / yellowish felty / steppes c Asia 25
palibinianum / R / 4" / free-flowering / Mongolia, Siberia 93
shinanense / R / to 4" / white-yellow gray / alpine Japan 25
sibiricum - see L. leontopodioides 25
souliei / R / 6" / grayish-white fls / w China 93
stracheyi / P / to 20" / silvery-gray bracts / sw China 93
wilsonii / R / 10"+ / stars of lvs, white-tomentose / w China 25

LEOPOLDIA - Liliaceae
comosa / Bb / 6-12" / greenish-brown fls / Medit. region, w Asia 186 / 185
comosa 'Alba' - the white form 185
comosa 'Plumosa' - mauve sterile fls only 185
plumosum - see L. comosa 'Plumosum' 185
tenuiflora / Bb / to 14" / fls bright bluish-violet / c Asia 4

LEPACHYS
columnaris - see RATIBIDA columnifera 25

LEPIDIUM - Cruciferae
cartilagineum ssp. crassifolium / 8" / white fls / e & c Europe 286
crassifolium - see L. cartilagineum ssp. crassifolium 286

LEPTARRHENA - Saxifragaceae
pyrolifolia / R / 8-16" / small white fls / Alaska to Washington 28

LEPTODACTYLON - Polemoniaceae
pungens / Sh / to 3'/ fls pink, lilac, yel., wht/ B.C. to Baja, Cal. 25

LEPTOLEPIA - Dennstaedtiaceae
novae-zelandiae / Gh F / to 1½' / 3-pinnate frond / New Zealand 15

LEPTOPTERIS
superba - see TODEA superba 15

LEPTOSPERMUM - Myrtaceae
boscawenii - see L. scoparium 'Boscawenii' 36
flavescens / HH T / to 25' / yellowish-white fls / Australia 45 / 25
flavescens v. grandiflorum - lvs & fls larger 25
flavescens v. obovatum / Sh / to 10' / white fls 36
humifusum / HH Sh / prostrate / small white fls / Tasmania 36 / 139
juniperinum / HH Sh / 6' / stiff, pungent lvs / e Australia 109 / 25
laevigatum / HH Sh / 20'+ / fls white / Australia 28
lanigerum / HH Sh / 5'+ / silvery lvs / Australia, Tasmania 109 / 139
nitidum / HH Sh / 3-10' / fls white or reddish / Tasmania 36

obovatum - see L. flavescens v. obovatum 36
pubescens / HH T / small / fls white / Australia 28
rotundifolium / HH Sh / 6' / fls pale pink / e Australia 25
rupestre - see L. humifusum 36
scoparium / HH Sh / to 12' / fls white to pink / N.Z., Tasmania 36 / 25
scoparium 'Boscawenii' - white fls, pink buds 36
scoparium 'Nanum' / 4-6" / lvs heather-like 134 / 129
scoparium 'Nichols' - fls carmine-red 184 / 139
scoparium 'Nichols Gloriosa' - pink flowered seedling 61
scoparium 'Nichols Nanum' / 4" / red fls 132
scoparium 'Nichols Pygmaeum' - see L. scoparium 'Nichols Nanum' 132
scoparium 'Prostratum' - see L. humifusum 139
scoparium 'Red Damask' - very dbl, dp red, long-lasting fls 139
scoparium 'Rotundifolium' - see L. rotundifolium *
sericeum / HH Sh / 6-10' / white fls / Tasmania 139
squarrosum / HH Sh / 9' / fls white, aging pink / se Australia 25

LEPTOSYNE
Stillmanii 'Sungold' - see COREOPSIS Stillmanii 25

LEPTOTAENIA
dissecta - see LOMATIUM dissectum 196

LESCHENAULTIA - Goodeniaceae
1 biloba / P / to 1½' / fls rich blue, purp, wht / Australia

LESPEDEZA - Leguminosae
bicolor / Sh / to 10' / fls purple / Japan 184 / 25
capitata / P / 18-38" / fls creamy-white / e & c United States 99
virgata /P/16-24"/stem purp; lvs dk grn; few fls/Jap., Kor., Manch. 200

LESQUERELLA - Cruciferae
alpina / R / 1-7" / yellow fls / Rocky Mountains 229
arctica / R / to 1' / bright yellow fls / Alaska to Greenland 209
englemannii ssp. ovalifolia / R / 5" / yel. fls / Col., Kan. to Tex. 229
fendleri / R / 4" / golden-yellow fls / Kansas, Colorado, Utah 229
kingii / R / 3-8" / yellow fls; gray lvs / Oregon, Calif., Utah 229
ludoviciana / R / to 16" / yellow fls / c United States 99
montana / R / stems on ground / yellow fls / Wyo. to New Mexico 198
occidentalis / R / to 8" / S-curved stems, yellow fls / n Calif. 229 / 228
2 ovalifolia - see L. engelmannii ssp. ovalifolie
recurvata / R / 4-16" / yellow fls / c Texas 226
3 tumulosa / R / congested dome / yellow fls / Utah

LESSINGIA - Compositae
leptoclada / P / 1-3' / lavender or bluish fls / s Sierra Nevadas 228

---

1 William E. Blackall and Brian J. Grieve, How to Know Westerm Australia's Wild Flowers, Parts, I, II, III, 1974.

2 Curtis Clark, Brittonia, 27:3.

3 Dwight Ripley, Quar. Bull. Alpine Garden Soc., 40:3.

LEUCAMPYX - Compositae
newberryi / P / 8-24" / white or cream daisies / Col., N. Mex. 229 / 227

LEUCANTHEMOPSIS - Compositae
alpina / 1-3" / fls white / mts. of Europe 289
radicans / P / dwarf runners/ ligule yel, turn org-red / s Spain 289

LEUCANTHEMUM - Compositae
alpinum - see LEUCANTHEMOPSIS alpina 289
atlanticum - large clumps dark grn lvs; wht daisy fls 154
atratum ssp. coronopifolium / P / 8-20" / basal lvs, spath-cun./Alps 289
catananche - see CHRYSANTHEMUM catananche 132
hosmariense - see CHRYSANTHEMUM hosmariense 132
Weyrichii - see CHRYSANTHEMUM Weyrichii

LEUCOCORYNE - Liliaceae
ixioides / HH Bb / 1' / fls white or pale blue / Chile 186 / 25
ixioides odorata - see L. odorata 186
narcissoides / HH Bb / ? / yellow fls, daffodil-like/Alacama, Chile (191)
odorata / HH Bb / ? / fragrant, smaller white fls / Valparaiso Prov. 186
purpurea / HH Bb / 1' / fls lavender & crimson-maroon / Chile 61

LEUCOGENES - Compositae
grandiceps / R / to 8" / dense heads, white-woolly bracts / N.Z. 132 / 25
leontopodium / R / to 8" / lanceolate bracts / New Zealand 129 / 25

LEUCOJUM - Amaryllidaceae
aestivum / Bb / 1½-2' / white fls / Europe, Asia Minor 192 / 129
aestivum 'Gravetye Giant' - tall with large fls 129 / 185
aestivum v. pulchellum / smaller / Majorca 279
aestivum vagneri - see L. vernum v. vagneri *
autumnale / HH Bb / 4-6" / fls white, tinged pink / sw Europe 212 / 129
autumnale v. oporanthum - taller, to 10" 185
autumnale v. pulchellum - with fls & lvs together 185
hiemale - see L. nicaense 25
longifolium / HH Bb / 8" / small white fls / Corsica (27) / 185
nicaense / HH Bb / 6" / white fls / Mediterranean region (31) / 25
pulchellum - var. pulchellum of L. aestivum or autumnale 185
roseum / HH Bb / 4" / pinkish fls / Corsica, Sardinia 267
trichophyllum / HH Bb / 10" / pink or white fls / Spain, Portugal 210 / 185
vernum / Bb / 8" / fls white / Europe 148 / 267
vernum v. carpathicum /fls greenish, yellow tipped /Poland, Roumania 269 / 129
vernum v. vagneri - robust form from Hungary 180 / 267

LEUCOPOGON - Epacridaceae
fraseri / Gh Sh / 6" /pinkish fls; frs orange-yellow / N.Z., Austrl. (480) / 25

LEUCOTHOE - Ericaceae
axillaris / Sh / to 6' / fls white to pink/ Va. to Fla & Miss. 25
catesbaei, dwarf form - see L. fontanesiana 'Nana' *
catesbaei minor - referrable to L. fontanesiana 'Minor'
davisiae / Sh / 5' / white fls / Oregon, California 36 / 222
editorum - see L. fontanesiana *
fontanesiana / Sh / to 6' / white fls / Va., Ga., Tenn. 36 / 25
fontanesiana 'Nana' - compact form 36

grayana / Sh / 1-3' / fls pale green / n Japan 200
keiskei / Sh / declining branches / white fls / Honshu 132 / 200
recurva / HH Sh / to 12' / white fls / Virginia to Georgia 25

LEUZEA - Compositae
conifera / B or P / 4-12" / fls purple / s Europe 28
rhapontica / P / to 28" / red or purple fls / Alps 314 / 289
scariosa - see L. rhapontica 289

LEVISTICUM - Umbelliferae
officinale / P / to 6" / fls greenish yellow / s Europe 25

LEWISIA - Portulacaceae
bernardina - see L. nevadensis 88
brachycalyx / R / rosette / white to pink fls / Utah, N. Mex., Calif 132 / 89
brachycalyx 'Alba' - selection for white fls *
bracycalix rosea - rose-colored form of L. brachycalyx
'Carroll Watson' - see L. cotyledon 'Carroll Watson' (403)
cantelovii / R / 6-15" / fls light pink / w Sierra Nevada Mts. 228 / 89
columbiana / R / 8-10" / white to light pink fls / Wash., B.C. 229 / 107
columbiana 'Alba' - the white phase *
columbiana v. columbiana/ R /5-12"/pnk-wht,pnk lines/B.C.,Wash,Vanc. 89 / 90
columbiana 'Rosea' - deep rose-red fls (433) / 132
columbiana v. rupicola / intermediate in size / Oregon 89
columbiana v. wallowensis - smaller in all parts 89
congdonii / R / 8-12" / rose fls / Fresno & Mariposa Cos., Calif. 89
cotyledon / R / 9" / variable pink fls / California 129 / 132
cotyledon 'Alba' - white fls 249
cotyledon 'Carroll Watson' - pure yellow fls (403)
cotyledon 'Drakes Strain' - see L. cotyledon 'Sunset Strain' (247) / 154
cotyledon v. finchae - included in the sp. 88 / 89
cotyledon v. heckneri - deep rose fls; fringed lvs 75 / 123
cotyledon v. howellii / lvs fluted / Siskiyou Mts. 89
cotyledon 'Jean Turner' - photograph in the reference 88
cotyledon millardii - included in the sp. 25
cotyledon 'Purdyi' - segregated for shorter,broader, red-flushed lvs 64
cotyledon 'Rose Splendour' - clear rose fls 184
cotyledon 'Sunset Strain' - wide range of new colors (481) / 89
X edithae - L. columbianum 'Rosea' x L. columbiana v. rupicola 88
finchae - see L. Cotyledon (481) / 89
heckneri - see L. cotyledon v. heckneri 88
howellii - see L. cotyledon v. howellii 129 / 89
'Jean Turner' - see L. cotyledon 'Jean Turner' 88
leana / R / 9" / white of light pink fls / Siskiyou Mts., Oregon 306 / 107
leana 'Alba' - the white form *
longipetala / Gh / 4" / rose-red fls / Eldorado Co., Calif. 89
millardii - see L. cotyledon 25
nevadensis / R / rosette / small white fls / Nevada 227 / 123
nevadensis 'Bernadina' / R / to 4" / fls white / Cal. to Wash. 25
oppositifolia / R / 6" / white fls / Oregon 228 / 132
purdyi - see L. cotyledon 'Purdyi' *
pygmaea / R / dense rosette / silvery-pink fls / Rocky Mts. 312 / 123
pygmaea ssp. longipetala - see L. longipetala 89
rediviva / R / tufts / pink fls / Montana & w-ward 229 / 107
rediviva 'Alba' - white fls 107

'Rose Splendour' / red and rose colored fls / cult by Weeks 88
rupicola - see L. columbianum v. rupicola 88
sierrae / R / tufts / pink or rose fls / California 89
stebbinsii / R / to 7" / white or rose fls / Mendocino Co., Calif. 89 / (192)
X 'Sunset Strain' - see L. cotyledon 'Sunset Strain' 83
X 'Trevosioa' - hyb. of L. cotyledon v. howellii x L. columbiana 184 / 89
triphylla / R / 2-4" / white or pink fls / nw United States 228 / 89
tweedyi / R / 4" / flesh-pink & pale apricot fls / Washington 129 / 132
tweedyi 'Alba' - white fls *
tweedyi 'Rosea' - rose-pink fls, yellow centers 24
wallowensis - see L. columbiana v. wallowensis 89

LEYCESTERIA - Caprifoliaceae
formosa / Sh / 6' / white fls, purplish bracts / Himalayas 132

LIATRIS - Compositae
aspera / P / to 6' / purple fls / c North America 224 / 25
aspera 'September Glory - fls rosy purple 95
aspera 'White Spire' - white-flowered clone (193)
borealis - see L. novae-angliae (193)

callilepis - see L. spicata 25
cylindracea / P / 18-24" / purplish fls / N.Y. to Arkansas 224
elegans / P / to 5' / purple or white fls / S. Carolina, Fla., Tex. (193)
gracilis / P / 1-3' / purple fls / Georgia, Alabama, Florida 28
graminifolia / P / 2'+ / rose-purple fls / coastal plain, N.J. & s 224 / 99
helleri / P / 4-16" / fls pink purp; lvs 1" / N.C., Va. 225 / 250
ligulistylis / P / to 3' / purple fls / c United States 224 / 25
novae-angliae / P / to 3' / fls rose-purple or white / N.Y. to N. Eng 25
punctata / R / to 1' / purple fls / c North America 224 / 25
pycnostachya / P / to 5' / purplish fls / c United States 224 / 25
pycnostachya 'Alba' - fls white 25
pycnostachya f. Hubrichtii - white fls 259
scariosa / P / 3' / purple fls / s Pa. to Ga., in mts. 224 / 25
scariosa alba / P /?/ heads 20-60 fld /white fls /Pa.,W.Va. to S.C. 99
secunda / P / to 32" / rose-purple fls / Florida to North Carolina 250
spicata / P / to 5' / fls reddish purple / ec & es United States 147 / 25
spicata albaflora / 1-3' / fls white / ne & s United States 99
spicata 'Kobold' / 18" / purple fls 300 / (194)
spicata f. montana / shorter inflorescence, dw. pl / Va., Ga., mts. 25
spicata 'White Spires' - see L. aspera 'White Spire' 95
squarrosa / P / to 3' / fls purple / Del., Ill., Mo., Ala. 224 / 25
squarrosa v. glabrata / plant glabrous / S.Dak., Tenn., Texas 25

LIBERTIA - Liliaceae
caerulescens / HH P / to 2' / fls blue & greenish-brown / Chile 25
formosa / HH P / 2-4½' / fls white & greenish-brown / Chile, in mts 279 / 25
grandiflora / HH P / to 3' / fls white & greenish / New Zealand 184 / 25
ixioides / HH P / to 2' / white & greenish fls / New Zealand 238 / 25
paniculata / HH P / 2' / white fls / e Australia, New Zealand 45 / 25
pulchella / HH R / 10"+ / white fls / N.Z., N. Guinea, Australia 109 / 25

LIBOCEDRUS
decurrens - see CALOCEDRUS decurrens 129 / 127

LIGULARIA - Compositae
clivorum - see L. dentata 25
clivorum 'Desdemona' - see L. dentata 'Desdemona' 268
dentata / P / 3-5' / orange-yellow fls / China, Japan 129 / 132
dentata 'Desdemona' / 3-4½' / bright orange-red fls (195) / 268
dentata 'Othello' / 3'+ / orange fls, lvs purplish-red beneath 25
glauca / P / to 3' / yellow fls / Carpathians, Bulgaria 289
X 'Hessei' / P/ to 6' / fls orange-yellow / Europe, Asia 25
hodgsonii / P / 12-32" / yellow fls / Honshu, Hokkaido 164 / 200
1 holmii / R / basal cordate lvs; fls yellow / Rockies 302
japonica / P / 4'+ / orange-yellow fls / Far East 129 / 25
'Othello' - see L. dentata 'Othello' 25
przewalskii / P / to 6' / yellow fls; deeply lobed lvs / n China 194 / 268
sibirica / P / to 3' / yellow fls / France to Himalayas, Japan 25
sibirica v. speciosa - larger & showier Far Eastern form 25
speciosa - see L. siberica v. speciosa
stenocephala / P / 5' / yellow fls / Far East 49 / 25
veitchiana / P / to 8' / bright yellow fls / China 25
Wilsoniana /P/to 6'/like L. veitchiana/disc.fls 10-12,ray 6-8/China 25

LIGUSTICUM - Umbelliferae
alpinum / P / to 12" / fls white or pink / Russian alps 287
grayi / P / to 2' / fls white, small / Cascade Mountains 228
lucidum / Sh or T / to 30' / fls white / China, Korea 25
mutellinoides / P/ to 12"/ fls pink or white/alps & arctic of Russia 287
scoticum / P / to 3' / dark green lvs; fls white / coasts of n Eu. 210 / 287
simplex - see L. mutellinoides

LIGUSTRUM - Oleaceae
japonicum / P / to 10' / evergreen, white fls / Japan, Korea 200 / 25
lucidum / HH Sh / to 20' / fls white / Japan, China 28
obtusifolium v. regelianum / low Sh / horizontally-branched / Japan 222

LILIUM - Liliaceae
albanicum / Bb/ to 16"/ yel. fls / Albania, nw Greece, sw Jugoslavia 290
Alexandrae - see L. nobilissimum 25
amabile / Bb / 2-3' / orange-red fls / Korea 75 / 267
amabile v. luteum / 2' / fls deep golden yellow 184 / 111
'Amber Gold' / to 3' / 10-12 Turks cap. yellow, spotted maroon 111
amoenum / fls deep pink, 1-3 / w China 61
auratum / Bb / to 8' / white & yellow fls, crimson spots / Japan 267 / 25
auratum Parkmannii / Bb / 2' / fls crimson, red papillae / Japan 268 / 311
auratum pictum / Bb / to 8" / fls densely spotted crimson / Japan 25
auratum v. platyphyllum / 6' / white fls heavily spotted 111
auratum v. rubrum - deep crimson band 267
autumnale - see L. michauxii 8 / 99
'Backhouse Hyb.' / 5-6' / cream, pink, wine or yellow fls, spotted (482) / 268

1 William A. Weber, Rocky Mountain Flora, 1976.

'Bellingham Hyb.' / 4-7' / yellow, orange, red fls, spotted	267	/	268
'Bellingham Hyb. Afterglow' / 6' / crimson-red fls			268
'Bellingham Hyb. Shuksan' / 4-5' / fls yel.-orange, crimson spots	129	/	268
'Black Beauty' - dark crimson-maroon fls			268
'Black Dragon' / 5-8' / white fls / purple-brown reverse	129	/	268
bolanderi / Bb / 3½' / deep crimson fls, purple spots / Ore., Calif	228	/	25
'Bright Star' / Aurelian hybrid / 4' / ivory-white fls	184	/	268
'Brocade' / 'Backhouse Hybrid' / orange-yellow fls / 5-6'			268
brownii / Bb / 3-4' / white & purple fls / wc China			111
bulbiferum / Bb / to 3' / orange-red fls / w & c Europe	148	/	267
bulbiferum v. croceum / 1-4' / light orange fls / w & c Alps	219	/	267
callosum / Bb / 3' / orange-red fls / Far East	166	/	25
canadense / Bb / 4-6' / yellow fls spotted maroon / e N. Am.	267	/	132
canadense 'Coccinea' - fls brick-red			25
canadense v. editorum / slender red fls / Pa. to Ala			99
canadense v. flavum - synonymous with the sp.			25
canadense v. rubrum - see L. canadense 'Coccinea'	(475)	/	25
candidum / Bb / 3-5' / white fls / n Greece	267	/	132
carniolicum / Bb / 4' / bright red fls / n Jugoslavia	148	/	267
carniolicum v. albanicum - see L. albanicum			290
carniolicum bosniacum / Bb / to 4"/ fls yel.spots/c & sw Jugoslavia			286
carniolicum v. jankae - see L. jankae	(528)	/	290
carolinianum - see L. michauxii			99
cattaniae - see L. martagon v. cattaniae			267
[1] caucasicum - see L. martagon			
centifolium 'Olympic Hyb.' - see 'Olympic Hybrids			268
centifolium 'Pink Perfection - see X 'Pink Perfection'			268
cernuum / Bb / 3' / rosy-lilac fls, purple spots / Korea, Manchuria	(483)	/	267
chalcedonicum / Bb / 3-4' / scarlet fls / Asia Minor	149	/	111
[1] ciliatum / 3-4½' / hairy; fls pale yel-crm, bl-purp center / Turkey			
'Cinnabar' / Asiatic hybrid / 2' / cinnabar-red fls	129	/	111
[2] 'Citronella' - a Fiesta hybrid			
columbianum / Bb / 4'+ / yellowish-orange fls / nw N. Am.	229	/	267
concolor / Bb / 3' / vermilion fls / China			25
concolor v. parthenion / usually spotted fls / Far East			25
concolor v. parthenion 'Coridion' - citron-yel., purp.-brown spots			25
cordatum - see CARDIOCRINUM cordatum			268
'Creelman Hybrid' - see L. regale 'Creelman'			*
X dalhansonii / to 6' / maroon fls, orange spots			25
dauricum / Bb / 2' / red fls, spotted black / ne China, Siberia			111
davidii / Bb / to 5' / orange to reddish-orange fls / w China			267
davidii 'Maxwill' / 7' / orange-red fls			267
davidii v. unicolor - spotting absent or reduced			267
davidii v. willmottiae / 7' / deep orange fls, dark spots	(475)	/	267
duchartrei / Bb / to 4' / white fls, purple spots / w China	(196)	/	267
'Edna Kean' / Preston Hybrid / 3-5' / orange-red fls			268
'Enchantment' / 2-3' / nasturtium-red fls	267	/	268
formosanum / HH Bb / 3-4' / white fls / Formosa			267
formosanum v. pricei - hardier mt. strain, fls flushed purple	100	/	267
giganteum - see CARDIOCRINUM giganteum			268

---

[1] Brian Mathew and Turhan Baytop, The Bulbous Plants of Turkey, 1984.

[2] R.H.S. The Lily Yearbook, 1967.

'Golden Clarion Str.' / 3-6' / Aurelian / golden & lemon-yel. fls 129 / 268
grayi / Bb / 2-4' / orange-yellow & dull red fls / Va. to N.C. 225 / 93
X 'Green Dragon' / Trumpet hyb. / 4-6' / white fls shaded chartreuse 184 / 111
'Green Mountain Strain' / 5-7' / fls white with green throat 268
hansonii / Bb / 4-5' / orange fls / Korea 111 / 93
'Harlequin Hybrids' / 3-5' / pink to tangerine reflexed fls 268
heldreichii / Bb / 16-24" / fls orange red, no spots /Greece, Albania 290
henryi / Bb / to 8' / bright orange fls / China 267 / 132
henryi 'Black Beauty - see L. 'Black Beauty' 268
X hollandicum / 2½' / red, orange, yellow fls 25
Humboldtii / Bb / to 6' / orange-red fls / Sierra Nevadas, Calif. 227 / 267
Humboldtii v. ocellatum / maroon spots, rimmed red / s Calif. 25
Humboldtii pricei / Bb / to 40" / 6-15 fls, red, orange spots / Cal. 61
'Imperial Silver Strain' /5-6' / white fls spotted vermilion 268
X imperiale 'George C. Creelman' - L. regale 'Creelman' *
jankae / Bb / to 32" / yellow fls, sometimes spotted / Balkans 290
japonicum / Bb / to 3' / rose-pink fls / Japan 129 / 25
kelleyanum / Bb / 2-3' / orange to yellow fls / California 268
kelloggii / Bb / 4' / cream to pink fls, aging purple / nw Calif. 25
X 'Kelmarsh' / 6' / white fls suffused purplish-maroon 268
kesselringianum / Bb / 2' / fls yellow / Caucasus 4
lancifolium / Bb / to 6' / fls orange, salmon-red, spotted / Far East 70 / 25
lancifolium v. flaviflorum / fls yel., purp spots, anthers red/Japan 25
lancifolium 'Splendens' - fls larger, rich red, bolder spots 25
lankongense / Bb / 3' / white fls, tinged yellow / Yunnan (484) / 25
leichtlinii / Bb / 3' / lemon-yellow fls, wine spots / Japan 111
leichtlinii v. maximowiczii - see L. leichtlinii v. tigrinum 25
leichtlinii v. tigrinum /8'/orange-red fls, brownish spots/Far East 25
leucanthum / Bb / to 4' / greenish-white & yellow fls / c China 25
leucanthum v. centifolium / to 9' / white fls / s Kansu 267
'Limelight' / Aurelian hybrid / 3-5' / lime-yellow fls 268
longiflorum / Bb / to 3' / white fls / Japan 208 / 25
longiflorum 'Slocum's Ace' - white trumpet lily to 3' 267
mackliniae / HH Bb / 2½' / white fls, flushed pink / Manipur (136) / 267
maculatum / Bb / 8-32" / orange-red to yellow fls / Honshu, Shikoku 200
maculatum v. dauricum / stems woolly / sandy soil/ Jap.,Kor.,Manch. 200
X maculatum 'Wilson' / Bb / 2' / fls apricot-orange 61
X 'Marhan' / 4-7' / rich orange fls, brown spotted 268
maritimum / Bb / 2' / crimson fls / w United States 228 / 111
martagon / Bb / to 6' / light purple to deep maroon fls / n Eu. 148 / 132
martagon v. album / 5'/ clean white fls (483) / 267
martagon v. cattaniae / 7' / deep wine to maroon fls / Balkans 267
martagon v. dalmaticum - see L. martagon v. cattaniae 25
martagon v. hirsutum - stems hirsute; purplish-pink fls 25
martagon v. sanguineo-purpureum / light purple fls, spotted / Albania 61
'Maxwill' - see L. davidii 'Maxwill' 267
medeoloides / Bb / 16" / fls apricot and red / Japan 273 / 267
michauxii / Bb / 1-4' / orange-red spotted fls / Va., to La. 225
michiganense / Bb / 5' / orange-red fls, spotted / e & c N. Am. 142 / 267
monadelphum / Bb/to 5'/waxy yel fls, maybe purp spots / n Caucasus 267 / 25
'Mrs. R. O. Backhouse' - Backhouse Hybrid / 5-6' / orange fls 268
myriophyllum - see L. sulphureum and L. regale 25
nanum / Bb /to 16" / lilac fls, speckled purple / Sikkim, Tibet 100 / 132
nepalense / HH Bb / yellow-green & purple fls / Bhutan & n-ward 144 / 267

nevadense / Bb / to 6' / orange-red fls / Sierra Nevadas, Calif		267
nobilissimum / Bb / to 2' / white / Ryukyu Islands		25
occidentale / Bb / 3-4' / orange-red fls / w United States	(67) /	111
'Olympic Hybrids' / 5-6' / pale ivory to sulphur fls		268
oxypetalum / Bb / 8" / greenish or bronzy-yellow fls / Himalayas		267
oxypetalum v. insigne / dwarf, mauvish-purple fls / nw Himalayas		267
X 'Paisley Strain' / 3-6' / yellow to mahogany fls	(483) /	268
pardalinum / Bb / to 8' / orange fls spotted red / w N. Am.	269 /	132
pardalinum 'Giganteum' / yellow fls, tipped red		25
X parkmanii - see L. auratum Parkmannii		
parryi / Bb / 5-6' / fls orange-red / w United States	227 /	111
parviflorum - see L. columbianum		142
parvum / Bb / 3' / yellow fls / w United States	175 /	111
pennsylvanicum / Bb / to 3' / fls red & yellow, spotted / ne Asia		25
'Perfection', pink strain - see L. X 'Pink Perfection'		*
philadelphicum / Bb / 1-4' / orange or orange-red fls / e U.S.	224 /	267
philadelphicum v. andinum / red-scarlet fls / nc North America	192 /	99
philippinense / Gh Bb / to 3' / white fls, tinged green / Phil. Isls.		25
philippinense v. formosanum - see L. Formosanum		25
X 'Pink Perfection' / 5' / fls in orchid shades		268
pitkinense / Bb / 3-6' / grenadine-red fls, purple spots / Calif.		268
pomponium / Bb / to 2½' / scarlet-red fls / Maritime Alps	149 /	268
ponticum / Bb / fls small, orange to yellow / Caucasus, Asia Minor		61
'Preston Hybrids' / 3-5½' / red & orange-red shades		268
pumilum / Bb / 1-2' / bright red fls / n China, Manchuria		267
pumilum 'Golden Gleam' - golden-yellow fls		267
pumilum 'Red Star' / 2' / scarlet fls, true from seed		311
pumilum 'Yellow Bunting' - canary yellow fls		184
pyrenaicum / Bb / 2-4' / lemon-yellow fls / Pyrenees	148 /	132
pyrenaicum 'Aureum' - fls deeper yellow		25
pyrenaicum 'Rubrum' - fls orange-scarlet, maroon spots		25
'Red Star' - see L. pumilum 'Red Star'		*
'Redstart' / 3' / mahogany-red fls, spotted		268
regale / Bb/ 6'/ wht fls, frag, lilac outside, yel. throat / w China	208 /	267
regale 'Album' - completely white fls		111
[1] regale 'Creelman' - see L. X imperiale 'George C. Creelman'		*
regale 'Pink Perfection' - see L. X 'Pink Perfection'		*
regale 'Sentinal Strain' - see L. 'Sentinal Strain'		268
rubellum / Bb / 2½' / deep pink fls / Japan	129 /	132
rubescens / Bb / 2-9' / pale pink & yellow fls / Ore., Calif.	228 /	267
sargentiae / Bb / 3-4' / white fls, yellow throat / w China	(408) /	111
'Sentinal Strain' / 3-5' / white fls, golden throat	184 /	268
shastense - see L. kelleyanum		25
'Shuksan' - see L. Bellingham Hyb. 'Shuksan'		268
speciosum / Bb / 4½' / red-pink & white fls, spotted / Japan	166 /	200
speciosum 'Gilrey' - large ruby-red fls		311
[2] 'Sterling Star' - white spotted		
sulphureum / Bb / to 8' / yel. inside, red outside / Burma, w China		25
superbum / Bb / to 8' / deep yellow & orange-red fls / e & c U.S.		267
superbum 'Rubrum' - pure red form collected by Mrs. Henry		98
'Sutton Court' / 5-6' / yellow fls, spotted purple		268
szovitsianum / Bb / 5' / deep yellow fls / s Caucasus	267 /	132

[1] R. H. S., Lily Yearbook, 1968.

[2] International Growers Exchange, Inc. Catalog, Fall, 1985.

taliense / Bb / to 4' / fls white with purple spots / w China 25
tenuifolium - see L. pumilum (484) / 25
tigrinum - see L. lancifolium (451) / 25
tigrinum flaviflorum - see L. lancifolium v. flaviflorum 25
trigrinum v. splendens - see L. lancifolium 'Splendens' 25
tsingtauense / Bb / 3' / fiery orange fls / e China, Korea 70 / 267
tsingtauense carneum / Bb / to 3' / fls rays / e China, Korea 25
vollmeri / Bb / 3' / yellow to reddish-orange fls / California 228
wallichianum / Bb / 6' / greenish & creamy-white fls / e Himalayas 25
wardii / Bb / 5' / pinkish-purple fls / se Tibet 261 / 267
washingtonianum / Bb / 2-7' / white fls / Wash., Ore., Calif. (475) / 228
washingtonianum v. purpurascens - pinkish-purple fls, vigorous (67) / 267
wigginsii / Bb / 3' / yellow fls, purple spots / Siskiyou Mts. 267
willmottiae - see L. davidii v. willmottiae 25
wilsonii / Bb / 3' / spotted orange-red fls / Japan 268

LIMNANTHES - Limnanthaceae
alba / A / 1' / white fls / California 228 / 206
douglasii / A / to 1' / yellow, pink, white fls / Oregon, Calif. 228 / 25
douglasii 'Lutea' - see L. douglasii v. sulphurea 182
douglasii v. sulphurea / all yellow fls / Point Reyes, Calif 25

LIMONIUM - Plumbaginaceae
auriculae-ursifolium / R / 8-16" / violet-blue fls / sw Eu., coastal 288
bellidifolium / R / to 1' / pale violet fls / Eur., saline soils 210 / 288
binervosum / P / to 18" / covella violet blue / w Europe 288 / 25
'Bonduellii' / A or B / to 2' / yellow fls / Algeria 25
cordatum / R / 6-12" / violet fls / w Italy, se France 288
cosyrense / R / 4" / lavender fls / s Europe 132
diffusum / HH P / to 16" / pale violet fls / s Portugal, sw Spain 288
dregeanum / Gh P / 10" / pink fls / arid reg., Cape of Good Hope 25
dumosum - see L. tataricum v. angustifolium 135
globularifolium - see L. ramosissimum 288
gougetianum / R / 6-10" / lavender fls / Italy 123
incanum - see L. tataricum v. angustifolium 28
incanum nanum - see L. tataricum v. nanum 28
latifolium / P / 20-32" / fls pale violet / se Europe, Balkans 182 / 288
1 latifolium 'Blue Cloud'
minutum / R / 6-9" / reddish-purple fls / se France (412) / 111
Mouretii / P / to 2½' / green, purplish & white / Morocco 25
nashii / P / 2'+ / fls lavender / sea-coast, Canada to Florida 225 / 99
paradoxum / like L. binervosum, fewer spikelets / Wales, Ireland 288
pectinatum / Gh / rosetted / rose fls / Canary Isls. 28
perezii / HH Sh / 3' / pale yellow fls / Canary Isls. 28
ramosissimum / HH P / 8-20" / pale pink fls / salt marshes, Medit. reg. 288
Suworowii - see PSYLLIOSTACHYS Suworowii 25
spathulatum / HH P / 24-32" / violet fls / n Africa 288
speciosum / P / 1' / rose or purplish-pink fls / s Russia 28
tataricum / P / 1' / ruby-red fls / s Europe to Siberia 28
tataricum v. angustifolium / P / 20" / silvery-lavender fls / w Eu. 93
tataricum v. nanum - dwarf strain 28
tetragonum - see L. dregeanum 25

---

1 Hillier's Hundred Catalog, Index, p. 53.

tomentellum / R / to 1' / pinkish fls / se Russia 288
vulgare / R / 6"+ / blue-purple fls / w Europe, n Africa 25

LINANTHUS - Polemoniaceae
nuttallii / R / 1' / fls white to cream / nw United States 229 / 25
nuttallii ssp. floribunda /P /to 28"/ whitish fls; filiform lvs / Ca. 196

LINANTHASTRUM - Polemoniceae
1 nuttalli / P / ? / showy roadside plant / w Rockies

LINARIA - Scrophulariaceae
aerugina / R / to 16" / fls yellow or brownish / Port., Spain 288
aeruginea v. nevadensis - see L. nevadensis 288
alpina / R / to 6" / blue & orange fls / Alps 210 / 25
alpina 'Rosea' - rose & orange-yellow fls 25
amethystea / A / 1'+ / fls bluish-violet / Spain, Portugal 212 / 288
amoi / P / to 1' / fls reddish purple / s Spain 288
angustissima / P / 2'+ / fls pale yellow / s & ec Europe 288
anticaria / HH P / to 18" / fls grayish-lilac / s Spain 288
atrofusca - see L. aerugina 288
bipartita - see L. incarnata 288
broussonetii - see L. amethystea 288
caesia / A, B, P /4-28"/ fls yellow, brown striped / w Iberian Pen. 288
canadensis / A or B/to 2½'/ corolla blue, white palate / N. Amer. 25
'Canon J. Went' - see L. purpurea 'Canon J. Went' *
crassifolia = see CHAENORHINUM origanifolium ssp. crassifolium 288
2 cuartanensis / R / ? / bluish-lavender fls / s Spain
cymbalaria - see CYMBALARIA muralis 25
cymbalaria alba-compacta - see CYMBALARIA muralis 'Alba-compacta' *
cymbalaria globosa alba - see CYMBALARIA muralis 'Globesa Alba' *
cymbalaria globosa rosea - see CYMBALARIA globosa rosea
dalmatica - see L. genistifolia ssp. dalmatica 227 / 288
faucicolia / A / 6"+ / violet fls / n Spain 25
genistifolia / P / 3' / yellow fls, orange beard / Italy to Russia 25
genistifolia ssp. dalmatica / 3'+ / yellow fls / Balkans 288
glareosa - see CHAENORHINUM glareosum 288
globosa rosea - see CYMBALARIA muralis 'Globosa Rosea' *
incarnata / A / to 2' / violet to red-purple fls / Spain 288
italica - see L. angustissima 288
japonica / P / 6-28" / fls pale yellow / Far East 200
lilacina /HH P / to 18" / orange & yellow & lilac fls / se Spain 288
macedonica - included in L. genistifolia ssp. dalmatica 288
maroccana / A / 9-15" / violet-purple & yellow fls / Morocco 132
melanantha - see L. aerugina 288
nevadensis / R / 8" / fls yellow & brown / Sierra Nevada, Spain 288
origanifolia - see CHAENORRHINUM origanifolium 288
pallida - see CYMBALARIA pallida 288
pallida alba - see CYMBALARIA pallida 288
purpurea / P / to 3' / lilac fls / e Europe 182 / 279
purpurea 'Canon J. Went' - pale pink fls, true from seed 129 / 282

---

1 William A. Weber, Rocky Mountain Flora, 1976.

2 Dwight Ripley, Quar. Bull. Alpine Garden Soc., 41:2.

repens / P / to 2' / fls white, purplish veins / Europe 149 / 25
supina / R / 6" / fls pale to deep yellow / Majorca 212 / 132
supina nevadensis - see L. nevadensis 288
triornithophora / P / 3' / deep lilac fls, yellow markings / sw Eu. 212 / 279
triphylla / A / to 1½' / white, orange, violet fls / se & ec Eu. 211 / 288
tristis / R / to 1' / yellow & brown fls / s Spain, Canary Isls. 25
tristis 'Lurida' / scree pl./ creamy-gray & heather fls / Atlas Mts. (529) / 64
tristis nevadensis - see L. nevadensis 288
tristis 'Toukbal' - yellow-green & maroon fls, self-sterile 154
verticillata / P / to 14" / yellow fls / Spain 288
villosa - see CHAENORHINUM villosum 288
vulgaris / P / 3'+ / bright yellow & orange fls / N. Am., Europe 224 / 220

LINDELOFIA - Boraginaceae
longiflora / P / 18" / fls gentian-blue / w Himalayas 194 / 279
pterocarpa / high mt. race of L. stylosa in Pamirs, c Asia/ red fls 11
spectabilis - see L. longiflora 279

LINDERA - Lauraceae
benzoin / Sh / 15' / early greenish-yellow fls / Maine to Texas 222

LINNAEA - Caprifoliaceae
borealis / R / trailing / rose or white fls / circumboreal 148 / 25
borealis v. americana / flower more tubular fl, to ½"/ N. America 25
borealis longiflora / R / trailing/ fls wht, purp tinge /B.C. s-Cal. 222

LINOSYRIS
vulgaris - see ASTER linosyris 160

LINUM - Linacea
alpinum - see L. perenne ssp. alpinum 287
alpinum 'Album' - see L. perenne ssp. alpinum 'Album' *
altaicum / P / to 2' / violet-blue fls / c Asia 25
anglicum - see L. perenne ssp. anglicum 287
angustifolium - see L. bienne 287
aquilinum / HH R / dwarf tuft / yellow fls, red tinged / Chile 64
arboreum / Sh / 1' / golden yellow fls / Crete 132
arboreum 'Gemmels Hyb.' - see L. X 'Gemmels Hybrid' 24
austriacum / P / to 2' / blue fls / c & s Europe 208 / 287
bienne / B or P / to 2' / blue fls / w & s Europe 287
californicum / A / 4-20" / fls pinkish, whitish / inner coast ranges 196
campanulatum / R / 10" / fls yellow / e Spain to c Italy 287
capitatum / R / to 14" / fls yellow / Balkans, Italy 287
catharticum / A / 6" / white fls, yellow claw / Europe 314 / 287
dolomiticum / R / 6" / yellow fls / on dolomite in Hungary 292 / 287
elegans / R / to 6" / fls yel; lvs thick / Balkan Peninsula 287
extraaxillare - see L. perenne ssp. extraaxillare 287
flavum / P / to 2' / fls yellow / c & se Europe 48 / 287
flavum 'Compactum' / HH R / 9" / bright yellow fls 300 / 268
X 'Gemmels Hybrid' / 6" / deep yellow fls (245) / 268
grandiflorum / A / to 2' / fls in shades of red / n Africa 208 / 25
grandiflorum 'Rubrum' - brilliant scarlet fls 129
hirsutum / P / 1½' / fls blue / ec & e Europe, Russia 287
julieum - see L. perenne ssp alpinum
lewisii - see L. perenne ssp. lewsii 227 / 25

macraei / R / ? / fls yellow / Chile 64
monogynum / HH P / to 2' / fls white / New Zealand 193 / 25
narbonense / P / to 2' / fls clear blue / c & e Europe 212 / 129
narbonense 'Gentianoides' - garden selection 46
narbonense 'Heavenly Blue' / 1½' / ultramarine-blue fls 268
neomexicanum / P / 8-12" / fls yellow / New Mexico, Arizona 227
olympicum / HH Sh / 4-6" / fls violet / Bithyrian Olympus 96
pallasianum / R / to 1' / pale yellow fls / Ukraine, Roumania 287
pallescens / R / 1' / lilac fls / Siberia 206
perenne / P / to 2' / blue fls / c & e Europe 147 / 287
perenne 'Alba' - fls white 25
perenne ssp. alpinum / 2-12"/fls 3/4" across, blue/Pyrenees, w N.A 287 / 25
perenne ssp. alpinum 'Album' - white fls *
perenne ssp. anglicum / stems decumbent or ascending / Britain 287
perenne ssp. extraaxillare / like ssp. anglicum / Carpathians 287
perenne ssp. lewisii / robust plant / w North America 228 / 25
perene montanum / P / 8-16" / fls blue / Jura, n Alps 287
1 perenne nanum 'Saphyer' / 8-10"/sky blue fls, wht eye;evergrn,bushy
rhodopeum / P / to 24" / stems few, fls yellow / Bulgaria, Greece 287
rigidum / P / to 20" / fls yel. / Me. to Alta., s to Mo., Tex., N.M. 25
salsoloides - see L. suffruticosum ssp. salsoloides 184 / 287
salsoloides 'Nanum' - see L. suffruticosum ssp. salsoloides 'Nanum' 189
strictum / A / to 18" / fls yellow / c Europe, Mediterranean 25
suffruticosum ssp. salsoloides / 10" / fls white / sw Europe 314 / 287
suffruticosum ssp. salsoloides 'Nanum' - 2" replica (378) / 189
sulcatum / A / 2½' / fls yellow / e North America 29
tenuifolium / P / 8-18" / pink or white fls / c & s Europe 287
usitatissimum / A / to 4' / fls blue 'Flax'/ probably Asia 25
usitatissimum 'Album' / 4' / the sometimes white Flax fls / Asia 25
viscosum / P / 2' / fls pink / s & sc Europe, in mts. 148 / 287
viscosum 'Album' - fls white *

LIPARIS - Orchidaceae
kumokiri / Bb / 6-12" / fls pale green to purple / Japan, Korea 200
lilifolia / R / to 10" / fls madder-purple / e & c U.S., China 224 / 66
Loeselii / R / to 10" / fls whitish or yel.-grn / N. Am., Eurasia 224 / 25
Makinoana / P / 4-8" / fls purple and wine red / Japan 200

LIPPIA - Verbenaceae
nodiflora / P / 1' / fls white / Medit. region 288

LIQUIDAMBAR - Hamamelidaceae
styraciflua / T / 50' / lvs a brilliant fall color / e & c U.S. 129

LIRIODENDRON - Magnoliaceae
tulipifera / T / 100' / green & orange fls / e North America 36/ / 129

LIRIOPE - Liliaceae
graminifolia / HH R / 8" / fls pale violet / China, Viet Nam 25
minor / R / 4-6" / fls pale purple / Far East 200
muscari / P / 1½' / fls dark violet / Japan, China 184 / 25
muscari 'Monroe White' / R / 6-12" / white fls (432) / (371)

---

[1] Park Catalog, 1986.

muscari 'Variegata' - see L. platyphylla 'Variegata' 25
platyphylla / P / 12-20" / pale purple fls / Far East 200
platyphylla alba / P / 12-20" / fls white / Japan 200
[1] platyphylla 'Variegata' / P / 12-18" / dark violet-blue fls
spicata / P / 16-18" / pale lavender fls / Japan 25
spicata v. minor - see L. minor 200

LISTERA - Orchidaceae
ovata / R / 4-12" / fls green & yellowish-green / Europe, Near East 194 / 290

LITHODORA - Boraginaceae
diffusa / HH Sh / prostrate / deep blue fls / s & w Europe 210 / 25
diffusa 'Grace Ward' / prostrate / large pale blue fls 110 / 36
diffusa 'Heavenly Blue' / dwarf / sky-blue fls 300 / 24
oleifolia / Sh / 1' / fls pale pink turning blue / Pyrenees 133 / 288

LITHOPHRAGMA - Saxifragaceae
bulbiferum - see L. glabrum 25
glabrum / R/ to 12" /fls pink, some wht/ B.C. to Cal. e to Rockies 25
parviflorum / R / 8-14" / fls white to purplish / nw N. America 228

LITHOSPERMUM - Boraginaceae
amelinii / P / to 12" / fls yellow / s United States 61
angustifolium - see L. incisum 25
buglossoides purpureo-caeruleum- see BUGLOSSOIDES purpureo-caeruleum 288
canescens / P / 4-18" / orange-yellow fls / N. Dakota to Oklahoma 229
carolinense / P / to 3' / orange fls / s & sc United States 225 / 99
croceum / differs but slightly from L. canescens / Oklahoma 229
diffusum - see LITHODORA diffus 184 / 25
diffusum 'Grace Ward' - see LITHODORA diffusum 'Grace Ward' 129 / 25
diffusum 'Heavenly Blue' - see LITHODORA diffusum 'Heavenly Blue' 25
doerfleri - see MOLTKIA doerfleri
X frobelii - see MOLTKIA X intermedia 'Frobelii' 25
graminifolium - see MOLTKIA suffruticosa 36
incisum / P / to 2' / bright yellow fls / e & c North America 224 / 25
multiflorum / P / to 2' / fls yel or orange / Wyo,Ariz.,w to Tex,Mex 25
officiale / P / to 3' / fls yel-wht, grn-wht/ Eur., Asia, nat. U.S. 25
oleifolium - see LITHODORA oleifolia 129 / 25
prostratum 'Grace Ward' - see L. diffusum 'Grace Ward' 36
purpureo-caeruleum - see BUGLOSSOIDES purpureocaerulea 288
ruderale / P / 8-20" / fls pale yellow / B.C. to Montana 228

LITHREA - Anacardiaceae
caustica / HH T / 20-30' / fls & frs whitish; lvs evergreen / Chile 25

LITTONIA - Liliaceae
modesta / Gh Cl / 6' / orange-red fls / Natal, n Transvaal 180 / 267

LLOYDIA - Liliaceae
graeca - see GAGEA graeca 149
[2] longiscapa / Bb / ? / fls white with orange base / Bhutan

---

[1] Henry Skinner, Jour. Royal Horticultural Society, Aug. 1971.

[2] Anonymous, Jour. Royal Horticultural Society, July 1963.

serotina / Bb / to 6" / fls yel.-white, purple veins / Eu., N. Am. 148 / 25
triflora / Bb / 3-8" / fls white / Japan 200

LOASA - Loasaceae
laterita - see CAJOPHORA laterita 93
nana / R / dwarf / yellow fls / Valdivian Andes 64
sigmoides / R / mass of small bristly lvs / fls orange-yel. / S. Am. 64
triphylla v. vulcanica / A / 1½' / yellow, red, white fls / Ecuador 194 / 25
vulcanica - see L. triphylla v. vulcanica 184 / 25

LOBELIA - Lobeliaceae
Cardinalis / P / 3' / scarlet fls / e & c North America 224 / 25
Cardinalis 'Alba' / P / to 3' / white fls / N.B. to Minn., s to Fla. 25
Cardinalis X fulgens / P / cultivars needing winter protection 268
Erinus / A or P / less than 8" / fls blue or violet 25
Erinus 'Crystal Palace Compacta' / A / 6" / royal blue fls 283
fulgens - see L. splendens (418) / 25
1 fulgens 'Bees Flame' / 3-4' / blazing red fls; beet red lvs
gattingeri / P / 8-16" / fls deep blue / Tennessee 250
georgiana / P / to 3' / fls pale blue / Virginia to Florida 99
ilicifolia / Gh / 3-6" / fls pink / South Africa 61
inflata / A / to 3' / fls not showy, pale violet / e & c N. Am. 224 / 99
kalmii / P / to 2' / blue fls, white-eyed / e & c North America 99
keniensis / Gh / 3' / fls bluish-violet / Kenya 156
linnaeoides / R / 3" / blue & white fls / New Zealand (202) / 96
puberula / P / 3'+ / blue to purple fls / Pa. to Ga. 226 / 99
roughii / R / 3" / blue & white fls / New Zealand 96
sessilifolia / P / 20-40" / blue-purple fls / Far East 200
siphilitica / P / 2-3' / fls blue or purplish / e United States 224 / 99
siphilitica 'Alba' - fls white 25
siphilitica 'Rosea' - pinkish fls *
siphilitica vedrariense - see L. X vedrariense 268
splendens / like L. cardinalis / Mexico 25
tupa / HH P / to 6' / red-brown fls / Chile 244 / 279
urens / P / 3-24" / fls blue or purplish / w Eur. n to England 289
X vedrariensis / HH P / 3' / crimson-violet fls 279

LOBULARIA - Cruciferae
maritima / P / low / small white fls / Europe, ne U.S. to Mich & s 99

LOISELEURIA - Ericaceae
procumbens / Sh / tufts / rose-colored fls / mts. of n hemisphere 148 / 200

LOMANDIA - Xanthorrhoeageae
sororia / P / rush-like / fls whitish / sw Australia 109

LOMATIA - Proteaceae
cuspidatum / P / ? / fls purple / Washington, mts 142
ferruginea / HH T / to 30' / fls tawny-red & yellow / Chile 36
minus / P / ? / fls yellow or white / Oregon 142
triternatum macrocarpum - see triternatum triternatum 142
triternatum triternatum / P / ? / fls yellow / Alta. to B.C. & s 142

---

[2] Wayside Catalog, Spring, 1986.

LOMATIUM - Umbelliferae
angustatum / R / 6" / fls yellow / Washington 195
angustatum flavum - see L. martindalei v. flavum 142
columbianum / P / to 2' / fls purple / nw United States 142
cous / R / to 1' / fls yellow; lvs sparse / Montana 64
dissectum / P / 3-5' / yellow or red fls / California to B.C. 228
grayi / P / 8-25" / fls yellow; lvs dissected / Colorado to Wash. 229
martindalei / R / 2-4" / fls white or cream / nw North America 228
martindalei angustatum - see L. angustatum 195
martindalei v. flavum - fls yellow 142
nudicaule / P / to 2' / yellow fls / Idaho, Calif., B.C. 229 / 25
suksdorfii / R / 2-8" / fls yellow / Klickitat Co., Washington 142
triternatum / P / to 2½' / yellow fls / Alberta, Wyo., Calif. 229 / 25
utriculatum / P / 4-20" / yellow fls / nw United States 228

LOMATOGONIUM - Gentianaceae
rotatum / R / 4-15" / white or blue fls / Rocky Mts. 229 / 198

LONAS - Compositae
annua / A / 4-12" / yellow fls for drying / Medit. Region 184 / 289
inodora - see L. annua 289

LONICERA - Caprifoliaceae
alpigena / Sh / 4-8' / yellow fls, tinged red / c & s Europe 148 / 36
X americana / Cl / yellow fls suffused reddish-purple / s & se Eu. 129 / 36
X bella 'Rosea' / Sh / to 10' / fls red / Japan, Russia 25
[1] X Brownii 'Dropmore Scarlet - billiant scarlet fls
Caprifolium / Cl / to 20' / fls white or purp. / Europe, w Asia 25
ciliosa / Cl / twining / fls yellow or orange-scarlet / w N. Am. 36
conjugialis / Sh / 4' / dark red fls / Washington to California 222
demissa / Sh / 12' / whitish fls; scarlet frs / Japan 222
dioica / Sh / to 9' / fls yellowish to purplish / ne N. Am. 28
fragrantissima / Sh / to 8' / creamy-white fls; red frs / China 208 / 25
glabrata / C. / ? / yellow fls, resembling L. japonica / Himalayas 233
Henryi / Sh / twining or prostrate / fls yel or red /fr black/China 25
involucrata / Sh/ to 3'/fls in pairs, yel; fr blk/Que to Alaska, Mex 25
Ledebourii / like L. involucrata /lower lf, pubescent / coast Calif. 25
microphylla / Sh / 3' / yellowish-white fls; orange-red frs / c Asia 25
periclymenum / Cl / to 20' / yellow, white, red fls / Eurasia 208 / 36
periclymenum v. serotina / Cl / tall / fls dark purple & yellow 110 / 25
X purpusii / Sh / 10' / white fls, fragrant 36
pyrenaiea / Sh / to 3' / fls white / Pyrenees 25
sempervirens / Cl / vigorous / orange-scarlet fls / Conn. to Texas 36
sempervirens 'Magnifica' - late flowering 25
standishii / Sh / 6-8' / fragrant white fls; red frs / China 36
syringantha / Sh / to 10' / fls pinkish or lilac / China 25
tatarica / Sh/ to 10'/ fls pink or white; fr. red / Russia-Turkestan 25
Xylosteum / Sh / to 10' / fls yel-white; fr. dk red / Eur., Asia 25

LOPEZIA - Onagraceae
coronata / Gh / to 3' / red & lilac-pink fls / s Mexico 25

[1] Wayside Catalog, Spring, 1986.

LOROPETALUM - Hamamelidaceae
chinense / HH Sh / to 12' / whitish fls / c & se China 28

LOTUS - Leguminosae
caucasica - see L. corniculatus 288
corniculatus / R / procumbent / yellow fls, turning orange / Eu. 147 / 129
corniculatus v. japonicus - fewer fls 200
creticus / HH P / yellow & purple fls / Medit. region 212 / 287
maritimus - see TETRACONOLOBUS maritimus 287
siliquosus - see TETRAGONOLOBUS maritimus 287
tenuifolius - see L. tenuis 287
tenuis / P / to 3' / yellow fls / Europe 287

LUCULIA - Rubiaceae
grandifolia / Gh Sh / 8'+ / fragrant white fls / Bhutan 100

LUDWIGIA - Onagraceae
alternifolia / Sh / 2-3' / yellow fls / e U.S., in wet places 28

LUETKEA - Rosaceae
pectinata / Sh / 2-6" / fls white / Calif. to Alaska, in mts 312 / 25

LUFFA - Cucurbitaceae
cylindrica / A Cl / high / Dish-cloth Gourd / Old World tropics 28

LUINA - Compositae
hypoleuca / R / 6-16" / straw-yellow fls / nw United States 228
Nardosmia / P / 2-3'/fls yel, several in heads / B.C. s to n Calif. 25

LUMA - Myrtaceae
apiculata / Gh Sh / 5-15' / fls creamy-white / Chile 25

LUNARIA - Cruciferae
annua / B / 1½-w' / violet-lilac to purple fls / Sweden 210 / 129
annua 'Alba' - white fls 28
annua 'Variegata' - with variegated lvs 129
biennis - see L. annua 25
rediviva / P / 3' / fls deep mauve / Europe 148 / 129
variegata - see L. annua 'Variegata' *

LUPINUS - Leguminosae
albifrons / HH Sh / 5' / blue, purplish, white fls / California 228 / 25
angustifolius / A / 8-32" / blue fls / s Europe 211 / 287
arboreus / HH Sh / 5' / clear yellow fls / California 227 / 279
arcticus / P / 3' / bluish fls / Alaska, Yukon to B.C. 84 / 25
arcticus v. subalpinus / P / to 2' / blue fls / Washington to Alaska 84
argenteus stenophyllus - see L. argenteus v. tenellus 84
argenteus v. tenellus / 16-28" / blue fls / Great Basin area, N. Am. 84
Benthamii / A / to 2' / fls blue / California 25
breweri / R / prostrate mats / fls violet & white / Oregon 228
caespitosus / R / 3-6" / pale blue fls; silvery lvs / w N. Am. 96
caudatus / P / 1-2' / blue fls / Idaho, New Mexico, California 229 / 84
diffusus / HH P / 1-3' / fls blue / Florida to North Carolina 225
latifolius / P / 1-4' / fls blue or purplish / w North America 229 / 228

latifolius v. subalpinus - see L. arcticus v. subalpinus 84
latifolius v. thompsonianus / 12" / fls not hairy / Wash., Oregon 142
lepidus / R / to 1' / violet-blue fls / B.C. to Oregon 229 / 25
lepidus v. aridus / lfts. longer, racemes 2"; fls smaller/ B.C.-Ore. 25
lepidus v. lepidus / racemes longer than lvs; fls ½" / B.C.-Ore. 25
lepidus v. lobbii / prostrate mats / sub-alpine / w North America 142
leucophyllus / P / 20-36" / fls pale lavender / B.C., Calif. 229 / 84
littoralis / P / to 2' / fls blue or lilac / n Calif. to B.C. 228
lobbii / R / 6" / fls blue & white / c Sierra Nevada, Calif. 228
luteus / A / to 2' / fls yellow / Mediterranean 25
lyallii / R / mats / fls blue / s Sierra Nevada Mts., Calif. 312 / 228
microcarpus / A / ? / blue-purple fls / South America 64
nootkatensis / P / 3' / fls white, blue, yellow / nw N. Am. 137 / 25
ornatus - see L. sericeus 142
[1] palaestinus / A / 16" / purple or lilac & cream fls / Israel, Egypt
parviflorus / P / 10-18" / fls light blue, small / n Rockies 229 / 84
perennis / P / to 2' / blue, pink, white fls / Maine to Florida 229 / 25
pilosus / A / to 3' / fls rose & red / s Europe 28
polyphyllus / P / 2½-4' / fls varii-colored / Wash. to Calif. 228 / 129
polyphyllus 'Roseus' - selection for pink fls *
[2] polyphyllus 'Russell hybrids mixed' / 3'+/ lg fls, many colors
sabinii / P / 2-4' / yellow fls, tinged purple / Ore., Wash. 142
sericeus / P / 1-2' / blue & white fls / B.C. to N. Mexico 229 / 142
stiversii / A / 1½' / yellow & rose fls / California 228 / 25
subalpinus - see L. latifolius 228
subcarnosus / A / 8-10" / deep blue fls, white spot / Texas 226 / 28
succulentus / HH P / 4-24" / fls deep purp-blue / coastal Calif. 227
sulphureus / P / to 3' / hairy; fls sm yel to blue or purp/B.C.-Cal. 25
texensis / A / to 1' / fls blue with dark blue wings / Texas 25

LUZULA - Juncaceae
albida - see L. luzuloides 93
arctica / R / 4" / sessile clusters of fls / arctic Europe 290
glabrata / R / tufted / brown fls / Calif. to B.C., Rocky Mts. 196
luzuloides / P / 2' / whitish fls / Europe 93
maxima - see L. sylvatica 93
multiflora / R / to 1' / brownish fls / e & c North America 99
nivea / P / 1-2' / fls white / w & c Europe, in mts 210 / 25
pedemontana / P / fls white, caespitose / sw Alps 290
pilosa / F / 6-12" / brown fls / Europe 301
spicata / P / 1-1½ / dense brown panicles / circumpolar 209 / 99
sylvatica / P / 18" / bracted panicles / Europe 28

LUZURIAGA - Liliaceae
erecta / HH Cl / 18" / white fls / Chile 61
parviflora / HH R / creeping / pale green fls / New Zealand 238
radicans / HH Sh / creeping / white fls; orange frs / Tasmania 139

LYCHNIS - Caryphyllaceae
'Abbotswood Rose' - see L. coronaria 'Abbotswood Rose' 129

---

[1] R. Benjamin, Bull. Am. Rock Garden Society, 29:1.

[2] Wayside Catalog, Spring, 1986.

alba - see SILENE alba	286
alpestris - see SILENE alpestris	160
alpina / F / 2-6" / fls pale purple / Alps, Pyrenees	148 / 286
alpina 'Alba' - fls white	25
alpina v. americana / P / 4-16" / coarse lvs, fls rosy/ Nfld, Gaspé	99
alpina 'Rosea' - fls rose pink	25
apetala - see SILENE wahlbergella	286
X arkwrightii / 9" / orange scarlet fls	46
atropurpurea - see L. viscaria ssp. atropurpurea	25
chalcedonica / P / 2-3' / brilliant scarlet fls / s Russia	147 / 129
chalcedonica 'Alba' - fls white	25
chalcedonica 'Rosea' - fls rose	25
cognata / R / ? / orange-scarlet fls / Korea in mts.	70
coronaria / P / 1-3' / fls purplish / Europe	147 286
coronaria 'Abbotswood Rose' / 1½' / vivid pink fls	129
coronaria 'Alba' - white fls	25
coronaria 'Atrosanguinea / 2½' / dark crimson fls	282
coronata / B / 1½' / fls brick-red to salmon / Far East	28
coronata v. siebodii - see L. sieboldii	200
'Dammeri' - name of no botanical standing	25
dioica - see SILENE dioica	286
flos-cuculi / P / to 3' / fls pale purplish / Europe	148 / 286
flos-cuculi ' Alba' - white fls	25
flos-jovis / P / 1-2' / fls purple or scarlet / c Alps	148 / 129
flos jovis 'Alba' - white fls	*
flos-jovis 'Horts Var.' / 6-12" / bright pink fls; gray lvs	268
flos-jovis 'Milkmaid' - selection for soft pink fls, dwarf	*
flos-jovis 'Nana' - dwarf	*
forrestii / R / 1' / fls white / ?	*
fulgens / P / to 30" / fls red / Siberia, Manchuria, Korea, Japan	25
furcata - see SILENE furcata	*
gracillima / P / 1½-3' / lvs thin; fls white / Japan	200 / 25
X haageana / R / 1' / fls scarlet to salmon shades	194 / 129
[1] kingii - see MELANDRIUM kingii	
kiusiana / P / 20-28" / red laciniate fls / Honshu, Kyushu	200
'Milkmaid' - see L. flos-jovis 'Milkmaid'	*
miqueliana / P / 20-32" / vermilion fls / Japan	200
nivalis / R / to 8" / purple or white / e Carpathians	286
presslii - see MELANDRIUM rubrum glaberrimum	61
sartorii - see L. viscaria ssp. atropurpurea	286
senno / P / 2' / fls scarlet, white / China, cult. in Japan	29 / 286
sibirica / P / 3-12" / fls pale yellow or white / arctic Russia	25
sieboldii / P / 16-28" / deep red fls / Japan	200
taimyrensis / R / to 1' / white or purplish fls / Siberia	209
tamyrense - see L. taimyrensis	209
viscaria / P / to 3' / fls dark or pinkish purple / Europe	149 / 286
viscaria 'Alba' - white fls	135
viscaria ssp. atropurpurea / 18" / purplish fls / Balkans	286
viscaria 'Splendens' - rose-pink fls	306 / 28
viscaria 'Splendens Plena' / 15" / doubled fls, carmine-pink	129
viscaria vulgaris - a synonym	286
vulgaris - see L. viscaria	*

[1] William A. Weber, Rocky Mountain Flora, 1976.

wilfordii / P / 20-30" / deep red fls / Far East 200 / 25
yunnanensis / R / 8" / white fls / sw China 25

LYCIUM - Solanaceae
chilense / HH Sh / 6-10' / yellowish & purplish fls / Chile 139
chinense / Sh / 12' / purplish fls, scarlet frs / China 28
intricatum / Sh / spiny 1½-6' / fls purple, pink or wht / s Europe 288

LYCOPODIUM - Lycopodiaceae
annotinum /R or P/to 12" trailing / lvs. firm, evergrn/n temp. zone 99 28
clavatum / R / trailing / Common Club-moss / n temperate zone 28
complanatum / R / trailing / evergreen / n hemisphere 28
obscurum / R / trailing / Common Ground-pine / N. America, Japan 28
volubile / Gh / scandent to 15' / flattened frs. / New Zealand 193 / 15

LYCORIS - Amaryllidaceae
radiata / HH Bb / 15" / bright orange-red fls / Far East 166 / 111
sanguinea / HH Bb / 18" / crimson-red fls / Japan 166 / 123

LYGODESMIA - Compositae
aphylla texana - see L. texana 226
juncea / P / 6-12" / fls rose-purple / prairie Wisconsin to Alberta 99
texana / R / ? / fls pale lavender or purplish / Texas 226

LYGODIUM - Schizaceae
palmatum / F / twining to 2' / Hartford Fern / Mass. to Florida 28

LYONIA - Ericaceae
ligustrina / Sh / to 12' / white fls / Maine, Florida, Texas 25
lucida / HH Sh / 6' / fls white to pink / Va., to Fla. & La. 222
mariana / Sh / to 6' / white or pinkish fls / e & s United States 36
ovalifolia /Sh or T/to 40'/lvs ellip.; fls wht-pink/w China, Himal. 25

LYSICHITON - Araceae
americanum / P / to 5' /yel. spathe,'West Skunk Cabbage'/ nw N. Am. 129 / 25
camtschatcense / P / 3' / white spathes / Japan 129

LYSIMACHIA - Primulaceae
ciliata / P / to 3½' / fls yellow / North America 25
clethroides / P / 3' / fls gray-white / China, Japan 147 / 279
ephemerum / P / 3' / fls white, tinged purple / sw Europe 212 / 129
japonica minuta - see L. japonicum v. minutissima 164
japonicum v. minutissimum - small yellow fls, creeper 164
lichiangensis / P / to 2' / fls pinkish-white / China 25
mauritiana / B / 4-16" / fls white to pinkish / Far Eastern shores 164 / 200
punctata / P / 3' / brassy-yellow fls. invasive / Eu., Asia Minor 147 / 279
quadrifolia / P / 1-3' / yellow fls / e United States 28
serphyllifolia / Sh / 4-16" / lvs ovate, obtuse; fls yel / Greece 288
terrestris / P / to 3' / yellow fls / e & c North America 224 / 99
thymiflora / P / 1-2' / yellow fls / n hemisphere 200

LYTHRUM - Lythraceae
salicaria 'Roseum Superbum' / P / 2½-3' / bright magenta-red fls 207

MAACKIA - Leguminosae
amurensis / T / 40' / fls whitish / Manchuria 28
amurensis v. Buergeri / lfts almost obtuse, pubescent beneath 25

MACFADYENA - Bignoniaceae
unguis-cati / Gh Cl / woody / bright yel. fls / Yucatan, Argentina 25

MACHAERANTHERA - Compositae
bigelovii / A or B / to 3' / purple to violet fls / Colo., N. Mex. 227 / 25
grindelioides / R / 10" / fls yellow / N. Mexico to Canada 142
tanacetifolia / A / 4-16" / fls bright red-violet / Mont. to Kans. 257 / 229

MACLEAYA - Papaveraceae
cordata / P / to 8' / lvs 8" across, white beneath / China, Japan 25
microcarpa / P / to 8' / apetalous fls / c China (418) / 25

MACROPIDIA - Haemodoraceae
fuliginosa / HH R / to 12" / fls black & yellowish-green / w Austrl. 194 / 45

MACROPODIUM - Cruciferae
pterospermum / P / 16-40" / large white fls / alps of Japan 165 / 200

MAGNOLIA - Magnoliaceae
acuminata / T / 60-90' / green & yellow fls / e North America (342) / 36
Campbellii /T/ to 80'/ fls pink outside, wht-pink inside/ Himalayas 25
Campbellii mollicomata /hardier, sev. basal trunks / fls not cupped 25
denudata / T / 30' / pure white fls / China 36
fraseri / T / to 50' / white or pale yellow fls / Va., Ga., Ala. 25
globosa / HH T / 10-20' / creamy-white fls / Sikkim to Yunnan 36
gloriosa - see H. grandiflora 'Gloriosa' 36
grandiflora / HH T / to 8' / fls white; evergreen lvs / se U.S. 208 / 129
grandiflora 'Gloriosa' / 60' / large creamy-white fls 36
X highdownensis / HH T / 15' / pendulous white fls 129
kobus / T / 18'+ / fragrant white fls / Japan 139
kobus v. borealis / T / to 75' / lvs to 6", vigorous / Japan 25
X loebneri / M. kobus x stellata / fls intermediate 25
X loebneri 'Leonard Messel' - taller tree 25
macrophylla / T / 50' / lg creamy-wht fls; lvs 3';frag / Ky. to Fla. 208 / 25
obovata / T / to 100' / white fls / China, Japan, in mts. 200
parviflora - see M. sieboldii 222
salicifolia / T / 20'+ / white fls; aromatic lvs / Japan 139
sieboldii / Sh / 10' / waxy white fls / Japan 184 / 129
sinensis / Sh / 15' / pendulous white fls / w China 110 / 139
X soulangiana / T / small, purplish & white fls 208 / 25
X soulangiana 'Lennei' / 20'+ / large white & deep purple fls 110 / 25
stellata / T / to 15' / fragrant white fls / c Japan 194 / 25
X thompsoniana - M. tripetala x virginiana / 6" fls 25
tripetala / T / to 40' / white fls / Pennsylvania to Mississippi 93 / 25
virginiana / T / to 60' / fragrant white fls / Mass. to Fla., Tex. 194 / 25
wilsonii / HH Sh / 10'+ / white fls / w China 110 / 25

MAHONIA - Berberidaceae
aquifolium / Sh / 3'+ / fls yellow / nw North America 129 / 25
aquifolium 'Heterophylla' - narrow, wavy-edged leaflets 139
bealei / HH Sh / to 7' / yellow fls; black frs, gray bloom / China 300 / 25

'Charity' - see M. X media 'Charity' 129 / (403)
japonica / Sh / 7' / lemon-yellow fls, long racemes / China (418) / 129
lomariifolia / HH Sh / 8-12' / yellow fls / Burma, w China 129 / 25
X media 'Charity' / HH Sh, to 15' / fls bright yellow 110 / (403)
napaulensis / Sh / to 12' / pale yel fls; shiny lvs / Himal, India 222
nervosa / Sh / 2' / yellow fls / British Columbia to California (206) / 107
pinnata / Sh / to 10' / yel. racemes; fr blk; blue fls / California 25
Piperana - like M. aquifolium, lvs gray-papillose beneath 25
pumila / Sh / 1' / yellow fls / w North America 107
repens / Sh / 1' / yellow fls / B.C. to New Mexico 107

MAIANTHEMUM - Liliaceae
bifolium / R / 9" / white fls / Eurasia 148 / 107
bifolium kamchaticum - see M. kamchaticum 25
canadense / R / 5" / starry white fls / e North America 22 / 107
canadense v. interius / larger fls / nc North America 25
dilitatum / R / 4-10" / white fls / Far East, w North America 200
dilitatum nipponicum - see M. kamtchaticum *
kamtchaticum / R / 14" / fls white; frs red / w U.S., ne Asia 25

MAIHUENIA - Cactaceae
poeppigii / Gh / caespitose / yellow fls / Chile 25

MAITENUS
boaria - see MAYTENUS boaria 206

MALESHERBIA - Malesherbiaceae
fasciculata / Gh Sh / 1' / white fls / Chile 206
linearifolia / HH A / 18" / purple or black fls / Chile 206

MALUS - Rosaceae
baccata / T / 16' / white fls; yellow frs / e Asia 208 / 25
florentina / T / 12' / white fls; red frs / Italy, Jugoslavia 288
floribunda / T / 10'+ / crimson buds; white fls / Japan 139
fusca / T / 35' / fls white or pinkish / Alaska to Calif. 137 / 222
hupehensis / T / 25' / fls white or pinkish / China (530) / 25
kansuensis / T / 25' / white fls; frs yellow to purple / nw China 25
X platycarpa / T / to 20' / fls white; fr yellow-green 25
'Red Jade' / T / 30-40' / drooping branches; fls wht; fr red / U.S. 36
rockii - w China crabapple related to M. baccata 222
sargentii / Sh / 6' / pure white fls; frs dark red 269 / 222
sieboldii v. zumi - see M. X zumi (530) / 25
toringoides / T / to 25' / fls wht; fr round yel., red tinge/w China 25
X 'Zumi' / T / to 20' / rosy buds, white fls 25

MALVA - Malvaceae
alcea / P / 1-4' / fls bright pink / most of Europe 210 / 287
alcea v. fastigiata / 3' / branches congested 129 / 25
moschata / P / 3' / fls rose or white / Eu., naturalized in U.S. 148 / 279
moschata 'Alba' - white fls 184 / 279
moschata v. heterophylla - lvs crenate or broadly lobed only 99
sylvestris / B / 2-3' / fls bluish-lilac / Europe 224 / 25
sylvestris 'Zebrina' / B gr. as A / fls rose-purple, striped deeper 28
trimestris - see LAVATERA trimestris *

MALVASTRUM - Malvaceae
coccineum - see SPHAERALCEA coccinea 25
hypomadarum / Gh Sh / to 10' / white fls, rose-purple eye / S. Afr. 28
lateritum / HH R / prostrate to 6" / brick-red fls / Uruguay 28

MAMMILLARIA - Cactaceae
multiceps / Gh / ? / pale yellow fls / Mexico 93

MANDEVILLA - Apocynaceae
laxa / HH Cl / 12'+ / fls white / Argentina, Bolivia 171 / 93
suaveolens - see M. laxa 128 / 93

MANDRAGORA - Solanaceae
autumnalis / R / to 1' / corolla violet; berry yel./ Medit & Port. 288
officinarum / R / to 1' / fls greenish-yellow / s Europe 194 / 25

MANFREDA - Agavaceae
maculosa / HH P / 3' / fls greenish or purplish white / Texas 28
virginica / P / to 2' / greenish-yellow fls / Ohio to Florida 25

MARGYRICARPUS - Rosaceae
setosus / Sh / 6-8" / green fls; white frs / Andes Mts. 184 / 123

MARIANTHUS - Pittosporaceae
bignonaceus / Cl / lvs soft; fls soft orange bells / se Australia 109

MARRUBIUM - Labiatae
candidissimum - see M. incanum 25
cyllenium / HH P / to 20" / yellowish fls / s Albania, s Greece 288
incanum / P / to 20" / white fls; white, woolly lvs / Italy, w Balk. 288
libanoticum / HH R / 1' / fls pale pink / Asia Minor 29
supinum / HH P / to 18" / corolla pink-lilac, pubesc. outside/Spain 288
velutinum / HH P / to 16" / corolla yel, pubescent outside / Greece 288
vulgare /P/to 18" corolla wht; lvs wht pubescent/Asia, Eur, n Africa 25

MARSHALLIA - Compositae
caespitosa / R / 6-16" / white fls / La., Texas n to Missouri 206
graminifolia / P / 6-40" / pink fls / Fla., La., N.C. 225
grandiflora / P / 1-3' / pink or white fls / Pa., N.C., Kentucky 224 / 225
mohrii / P / 1-2' / pink or wht / interior U.S., plateau Ga., Ala. 225
obovata / P / 4-28" / white fls / Florida to North Carolina 225
obovata v. scaposa - leaves basal only 216

MASSONIA - Liliaceae
pustulata / B / 5-6" / fls 10-20, white / South Africa 61

MATRICARIA - Compositae
ambigua / R / 1' / pure white fls / circumpolar (325)
aurea - see CHRYSANTHEMUM parthenium 'Aurea'
capensis - see CHRYSANTHEMUM parthenium
'Lemon Ball' - see CHRYSANTHEMUM parthenium 'Lemon Ball'
maritima - see TRIPLEUR ospermum maritimum 25
matricarioides / A / to 1' / white fls / nw North America, e Asia 200
tchihatchewii - see TRIPLEUROSPERMUM tchihatchewii 25

MATTEUCCIA - Polypodiaceae
pensylvanica / F / 3-5' / Ostrich Fern / North America 159
struthiopteris / F / to 5' / moist soil / Eurasia 147 / 77

MATTHIOLA - Cruciferae
bicornis / see M. longipetala ssp. bicornis 25
fruticulosa ssp. perennis / P / 2' / fls yellow to purp-red / s Eu. 286
fruticulosa ssp. valesiaca / P / 2' / fls yel to red / s Europe 286
incana / B / to 2½' / varii-colored fls / s Europe 210 / 25
longipetala ssp. bicornis / A / to 18' / fls pink or purple / se Eu. 25
scapifera / R / close tufts / fls cream to deep purple / Atlas Mts. 154 / 64
sinuata / B / to 2' / pale purple fls; whitish lvs / s & w Europe 149 / 286
valesiaca - see M. fruticulosa ssp. valesiaca 286

MAURANDIA
antirrhinifolia - see ASARINA antirrhinifolia 227 / 25
barclaiana - see ASARINA barclaiana 25
erubescens - see ASARINA erubescens 128 / 25
lophospermum - see ASARINA lophospermum 25
scandens - see ASARINA scandens *

MAYTENUS - Celastraceae
boaria / HH Sh or T / 20'+ / greenish-white fls; evergreen / Chile 206 / 36

MAZUS - Scrophulariaceae
japonicus 'Albiflorus' - see M. miquelii f. albiflorus 127 / 200
pumilio / R / 1-3" / mauve & white fls / N.Z., Australia 109 / 123
miquelii / R / 3-6" / fls pale purp, rarely white / Honshu, Japan 200
miquelii f. albiflorus / stoloniferous / white fls / Japan 200
radicans / R / creeper / white & gold fls / N.Z., Australia 193 / 123
reptans / R / prostrate / blue-purple & gold fls / Himalayas (133) / 123

MECONOPSIS - Papaveraceae
aculeata / P / to 2' / fls blue, purp., red; monocarpic / w Himal. 194 / 25
baileyi - see M. betonicifolia f. baileyi 25
X beamishii / to 4' / fls yellow, purple blotched 275
bella / R / 3' / pale blue fls / Himalayas (531) / 25
betonicifolia / P / 2-5' / rich sky-blue fls / Tibet, Yunnan 275 / 129
betonicifolia 'Alba' - white fls *
betonicifolia f. baileyi - only slightly differing (210) / (209)
betonicifolia pratensis - referrable to M. baileyi 275
cambrica / R / 1' / fls yellow or orange / w Europe 275 / 129
cambrica 'Aurantiaca' - fls orange 25
cambrica 'Flore Pleno' - fls double, orange or yellow 25
chelidonifolia / P / to 3' / fls yellow to 1" / w China 25
X cookei / P / 1' / fls ruby-red 268
'Crewdson Hybrid' - similar to M. X sheldonii (211)
delavayi / monocarpic / 5-9" / fls deep purple-blue / Yunnan 28
dhwojii / B / 2-2½' / yellow fls / Nepal 275 / 208
discigera / P / to 7" / many fl stalk, red, purp, blue/ Nepal, Tibet 61
gracilipes / P / 2' / yellow fls / Nepal 268
grandis / P / to 3' / fls deep blue / Nepal, Tibet 275 / 129
grandis 'Alba' - white fls *
grandis beamishii - see M. X beamishii *

grandis 'Branklyn' - see M. X sheldonii 'Branklyn' 279
grandis X betonicifolia - see X Sheldonii 25
horridula / B / 3' / purple or wine-red fls / c Asia 194 / 208
integrifolia / monocarpic / 1½ / yellow fls / Tibet 194 / 129
integrifolia alba - white form 64
latifolia / monocarpic / 3-5' / pale blue fls / Kashmir 275
longipetiolata / monocarpic / 18" / yellow fls / Nepal 139 / 275
lyrata - fls small, blue or purple 96 / 64
napaulensis / monocarpic / to 8' / red, purple, blue fls / Nepal (304) / 275
paniculata / monocarpic / to 6' / fls yellow / Nepal, Assam (532) / 275
paniculata 'Alba' / white fls *
punicea / P / 2½ / pendulous red fls / ne Tibet, w China 275
quintuplinervia / P / 1-1½ / fls lavender-blue / Tibet 275 / 129
regia / monocarpic / 4-5' / fls yellow / Nepal 275 / 129
robusta / large spires of yellow fls 96
X sarsonii / P / 3' / fls sulphur-yellow 279 / 275
X sheldonii / P / 3' / fls deep blue 279 / 268
X sheldonii 'Branklyn' - fls 8" across 129 / 279
sherriffii / P / 1-2' / pink to rosy-red fls / se Tibet, Bhutan 100 / 268
simplicifolia / A / 18" / blue or violet-blue fls / c Asia 275 / 208
superba / monocarpic / to 3½' / white fls / Tibet, Bhutan 275
villosa / P / 15-24" / yellow fls / Nepal, Sikkim, Bhutan 275
wallichii - see M. napaulensis 25

MEDEOLA - Liliaceae
virginiana / P / to 2½' / gr-yellow fls; dark purp. frs / N. Am. 275 / 25

MEDICAGO - Leguminosae
intertexta / A / to 20"/glabrous procumbent or ascend./Medit.& Port. 287 / 77
marina / P / 8-20" / white-tomentose plant / Medit. region 210 / 287
minima / A / to 16" / spiny, sev. seeded, pubescent / Europe 200 / 287
orbicularis / A / to 3', procumbent / yellow fls / s Europe 210 / 287
sativa / P / 1-3' / green & purple fls: Alfalfa / Europe 28

MEGACARPAEA - Crucierae
polyandra / P / 6' / yellowish white fls; large lvs / Kumaon 184 / 279

MELALEUCA - Myrtaceae
acuminata / HH Sh / to 15' / white or pink fls / Australia 25
armillaris / HH T / to 30' / white bottle-brush fls / Australia 190
decora / HH T / to 40' / whitish bark; fls white / New South Wales 25
decussata / HH T / to 20' / fls lavender / s Australia 109/ / 25
densa / HH Sh / 5' / lvs opposite / w Australia 25
diosmifolia / HH Sh / 4' / reddish fls / Australia 206
fulgens / HH Sh / 4' / red fls / w Australia 45
gibbosa / HH Sh / 5'+ / light purple fls / Australia, Tasmania 139
hypericifolia / Gh Sh / to 6' / fls crimson-red / N.S. Wales (533) / 25
laterita / HH Sh / 10' / fls scarlet-red / w Australia (533) / 25
nesophylla / HH T / 20' / lavender or rose-pink fls / w Australia 45 / 25
[1] scabra / P / to 3' / fls magenta; lvs obtuse / w Australia
squamea / HH Sh / 6-10' / purplish fls / Tasmania, Australia 139

---

[1] William E. Blackall and B. J. Grieve, How to Know Western Australian Wildflowers, Part I, II, II.

squarrosa / HH T / to 20' / whitish fls / Australia, Tasmania <u>109</u> / 25
wilsonii / HH Sh to 4' / mauve to crimson fls / s Australia 45

MELAMPODIUM - Compositae
cinereum / R / 10"+ / white daisies / s Texas 64
leucanthum / R / 6-8" / large white fls / Tex., Kan., Ariz. 226

MELAMPYRUM - Scophulariaceae
cristatum / A / 6-20" / pale yellow fls; purple bracts / Europe <u>314</u> / 288
nemerosum / P / 6-20" / bright yellow fls / n & c Europe 288

MELANDRIUM - Caryophyllaceae
affinis - see SILENE furcata 286
album - see SILENE alba 286
apetalum - see SILENE uralensis ssp. apetala 25
apetalum f. okadie / R / 5-8" / one fl; seed flat, red-brn / Honshu 200
diurnum - see SILENE dioica 286
elisabethae - see SILENE elisabethae 286
keiskeii - see SILENE keiskeii 107
keiskei v. akaisialpinum / R / tufted low / 1-2 rose fls / Honshu 200
keiskei v. minor - see SILENE keiskei v. minor 25
[1] kingii / R / to 4" / fls erect / Rocky Mountains
rubrum - see SILENE dioica 286
rubrum glaberrimum / P / to 2' / fls sm, red-purp; many stems / Czech. 61
sakalinense - see SILENE sachalinensis
zawadskii - see SILENE zawadskii 286

MELANTHIUM - Liliaceae
hybridum / P / to 3 1/4' / fls green / Connecticut to Georgia 25
virginicum / P / to 5' / fls grn-yel, dark in age / N.Y. to Ind. & s 25

MELASPHAERULEA - Iridaceae
graminea - see M. ramosa <u>184</u> / 25
ramosa / HH C / 18-30" / starlike white fls, aging greenish / S Afr. 25

MELICA - Graminaea
altissima / P Gr / 3' / tawny to purple spikes / Eurasia 25
altissima atropurpurea / P Gr / 2-7' / 2 ftle. florets / Czech, Rus, Bulg. 290
caerulea - see MOLINIA caerulea 124
ciliata / P Gr / 8-30" / panicles purplish or whitish / Eurasia <u>210</u> / 288
transilvanica / Gr / to 3' / sim. to M. ciliata, pancls short / Eurasia 25

MELICOPE - Rutaceae
ternata / T / to 20' / fls greenish; seed glossy black / New Zealand 25

MELISSA - Labiatae
officinalis / P / to 3' / Lemon Balm; yellow fls / Europe <u>184</u> / 288

MELITTIS - Labiatae
melissophyllum / 8-28" / fls white, pink, purple / w, s & c Europe <u>93</u> / 288

---

[1] William A. Weber, <u>Rocky Mountain Flora</u>, 1976.

MENODORA - Oleaceae
heterophylla / R / 6" / fls reddish & yellow / Texas 226

MENTHA - Labiatae
arvensis / P / to 2' / lilac to white fls, sickly scent / Europe 288
arvensis v. canadensis - see M. arvensis v. villosa 99
arvensis v. villossa / more pubescent / Nfld to Alaska, s to U.S. 99
insularis - see M. suaveolens 288
microphylla / P / 16-44" / lilac or white fls / Italy, Balkans 288
X piperita v. citrata / P / 3' / Lemon Mint 201 / 25
requienii / HH R / creeping / pl. lilac fls, pungently minty / Cors. 288
X rotundifolia - M. longifolia X M. suavolens, variable 288
suaveolens / P / 16-40" / whitish or pink fls / s & w Europe 288

MENTZELIA - Loasaceae
albicaulis / to 16"/stems gr-wht;fls stalkless, yel/Cascad, Sierra N 229
decapetala / B or P / to 4' / white or yellowish fls / c N. America 229 / 25
laevicaulis / P / to 5' / fls pale yellow / Calif. to B. C. 229 / 208
multicaulis / P / 8-12" / st wht, shiny; fls yel / w Colo. 229
multiflora / to 3' / sm yel-wht stems; fl narrow bract / s Cal.-Wyo. 229
nuda / B or P / to 2' / fls creamy white / Neb. & Colo, s to Texas 25
pumila / P / ? / yellow fls / Wyoming to Mexico 227

MENYANTHES - Gentianaceae
trifoliata / aq P / to 10" / white or purplish fls / n temp. zone 148 / 25

MENZIESIA -Ericaceae
ciliicalyx / Sh / to 3' / fls pale yellow / Honshu 132 / 200
ciliicalyx v. multiflora - purplish fls / n Japan 200
ciliicalyx v. purpurea - fls purple 200
ferruginea / Sh / 6' / fls white, tinged pink / Alaska to Oregon 36
lasiophylla - see M. ciliicalyx v. purpurea 200
pentandra / Sh / ? / many branched; fls yellowish / Japan 200
pilosa / Sh / 20-6' / yellowish fls tinged red / Pa. to Georgia 28
purpurea / Sh / to 8' / fls rose-purple / Kyushu 200
tubiflora - see M. ciliicalyx 200

MERENDERA - Liliaceae
attica / Bb / to 2½" / white to rose / Greece, low altitudes 185
bulbocodium - see M. pyrenaica 290
filifolia / Gh C / 1'+ / bright pink fls / s France, n Africa 185
kurdica / C / to 4" / bright pink fls / Kurdistan 185
[1] manissadjhanii / Bb / lvs 2-3; fls 1-3 wht or mauve / Turkey, Iran
montana - see M. pyrenaica 212 / 290
pyrenaica / C / to 4" / solitary fls, pinkish-purple / c Pyrenees 290
sobolifera / C / 3" / pale rosy-lilac to white fls / Asia Minor 267
trigyna / C / to 5" / fls pale to dark pink / Caucasus 4

MERREMIA - Convolvulaceae
dissecta / Cl / ? / white funnel, red throat / Texas to Florida 226

---

[1] Brian Mathew and Turhan Baytop, The Bulbous Plants of Turkey, 1984.

MERTENSIA - Borginaceae
alpina / P / to 8" / fls double, blue / Rocky Mountains 25
asiatica / P / to 3' / fls blue / ne Asia & Aleutians 200
bella / P / 4-28" / blue fls / w Oregon, Idaho 228
ciliata / P / 2' / fls light blue / Rocky Mts. 229 / 279
echioides / R / 15" / deep blue fls / Himalayas 25
lanceolata / P / 18" / blue fls / Sask. to New Mexico 229 / 25
macdougalii / P / 3-10" / fls blue / Arizona 224
maritima / R / mats / turquoise & mauve fls / northern seashores 261 / 25
oblongifolia / R / to 1' / corolla blue crested/ Mont-Wash, n Calif 25
paniculata / P / 3' / pinkish fls turning blue / n N. America 229 / 25
paniculata borealis /P/to 3'/lvs smooth; fls blue-pink/Que to Alaska 25
primuloides / R / 6" / blue, yellow, white fls / Himalayas 25
pterocarpa / R / 9"+ / sky-blue fls / Far East 46
rivularis - see M. pterocarpa 154
sibirica / P / 1-5' / purplish-blue fls / e Siberia 28
virginica / P / 2' / fls purplish-blue / e U.S. 224 / 129
virginica 'Alba' - white fls 99
viridis / R / to 14" / blue fls / Mont., Col., Utah, in mts. 229 / 25

MESEMBRYANTHEMUM - Aizoaceae
aureau - see M. chrysum 156
chrysum / R / prostate 2-8" / fls yellow / Cape Province 156
cooperi - see DELOSPERMA cooperi 99 / 61

METANARTHECIUM - Liliaceae
luteoviride / R / 8-16" / yellow-green fls / Japan in mt. meadows 166 / 200

MEUM - Umbellliferae
anthamanticum / P / to 2' / white or purplish fls / w & c Europe 148 / 287

MIBORA - Gramineae
minima / A Gr / 2" / recommended association plant / Alps 123
verna - see A. minima *

MICHAUXIA - Campanulaceae
campanuloides / B / 4-8' / white & purplish fls / Near East (418) / 30

MICROCACHRYS - Podocarpaceae
tetragona / Gh Sh / low / bright red cones / Tasmania 127 / 25

MICGROGLOSSA
albescens - see ASTER albescens 268

MICROMERIA - Labiatae
corsica - see TEUCRIUM marum 24
croatica / Sh / to 8" / purplish fls / Jugoslavia 288
dalmatica / P / to 20" / fls pale lilac / Balkan Peninsula 288
graeca / Sh / 15" / fls purple / Medit. region, Portugal 149 / 25
hispida / HH Sh / to 1" / purple fls / Crete 288
juliana / HH Sh / 4-16" / purplish fls / c Portugal, Medit. reg. 288
piperella / R / 6" / fls purplish / s Europe 184 / 25
pygmaea - see SATUREIA montana 46
rupestris - see M. thymifolia 288
teneriffae / HH Sh / 6-12" / fls small, purple / Teneriffe 81

thymifolia / P / 18-20" / violet & white fls / w Balkans to Italy 288

MICROSERIS - Compositae
alpestris - see NOTHOCALAIS alpestris 196
nutans / P / 4-30" / yellow fls / California & northward 229 / 228
troximoides - see NOTHOCALAIS troximoides 196

MICROSORIUM scandens - see PHYMATODES scandens 15

MILIUM - Gramineae
effusum 'Aureum' / P Gr / 3' / yellow-leaved 212 / 268

MILLA - Liliaceae
biflora / HH Bb / to 1' / fls waxy-white / sw US to Guatemala 227 / 25

MIMOSA - Leguminosae
pudica / Gh Sh / 3' / mauve fls; sensitive lvs / tropical America 208 / 25

MIMULUS - Scrophulariaceae
alatus / R / 5" / blue-violet or pink fls / e, se & c US 224 / 99
alsinioides / R / 2-12" / yellow & red fls / California to Canada 228
aurantiacus / HH P / 2-7' / orange fls / coastal California 228
'Bonfire' / P gr as A / 9" / orange-scarlet fls 282
X burnetii / 9-12" / yellow & bronze fls 290 / 129
cardinalis / R / 1' / scarlet & yellow fls / Oregon to Arizona 229 / 107
coccineus / A / to 7" / red-purple fls / California 196
cupreus / R / 8-10"/ yellow fls aging to copper / Chile 123
cupreus 'Red Emperor' - compact, bright crimson-scarlet fls 282
cupreus 'Whitecroft Scarlet' / 4" / vermilion fls 129
douglasii / R / 3" / purple fls / n California, s Oregon 228
glutinosus / HH Sh / to 3½' / fls salmon-yellow / c & s California 208
glutinosus aurantiacus - see M. aurantiacus *
guttatus / P / 2-36" / yellow fls, spotted red / w N. Am. 224 / 228
inflatus / A? / 4-12" / yellow fls / Japan 164
langsdorfii - see M. guttatus 142
lewisii / P / to 2½' / fls rose-red to pink / nw N. Am. 229 / 25
lewisii guttatus - see M. guttatus 228
luteus / R / prostrate / fls yellow, spotted red / Chile 129 / 25
luteus guttatus - see M. guttatus 288
moschatus / R / 1' / pale yellow fls / nw & ne N. Am. 224 / 25
nanus / R / 2-6" / purple, yellow, white fls / ne California 229 / 228
nasutus / P / 4-30" / yellow fls / California, Arizona, Mexico 227
primuloides / R / mat to 5" / yellow fls / Calif., Oregon 129 / 228
primuloides f. linearifolius / linear lvs / Siskiyou Mts. 196
ringens / P / 2½-3½' / violet fls / e N. Am. 224 / 208
tilingii / R / 8" / yellow & red fls / Baja California to B. C. 312 / 228

MINUARTIA - Caryophyllaceae (Some authorities consider this to be Arenaria)
artica / P / 1-2' / lvs linear; fls solitary wht / n Urals 286
austiaca / R / to 8" / white fls / e Alps 314 / 286
biflora / R / mats / white to purplish fls / circumpolar 314 / 276
capillacea / R / to 1' / white fls / France to Jugoslavia 286
caucasica - see M. circassica 77
circassica / R / to 8" / white fls / Caucasia 77
condensata - see M. recurva ssp. juressi 286

flaccida - see M. villarsii 286
gerardii - see M. verna 286
graminifolia / R / to 6" / white fls / s Europe 148 / 286
graminifolia clandestina / P/ 4-12"/ lvs rigid; fls wht/Sic., Alban. 286
imbricata / R / mats to 4" / white fls / w & s Caucasia 77
juniperina / R / to 10" / white fls / s & w Greece 286
laricifolia / R / to 1' / white fls / c Spain to Carpathians 148 / 286
laricifolia ssp. kitabelii / robust form / e Austrian Alps 286
mutabilis / R / 4-8" / white fls / sw & sc Europe, in mts. 286
recurva / R / 4" / white fls / Portugal to s Carpathians 286
recurva ssp. juressi / 6" / Sicily, Italy, w Balkans 286
rosanii / R / 1-2" / fls white / Italy 28
rubella / R / caespitose / high montane variant of M. verna / Eu. 286
sedoides / R / cushion / solitary white fls / Pyrenees, Alps 286
setacea / R / to 8" / white fls / France, Russia, Greece 286
stricta Kitaibelii - see M. laricifolia ssp. Kitaibelii 286
verna / R / to 8" / white fls / s, w & c Europe 148 / 286
verna 'Aurea Nana' - see M. v. 'Caespitosa Aurea' *
verna 'Caespitosa', - bright green turf *
verna 'Caespitosa Aurea' - a dense yellow-green moss in effect 107
villarsii / R / 8" / white fls / mts. of s Europe 286

MIRABILIS - Nyctaginaceae
greenei / P / 1-2' / rose-purple fls / inner coast ranges, w US 196
jalapa / HH A / 1½-3' / varii-colored fls / C. Am. to Peru 129 / 208
multiflora / P / to 3' / fls purplish-red / Utah, Col. to Texas 229 / 25
nyctaginea / P / to 3' / fls white or pale pink / Mont. to Mexico 308 / 25

MISCANTHUS - Gramineae
sacchariflorus / P Gr / 7' / silky panicles / Asia 184 / 25
sinensis 'Gracillimus' / Gr / 8' / leaf with silvery mid-rib 268

MITCHELLA - Rubiaceae
repens / R / creeping / pinkish fls / Nova Scotia to Mexico 224 / 107

MITELLA - Saxifragaceae
breweri / R / 1' / greenish-yellow fls / Sierra Nevada Mts., Montana 229 / 228
caulescene / R / 4-16" / greenish fls / Calif., B.C., Montana 229 / 228
diphylla / R / 8"+ / lacy white fls / ne N. Am. to Missouri 224 / 25
ovalis / R / 6-12" / creamy fls / Oregon, Washington, B. Columbia 228
pentandra / R / 4-12" / greenish-yellow fls / Alaska, Calif., Col. 229 / 228
stauropetala - see M. trifida 235
trifida / R / 6-14" / white fls, tinged purple / Cascades to Rockies 229 / 228

MITARIA - Gesneriaceae
coccinea / Gh Sh / scandent / bright scarlet fls / Chile 28

MOEHRINGIA - Caryophyllaceae
glaucovirens / R / 6" / white fls / s Alps 286
muscosa / R / mossy mat / white starry fls / s Europe 314 / 123

MOLINIA - Gramineae
caerlulea / P Gr / to 3' / Moor Grass / Eurasia, intro. Maine to Pa. 220 / 25
caerlulea v. variegata - lvs striped green & cream 184 / 25

MOLOSPERMUM - Umbelliferae
cicutarium - see M. peloponnesiacum 148
peloponnesiacum / P / to 6' / white fls / Pyrenees, Alps 314 / 123

MOLTKIA - Boraginaceae
caerulea / R / suffruticose / blue fls / Asia Minor 28
doerfleri / P / 18" / tubular violet fls / ne Albania 25
X intermedia / Sh / to 15" / violet-blue fls 129 / 132
X intermedia 'Froebelii' / 6" / shrubby, bright blue fls 25 / 132
petraea / Sh / 6" / pinkish-blue fls which darken / Greece 36 / 123
suffruticosa / Sh / trailing to 10" / blue fls / Italy, in mts. 148 / 288

MOLUCELLA - Labiatae
laevis / A / to 3' / white fls in inflated calyx / Asia Minor 129 / 25

MOMORDICA - Cucurbitaceae
charantia / A Cl / to 10' / fls orange-yellow / tropics 128 / 25

MONARDA - Labiatae
'Adam' - 3' / cerise-red fls 268
'Croftway Pink' - see M. didyma 'Croftway Pink' 25
didyma / P / 2-3' / scarlet fls / N. Y. to Ga., w to Mich. 147 / 129
didyma 'Croftway Pink' - fls pink 300 / 25
fistulosa / P / to 4' / fls bright lavender / e North America 308 / 25
fistulosa v. menthifolia - western ranging var., shorter petiole 25
media / P / 3' / deep purple-red fls / N.Y., Ind., Tenn. 99

MONARDELLA - Labiatae
ordoratissima / R / 9-12" / bracts whitish to purplish / w US 28
ordoratissima v. discolor / lvs ± densely sh.-hairy beneath/Wis,Ore. 142

MONESES - Pyrolaceae
uniflora / R / 2-6" / fls white or rosy / n hemisphere 148 / 200

MONOTHROPA - Pyrolaceae
hypopitys / R / to 8" / yellowish saprophyte / Europe, N. America 288
uniflora / woodland pl / to 1' / white saprophyte / Asia, N. Am. 22 / 202

MONSONIA - Geraniaceae
speciosa / Gh / 1' / fls cream to pink / South Africa 25

MONTANOA - Compositae
grandiflora / Gh Sh / to 12' / large heads, white fls / Mexico 254

MONTBRETIA - Iridaceae
solfaterre - see M. 'Solfatare' 268
'Solfatare' / HH C / 1½-3' / apricot-yellow fls; bronze lvs 268

MONTIA - Portulacaceae
australasica - see CLAYTONIA australasica 15
parvifolia / R / to 12" / fls pink to white / w N. Am. 228 / 142
perfoliata / A / 4-12" / fls white / B.C. to Mexico 25
saxosa / R / dense tufts / pink fls / n coast ranges of California 228
sibirica / A / to 15" / fls white to pink / nw North America 229 / 25

MORAEA - Iridaceae
catenulata / HH C / 15" / fls white, tinged blue / Mauritius 29
ciliata / HH C / 4" / lilac fls / Cape Province 185
edulis / HH C / 6"+ / fls variable in color / South Africa 212
framesii / HH C / 6-9" / white fls / South Africa 121
grandiflora - see HOMERIA oreyniana 25
huttoni - see M. spathulata 61
iridoides - see DIETES vegeta 128 / 25
ixioides - see LIBERTIA ixioides 160
juncea / HH C / 1' / yellow-brown fls, marked blue / South Africa 25
moggii / HH C / 18" / fls yellow / Transvaal 268
natalensis / HH C / 12-18" / lilac fls / Natal 121
papilionacea / HH C / 2' / yellow-brown fls / Cape Province 534 / 111
polystachea / HH C / to 3½' / lilac fls / South Africa 535 / 25
ramosa - see M. ramossissma 111
ramosissima / HH C / to 5' / yellow fls / South Africa 223
setacea / HH C / 6" / yellow fls / Cape of Good Hope 206
spathacea - see M. spathulata 267
spathulata / HH C / 2' / fls bright yellow / South Africa 279 / 267
stricta / HH C / 6' / purple fls / e Cape, Lesotho 185
tricuspis / HH C / 1-2' / fls wht-lilac, purp. spots / Cape Colony 28
tricuspis 'Lutea' - yellowish fls *
tripetala / HH C / 18" / lilac or pale blue fls / South Africa 25
tristis - see M. juncea 25
villosa / HH C / to 1' / bright purple fls / South Africa 194 / 25

MORICANDA - Cruciferae
arvensis / B / 12-16" / violet fls; fleshy lvs / w Medit. region 194 / 25

MORINA - Dipsaceae
alba / P / ? / creamy-white fragrant fls / China (213)
betonicoides / P / to 1½' / fls bright rose-red / Sikkim 25
kokanica / P / 2-3' / bright violet fls / Turkestan 93
longifolia / P / 2-3' / white fls, aging crimson / Nepal 194 / 129
persica / P / 24-40" / fls dark rose / Iran, Near East (536) / 93

MORUS - Moraceae
alba / T / 30-45' / white or pinkish frs / China 210 / 36

MOSLA - Labiatae
punctulata / A / to 2' / rose fls / Far East 164 / 200

MUEHLENBECKIA - Polygonaceae
axillaris / HH Sh / prostrate to 3' / black frs / New Zealand 194 / 15

MUILLA - Liliaceae
transmontana / R / 4-20" / white fls / ne Calif., Nevada 229

MULGEDIUM
alpinum - see LACTUCA alpina 25

MUSCARI - Liliaceae
ambroisiacum - see M. moschatum 290
'Argaei Album' / late-flowering white / uncertain origin 25
armeniacum / Bb / 4-8" / bright blue fls / Armenia 269 / 129

armeniacum 'Album' / 6" / white fls / Turkey 17
armeniacum 'Cambridge Blue' - probably same as M. armeniacum 'Cantab' *
armeniacum 'Cantab' / 6" / clear Cambridge-blue fls 17
armeniacum 'Early Giant' - like 'Heavenly Blue' but earlier 267
armeniacum 'Heavenly Blue' - blue selection 268
armeniacum v. szovitsianum - purple-blue fls (16)
atlanticum - see M. neglectum 148 / 290
atropatanum - see M. tenuiflorum 268
aucheri / Bb / to 6" / sky-blue, globose fls / Anatolia 268
azureum - see PSEUDOMOSCARI azureum 185
azureum album - see HYACINTHELLA azurea 'Alba' 268
botryoides / Bb / 6-9" / fls sky-blue / c & s Europe 129 / 268
botryoides 'Album' - white fls 129 / 268
botryoides 'Heavenly Blue' - see M. armeniacum 'Heavenly Blue' 268
botryoides v. heldreichii - inflorescence dense / Balkans 268
botryoides v. kerneri - inflorescence loose, linear lvs 268
botryoides 'Pallidum' - pale blue fls 268
botryoides 'Plenum' - doubled fls *
bourgaei / Bb / 6" / brilliant blue & white fls / Anatolian alps 268
caucasicum / Bb / to 14" / brown & buff fls / Turkey, Iran 268
chalusicum - see PSEUDOMUSCARI chalusicum 185
colchicum - see M. armeniacum 268
commutatum / Bb / 6" / deep blackish-blue fls / Italy, Balkans 211 / 268
comosum - see LEOPOLDIA comosa 185
comosum 'Album' - see LEOPOLDIA comosa 'Alba' 185
comosum 'Monstrosum' - see LEOPOLDIA comosa v. plumosa 185
comosum v. plumosum - see LEOPOLDIA comosa v. plumosa 185
conicum / Bb / 9" / blue fls / Iran 111
dolychanthum / Bb / 1' / bluish-violet fls / Caucasus 25
'Early Giant' - see M. armeniacum 'Early Giant' 267
elegans / Bb / ? / fls bright blue-voilet / Caucasus 25
graecum - see M. comosum 268
grandifolium / Bb / 15-18" / fls livid blue / ? 61
grossheimeri - see M. armeniacum 268
heldreichii - see M. botryoides v. heldreichii 268
holzmannii - LEOPOLDIA comosa *
kerneri / narrow-leaved variant / near M. botryoides / n Italy 290
latifolium / Bb / 1' / fls dark dusky-blue / Asia Minor 267 / 129
leucostomum - see M. neglectum 268
macrocarpum / Bb / 4-6" / fls purp. in bud, bright yellow / s Aegean 290
moschatum / Bb / 4-6" / fls purp. early, to green, whitish / Anat. 290
moschatum 'Album' - fls white *
muscarima - see MUSCARIMA moschatum *
neglectum / Bb / to 1' / fls deep blue / Eu., n Africa, Orient 267 / 268
neglectum v. pulchellum / 6" / without offsets / Greece 268
neglectum racemosum - a synonym 268
pallens / Bb / to 10" / fls light blue / Caucasus, Turkey 268
paradoxum - see BELLEVALIA paradoxa 268
pinardi / Bb / to 2½' / fls yellow & pale violet / sw Asia Minor 454 / 25
plumosum - see LEOPOLDIA comosa v. plumosa 185
polyanthum / Bb / to 1' / deep blue fls / Asia Minor 25
polyanthum 'Album' - see M. armeniacum 'Album' 17
pulchellum / diploid forms of M. neglectum / under 6" / Greece 290
racemosum - see M. neglectum 268
sosnovskyi - see armeniacum 268

steupii / Bb / 1'+ / fls violet / c Europe, Ukraine 2
szovitsianum - see M. armeniacum v. szovitsianum 268
tenuiflorum / Bb / to 10" / fls violet / c Europe, Ukraine 268
tubergenianum / Bb / 4-6" / fls dark & light blue / nw Iran 267 / 129
turkewiczii / Bb / to 4" / brownish fls / ne Turkey 268
woronowii / Bb / to 16" / bluish-violet fls / c Asia, Iran 4

MUSCARIMA - Liliaceae
macrocarpum / Bb / 6-8" / fls yellow & brown / Greece, Turkey 185
moschatum / Bb / 4-6" / purplish fls, aging yellow / w Turkey 185

MUSINEON - Umbelliferae
tenuifolium / R / stemless / yellow fls / Rocky Mts. 65

MUTISIA - Compositae
decurrens / Sh / vine / fls yellow, orange / Chile, Argentina 25
ilicifolia / HH Cl / to 10' / fls bright pink or mauve / Chile 194 / 128
latifolia / Gh Cl / 8' / pink daisies / Chile 128
oligodon / Gh Cl / 1½' / pink fls / Chile, Argentina 25
pulchella - see M. spinosa v. pulchella 36
retusa / HH Cl / 10-20' / pink fls / Chile 61
spinosa / Gh Cl / 20' / pale rose fls / Chile, Argentina 25
spinosa v. pulchella - lvs white-tomentose beneath 36
subulata / HH Cl / 6-9' / fls bright orange-scarlet / Chile 130 / 93

MYOPORUM - Myopoaceae
debile / HH Sh / to 3' / pink fls, rosey edible frs / Australia 25

MYOSITIDUM - Boraginaceae
hortensia / HH P / 1-2' / fls drk blue, pale outside /Chatham Isls. 93 / 25
nobile - see M. hortensia 25

MYOSORUS - Ranunculaceae
minimus / A / to 5" / fls pale greenish-yellow / Europe, weedy 210 / 286

MYOSOTIS - Boraginaceae
alpestris / R / 6" / light blue fls, yellow eye / se Europe 148 / 128
alpestris 'Alba' - see M. sylvatica 'Alba' 25
alpestris 'Nana' - probably M. sylvestris 'Compacta' *
alpestris 'Rosea' - pink form 107
alpestris v. rupicola / R / 1" / brilliant blue fls / s Eu., in mts 24
alpina / R / mat / bright blue fls / Pyrenees 212
australis / R / 6" / yellow fls / New Zealand 123
australis v. conspicua - fls larger 15
[1] colensoi / R / trailing / white fls / Broken River Basin, N. Z.
eximia / R / to 10" / white fls / limestone cliffs / New Zealand 15
explanata / HH R / to 1' / fls white / New Zealand 238 / 25
lyallii / HH R / compact / fls white / New Zealand 15
macrantha / R / 1'+ / fls tawny orange / S. Island, New Zealand 96
palustris - see M. scorpioides 288
pseudoproquinqua / A or P /4-12" / pale blue fls / Caucasus 11
pulvinaris / R / 4" / white fls / alps of Otago, New Zealand 193 / 15

---

1 Adams, Mountain Flowers of New Zealand, 1965.

pygmaea / HH R / 4" / white fls / New Zealand 15
rakiura / HH R / to 1' / white fls / coastal New Zealand 15
rupestris - see ERETRICHUM rupestre 11
rupicola - see M. alpestris v. rupicola 24
saxosa / R / 2-3" / fls white / New Zealand 123
scorpioides / B or P / to 3' / blue fls / c & n Europe 288
sylvatica / A or B / 8-20" / pale blue fls / Eu., escape in U.S. 229
sylvatica 'Alba' / to 2' / white fls 25
sylvatica 'Compacta' - dense form 25
sylvatica 'Rosea' - fls rosy 25

MYRCEUGENIA
apiculata - see LUMA apiculata 93

MYRICA - Myricaceae
cerifera / Sh / to 35' / grayish-white frs / New Jersey to Florida 25
gale / Sh / 1-5' / brownish catkins / Maryland to Florida 28
pensylvanica / Sh / to 6' / waxy-gray frs / e North America 93 / 99

MYRIOCEPHALUS - Compositae
stuartii / A / 6" / yellow & white fls / Australia 45

MYRRHIS - Umbelliferae
odorata / P / 2-3' / fls whitish, small / Europe 148 / 287

MYRSINE - Myrsinaceae
nummularia / HH Sh / prostrate / minute fls; blue-purple frs/ N.Z. 238 / 15

MYRTUS - Myrtaceae
communis / HH Sh / 10-12" / white fls / Levant, originally 110 / 36
communis var. tarentina / HH Sh / 6' / lvs small / s Europe 287
luma / HH Sh / 20'+ / white fls / Chile 129 / 93
nummularia / Sh / 1-2" / white fls; pink frs / Falkland Isls. 36 / 123
ugni / HH Sh / 4'+ / white fls; purple frs / Chile 93

NANDINA - Berberidaceae
domestica / HH Sh / to 8' / wht. fls, bright red frs / India, e Asia 167 / 25

NARCISSUS - Amaryllidaceae
alpestris - see N. pseudo-narcissus ssp. moschatus 290
asturiensis / Bb / 3-5" / deep gold fls / nw Spain, in mts. 212 / 129
atlanticus / Bb / 12" creamy-white fls / Atlas Mts., Morocco (539)
'Baby Moon' / 9" / rich buttercup-yellow fls / Jonquil 111
bulbocodium Bb / 2-5" / light yellow fls / Spain, Portugal 212 / 129
bulbocodium ssp. albidus / HH Bb / whitish-yellow fls / Morocco (539)
bulbocodium ssp. albidus f. zaianicus / whitish fls / Morocco (539)
bulbocodium v. citrinus / 6" / pale lemon-yellow fls 306 / 267
bulbocodium v. conspicuus - deep yellow fls 267 / 129
bulbocodium Graellsii /B / to 15" / fls wht or primrose yel./sw Eur. 25
bulbocodium mesatlanticus-see N.b. ssp. romieuxii f. mesatlanticus 267

1	bulbocodium v. monophyllus - see N. cantabricus ssp. monophyllys		/ 25
	bulbocodium v. nivalis / 2½-3" / fls pale yellow	212	/ 185
	bulbocodium ssp. obesus / fls deep rich yellow, robust	(453)	/ 267
1	bulbocodium obvallaris - see N. pseudo-narcissus ssp. obvallaris		25
	bulbocodium petunoides - see N. cantabricus v. petunoides		
	bulbocodium v. rifanus - pale yellow fls, longer trumpet		117
	bulbocodium ssp. romieuxii / lemon-yellow fls in winter / n Africa		129
	bulbocodium ssp. romieuxii v. mesatlanticus - deeper yellow fls		267
	bulbocodium tanaticus / nearly white fls / Monacco	(453)	/ 267
	bulbocodium v. tenufolius - rich yellow fls; narrow prostrate lvs		267
	calcicola / Bb / 5-6" / fls bright yellow / Portugal	(453)	/ (539)
	canaliculatus - see N. Tazetta		25
	cantabricus / Bb / nearly sessile fls; fragrant white / Spain	212	/ (539)
	cantabricus v. cantabricus - lvs spreading; fls white		25
	camtabrocus clusii - see N. cantabricus v. cantabricus		25
	cantabricus v. foliosus - lvs erect		25
	cantabricus ssp. monophyllus / white fls / s Spain, n Africa	267	/ (539)
	cantabricus v. petunoides - corolla widely expanda	267	/ (539)
	cuatrecasasii / Bb / fls deep yel., fragrant / c Spain, n Portugal		290
	cyclamineus / Bb / 4-8" / deep yellow fls / Spain, Portugal	132	/ 129
	cyclamineus 'Beryl' / 6-9" / butter-yellow & orange-red fls		117
	cyclamineus 'February Gold' - see N. 'February Gold'		*
	cyclamineus 'Woodcock' - see N. 'Woodcock'		*
	dubius / HH Bb / 8-18" / white fls / s France, ne Spain		290
	elegans / HH Bb / 6" / pure white fls / Morocco, Medit., Isls.	(538)	/ (539)
	'February Gold' - Cyclamineus Hybrid, larger earlier fls		184
	fernandesii / Bb / 4" / yellow, 2-flowered / Portugal	453	
	'Fortune' - orange & yellow large-cupped fls		(539)
	gaditanus / Bb / 5-8" / bright yellow, fragrant fls / s Spain, Port.		212
	hedraeanthus / Bb / horizontal scape / pale yellow fls / se Spain		(539)
	hispanicus - see N. pseudo-narcissus ssp. major		290
	X 'Jessamy' / Bb / 4" / white fls, bulbocodium hyb.		17
	jonquilla / Bb / 8-12" / fls yellow, small-cupped / s Europe	186	/ 129
	jonquilla 'Baby Moon' - see N. 'Baby Moon'		*
	jonquilla v. henriquesii / longer fls, orange-yellow / Portugal		(539)
	jonquilla scaberulus - see N. scaberulus		
	X juncifolius / Bb / to 6" / deep yellow fls / Spain, Portugal	148	/ 129
	X juncifolius calcicola - see N. calcicola		(539)
	'Little Beauty' - hybrid bicolor trumpet, dwarf		(217)
	'Little Gem' - selected form of N. minor		(217)
	lobularis - see N. pseudo-narcissus ssp. obvallaris		267
	loiseleurii - see N. triandrus v. loiseleurii		267
	longispathus / Bb / 12-60" / medium yellow fls / se Spain		290
	marvieri - see N. rupicola v. marvieri		
	minimus - see N. asturiensis		129
	minor / Bb / to 8" / variable bicolor trumpet / Portugal	306	/ 267
	minor v. conspicuus - varient of garden origin		184
	minor v. pumilus / to 6" / solid yellow fls		(217)
1	minutiflorus / Bb / 5-8" / uniform yellow fls / Spain, Portugal		
	nanus - synonymous with N. minor		185
	nanus 'Little Beauty - hybrid bicolor, dwarf		(217)

---

1 Frederick G. Meyer, American Horticultural Magazine, Jan. 1966.

nanus 'Little Gem' - selected form of N. minor		(217)
nevadensis - see pseudo-narcissus ssp. nevadensis		(539)
nivalis - see N. bulbocodium		290
nobilis - see N. pseudo-narcissus ssp nobilis		(539)
X 'Nylon' / Bb / 4" / milk-white fls / bulbocodium hyb.		(539)
obvallaris / Bb / 12" / golden yellow fls, cultivar ? / sw Wales		290
X odorus 'Rugulosus' / Bb / yellow fls, shorter segments		25
pachybolus - see N. tazetta ssp. pachybolus		(539)
pallidiflorus - see N. pseudo-narcissus ssp. pallidiflorus		(539)
panizzianus - see N. papyraceus ssp. panizzianus		
papyraceus / HH Bb / 1' / pure white fls / Medit. region, sw Europe		290
papyraceus ssp. panizzianus / fls smaller / se Fr., n Italy & w-ward		290
poeticus/ Bb / 1' / white fls / s Europe	148 /	111
poeticus v. hellenicus / 12-18" / sm. fls / Mt. Pindus, Greece	149 /	(539)
pseudo-narcissus / Bb / 1' / yellow fls / British Isles, Europe	148 /	111
pseudo-narcissus ssp. longispathus / longer spathes & pedicels/ Spain	/	212
pseudo-narcissus ssp. major / to 1' / deep yellow fls / Iberian Pen.		290
pseudo-narcissus ssp. nevadensis / 6"+ / fls yellow & gold	212 /	(539)
pseudo-narcissus ssp. nobilis / larger plant / yellow & gold fls		(539)
pseudo-narcissus ssp. obvallaris / 12" / golden yellow fls		212
pseudo-narcissus ssp. pallidiflorus / creamy fls / w Pyrenees		(539)
pumilus - see N. minor v. pumilus		25
pumilus v. minutiflorus - see N. minutiflorus		
requieni / Bb / fls frag, deep yel. / s France, se Spain		290
rupicola / Bb / 6" / rich deep yellow fls / Spain, Portugal	212 /	267
rupicola v. marvieri / fls soft yellow / Atlas Mts.		267
scaberulus / Bb / 6" / deep yellow & orange fls / Portugal	453 /	(539)
serotinus / HH Bb / 8" / white & yellow fls in fall / Medit. reg.	212 /	267
tanaticus - see N. bulbocodium tanaticus		267
tazetta / HH Bb / 1' / fls white, bunch-flowered / se Eu. & e-ward	149 /	111
tazetta aureus / Bb / 8-16" / bicolor brn & yel / se Fr., It., Sard.		290
tazetta ssp. pachybolus / Gh Bb / large bulbs / white fls / Algeria		(539)
tazetta ssp. panizzianus - see N. papyraceus ssp. panizzianus		290
tazetta ssp. papyraceus - see N. papyraceus		290
'Tete-a-Tete' - yellow cyclamineus hybrid		268
triandrus / Bb / 4-6" / creamy white fls / nw Spain, Portugal	148 /	129
triandrus v. albus - synonymous with the sp.		25
triandrus calathinus - see N. triandrus Loiseleurii		25
triandrus v. cernuus / fls larger yellow bicolor / Portugal, Spain		25
triandrus v. concolor - pale golden-yellow fls		267
triandrus v. concolor 'Aurantiacus' - darker fls, earlier		95
triandrus v. Loiseleurii / fls white or light cream / Glenan Is		267
triandrus pallidulus / fl usually one, cream-yellow / Port, Spain		290
triandrus pulchellus / Bb / to 1' / fls yel., corona wht/Spain, Port		25
triandrus triandrus / Bb / to 1' / fls pure wht / Spain, Portugal		25
vividiflorus / B / to 1½' / fls green / Morocco		25
watieri / HH Bb / 3" / ice-white fls / n Africa	267 /	132
'Wee Bee' / 5" / soft yellow fls / sport of N. minor		117
wilkommii / lvs smooth; fls yellow to orange / sw Spain, s Port.		290
'Woodcock' - yellow cyclamineus hybrid		(539)

NARDOPHYLLUM - Compositae

bryoides / Sh / few inches / small yellow fls / Patagonia		154

NARDOSTACHYS - Valerianaceae
jatamansi / R / 6" / rose, lilac, red-purple fls / Kashmir 64

NARTHECIUM - Lilaceae
asiaticum / P / 2' / greenish-yellow fls / Japan 166 / 25
ossifragum / P / 2-18" / yel. inside, grn out / Ireland, e Europe 194 / 301

NASSAUVIA - Compositae
revoluta / R / to 6" / white fls / Andes Mts. 64

NASTURTIUM - Cruciferae
officinale / Aq. P / fls white, 'Water Cress' / Eur, nat U.S. 25

NAUMBERGIA
thysiflora - see LYSIMACHIA thysiflora 25

NECTAROSCORDUM - Lilaceae
dioscorides - see N. siculum ssp. bulgaricum 290
siculum / Bb / 2-4' / multi-colored fls / w Medit. region 290
siculum ssp. bulgaricum / fls gr-white, tinged pale pink / se Eu. 290

NEILLIA - Rosaceae
thibetica / sh / to 6' / pink or white fls / China 25

NEMASTYLIS - Iridaceae
acuta / Bb / 6-18" / fls bright blue / Louisiana to Texas 257 / 25
geminiflora - see N. acuta 25
purpurea / Bb / 1-2' / red-purple fls / Louisiana., e Texas 225

NEMESIA - Scrophulariaceae
frutescens / Sh / 2½' / fls gold / Cape of Good Hope 206

NEMOPANTHUS - Aquafoliaceae
mucronatus / Sh / to 10' / fls yel.; fr dp red / Nfld. to Minn. 25

NEMOPHILA - Hydrophyllaceae
insignis - see N. menziesii 93
maculata / A / trailing / fls white, blotched purple / California 175 / 228
Menziesii / A / 8" / cornflower-blue fls / California, Oregon 228 / 93
Menziesii 'Alba' - white fls 25

NEOBESSEYA
missouriensis - see CORYPHYANTHA missouriensis 25

NEOMARICA - Iridaceae
caerulea / P or Gh / 2-3' / fls blue or lilac / Brazil 90 / 93

NEOPANAX - Araliaceae
colensoi / Gh Sh / 15' / almost black frs / New Zealand 184 / 15
colensoi montanum / Sh / to 10' / smaller than type / New Zealand 15

NEOPORTERIA - Cactaceae
subgibbosa / Gr / 3' / fls pink or red / Chile 25

NEOTINEA - Orchidaceae
intacta - see N. maculata 85 / 290
maculata / Tu / to 10" / fls dull pink or gr-white / Medit. reg. 290

NEOTTIA - Orchidaceae
intacta - see NEOTINEA maculata *
nidus-avis / P / 8 -30" / fls brownish-yellow / Eurasia 148 / 93

NEPENTHES - Nepenthaceae
hookerana / Gr / has pitchers of 2 shapes / Malay Peninsula 194 / 25
rajah / P / very large petioles to 6" / Borneo 61

NEPETA - Labiatae
cataria / P / to 3' / fls white, purple spotted / Europe 210 / 25
grandiflora / P / 16-32" / blue fls / Caucasus 288
Mussinii / B / to 1' / fls blue / Caucasus, Iran 25
nepetella / P / to 4' / fls pink, white, blue / sw Europe 148 / 25
nervosa / P / 1-2' / clear blue fls / Kasmir 130 / 207
nuda / P / 2-4' / fls pale violet to white / s, e, & ec Europe 288
subsessilis / P / 20-40" / purple fls / Honshu, Shikoku 200
tuberosa / P / to 2½' / purple or violet fls / Port., Spain, Sicily 25
violacea / see N. nuda 288

NERINE - Amaryllidaceae
appendiculata / HH B / to 2' / fls pale pink / South Africa 267
bowdenii / HH Bb / 2' / fls in shades of pink / South Africa 194 / 129
bowdenii 'Quinton Wells' - dark-colored fls, crinkled segments 279
bowdenii wellsii - see 'Quinton Wells' 279
filifolia / Gr Bb / to 8" / bright red fls / South Africa 25
flexuosa / Gr Bb / 3' / fls pale pink / South Africa 25
flexuosa 'Alba' - fls white 25
masonorum / Gh Bb / 9" / rose-pink fls, darker stripe / S. Africa 25
undulata / Gh Bb / 1½' /pale pink fls / South Africa (332) / 25

NERIUM - Apocynaceae
Oleander / Sh / to 20'/fls yel, pink, purp., wht. / Medit., Japan 129 / 25

NERTERA - Rubiaceae
balfouriana / HH R / dense patches / frs yellow, lt orange / N.Z. 193 / 15
ciliata / HH R / patches / frs orange / New Zealand 15
depressa - see N. granadensis 25
dichrondraefolia / R / mats / frs red or orange / New Zealand 15
granadensis / R / mat / fls greenish, frs orange / N. Z., Andes Mts. 184 / 123

NICANDRA - Solanaceae
physaloides / A / 3'+ / blue fls / Peru 129 / 25

NICOTIANA - Solanaceae
'Langsdorffii / A / to 5' / fls green or yellow green / Brazil 61
'Lime Green' / 30" / greenish-yellow fls 129 / 282
rustica / A / 1½-3½' / fls greenish-yellow / South Africa 210 / 25
suaveolens / A / to 3½' / fls wht inside, grn-purp out / se Austr. 25
sylvestris / HH P / 3'+ / greenish-white fls / Argentina 25

NIEREMBERGIA - Solanaceae
caerulea - see N hippomanica v. violacea 184 / 25
frutescens / HH P / 1-3' / white fls tinted blue / Chile 28
hippomanica / F / 6-12" / fls bluish / Argentina 25
hippomanica v. violacea / HH R / to 1' / violet fls, / Argentina (327) / 25

NIGELLA - Ranunculaceae
damascena / A / to 18" / white to light blue fls / s Europe 210 / 25
damascena 'Miss Jekyll' / A / 18" / fls semi-dbl., Cambridge-blue 129 / 268
'Miss Jekyll' - see N. damascena 'Miss Jekyll' 268
damascena ' Persian Jewels' - varii-colored fls 283
sativa / A / 18" blue fls / Medit. region 210 / 25

NIGRITELLA - Orchidaceae
nigra / R / to 1' / green or purple fls / n Europe, in mts. 210 / 85

NIVENIA - Iridaceae
corymbosa / HH Sh / 5' / blue fls; evergreen lvs / Cape of Good Hope 223

NOCCAEA
stylosa - see HUTCHINSIA stylosa

NOLANA - Nolanaceae
paradoxa / P gr. as A / 6-8" / dark blue fls / Chile 25

NOLINA - Agavaceae
georgiana / HH P / 2-3' / whitish fls / Georgia 64
microcarpa / HH P / stemless / pale yellow fls / N. Mexico, Arizona 29
texana / HH P / 2-4' / whitish fls / c & s Texas 226

NOMOCHARIS - Lilaceae
aperta / Bb / 3'' / pale pink fls / Yunnan, se Tibet (218) / 267
farreri / Bb / 3' / white fls, marked red / ne Upper Burma (540) / 267
mairei / Bb / 2½' / white fls, spotted purple / Yunnan 132 / 267
nana - see LILIUM nanum 25
oxypetala - see LILIUM oxypetalum 268
pardanthina / Bb / to 3' / pink fls / n Yunnan 129 / 25
saluenensis / Bb / 3' / rose-pink, white fls / China, Burma 267 / 129

NOTHOCALAIS - Compositae
alpestris / R / to 8" / yellow fls / Calif. to Washington, in mts. 196
troximoides / R / to 1' / yellow fls / Rocky Mts., Cal. to B. C. 196

NOTHOCHELONE
nemerosa - see PENSTEMON nemerosus (219)

NOTHOFAGUS - Fagaceae
antarctica / HH T / 90' / deciduous / Chile 222
obliqua / HH T / 100' / deciduous / Chile, Argentina 25

NOTHOLAENA - Polypodiaceae
marantae / F / 1' / winter-hardy / Medit. region, Himalayas 90

NOTHOLIRION - Lilaceae
bulbuliferum / Bb / 5' / pale pinkish-lilac fls / Tibet (67) / 267

campanulatum / Bb / 4' / deep crimson fls / w China, Burma, Yunnan 267
hyacinthinum - see N. bulbuliferum 25
macrophyllum / B / to 3' / fls rose / Himalayas 25
thomsonianum / Bb / 4' / pale pinkish-lilac fls / Himalayas 267

NOTHOPANAX
colensoi - see NEOPANAX colensoi 15

NOTHOSCORDUM - Lilaceae
bivalve / Bb / 16" / yellowish or white fls / Va. & s-ward 224 / 25
fragrans - see N. inodorum 25
inodorum / Hh Bb / to 2' / white or lilac fls / s US, introduced ? 184 / 25
striatum - see N. bivalve 28

NOTOSPARTIUM - Leguminosea
carmichaeliae / HH Sh / 4-10' / purplish-pink fls / New Zealand 255 / 36
glabrescens / Sh or T / 30' / slen. pend. br.; fls purp-flush / N.Z. 15
torulosum / Sh / to 12' / drooping brs; fls purp-flush / N.Z. 15

NOTOTHLASPI - Cruciferae
rosulatum / R / 3-6" / white fls; felty lvs / New Zealand 220 / 123

NUPHAR - Nymphaeaceae
polysepalum / Aq / lvs & fls floating / fls yellow / nw U. S. 28

NYMPHAEA - Nymphaeaceae
capensis / Aq P / floating / fls sky blue / South Africa 25

NYSSA - Nyssaceae
sylvatica / T / 25'+ / brilliant fall foliage / e N. America 36 / 129

OAKESIELLA
sessilifolia - see UVULARIA sessilifolia 99

OCHNA - Ochnaceae
atropurpurea / Gh Sh / 4' / yellow fls; purple frs / South Africa 194 / 25

OCIMUM - Labiatae
basilicum / A / to 2' / white or purplish fls / old world tropics 201 / 25

ODONTOSTOMUM - Liliaceae
hartwegii / HH P / 1-3½' / yellowish or white fls / California 228

ODONTOSPERNUM
maritimum - see ASTERICUS maritimus 184 / 25

OEMLERIA - Rosaceae
cerasiformis / Sh / to 15' / fls white, fr. blue-black / B.C.-Calif 25

OENOTHERA - Onagraceae
acaulis / P or B / stemless / white fls, fading pink / Chile 107
acaulis v. aurea - yellow fls 123
acaulis lutea - see O. acaulis v. aurea 123

acaulis taraxifolia - included in the sp. 93
albicaulis / A or B / to 1' / white fls aging rose / mid-US & w-ward 28
argillicola / P / 2-5' / fls bright yellow / e Alleghenies 313
argophylla - see O. missouriensis v. incana 64
biennis / B / 1-6' / large yellow fls / e N. America 224 / 25
brachycarpa / R / stemless / fls purplish / Mont., Kan., w & s-ward 28
caespitosa / R / 2-4" / white to pink fls / mid-western U.S. 229 / 107
caespitosa eximia / R or B / to 1' / fls white; with stems / w N.A. 142 / 61
caespitosa f. macroglottis - thick-ribbed frs *
chenanthifolia / P / prostrate / fls yellow / coastal s California 25
coronopifolia / P / 2-10" / fls few, white / Ida., S.Dak., Utah 229
deltoides / R / 2-10" / fls white, aging pink / Oregon to Nevada 229
deltoides v. piperi - A / 4" / white fls aging pink / Siskiyous 196
depressa - see O. strigosa 287
drummondii / A / 1-2' / fls bright yellow / Texas coast 28
erythosepala / B / 1-5" / yellow fls, red calyx / w & c Europe 182 / 287
'Fireworks' - see C. tetragona v. 'Fireworks' 129
flava / R / tufted / yellow fls / N. Mexico, Arizona, Mexico 229 / 227
fraseri - see C. tetragona v. fraseri 93
fremontii / R / 2-6" / yellow fls / Kansas, Nebraska 229
fruticosa / P / 1-2' / yellow fls / e United States 120 / 25
fruticosa v. youngii - see O. tetragona 25
glabra - an impoverished O. caespitosa 64
glauca fraseri - see O. tetragona v. fraseri 93
heterantha / R / tufted / yellow fls / Sierra Nevada Mts. 229 / 228
hookeri / B or P / yellow fls aging red or orange / w. United States 226 / 25
kunthiana / R / to 8" / pink fls / Texas, Mexico 226 / 259
lavandulifolia / P / 6-8" / fls yellow / s United States 25
longiflora / B or P / 1-3' / yellow fls, aging red / South America 25
longissima / P / 4-10' / pale yellow fls, aging reddish / Cal., Col. 229
mexicana - see O. kunthiana 259
mexicana childii - see O. kunthiana 259
mexicana rosea - see O. kunthiana 259
missouriensis / R / 6-9" / lemon-yellow fls / se United States 147 / 129
missouriensis v. incana - bright yellow fls, 3" across (221)
[1] mutica / R / rosette / white or pink fls / Chile
nuttallii / P / 16-40" / white fls / Montana to Colorado 229
odorata / HH P / 1½' / yellow fls aging red / s North America 207
pallida / P / 4-20" / fls white / Texas, S. Dakota, Washington 229 / 227
perennis / P / 4-20" / small yellow fls / North America 224 / 25
perennis pumila - a synonym of the sp. 99
perennis v. rectipilis - with straight hairs / e N.B., Ont. 99
pilosella / P / 6-20" / yellow fls / c United States 224 / 25
pumila - see O. perennis 123
riparia - see O. tetragona v. riparia 129
rosea / P / 6-20" / rose to purplish-red fls / s U.S., S. America 210 / 25
serrulata / B or P / yellow fls / Minnesota, w & s-ward 28
speciosa / P / 2' / white, pink, lavender fls / Missouri & s-ward 224 / 226
speciosa childsii - see O. kunthiana 259
stricta / P / 8-24" / yellow fls / Chile; introduced in California 228
strigosa / to 9' / plant greyish, calyx red; fr strigose / c Europe 287
subacaulis - see O. heterantha 259

---

[1] Modic, Bull. Am. Rock Garden Soc., 19:2.

taraxifolia 'Lutea' - see O. acaulis v. aurea *
taraxifolia - see O. acaulis 123
tetragona / P / 2' / yellow fls / N. Y., Ill., to Georgia 225 / 99
tetragona 'Fireworks' / 18" / yellow fls, red in bud 147 / 129
tetragona v. fraseri - glaucous plant of higher elevations 184 / 99
tetragona v. riparia / non-glandular & gland-tipped hairs / N/S Car. 129 / 25
tetraptera / A / to 16" / petals white / Texas to South America 25
triloba / B / rosette / fls pale yellow to roseate / W. Va. to Texas 229 / 99
triloba flava - superfluous through involved synonymy 235

OLEARIA - Compositae
alpina / HH Sh / 3' / white fls, glossy lvs / Tasmania 109
arborescens / Gh Sh / to 12' / large corymbs / New Zealand 15
Cheesemanii / Sh / bark flaking; fls yellow & white / New Zealand 25
colensoi / Gr Sh / to 9' / coriaceous lvs, buff or white below / N.Z. 15
cymbifolia - see O. nummularifolia f. cymbifolia 139
forsteri - see O. paniculata 15
gunniana - see O. phlogopappa 184 / 139
gunniana 'Splendens' - see O phlogopappa 'Splendens' 36
haastii / HH Sh / 9' / evergreen; fls white, small / New Zealand 129
ilicifolia / HH Sh / to 15' / fragrant white fls / New Zealand (223) / 25
insignis - see PACHYSTEGIA insignis 139
insignis minor - see PACHYSTEGIA 'Minor' 139
lyrata / HH Sh / 9' / evergreen; white fls / Australia, Tasmania 37
macrodonta / HH Sh / 15' / white & yellow fls / New Zealand 36 / 129
macrodonta 'Major' - larger lvs & fls 129 / 36
megaphylla / HH Sh / 5' / fls white / Australia 45
moschata / Sh / to 12' / musky fragrance / New Zealand 15
myrsinoides / Sh / low / 3-5 fls head, white-yellow / Tasmania 61
nummularifolia / HH Sh / 6-10' / creamy-white fls / N. Zealand 139
nummularifolia f. cymbifolia - lvs strongly revolute 139
X oleifolia / HH Sh / 6-10' / white daisies 139
paniculata / Sh or T / to 18" / fls wh;evergrn lvs, glab. above/N.Z. 15
pannosa / HH Sh / 4' / lvs tomentose beneath / Australia 29
phlogopappa / HH Sh / 6-10' / white fls / Tasmania, se Australia 36 / 139
phlogopappa 'Splendens' / wide color range / Tasmania 36
phlogopappa v. subrepanda - denser habit 139
purpuracea / Sh or T / to 15' / fls yellow & white / New Zealand 25
X scillionesis / HH Sh / 5' / pure white fls 110 / 36
semidentata / HH Sh / 1-3' /light to violet-purple fls / Chatham Is. (223) / 93
solandri / Sh or T / 12' / stiff spreading brs; yel tomentum / N.Z. 15
stellulata / taller, less compact than O. phlogopappa / e Australia 36
stellulata 'Splendens' - see O. phlogopappa 'Splendens' 109 / 36

OLYMPOSCIADUM - Umbelliferae
caespitosum / R / to 1' / white fls/ Turkey, on limestone 77

OMPHALODES - Boraginaceae
'Anthea Bloom' - see O. cappadocica 'Anthea Bloom' 184
Cappadocica / R / 6-8" / sky-blue fls / Asia Minor, Turkey 269 / 29
Cappadocica 'Alba' - white fls *

[1] Cappadocica 'Anthea Bloom' / 8" / fls intense blue		
Krameri / P / 10-16" / fls blue / Japan		200
kuzinskyana / A / 2-6" / corolla blue, rarely white / Spain, Port.		288
linifolia / A / 16" / fls white or bluish / sw Europe	210 /	288
linifolia 'Alba' - selection for white fls		*
lojkae - neater version of O. Cappadoica		64
luciliae / P / 4-8" / fls rose, age-blue / Greece, Asia Minor	152 /	129
verna / R / 8" / blue fls / Europe	210 /	25
verna 'Alba' - fls white		25
verna 'Anthea Bloom' - selection for sky-blue		25
OMPHALOGRAMMA - Primulaceae		
elegans / P / 6-10" / hairy; fls deep purple / Tibet		61
minus / R / 4" / fls purple / China		224
soulei / R / to 8" / violet-purple fls / w China		93
vinciflorum / R / 6-8" / strong blue-violet fls / w China	132 /	129
ONCIDIUM - Orchidaceae		
bifolium / Gh P / ? / fls yellow, barred brown / Argentina		*
ONOBRYCHIS - Leguminosae		
arenaria / P / 4-32" / pink fls, veined purple / c, e & se Europe		287
montana / P / to 20" / pink fls, usually purple veins / s Europe	148 /	287
pyrenaica / P / dwarf / corolla pink / Pyrenees		287
saxatilis / R / caespitose / pale yellow fls, pink veins / w Medit.		287
supina / P / 16-24" / corolla pink, or wht veined / s Eur., Calabria		287
viciifolia / P / to 2' / fls pink, rarely white / Eurasia	132 /	25
ONOCLEA - Polypodiaceae		
sensibilis / F / 1-2' / fertile fronds separate / N. Am.	424 /	159
ONONIS - Leguminaceae		
aragonensis /Sh/ 6-12"/hairy stems; fls yel; seeds grn-brn/ Pyrenees		287
arvensis - see O. spinosa		25
cenisia - see O. cristata		287
cristata / R / procumbent / fls pink / Alps, Pyrenees	314 /	287
fruiticosa / Sh / 10-40" / pink fls / Spain, se France	212 /	287
hircina / HH Sh / to 2' / fls rose & white / s Eu., n Africa		25
minutissima / Sh / to 1' / yellow fls / w Medit. reg., Jugoslavia		287
mitissima / A / 1-2' / pink fls / Medit. region, Portugal		287
natrix / Sh / 8-24" / yellow fls / s & w Europe	148 /	287
pubescens / A / 1' / yellow fls / Mediterranean region		287
pusilla / R / 10" / yellow fls / s Europe, Czecho-slovakia		287
reclinata / A / procumbent to 6" / fls pink or purple / s & w Eu.		287
repens / P / 16-18" / pink or purple fls / w & c Europe		287
rotundifolia / Sh / 14-20" / pink or whitish fls / Spain, Italy	148 /	287
spinosa / P / to 2' / pink fls / Europe	314 /	25
ONOPORDUM - Compositae		
acanthium / B / 3-10' / purplish fls / Europe	182 /	225
acaule - see O. acaulon		225
acaulon / R / large rosette / fls white / Spain, n Africa	212 /	225

[1] Anonymous, Jour. Royal Hort. Soc., Nov. 1965.

arabicum - see O. nervosum
bracteatum / B / to 6' / stems yel.,hairy;corolla purp/s Balkan Pen. 289
illyricum / B / 3-9' / fls purple / Morocco, Near East 210 / 225
nervosum / B / 9' / rose fls / Spain, Portugal 25
tauricum / B / tall / fls purple / Mediterranean 25

ONOSMA - Boraginaceae
albopilosum / HH Sh / ? / fls white, aging rose / Asia Minor 61
alboroseum / R / to 8' / fls white, coloring in age / Asia Minor 24 / 25
aucheriana - see O. montana 288
cinerara - see O. alboroseum 24
decipiens / R / 4" / white fls / Cappadocian alps 96
dicroanthum - see O. setosum v. dicroanthum 204
echioides / R / to 1' / pale yellow fls / Italy, w Balkans 148 / 288
erecta / R / 6-10" / fls pale yellow / s Greece, Crete 288
euboica / R / 8-12" / pale yellow, puberulant; nuts smooth/Greece 288
frutescens / R / 10" / pale yellow fls, tinged purple / Greece 288
fruticosa / R / 1' / white to yellow fls / Cyprus 288
helvetica / P / 8-20" / fls pale yellow / sw Alps 148 / 288
montana / R / to 1' / fls pale yellow / s Balkans 149 / 288
nanum / R / 5" / fls yellow, aging blue / Asia Minor 25
pyramidale / HH P / to 2' / fls scarlet, turning lilac / Himalayas 25
sericeum / R / 1' / creamy-yellow fls / Caucasus 107
setosum / B / to 1' / fls yellowish-white / Iran 204
setosum v. dichroanthum - fls blue-violet 204
stellulata helvetica - see O. helvetica 288
stellulata taurica - see O. taurica 288
stellulatum / R / 10" / pale yellow fls / w Jugoslavia 75 / 288
taurica / R / 4-16" / fls pale yellow / se Europe 288
tornensis / R / 6-12" / fls pale yellow / se Czechoslovakia 288

ONOSMODIUM - Boracivaceae
molle / P / 1-2' / fls greenish-yel white / lime barrens, Tenn., Ky. 99

OPHIOPOGON - Liliaceae
Jaburan / P / to 2'+/ fls white, fr. blue-violet / Japan 25
Jaburan 'Variegatus' / similar to O. Jaburan / variegated lvs 25
ohwii / P / 12-16" / fls white / Japan 200
planiscapus 'Arabicus' / P / 20" / lvs dark purp, fls white, pale purp 25
planiscapus 'Nigrescens' - see O. planiscapus 'Arabicus' 25

OPHRYS - Orchidaceae
apifera/ P / 6-20" / pink & brownish-purple fls / Eurasia 149 / 85
bombyliflora / R / 3-10" / fls multi-colored / Mediterranean 290
fusca / R / to 16" / dark puce & yellow fls / Medit. region 212 / 25
insectifera / P / to 2' / green & blackish-violet fls / Europe 290
lutea / R / to 1' / greenish-yellow fls / Eurasia, n Africa 211 / 85
muscifera - see O. insectifera 85 / 290
sphegoides / P / to 2½' / brown to buff fls / w, s & c Europe 212 / 85
tenthredinifera / P / 4-18" / fls pink & purplish-red / Medit. reg 211 / 85

OPITHANDRA - Gesneriaceae
primuloides / P / 3-4" / fls lilac & white / Japan 25

OPLOPANAX - Araliaceae
horridus / Sh / 3-10' / scarlet frs / nw N. Am. 541 / 25
japonicus / Sh / 3' / fls greenish; frs red / Japan 200

OPUNTIA - Cactaceae
aurea - see O. basilaris v. aurea 25
basilaris / HH P / low / fls reddish / sw US 229 / 25
basilaris v. aurea / fls yellow / s Utah, n Arizona 25
compressa - see O. humifusa v. austrina 25
cymochila / horicultural segregate of O. compressa 28
engelmannii / HH P / 2-5' / fls yellow, fading red / US, Mexico 28
erinacea / Gh P / low fls white, yellow, red / Calif., Arizona 25
erinacea v. utahensis / P / low spreading / fls wht-yel-red/ sw U.S. 25
fragilis / R / prostrate / greenish-yellow fls / Wisc. to Rockies 28
humifusa / R / prostrate / yellow fls / e & c US 300 / 25
humifusa v. austrina / erect to 3' / white aging gray / Del. to Fla. 25
phaeacantha v. rufispina / low / yellow stiff-spined lvs / Cal., Tex. 25
polyacantha / R / low fls mostly yellow / c & w US 229 / 25
polyacantha v. polyacantha/ P/low spread/ fls yel, joint 1-4"/Colo. 25
polyacantha Schweriniana / P / low spread/ fls yel, joint 1-2" /Colo. 25
Rafinesquii - see O. humifusa v. austrina 25
rhodantha - see O. erinacea v. utahensis 25
rutila - see O. phaeacantha v. rufispina 25
Schweriniana - see O. polyacantha Schweriniana 25
1 'Smithwick' / twice size O. fragilis / yellow fls
vulgaris / P,short/ to 20'/ fls yel or red; fr red/ Brazil, n Argen.
whipplei / HH Sh / 2' / greenish-yellow fls / sw US, Mexico 28

ORCHIS - Orchidaceae
aristata / P / to 16" / fls pale to deep purple / Alaska, Far East 166 / 66
collina - see O. saccata 290
coriophora / P / 8-16" / brown & wine-purple fls / Medit. region 148 / 211
cruenta / R / to 1' / purple-red bracts / n & c Europe, Siberia 85
elata - see DACTYLORHIZA elata 232 / 290
elodes - see O. ericetorum 85
ericetorum / P / 8-16" / varii-colored fls / acid soils, Europe 85
ericetorum 'Heath Spotted Orchid' / 5-7 lvs, sharp tip/ n & w Brit. 85
foliosa - syn. of O. maderensis 61
fuchsii / P / 8-16" / varii-colored fls / limy soils, Europe 85
fuchsii 'Common Spotted Orchid' / lvs broad, round tip/ s & e Brit. 85
graminifolia / P / 3-6" / fls rose-purple / Japan 200
incarnata / P / 6-24" / varii-colored fls / marshes, Europe 292 / 85
italica / P / 8-16" / lilac & pink fls / Europe, Asia Minor 212 / 85
lactea / R / 3-8" / fls white or grn-pink / Medit. & Balkans 290
latifolia - may be O. incarnata, majalis or purpurella 85
laxiflora / P / 12-20" / reddish-purple fls / Medit. reg. 149 / 85
longicornu / R / 4-15" / white fls, purplish spots / s Eu., n Afr 85
macrostachys / P / to 20" / pink fls, purple spots / Spain to Balkans 85
maculata / P / to 24" / vari-colored / Europe 290
maculata ssp. maculata / to 2' / solid; fls pink, lilac, red or purp 290
maculata ssp. macrostachys /lvs no spots/ labellum dp, 3 lobe/s Eur. 290
maderensis / HH P / 12-18" / bright reddish-purple fls / Madeira 129

---

1 Claude A. Barr, Jewels of the Plains, 1982.

majalis / P / 8-20" / lilac-purple fls / Europe, in damp areas 85
mascula / P / 12-22" / violet fls, white patch / Eurasia 148 / 85
militaris / P / 8-24" / pink fls, purple spots / c & w Europe 182 / 85
morio / R / 4-16" / varii-colored fls / Europe 85 / 211
morio ssp. champagneuxii - slender tufts, long-spurred fls 85
morio v. longicornu - see O. longicornu *
morio ssp. picta / 6-10" / pale red-violet fls / widespread in Eu. 85
pepilonacea grandiflora / Tu/ 8-16"/ fls purp., rare red-brn/ Spain 290
picta - see O. morio ssp. picta 211
praetermissa / P / to 30" / lilac & magenta fls / n Europe 182 / 85
provincialis / R / to 1' / yellow fls / Switzerland , s Europe 149 / 85
purpurea / P / 8-16" / varii-colored fls / Medit. region 85
purpurella / R/ 8"/purple or magenta fls / w Europe, Norway, Holland 85
quadripunctata / R / compact / pink fls / s Eur., Asia Minor 149 / 85
romana / R / to 1' / red or yellow fls / Medit. region 85
saccata / R / 5-15" / fls brownish or greenish-purple / Medit. reg. 290
saccifera - see O. macrostachys 85
sambucina / R / 4-12" / red or yellow fls / Europe 184 / 85
simia / P / 8-18" / white & pink fls / c & s Eu., Near East 149 / 85
spectabilis / P / to 10" / pink to mauve, rarely white / N. America 99 / 25
traunsteineri / P / 8-18" / purple fls / n Europe 85
tridentata / P / 8-18" / white to pale violet fls / c Eur, Medit. reg 210

OREOPHILA
glacialis - see HYPOCHARERIS glacialis *

ORIGANUM - Labiatae
amanum / HH Sh / 2-4" / deep rose fls / Greece 129 / 123
dictamnus / HH Sh / to 8" / pink fls, purple bracts / Crete 184 / 288
laevigatum / HH Sh / 8-20" / deep purplish-pink fls / Turkey 268
majorana / P gr. as A / 1-2' / Sweet Marjoram / Europe 28
Onites / P / ? / fls purple or white / se Europe, Turkey, Syria 25
pulchellum / HH Sh / 8-6" / fls rose / se Europe 25
pulchrum - see O. scabrum ssp. pulchrum 288
rotundifolium / Sh / 4-8" / pale pink fls / Turkey 79 / 279
scabrum ssp. pulchrum / HH P / 1½' / pink fls / Mt. Evvoria, Greece 288
tyttanthum / 1-2½' / tiny white to rosey-violet fls / Soviet Union 95
vulgare / Sh / 3' / white or red fls / Europe 210 / 288

ORLAYA - Umbelliferae
grandiflore / A / 1' / fls pink or white / Medit. region 24

ORNITHOGALUM - Liliaceae
arabicum / Gh Bb / 2' / white fls / Arabia 111
arcuatum / 8-18" / milky-white fls / Caucasus 226 / 4
balansae / Bb / 6" / white fls, striped gray-green / Middle East 111
caudatum / Gh Bb / 2-4" / greenish-white fls / n Africa 181
chionophyllum / Bb / 2-4" / white fls, green-striped / Greece, Cyprus 185
collinum / Bb / 3-8" / white fls, green-striped / Medit. region 290
comosum / Bb / to 6" / greenish-white fls / s Eu., e Medit. region 267
concinnum / Bb / to 1' / white fls, green banded / c & n Portugal 212
exscapum / Bb / 6" / green & white fls / s Europe 28
fimbriatum / Bb / 5" / starry white fls, green banded / Asia Minor 111
flavescens - see O. pyrenaicum v. flavescens 181
flavissismum / Gh Bb / to 1' / orange-yellow fls / S. Africa 223

gussonei - see O. collinum 290
longibracteatum / Gh Bb / to 2' / fls greenish-white / South Africa 25
miniatum / Gh Bb / 1' / white, yellow, orange fls / South Africa 25
montanum / Bb / 18" / green & white fls / se Europe 194 / 111
nanum - see O. sibthorpii 149 / 290
narbonense / Bb / 18" / green & white fls / Medit. region 149 / 267
nutans / Bb / 1' / soft jade-green fls / s Europe 211 / 129
orthophyllum - like O. umbellatum, offsets lacking or fewer/s&c Eur. 290
ponticum /Bb/to 3'+/per. wht, grn stripe back,anth. red-brn/sw Asia 290
pyramidale / Bb / to 2' / milky-white fls, green-striped / c Eu 290
pyrenaicum / Bb / 2' / greenish-yellow fls / s Europe 210 / 267
pyrenaicum v. flavescens / pale yellow fls / Atlas Mts. 181
reverchonii / Bb / to 16" / pure white fls / s Spain 212
saundersiae / Gh Bb / 3'+ / white fls / South Africa 186 / 25
schelkovnikovii / Bb / to 2' / white fls / Armenian S.S.R. 4 25
sibthorpii / Bb / to 6" / white fls, pale green stripe / Balkans 290
subcucullatum / Bb / 18" / white fls / sw Europe 111
tenuifolium - see O. collinum 290
thrysoides / Gh Bb / 1½' / white to golden-yellow fls / S. Africa 129
umbellatum / Bb / 1' / white fls, weedy plant / Europe 212 / 25
umbellatum v. algeriense / HH Bb / to 1' / white fls 181
umbellatum baeticum - see O. umbellatum v. algeriense 181
unifolium concinnum - see O. concinnum 212

OROBANCHE - Orobanchaceae
amethystea - parasitic on Compositae / whitish fls / Europe 149 / 288
caryophyllacea / P/ 6-20"/frag. fls yel. or pink, tinged purp /Eur. 288
elatior / parasitic / 28" / yellow fls / Eng. to Bulg., on clover 288 206
hederae - parasitic on English Ivy / cream or purple fls / Eur. 288
major - see O. elatior
ramosa / parasitic to hemp,tobacco,etc./from Asia, nat. N.Y. to Ill. 99
rapum - see O. rapum-genistae
rapum-genistae / ?/ 8-32"/ fls fetid, parasitic/ w Eur. to Scotland 288
teucrii / parasitic on Teucrium / fls pink or yellow / Europe 288
uniflora / parasitic / 6-24" / white to purple fls / nw US 228

OROBUS - Leguminosae
aurantiacus - see LATHYRUS aureus 279
aureus - see LATHYRUS aureus 287
owerinii / R / acaulescent / purple fls / Caucasus 9
pannonicus - see LATHYRUS pannonicus 287
vernus - see LATHYRUS vernus 287
vernus albo-roseus - see LATHYRUS vernus 'Albo-roseus' 287

ORONTIUM - Araceae
aquaticum / Aq / 2' / yellow & white fls / e United States 99

OROSTACHYS - Crassulaceae
fimbriata / B / 4-6" / fls reddish / se Siberia 25
furusei - see SEDUM furusei 200
spinosa / B / to 12" / fls greenish-yel., sessile / e Rus., nc Asia 286 / 25

ORPHANIDESIA - Ericaceae
gaultheriodes / Sh / prostrate / pink fls / Asia Minor 129 / 25

ORPHIUM - Gentianaceae
frutescens / Gh Sh / to 2' / pink fls / South Africa 233 / 25

ORTHILLA - Pyrolaceae
secunda / R / 2-10" / petals green-white, erect / Europe 288

ORTHOCARPUS - Scrophulariaceae
copelandii / R / 4-14"/ rose-purple & white fls / Calif., Oregon 228
imbricatus / A / to 1' / purple & white fls / Wash. to n Calif. 228 / 25

ORTHROSANTHUS - Iridaceae
chimboracensis / HH P / 1½' / dark blue fls / Mexico, Peru 293 / 93
laxus / HH P / 2' / pale blue fls / w Australia 45
multiflorus / HH P / 1-2' / sky-blue fls / s & w Australia 28

ORYCHOPHRAGMUS - Cruciferae
violaceus / P / 1-2' / fls lavender-blue / Asia 93

OSMANTHUS - Oleaceae
delavayi / Sh / to 6' / fragrant white fls / w China 269 / 25
fragrans / T / to 30' / fls white, fragrant / e Asia 25

OSMORHIZA - Umbelliferae
Claytonii / P / to 3' / plant villous-pubescent / c N. Am. 225 / 25
longistylis / P / to 3' / glabrous to pubescent / e to wc N. Am. 224 / 25

OSMUNDA - Osmundaceae
cinnamomea / F / 5' / cinnamon-brown young fronds / N. & S. Am. 25
Claytoniana / F / 4' / fertile & sterile fronds on 1 stem / N. Am. 25
japonica / F /to 3'+/sporophyllis cinnamon brn / Japan, China-Himal. 200
regalis / F / to 6' / tolerant waterside plant / temperate zone 184 / 25

OSTEOSPERMUM - Compositae
Barberae / HH Sh / gr. as A / to 1' / purple daisies / South Africa 25
Barberae 'Compact' / to 9" / hardier as DIMORPHOTHECA 129

OSTROWSKIA - Campanulaceae
magnifica / P / 5-8' / pale lilac fls / Turkestan 194 / 25

OSTRYA - Betulaceae
virginica / T / 30'+ / hop-like frs / e North America 25

OSYRIS - Santalaceae
alba / HH Sh / 4' / red frs / s Europe 210 / 286

OTANTHUS - Compostae
martimus / R / 1' / yel. fls, wht felty lvs / w Eur. to Near East 211 / 25

OTHONNOPSIS
cheirifolia - see HERTIA cheirifolia 184 / 93

OURISIA - Scrophulariaceae
alpina / R / neat rosette / carmine-crimson fls / Peru 64
breviflora / R / trailing / blue, pink, white fls / Aucklands 194 / 64

caespitosa / R / creeper / white fls / New Zealand 238 / 268
caespitosa v. gracilis - smaller and slenderer 268
calycina - see O. macrocarpa var. calycina 238
coccinea / R / 6-12" / scarlet-red fls / Chilean Andes 19 / 93
Crossbyi / P / prostrate / bracts in whorls / New Zealand 15
elegans - darker fls than O. coccinea, perhaps a var. 64
fragrans / R / close creeper / white & blush-pink fls / Argentina 64
integrifolia / P / prostrate / fl to 3", white & pink/ Tasmania 15
macrocarpa / P / to 2' / white fls / New Zealand 542 / 25
macrocarpa v. calycina - colonial form from creeping rhizomes 238
macrophylla / R / 8-10" / white fls / New Zealand 238 / 123
microphylla / R / crevice-plant / rosy-lilac fls / Argentina 64
Poeppigii / R / 4-8" / crimson-scarlet fls / S. America 64
pygmaea / R / ? / single rose fls / Andes Range 64
racemosa - allied to O. Poeppigii 64
sessilifolia / R / 2-6" / white fls, purple base / New Zealand 96
X 'Snowflake' / mats / white fls 83
vulcanica / HH R / leafy mat / New Zealand 15

OXALIS - Oxalidaceae
acetosella / R / 3" / fls white, purple veins / Eurasia 165 / 25
acetosella ssp. griffithii - see O. griffithii 200
acetosella f. rosea - see O. montana f. rhodantha 99
adenophylla / R / 4" / lilac-pink fls / Chile 90 / 129
braziliensis / Gh Tu / 6-10" / rose-purple fls / Brazil 25
comberi / HH R / creeping / yellow fls, veined red / Argentina 64
corniculata / R / prostrate / fls yellow / Europe 28
corniculata v. atropurpurea - lvs red-purple 28
corniculata purpurea - see O. corniculata v. atropurpurea 28
depressa / HH Tu / 4" / fls bright rosy-pink / South Africa 129 / 111
enneaphylla 'Rosea' / R / 3" / rose-pink fls / Falkland Isls. 24 / 123
grandiflora / P / stemless / fls 3, white / Venezuela 61
griffithii / R / 6"+ / white or pale rose-purple fls / Far East 200
hirta / HH Tu / weak-stemmed / deep rose to lavender fls / S. Africa 111
inops / R / 2-3" / shell pink fls / Basutoland 129
laciniata / HH R / low / varii-colored fls, scented / Patagonia 129 / 268
luteola / Gh Tu / stemless / brilliant yellow fls / Cape Province 111
magellanica / R / carpeter / white fls; bronzy lvs / Magellan Str. 46
montana f. rhodantha / creeping American woodland plant / rose fls 99
obtusa / Gh Tu / ? / orange-red fls / Cape of Good Hope 185 / 223
oregana / R / 2-6" / white or pink fls / Calif., Oregon 229 / 228
polyphylla / P / 4" / fls pale purp / Cape of Good Hope 206
purpurea / Gh Tu / 6" / rose, violet, white fls / Cape of Good Hope 223 / 25
Simsii / Gh Tu / 1' / deep red fls / Chile 28
speciosa - see O. variabilis 'Rosea' 28
trillifolia / R / 5-12" / fls pink or white / n Calif. to Wash 25
valdivensis / A / 2-8" / fls bright yellow / Chile 543 / 25
variabilis 'Rosea' / Gh / 6" / deep rose fls / South Africa 93

OXYDENDRON - Ericaceae
arboreum / T / 20' / white fls / se US 269 / 129

OXYPETALUM - Asclepiadaceae
caeruleum / P gr. as A / to 3' / pale blue fls / Brazil, Uruguay 25

OXYRIA - Polygonaceae
digyna / R / to 1' / pink winged frs / n hemisphere 148 / 107

OXYTROPIS - Leguminosae
ambigua / R / to 1' / purple fls / se Russia, n Asia 287
amethystea / R / acaulescent / fls pale purple / sw Alps 287
argentea / P / 6" / Siberia 206
campestris / R / to 8" / purple fls / se Russia, n Asia 148 / 287
campestris gracilis / P / to 1'/lfts 17-33, fls wht / B.C.-Wash & sw 25
campestris v. johannensis - see O. johannensis 99
campestris ssp. sordida /fls varying to light viol. / Finland, Rus. 287 209
campestris v. terrae-novae / P / dwarf of sp./ Hudson Straits, Nfld. 99
carpatica / P / acaulescent, corolla bright blue / Carpathians 287
deflexa / R / 8" / white to bluish-purple fls / N. Am., Siberia 229 / 142
fetida / R / acaulescent / yellowish fls / sw & wc Alps 314 / 287
foucandii / P / acaulescent, corolla lilac; legume ellip./ Pyrenees 287
Halleri / R / to 6" / fls pale purple / Scotland, Alps 25
japonicus / R / 4" / purple fls / Honshu 165 / 200
japonicus v. sericea - see O. yezoensis 200
jacquinii / R / 2-8" / corolla purplish violet / alps, French Jura 287
johannensis / P / to 12" / fls purple-violet / Nfld, N.B., Me. 99
lambertii / R / 4-6" / creamy fls, varii-tinged / wc N. Am. 224 / 220
lapponica / R / to 4" / violet-blue fls / Scandinavia, in mts. 314 / 287
lehmannii /R / to 8" / violet fls / U. S. S. R. 9
megalantha / R / 8" / fls bluish-purple / Hokkaido 165 / 200
montana jacquinii - see O. jacquinii 287
montanus - see O. amethystea 243 / 287
multiceps / P / lvs paired; 1-4 fls, pink / Rockies [1]
nigrescens / R / mats / purple or white fls / n Arctic regions 209
olympia - see O. purpurea 287
owerinii - see OROBUS owerinii 9
pallasii / P / 8-20" / light yellow fls / arctic Norway, Russia 287
pilosa / P / to 20" / light yellow fls / c & e Europe 148 / 287
podocarpa / R / dwarf / purple fls / California & n-ward 312 / 96
purpurea / R / 6" / purple fls / Albania, Greece 287
pyrenaica / R / to 8" / fls purplish or blue violet / s & sc Eu. 212 / 287
rishiriensis / R / 4-6" / pale yellow fls / Hokkaido 200
sericea / R / to 10" / fls cream to white / Montana to Texas 229 / 142
sericea v. spicata / fls lemon to sulphur-yellow / B.C., Ida., Wyo. 142
shokanbetsuensis / R / 4-6" / fls reddish-purple / Hokkaido 200
splendens / P / 18" / rose or carmine fls / nw N. Am. 229 / 25
submutica / R / 4" / red fls / c Asia 9
terrae-novae / P/ dwarf/ fls purple or violet /Hudson Strts to Nfld. 99
uralensis / P/to 1'/blue-purp fls; more vigor than O. Halleri/Urals 287 / 25
viscosa / see O. fetida 287
yezoensis / R / 4" / purple fls; white-woolly lvs / Hokkaido 200

OZOTHAMNUS - Compositae
ledifolius / HH Sh / 3'+ / white fls Tasmania 139
thyrsoides / HH Sh / 6' / white fls / Australia, Tasmania 139

---

1 WILLIAM A. WEBER, Rocky Mountain Flora, 1976.

PACHISTIMA - Celastraceae
canbyi - see PAXISTIMA canbyi
myrsinites / Sh / 2'+ / glossy lvs / nw North America 142

PACHYLAENA - Composite
atriplicifolia / R / rosetted / carmine-pink daisies / Andes Mts. 26

PACHYSANDRA - Buxaceae
procumbens / P / to 1' / fls greenish or purplish / to Fla. & La. 25
terminalis / P / 1' / fls whitish, evergreen lvs / Japan 280 / 25

PACHYSTEGIA - Compositae
insignis / HH Sh / 9' / white & yellow fls; evergreen lvs / N.Z. 238 / 139
insignis 'Minor' - rock-garden size, small & slender 238 / 25

PAEDEROTA - Scrophulariaceae
bonarota / R / to 8" / fls violet-blue / e Alps 148 / 288
lutea / P / 2½-8" / less hairy than P. bonarota / fls yel./ e Alps 288

PAEONIA - Ranunculaceae
abchasica / P / lvs pale, not grey beneath; fls yel / Caucasus 5
albiflora f. hortensis - see P. lactiflora 61
anomala / P / 2-3' / red fls/ ne Russia, n Asia 286
anomala v. intermedia - pistils villous 25
arietina - see P. mascula ssp. arietina 25
banatica - see P. officianalis banatica 25
broteri / HH P / 14" / fls carmine or purplish / sw Spain, Portugal 212 / 211
brownii / P / 12-15" / fls deep red & yellow / w N. America 229 / 107
cambessedesii / HH P / 1½' / rose-pink fls / Balearic Isls. (169) / 129
caucasica - see P. mascula 25
1 'Chameleon' - yellow to red crepy fls, near P. mlokosewitschii
clusi / HH P / 1' / white fls / Crete 310
coriacea / HH P / 20" / rose-pink fls; leathery lvs / s Spain 211
daurica / P / 3' / red fls / se Europe / 25
delavayi / Sh / 4-6' / dark maroon-red fls / w China 107 / 25
delavayi v. angustiloba - lvs more finely divided 222
delavayi lutea - see P. lutea 28
emodi / P / 1-3' / white fls / Kashmir 279 / 25
humilis - see P. officinalis ssp. humilis 286
japonica / P / 16-20" / white fls / Japan, China, Manchuria 200
kevachensis - see P. mascula 77
lactiflora / P / 30-32" / white to rose-purple fls / Asia 147 / 200
lobata / may be P. lactiflora, broteri, peregrina or villosa 25
lobata alba - a confused name applied to several sub species 25
ludlowii / Sh / to 6' / large golden-yellow fls / Tibet 139
lutea / Sh / to 5' / yellow fls / w China 208 / 25
lutea v. ludlowii / 8-9' / fls 5" across / Tibet 194 / 25
mascula / P / to 3' / red fls / s Europe 182 / 286
mascula ssp. arietina / lvs pubescent beneath / e Europe 286
mascula ssp. russii / with ovate leaflets / islands of Medit. Sea 286
mascula ssp. russii v. reverchonnii - carpels glabrous 310

---

1 Carl Worth, in correspondence.

mlokosewitschii / P / 2' / citron-yellow fls / Caucasus 269 / 129
mollis / P / 1-1½' / red or white fls / garden origin? 25
Moutan - see P. Suffruticosa 25
Moutan 'Hanadajai' - see P. suffrutcosa 'Hana-daigin'
obovata / P / 16-20" / pale rose fls / Far East 200
obovata 'Alba' / 18" / white fls / Siberia, China 184 / 279
obovata v. japonica / P / to 2' / white fls / China,Manchuria,Japan 25
obovata v. willmottiae - white fls; lvs densely hairy beneath 93
officinalis / P / 1½-2' / solitary red fls / France to Albania 148 / 129
officinalis 'Anemonaeflora Rubra' - crimson fls, enlarged stamens 268
officinalis banatica /P/1½-2'/fls red/ ctr lft segmented/Hung,Jugo. 25
officinalis ssp. humilis - segmented leaflets, glabrous pods 212 / 286
officinalis 'Lobata' - lvs distinctly lobed 28
officinalis ssp. villosa - segmentd leaflets, floccose stems & pods 25
peregrina / P / 2' / fls red / Italy, Roumania, Balkans 138 / 77
Potaninii / Sh / stoloniferous / fls 2½", deep red-white/Szechwan 25
Potaninii 'Alba' / 1½' / white fls / China 255
rhodia / P / to 14" / lfts. narrow; fls white / Rhodes 61
russii - see P. mascula ssp. russii 25
russii v. reverchonii - P. mascula ssp. russii X P. coriacea 286
suffruticosa / Sh / 3-4½' / fls in showy colors / nw China, Tibet (228) / 200
suffruticosa 'Hana-daigin' - deep purple fls 197
suffruticosa v. spontanea - leaflets smaller than typical Tree Peony 25
tenuifolia / P / 1-1½' / dark crimson fls / se Europe 286
Veitchii / P / 1' / deep magenta fls / China 25 / 279
Veitchii alba / P / to 1 3/4' / fls white / w China 25
Veitchii v. Woodwardii - dwarf, clear light pink fls (349) / 279
[1] 'White Wings'
Wittmanniana / P / 2-3' / solitary white fls / nw Caucasus 25 / 279
Woodwardii - see P. veitchii v. woodwardii 25 / 93

PALIURUS - Rhamnaceae
Spina-Christi / T / to 20' / fls grn-yel / s Austral. to n China 25

PANAX - Araliaceae
quinquefolius / P / 12-18" / white fls; red frs / e U.S. 224 / 107
schinseng v. japonica / P / to 3' / fls yellow-green / Korea, Manch. 200
trifolius / P / to 9" / wht fls, tinged pnk; frs yellowish / e N.A. 224 / 99

PANCRATIUM - Amaryllidaceae
illyrcium / HH Bb / 1' / white fls / Medit. region 186 / 129
maritimum / HH Bb /lvs to 20" / white fls / Medit. region, coastal 212 / 290

PANDOREA - Bignoniaceae
pandorana / HH Cl / twining / yellow or pinkish-wht fls / Australia 128

PANICUM - Gramineae
virgatum / P Gr / 2-6' / decorative panicles / N. Am. 71 / 268

PAPAVER - Papaveraceae
aculeatum / A 1-4' / fls scarlet-orange / Australia, South Africa 28
alaskanum - see P. radicatum 64 / 209

---

1 Hillier's Hundred Catalog, Index, p. 53.

alboroseum / R / to 8" / white or rose fls / Kamtchatka	13	/ 209
alpinum / A or B / 4-8" / varii-colored fls / e Alps, Carpathians	90	/ 129
alpinum 'Album' - white fls		*
alpinum burseri - see P. burseri		286
alpinum corona-sancti-stephani - see P. corona-sancti-stephani		286
alpinum ssp. ernesti-mayeri / R / 6" / white fls / se Alps		286
alpinum ssp. tatricum / white fls / w Carpathians		286
alpinum kerneri - see P. kerneri		286
anomalum / P / to 28" / white fls, glabous capsules		(229)
apokrinomenon / R / 15" / reddish-orange fls / w Turkey	77	/ (229)
arenarium / A / 8-10" / bright red fls / Caucasus, n Iran		5
atlanticum / HH P / 1-2' / dull orange fls / Morocco		286
bracteatum / P / to 3' / purplish-crimson fls / Iran, Caucasus		77
Burseri / R / 8" / white fls / n Alps, Carpathians	148	/ 286
cambricum - see MECONOPSIS cambrica		*
caucasicum - see P. fugax		25
chibinense / resembles P. radicatum, finely dissected lvs		286
commutatum / A / 2' / red fls spotted black / Crete, Asia Minor	77	/ 286
commutatum 'Ladybird' / 18" / brilliant scarlet fls marked black		59
corona-sancti-stephani / R / 5" / yellow fls / e & s Carpathians		286
croceum - see P. nudicaule f. croceum		28
dahlianum / R / caespitose / fls yellow or white / arctic Norway		286
degenii - see P. pyrenaicum ssp. degenii		286
dubium / A / to 2' / red fls; weedy / Europe		286
Fauriei / R / 4-8" / greenish-yellow fls / n Japan	200	/ 25
floribundum - see P. fugax		77
fugax / B / to 2' / orange fls / Caucasus		25
glaucum / A / to 4' / fls red / Syria, Iraq, Iran		25
Heldreichii - see P. spicatum		25
horridum - see P. aculeatum		28
kerneri / R / 8" / yellow fls / se Alps, c Jugoslavia	148	/ 286
'King George' - see P. orientale 'King George'		208
kluanense / R / tiny / pale yellow fls / Colorado	229	/ 312
lapponicum / R / 6-8" / yellow fls / arctic Norway & Russia		88
lateritium / P / 1-2' / brick-red or purplish / Armenia	77	/ 286
macounii / R / 1'+ / yellow fls / arctic N. Am. & Eurasia	22	/ 209
macrostemum / A / 8-20" / fls red or purplish / Asia Minor, Caucasus	77	/ 279
mairei / R / long stems / reddish fls / Atlas Mts		64
Miyabeanum / R / 8" / lemon yellow fls / Japan		24
Miyabeanum fauriei - see P. fauriei		200
Miyabeanum 'Takewoki' - minature with silvery lvs		(230)
monanthum / R / stemless / orange to pink fls to 1'+ / Cauc., Alps		5
Miyabeanum v. tanewaki - see P. nudicaule		286
Nordenhagenianum / R / to 1' / yellow fls / Iceland, Scandinavia		25
Nordenhagenianum ssp. islandicum / P/to 12"/ fls yel./Iceland,Scand.		25
nudicaule / B / 1' / white or yellow fls / arctic regions	147	/ 25
[1] nudicaule ssp. album - white fls		
nudicaule aurantiacum - see P. nudicalue ssp. xanthopetalum		*
[2] nudicaule 'Champagne Bubbles' - lg. stem; lg. fls, mix of pastels		
nudicaule 'Croceum' - yellow color selection		25

---

[1] Cullen, Baileya, 16:3.

[2] Thompson and Morgan Seed Catalog, 1986.

nudicaule 'Rubro-aurantiacum' - reddish-yellow selection 25
nudicaule ssp. xanthopetalum - yellow fls (229)
oreophilum / P / 2' / orange fls / Caucasia 77
orientale / P / 2-3' / scarlet fls / Armenia 147 / 129
orientale 'King George' - scarlet fls, fringed petals 208
orientale 'Mrs. Perry' - single pink fls 49
orientale 'Olympia' / 2½' / orange-scarlet double 268
pavonium / A / 1' / scarlet fls / Turkestan, Afghanistan 28
pilosum / P / 2'+ / orange-red fls / Bithynia, Galatia 194 / 77
'Pink Chiffon' - see P. somniferum 'Pink Chiffon' *
pseudocanescens / R / 6" / yellow fls / Siberia, Mongolia 25
pyrenaicum - see P. rhaeticum 25
pyrenaicum ssp. degenii / small orange or yellow fls / sw Bulgaria 286
pyrenaicum ssp. corona-sancti-stephani - see P. corona-sancti-s. 286
pyrenaicum ssp. rhaeticum - see P. rhaeticum 286
pyrenaicum sendtneri - see P. sendtneri 286
radicatum / R / to 1' / fls yellow, pink, white / nw Europe 194 / 286
radicatum 'Album' - the white phase *
radicatum lapponicum - see P. lapponicum *
rhaeticum / R / to 8" / golden yellow fls / s, w, & e Alps 148 / 286
rhaeticum ssp. degenii - segregate of P. alpinum 95
Rhoeas / A / 1½-2' / scarlet fls / n temperate zone 229 / 129
rupifragum / HH P / to 18" / brick-red fls / s Spain 286 / 25
rupifragum atlanticum - see P. atlanticum 286
sanctae-coronis - see P. corona-sancti-stephani *
sendtneri / R / 6" / white fls, lf base compact tunic / c & e Alps 148 / 286
[1] somniferum / A / 3'+ / white to purple fls / w & c Medit. region 210 / 286
somniferum 'Pink Chiffon' - double pink fls *
spicatum / P / 2½' / orange-red fls; white-hairy lvs / Asia Minor 25
[2] splendissimum - synonym of P. orientale
strictum / R / ? / orange fls; deeply divided lvs / w Turkey (229)
suaveolens / R / 4" / fls orange to pink / s Spain, Pyrenees 148 / 212
syriacum / A / to 1' / deep crimson fls / Cilicia, Syria 77
'Takewoki' - see P. miyabeanum 'Takewoki'
tauricola / B / 12-16" / fls red to orange; branched / Turkey 77
tianschanicum / R / tufted / orange fls / c Asia, in alpine meadows 5
triniifolium / B / to 1' / fls pale red / Asia Minor 77 / 25

PARADISEA - Liliaceae
liliastrum / P / 1½-2' / pure white fls / Alps, Pyrenes 148 / 267
liliastrum 'Major' / to 3' / larger fls 184 / 25
lusitanica / HH P / to 5' / white fls / Portugal 25

PARAGEUM
cathifolium - see GEUM calthifolium 200

PARAHEBE - Scrophulariaceae
Bidwillii / HH Sh / 6" / white fls / South Isl., New Zealand 134

---

[1] NOTE: Cultivation of P. somniferum prohibited in United States.

[2] Cullen, Baileya, 16:3.

x Bidwellii - procumbent, leathery lvs 25
catarractae / HH Sh / to 2' / white fls, purplish veins / N. Zealand 25
catarractae 'Diffusa' / densely matting / fls white veined rose-pink 139
decora / HH Sh / to 2' / fls white or pink / New Zealand 25
Hookeriana / HH Sh / dwarf / white fls, pink veins / New Zealand 1
linifolia / Sh / to 10" / 2-4 racemes, fls white to pale rose / N.Z. 25
Lyallii / HH Sh / prostrate / white fls, veined pink / New Zealand 139
perfoliata / HH Sh / 2' / spode-blue fls / Australia 279
plano-petiolata / HH Sh / prostrate / pink, lavender, wht fls / N.Z. 15
spathulata / HH Sh / dwarf / white fls, gray lvs / Ruapehu, N.Z. 1

PARAQUILEGIA - Ranunculaceae
grandiflora / R / small / lavender-blue fls / c Asia (543) / 25

X PARDANCANDA - (intergeneric hyb.)
1 'Norrissii' / 3' / 2" fls, many colors with markings

PARIS - Liliaceae
japonica / P / 12-32" / white fls / Honshu, in high mts. 200
obovata - see P. verticillata 200
polyphylla / P / to 3' / fls yellow, seeds scarlet / Himalayas 25
quadrifolia / R / 9-12" / yellowish-grn fls; bluish-black frs /Eur. 148 / 290
verticillata / P / 8-16" / yellow fls / Far East 200

PARNASSIA - Parnassiaceae
alpestris - see P. palustris ssp. alpestris 96
asarifolia / R / 10-16" / white fls / Virginia & North Carolina 60 / 225
caroliniana - see P. glauca 99
fimbriata / P / 6-12" / white fls / nw N. America 229 / 123
fimbriata v. hoodiana / differing slightly / n. Cascades to Alaska 142
foliosa nummularia / P / 6-12" / fls white / Honshu 200
glauca / R / 3-12" / white fls/ New Brunswick to Virginia 224 / 107
grandifolia / R / 1' / white fls / Florida, Texas, Missouri 225 / 99
kotzelbuei / R / to 10" / white fls / n N. America, ne Asia 99
nubicola / R / 12" / white fls / Himalayas 25
palustris / P / to 1'/ white fls, 1" across / N.Amer, n Asia & Eur. 148 / 286
palustris ssp. alpestris / taller, larger fls / Alps 96
palustris v. neogea - N. Am. var. with lvs cordate-based 132

PAROCHETUS - Leguminosae
communis / HH R / 1-2" / vivid blue fls / Himalayas 75 / 123

PARONYCHIA - Caryophyllaceae
argentea / R / mats / fls concealed by silvery bracts / s Europe 211 / 286
argyrocoma / R / 3-8" / silvery bracts / Me. to Tenn. 107
canadensis / A / to 1' / minute lvs; green fls / n & c US 107
capitata / R / 6" / conspicuous bracts / s Europe 212 / 286
cephalotes / R / 6" / silvery bracts / e, c & se Europe 286
kapela / P / to 6" / fls tiny, greenish /Spain, Pyrenees 148 / 286
kapela ssp. serphyllifolia / stems procumbent / Alps 286
nivea - see P. capitata 286
pulvinata / R / low mat / lvs rounded; fls sessile / high plains, US 61

---

[1] Wayside Catalog, Spring, 1986.

serpyllifolia - see P. kapela ssp. serpyllifolia 286
sessiliflora / R / 1" mats / calyx brownish-yellow / c N. Am. 229 / 75
virginica / R / 4-16" / bristles replacing petals / se & sc US 225 / 99

PARROTIA - Hamamelidaceae
Jacquemontania - see PARROTIOPSIS Jacquemontia 25
persica / Sh or T / to 30' / crimson stamens / Iran, Caucasus 36 / 139

PARROTIOPSIS - Hamamelidaceae
Jaquemontia / T / to 20' / lvs yel. in autumn; fl bracts wht /Himal. 25

PARRYA - Cruciferae
macrocarpa / P / 1, 2-3" / fls lilac, 1" across / Arctic 61
Menziesii - see PHOENICAULIS cheiranthoides 93
nudicaulis / P / to 16"/petals wht or purp; fr siliqua/ arctic Eur. 286

PARSONIA - Apocynaceae
capsularis / HH Cl / high / white yellow to dark red fls / N. Z. 15
heterophylla / HH Cl / robust / white fls / New Zealand 238

PARTHENOCISSUS - Vitaceae
incerta / Cl / trailer / lvs lustrous / North America 99
quinquefolia / Cl / high / black frs / ne US to Mexico 184 / 25
tricuspidata / Cl / high bloomy dark blue frs / Far East 139
vitacea - see P. incerta 99

PASITHEA - Liliaceae
caerulea / HH P / 3' / blue fls / Chile 279

PASSIFLORA - Passifloraceae
caerulea / Gh Cl / moderate / white, purple & blue fls / S. America 187 / 25
cinnabarina / Gh Cl / high / fls reddish-brown / Australia 45
edulis / Gh Cl / vigorous / white & purple fls / Brazil 128 / 25
incarnata / Gh Cl / high / pink to purple fls, edible frs / s US 229 / 25

PATERSONIA - Iridaceae
1 Drummondii / lvs very narrow, hairy / w Australia
glabrata / Gh / 3-6" / purple fls / Australia 64
glauca / Gh / 18" / fls blue / Australia, Tasmania 544 / 25
longiscapa / Gh / 6-20" / fls blue / Australia 206
1 occidentalis / P / tall, stout / very variable / w Australia
sericea / Gh / 18" / blue fls / New South Wales, Australia 109 / 45
umbrosa / Gh / 2' / blue fls / w Australia 121
1 xanthina / R / lvs narrow, grass-like; fls yel / w Australia

PATRINIA - Valerianaceae
gibbosa / P / 20-28" / yellow fls / n Japan 200
heterophylla / R / 1' / yellow fls / China 206
intermedia / R / 1'+ / fragrant yellow fls / Siberia 107
palmata - see P. triloba v. palmata 200
rupestris - see P. intermedia 93

---

1 William E. Blackall and B.J. Grieve, How to Know Your Western Australian Wildflowers, Parts I, II, III.

scabiosifolia / P / to 3'+ / fls yellow / temperate e Asia 25
sibirica / R / to 6" / yellow fls / Hokkaido, Siberia 164 / 200
triloba / P / to 2' / yellow, short-spurred fls / Honshu 93 / 200
triloba v. palmata / long-spurred fls / Japan 164 / 200
villosa / P / 20-40" / white fls / Far East 164 / 200

PAULOWNIA - Scrophulariaceae
tomentosa / T / 50' / pale lilac-mauve fls / China 129 / 222

PAXISTIMA - Celastraceae
Canbyi / Sh / to 16" / fls small bluish white / Va., W.Va. 25

PECTEILIS - Orchidaceae
radiata / P / 16" / white fls / Japan 166 / 25

PEDICULARIS - Scrophulariaceae
aposochila / R / to 6" / red fls / alps of Japan 273 / 200
attollens / P / 6-16" / purple or lavender fls / Oregon 228
bracteosa / P / 1-4' / yellow to red fls / Calif. to B. C. 228
canadensis / R / 1'+ / yellow fls / e N. Am. 224 / 99
carpatica - see P. hacquetii 288
chamissonis v. japonica / 8-24" / reddish fls / n Japan 200
contorta / P / to 2'/ fls wht or yel., purple spots / B.C. to Calif. 25
crenulata / R / 4-16" / fls rose / Wyoming, Nevada, Colorado 229
densiflora / P / 4-20" / bright red to purplish fls / Calif. to Ore. 228
elegans / R / decumbent / fls pinkish-red / c & s Apennines 288
elongata / R / 6-16" / pale yellow fls / s Alps 148 288
foliosa / P / 8-20" / fls pale yellow / s & sc Europe, in mts. 288
grayi / P / 1½-4' / fls greenish-yellow to reddish / Col., N. Mex 229 / 28
groenlandica / P / 1-2' / fls rose to red / n N. Am. 312 / 107
groenlandica v. surrecta / 2' / red or purple fls / N. Am. 25
hacquetii / P / 1-4' / fls pale yellow / Alps, Carpathians 288
kerneri / R / 4" / deep rich rose fls / European Alps 123
lanceolata / P / to 3' / yellow fls / c N. Am. 224 / 25
lapponica / P/ 4-10"/ fls yellow or cream / arctic Eur., N.A. & Asia 288
oederi / R / to 6" / yellowish fls / northern hemisphere 200
ornithorhyncha / R / 2-12" / purplish fls / Cascade Mts. to Alaska 228
palustris / P / 4-32" / rose-purple fls / arctic Europe 148 / 209
recutita / P / to 2' / greenish-yellow fls, tinged crimson / Alps 148 / 288
resupinata / R / 12" / purple fls / Siberia 164 / 206
rostratocapitata / R / to 8" / pink to purplish-red fls / e Alps 148 / 288
rostratospicata / P / 6-18" / fls pink to purplish-red / Alps 314 / 288
sudetica / R / 4-10" / pink-purp-red fls / n & c Eu., nw N. Am. 229 / 288
sylvatica / R / to 10" / pink or red fls / w & c Eu. to c Sweden 148 / 288
tuberosa / R / 4-10" / pale yellow fls / Alps, Pyrenees 148 / 288
verticillata / R / to 6" / red fls / n hemisphere 148 / 288

PEDIOCACTUS - Cactaceae
simpsonii / R / 4" / fls yellow-green to purple / Col., Wyo. 28

PEGANUM - Zygophyllaceae
harmala / P / 1-2' / fls 5-mawes, greenish white / Med & se Eur. 287

PELARGONIUM - Geraniaceae
alchemilloides / P / 6-18" / lvs kidney shaped; fls wht & rose 61

australe / HH R / 9" / white to rose fls / Australia 45 / 25
endlicheranum / P / 18" / fls deep rose / Asia Minor 25
grossularioides / P / ? / fls deep rose purple / nat. in California 25
odoratissimum / P / to 1½' / fls wht, sometimes veined red/S.Africa 25
pulverulentum / P / 1' / gray, blood color / Cape of Good Hope 206
tomentosum / Sh / to 3' / fls white blotched red / S. Africa 25
vitifolium / P / 1'+ / petals rose / South Africa 25

PELLAEA - Polypodiaceae

andromedaefolia / F/to 2½'/ petioles rosy flesh color/ Cal.to Baja 25
atropurpurea / F / 15" / gray to blue-green fronds / N. Am. 191 / 159
Brewerii / F / lvs to 10" / petioles slender, brown/ Wash. to Calif. 25
Bridgesii / F / to 1' / fronds blue-green /Sierra Nev. to Ore & Ida. 25
calomelanos - see PITYROGRAMMA calomelanoa 215
glabella / F / 9" / smaller plant / Que. to Va. & Colorado 99
rotundifolia / Gh F / 12" arching fronds / round leaflets / N. Z. 93 / 128

PELTARIA - Cruciferae

turkmena / Sh / 20-28" / white fls / c Asia 11

PELTIPHYLLUM - Saxifragaceae

peltatum / P / 2-3' / white to pale fls / California 147 / 129

PELTOBOYKINIA - Saxifrageceae

tellimoides / P / 1-2' / creamy-white fls / Honshu 165 / 200
watanabei / P / 10" / fls pale yellow / Japan 200

PENNISETUM - Gramieae

alopecuroides / P Gr / 12-32" / purple or green fls / Far East 129 / 200
japonicum - see P. alopecurioides 200
orientale / HH P Gr / ½-3' / silky purplish inflorescence / Asia 419 / 268
setaceum / P / to 3½' / inflorescent pink or purple / ? 25

PENSTEMON - Scrophulariaceae

acuminatus / P / to 2' / lilac to violet fls / Neb., Minn., s&e-ward 28
adamsianus - see P. fruticosus 41
aggregatus - see P. rydbergii v. varians 142
albertinus / R / 6-8" / fls light blue / nw North America (546) / 107
albidus / P / to 2½' / fls blue to blue-purple/ Rockies to N. Mex. 25
alpinus /R / to 15" / dark blue to purple fls / Wyoming, Colorado 229 / 42
ambiguus / P / 8-20" / fls whitish to bright pink / Kans., Col. 229
angustifolius / P / 8-20" / blue to pinkish-lavender fls/ N. Dak.& s 229 / 227
angustifolius 'Pygmeus' - of the 8" range? *
antirrhinoides / HH Sh / to 6' / yellow fls / s Calif., Baja Calif. 227 / 25
arenicola / R / 8-12" / sky-blue fls / n Rockies (236)
aridus / P / 4-10" / blue or blue-purple fls / Mont., Idaho (545) / 25
arizonicus - see P. virgatus ssp. arizonicus 41
arkansanus / P / to 2' / fls white with violet lines /Mo., Ark.,Tex. 25
attenuatus / P / to 2' / pale yellow to blue-purple fls /Ida., Wash. 229 / 25
azureus / HH P / 2½' / fls deep blue-purple / c California 228 / 25
barbatus / P / to 6' / red fls / Utah to Mexico 129 / 25
barbatus 'Nanus' / 15" / listed as 'Praecox Nanus' 283
barbatus 'Rose Elf' / 18" / clear shell-pink fls 207
Barrettiae / Sh / 1' / lilac-purple fls / Columbia River Gorge (547) / 107

1	'Blue of Zurich' - see P. heterophyllus 'Blue of Zurich'		
	brachyanthus - see P. procerus ssp. brachyanthus		142
	brachyanthus f. albus - see P. procerus ssp. brachyanthus 'Albus'		*
	Bradburyi - see P. grandiflorus		25
	brandegei / P / to 2½' / blue to reddish-purple fls / Col., N. Mex.		25
	brevisepalus / P / 1-3' / purple or violet fls / Va. to Tenn.	224 /	99
	Bridgesii / Sh / 1' / red fls / Colorado, Arizona, New Mexico		227
	caespitosus / R / mats / lilac-purple fls / Rocky Mts.	229 /	107
	californicus / HH R / 6" / blue or purplish fls / s & Baja Calif.		227
	calycosus / P / to 3' / fls white or purplish / c United States	224 /	99
	campanulatus / P / 18" / fls purple, rose, blue / Mexico		279
	campanulatus pulchellus - included in the sp.		41
	canescens / P / 1-3' / purple or violet fls / Pa. to Ala.	224 /	99
	X 'Cardinal' / 12" / P. rupicola x cardwellii / red fls		(237)
	Cardinalis / P / 16-28" / fls dull red or crimson / New Mexico		28
	Cardwellii / Sh / 10" / purplish fls / Oregon, Washington	228 /	107
2	Cardwellii v. albus - white form		
	Cardwellii 'John Bacher' - strong white form		(192)
	caryi / 4-10" / 1 side inflor. purplish; blue-grn lvs/Bigttorn Co,US		229
	centranthifolius / HH P / 1-4' / fls scarlet / coast ranges, Calif.		228
	cinicola / P / 2' / purplish fls / n California, se Oregon		228
	clutei / HH P / to 3½' / fls pink to rose / Arizona		25
	cobaea / P / 2' / large purple fls / Mo., Neb. & s-ward		107
	comarrhenus / P / 12-24" / pale blue fls / Col., Utah, Arizona		42
	confertus / R / 6-20" / cream to yellow fls / Rocky Mts.	229 /	123
	confertus albus / P / 8-24" / many sm. fls, yel & wht / nw N.Am.		229
	confertus v. caerulo-purpurascens - see P. procerus		41
2	confertus 'Kittitas' - mats of olive-green lvs; sulphury fls		
	cordifolius / HH Sh / to 2' / red fls / California	227 /	25
	corymbosa / Sh / to 2' / fls red, yel beard / California		25
	Crandallii / Sh / to 4" / dark blue fls / Colorado, Utah		229
	Crandallii ssp. glabrescens / 6-8" / bluish fls, New Mexico		25
	Crandallii ssp. procumbens / P / to 8" / drk blue-purp, decumb/Colo.		25
	cristatus - see P. eriantherus		25
	Cusickii / P / to 1½' / purple to blue / Ore & w Idaho		25
	cyananthus / P / 12-32" / deep blue fls / e Idaho, Wyo., Utah		229
	cyananthus brandegei - see P. brandegeei		41
	cyaneus / R / mat, 2-6" / bright blue fls / Mont., Ida., Wyoming		(238)
	dasyphyllus / P / ? / violet-blue fls / New Mexico		227
	Davidsonii / R / 6-9" / ruby-red fls / California	312 /	123
	Davidsonii v. davidsonii / P / lvs sm; fls blue to purp /Cal. & Ore.		142
	Davidsonii ssp. menziesii / finely-toothed lvs / Wash., Ore., B.C.		25
	Davidsonii ssp. menziesii f. albus 'Martha Raye' - wide-mouthed cv		(237)
	Davidsonii ssp. menziesii 'Microphyllus' - 4" miniature		123
	deustus / HH Sh / 8-24" / yellowish-white fls / California	229 /	228
	diffusus - see P. serrulatus		25
	Digitalis / P / to 4½' / fls white or whitish / e North America	224 /	99
	Digitalis alba /to 5'/ smooth stems, cor. bell showy wht/S.Dak-Okla.		229

---

[1] Bennett, Bull. Am. Penstemon Soc., Vol. 30, 1971.

[2] Manton, Bull. Am. Rock Garden Soc., 20:1.

Entry	Ref.
[1] Digitalis 'White Queen' - maximum-sized plant, fls pure white	
diphyllus / P / to 2½' / fls lavender to lilac-blue /Mont. to Wash.	142
dissectus / P / to 16" / fls purple, yellow; bearded / Georgia	25
eatonii / P / 1-2' / scarlet fls / s California, Nevada, Utah	227
X Edithae / P / to 10" / fls pink / Garden hybrid	25
[2] 'Elfin Pink' - clear pink fls	
ellipticus / P / procumb. to 6" / light violet / N.Dak nw to Alta.	25
eriantherus / P / to 16" / purple fls / Dakotas n & w-ward	107
euglaucus / P / to 20" / blue fls / Oregon, Washington	25
[3] 'Evelyn' - P. campanulatus cultivar	
Fendleri / P / 8-20" / fls blue with purple lines / sw Kans-Mex.	227 / 25
[3] 'Flathead Lake' - see P. X johnsoniae	
floridus / HH P / to 4' / rose-purple fls / Calif., Nevada	227
frutescens / R / 4-8" / pale purple fls / Far East	273 / 200
fruticosus / Sh / variable / purple fls / nw North America	229 / 107
fruticosus confertus - see P. confertus	*
fruticosus crassifolius - included in the sp.	129 / (237)
fruticosus v. fruticosus /P/6-16"/shruby; fls blue-lav/Wash,Ida,B.C.	142
fruticosus ssp. scouleri - longer fls; lvs linear-lanceolate	207 / 25
fruticosus ssp. scouleri 'Albus' - good white form	123
fruticosus ssp. scouleri 'Six Hills' / 6-9" / purplish-pink fls	123
[3] fruticosus ssp. serratus - best alpine form, wide color range	
Gairdneri / R / to 1' / fls lavender-blue / e Oregon	229 / 25
Gairdneri ssp. oreganus - lvs opposite-appearing; fls bluish to wht.	25
'Garnet' / 1½-2' / deep red fls / P. hartwegii X. P. cobaea	268
Garrettii / P / to 2½' / fls blue, yellow bearded / Utah	25
gentianoides / HH P / 4-5' / bluish-purple fls / Guatemala, Mexico	25
glaber / P / 2' / blue to purple fls / Missouri River reg. & w-ward	229 / 268
glaber v. alpinus - see P. alpinus	25
glabrescens - see P. crandallii ssp. glabrescens	41
glaucus v. stenosepalus - see P. Whippleanus	25
glaucus Whippleanus - see P. Whippleanus	25
globosus / R / 10-16" / bright blue fls /c Idaho, Wallowa Mts., Ore.	229
gloxinioides - name of no botanical standing	25
[4] 'Goldie' / P. confertus 'Kittatas' x pink f. P. euglaucus / yellow	
Gormanii / R / to 2' / fls blue-purple / Alaska, B.C.	13 / 25
gracilis / P / to 20" / fls pale violet / c North America	224
grandiflorus / P /to 4' /fls lilac or blue-lavender/Ill., Wyo., Tex.	224 / 25
hallii / R / 6-8" / violet fls / Colorado, in mts.	312 / 123
Hartwegii / P / to 4' / corolla scarlet 2" / Mexico	25
Haydenii / P / to 2' / fls blue / Nebraska	25
heterodoxus / R / 3-10" / deep rose-purple fls / California, Nev.	228
heterophyllus / R / to 2' / blue fls, pink tinged / California	227 / 123
heterophyllus 'Blue Gem' - gentian-blue fls	207

---

[1] Davidson, Bull. Am. Rock Garden Soc., 27:4.

[2] Rocknoll Catalog, Spring, 1986.

[3] Bennett, Bull. Am. Penstemon Soc., Vol. 30, 1971.

[4] Alpines of the Americas. Report of the First Interim International Rock Garden Plant Conference, 1976.

1 heterophyllus 'Blue of Zurich' - gentian-blue fls
1 heterophyllus ssp. purdyi - useful as annual bedder
heterophyllus 'True Blue' - azure-blue fls 123
hirsutus / P / to 3' / fls purplish or violet / Me. to Va., & Wisc. 224 / 25
hirsutus f. albiflorus - white fls 41
hirsutus albus - see P. hirsutus f. albiflorus 41
hirsutus 'Minimus' / 4-6" / stems stiffly erect (239)
hirsutus 'Pygmaeus' / 4" / pale violet or purp. fls / 50% true (378) / (237)
'Holly" / cv. of P. fruticosus ssp. serratus / 10" / fls lilac-blue (237)
humilis / R / to 1' / blue or bluish-purple fls / Calif., Col. 229 / 228
2 humilis ssp. brevifolius / 4" / azure-blue fls
Jamesii / P / to 20" / lavender or blue-purple fls / Col. 227
X Johnsoniae / P / 16-40" / fls in shades of red 41
'Kobolt' / P / 24-28" / coral-red sel. of P. barbatus, presumably *
kunthii / HH P / to 3' / fls red or pink / mts. w & s of Mexico City (238)
labrosus / HH P / 12-28" / scarlet fls / c & s California 227
laetus / P / 8-32" / blue-lavender & darker fls / Calif., Oregon 228
laetus ssp. roezlii - shorter fls 25
laevigatus / P / 16-40" / fls white & violet / N.J. to Florida 225
laevigatus digitalis - see P. digitalis *
lanceolatus / HH P / 16-20" / scarlet fls / Arizona, Texas (238)
laricifolius / R / 10" / purple fls / Wyoming 229 / 25
laricifolius ssp. exilifolius / 6-8" / white fls / Wyoming (238)
linarioides / HH P / 18" / fls lilac to purple to w New Mexico 227 / 25
linarioides v. coloradoensis / P / to 15"/fls purp, yel beard/w N.M. 25
Lyallii / P / to 2½' / lavender fls / Idaho, Montana 138 / 25
Menziesii - see P. davidsonii ssp. menziesii 44
Menziesii alba 'Martha Raye'-see P.davidsonii ssp. m.f. albus 'M.R.' (237)
Menziesii Davidsonii - see P. davidsonii 25
Menziesii 'Microphyllus' - see P. davidsonii ssp. m. 'Microphyllus' 123
3 'Manito' / see P. Edithiae / blue fl / dwarf
montanus / R / 8-10" / pink-purple fls / Ida., Mont., Wyo. 229 / 107
Murrayanus / P / to 3' / scarlet fls / e Texas, Oklahoma (546) / 25
nemerosus / P / 12-32" / pink-purple fls / Calif. to B.C. 228
Newberryi / R / 9-12" / rosy-purple fls / California 133 / 123
Newberryi f. humilior / 6-9" / fls cherry-red 123
Newberryi v. rupicola - see P. rupicola 41
nitidus / R / to 1' / clear blue fls / Sask. to Mexico 107
neotericus /8-24"/lvs leathery, bl-wht;fls bl-purp, bells/s Nev&Cal 228
oreocharis / P / to 28" / blue-purple fls / nw United States 228
ovatus / P / 12-40" / blue fls; glandular hairy lvs / Ore., B.C. 174 / 228
Palmeri / HH P/ to 5'/fls white, tinged pink / Calif., Utah, Ariz. 229 / 25
Parryi / P / to 4' / rose-magenta fls / Arizona, Mexico 227
parvulus / R / to 14" / blue-purple fls / nw Calif., se Ore. 228
parvus / R / 2-4" / blue fls / sc Utah, in mts. 42
Peckii / R / ? / pink or bluish fls / s Oregon 25
pennellianus / P / 18-20" / fls deep blue / Washington, Oregon 42
pinifolius / R / to 1' / scarlet fls / Arizona, New Mexico 227 / 107

---

[1] Bennett, Bull. Am. Penstemon Soc., Vol. 30 1971.

[2] Carl Worth, Bull. Am. Rock Garden Soc., 19:1.

[3] Alpines of the Americas. The American Rock Garden Society, 1976.

'Prairie Dawn' / multi-species hybrid / 12-18" / pale pink fls (237)
procerus / R / to 1' / purplish-blue fls / Colorado, w- & n-ward 229 / 107
procerus ssp. brachyanthus - fls larger 142
procerus ssp. brachyanthus 'Albus' - fls white *
procerus ssp. formosus / densely caespitose / Oregon & California 196
procerus ssp. tolmiei / alpine form / lighter-colored fls 64 / 41
procumbens - see P. suffrutescens 41
pruinosis / R / 4-16" / deep blue to lavender fls / c Wash, B.C. 25
pseudospectabilis / HH P / 3'+ / rose-purple fls / s California 227
pubescens - see P. hirsutus 41
pulchellus - see P. campanulatus 41
purpusii / R / 4-8" / violet to blue fls / California 228 / (238)
pygmaeus - see P. hirsutus 'Pygmaeus' 41
Rattanii / P / to 4' / lavender to purplish fls / Ore., California 228 / 25
Rattanii ssp. kleei / HH P / 1-4' / lav.-purp. fls / Santa Cruz Mts. 196
Richardsonii / P / 8-32" / lavender fls / Columbia R., n- & e-ward 229 / 228
Richardsonii v. dentatus - with lvs dentate only (238)
Richardsonii v. richardsonii /P /lvs nar; fls bearded /Wash,B.C.,Ore 142
roezlii - see P. laetus ssp. roezlii 25
'Royal Beauty' / 2' / wine-purple fls 172
rupicola / R / to 4" / deep rose fls / Wash. to n Calif. 19 / 25
rupicola 'Albus' - white fls (546) / 25
Rydbergii / P / 18"+ / deep indigo-blue fls / Wyo. to N. Mex. 25
Rydbergii v. varians - northwestern form, lighter in color 142
Scouleri - see P. fruticosus ssp. scouleri 129 / 41
Scouleri 'Albus' - see P. fruticosus ssp. scouleri 'Albus' 133 / 123
Scouleri 'Six Hills' - see P. fruticosus ssp. scouleri 'Six Hills' 123
secundiflorus / P / to 20" / lavender fls / N. Mex., Col., Wyo. 227
'Seeba Hybrid' / P. grandiflorus x murrayanus / varii-col. fls (237)
sepalulus / R / to 1' / pale mauve fls / California 64
serrulatus / P / 8-28" / deep blue to dark purple fls / Ore., B.C. 129 / 228
'Six Hills' - see P. fruticosus ssp. scouleri 'Six Hills' 46
Smallii / P / 16-30" / purplish or crimson fls / N. & S. Carolina (395) / 225
speciosus / P / 8-32" / fls in blue shades / Wash., Calif. 42
speciosus ssp. kennedyi / 2-16" / e California, Nevada 42
spectabilis / HH P / to 4' / lavender-blue fls / s & Baja Calif. 227 / 25
strictus / P / 15-48" / deep violet-blue fls / Col., N. Mex. 229 / 42
subserratus / R / ? / blue fls / Oregon 228
suffrutescens / R / mats / green lvs; purplish fls / Utah 235
tenuiflorus / P / to 2' / white fls / Alabama, Tennessee, N. Mexico 99
tenuifolius[1] / HH Sh / to 15" / lavender & pale purp. blue / Mexico
tenuis / P / 18-24" / violet-purp. fls / marshes, Ark., La., se Tex. 226
teucreides / R / mat, 2-4" / fls blue to blue-purple / Colorado 229 / (238)
Thompsonii - see P. Davidsonii 41
Thurberi / HH Sh / to 30" / blue or bluish-rose fls / California 227
Tolmiei - see P. procerus ssp. tolmiei 41
Tolmiei brachyanthus - see P. procerus ssp. brachyanthus 142
tolmiei brachyanthus 'Albus' - see P. procerus brachyanthus 'Albus' 41
tracyi / 3-4" / fls pink; lvs leathery / Trinity Co., California 228
triphyllus / Sh/ 2½'/ pale lavender to lilac-blue fls / Ida., Wash. 25
uintahensis / R / 4-16", usually low / lav., blue, purp. fls / Utah 42
unilateralis / P / 18-24" / reddish-purple fls / Wyo., Colorado 42

---

[1] Lodewick, Penstemon Field Identifier, 1970.

utahensis / P / 1-2' / fls crimson-red to carmine / Nevada, Utah 229
venustus / Sh / to 2½' / light purple fls / Idaho, Oregon, Wash. 28 / 25
virens / R / 4-14" / blue-violet or blue fls / se Wyo., Col. 229
virens 'Albus' - albino form *
virgatus ssp. arizonicus / 18" / pale violet fls / se Arizona 42
Watsonii / P / 12-28" / blue or blue-purple fls / Ida., Nev., Col. 229
Whippleanus / P / to 2' / purple or lavender fls / Mont. to N. Mex. 229 / 25
Wilcoxii / P / 3'+ / bright blue-bluish-purp. fls / Wash., Ore., & e 229 / 25
Wrightii / P / to 2' / fls bright red / w Texas, Arizona 28

PENTACHRONDA - Epacridaceae
pumila / HH Sh / procumbent / white fls; red frs / New Zealand 238 / 123

PENTAGLOTTIS - Boraginaceae
sempervirens / P / 1-3' / fls rich blue / Europe 25

PENTAPERA - Ericaceae
sicula - see ERICA sicula 288
sicula libanotica / Sh / dwarf / fls white or pink / Sicily, Syria 139

PENTHORUM - Saxifragaceae
sedoides / P / to 3' / yellowish-green fls / e North America 224 / 99

PEPEROMIA - Piperaceae
rotundifolia / Gh / creeping / fls in terminal spikes / Trop. Am. 25

PEREZIA - Compositae
bellidifolia / R / 9" / fls pale blue or lilac / South America 64
fenckii - brilliant blue South American alpine 64
linearis / R / to 1' / pure, brilliant blue fls / South America 64
multiflora / A / erect / reddish pappus / Brazil 182 / 93
pilifera / R / ? / creamy or blue fls / South America 64
recurvata / R / 2-3" / deep blue fls / Fuego, Falkland Isls. 194 / 123

PERILLA - Labiatae
frutescens / A / to 4' / white or reddish fls / Far East 28
frutescens var. crispa / 1½' / deep purple lvs 200
frutescens var. laciniata - cut-leaved purple form 28
laciniata - see P. frutescens var. laciniata 28

PERIPLOCA - Asclepiadaceae
graeca / Cl / 40' / green or purplish fls / sw Eu., w Asia 194 / 25

PERNETTYA - Ericaceae
ciliata / HH Sh / dwarf / evergrn lvs, possibly hardy / Mex., in mts 36
furens / HH Sh / 3' / white fls / Chile 209
lanceolata / Sh / to 18" / fls wht; fr pink; lvs shiny / Australia 109
leucocarpa - see P. pumila v. leucocarpa 75
macrostigma / trailing Sh / to 3' / rosy pink fls / N. Z. 15 / 25
mucronata / HH Sh / 3'+ / white fls / Chile 269 / 288
nana / HH Sh / 4" / white fls, red frs / New Zealand 107
prostrata / HH Sh / dwarf to prostrate / black frs / Venez. to Chile 139
prostrata ssp. pentlandii /Sh/to 1'/frs purp-black/Costa Rica-Chile 25
pumila / Sh / dwarf / lvs tiny; frs white / Magell.Str., Falklands 139
pumila v. leucocarpa / stragling shrub; frs white / Chile 25

tasmanica / HH Sh / prostrate / red frs / Tasmania 132
tasmanica 'Fructo-alba' - white-fruited form *

PEROWSKIA - Labiatae
atriplicifolia / P / to 3'+/ fls lavendar-blue / w Pakistan 25

PERSEA - Lauraceae
lingue / HH T / to 60' / reddish hairy frs / s Chile 46

PERTYA - Compositae
robusta / P / creeping rhizomes / fls in spikes / Honshu 200
sinensis / Sh / to 4½' / bitter lvs; fls purple-pink / China 139

PETALOSTEMON - Leguminosae
candidum / P / to 2½' / fls white / c North America 229 / 25
foliosum / P / 1-3' / rose or purplish fls / Tenn., Ill. 225
gattingeri / P / 8-20" / rose or purplish fls / Tenn. Alabama 225
purpureum / P / 1-3' / roseate to crimson fls / c N. Am. 229 / 99
searlesiae / R / 10" / rose-pink fls / e Arizona, Utah, Calif 227
villosum / P / 2' / rose-purple fls / Sask., Mich. to Texas 229 / 25

PETROCALLIS - Cruciferae
pyrenaica / R / matlike to 4" / fls wht to pink / c & s Europe 212 / 25
pyrenaica 'Alba' - the rarely white form 286
pyrenaica v. leucantha - not recognized as var., fls rarely white 286

PETROCOPTIS - Caryophyllaceae
crasifolia / P / rosette lvs; fls white / Pyrenees 286
glaucifolia / R / 3" / fls purplish / n Spain, in mts 286
grandifolia / R / tufts / fls purplish / nw Spain 286
hispanica / R / caespitose / white fls / wc Pyrenees 286
lagascae - see P. glaucifolia 286
pyrenaica / R / 3" / fls white or pale purple / w Pyrenees 314 / 286
pyrenaica 'Grandiflora' - larger-flowered selection *
pyrenaica 'Rosea' - roseate color form *

PETROMARULA - Campanulaceae
pinnata / HH P / 3' / fls pale blue / Crete 149 / 289

PETROPHILA - Proteaceae
serruriae / Gh Sh / 3-4' / white fls / Australia 184

PETROPHYTUM - Rosaceae
ceaspitosum / Sh / mat / white fls, blue-gray lvs / w United States 229 / 107
cinerascens / Sh / prostrate / lvs gray; fls white / Washington 25
hendersonii / Sh / 6" / fls pale greenish-yellow / Olympic Mts. 132 / 268

PETRORHAGIA - Caryophyllaceae
illyrica ssp. haynaldiana / R / to 16" / wht, pale yel fls / Balkans 25
nanteulii / A / to 20" / pink or purplish fls / w Europe 286
prolifera / A / 20" / fls pink or purplish / c Europe 286
saxifraga / P / to 1½' / white or pink fls / c & s Europe 194 / 286

PETTERIA - Leguminosae
ramentacea / Sh / to 7' / yellow fls / Albania, Jugoslavia 47

PEUCEDANUM
graveolens - see ANETHUM graveolens 25
utriculatum - see LOMATIUM utriculatum 228

PEUMUS - Monimiaceae
Boldus / HH T / 25' / fls whitish; edible frs / Chile 25

PHACELIA - Hydrophyllaceae
bipinnatifida / B / 1-2' / fls violet, blue / W. Va., Ark. & s-ward 224 / 25
campanularia / A / to 20" / bright blue fls / s Calif., Col. 227 / 25
dubia / A / to 20" / fls white to lilac / N. Y., Ohio, Tenn., Ga 224 / 99
franklinii / A / 6-18" / blue or bluish-white fls / w & c Canada 58
hastata / P / to 20" / fls white to lavender / B.C. to n Cal. & e 25
hastata leucophylla - see P. leucophylla 178
hastata v. compacta / 8" / densely wh. hirsute pl / Cal., Ore 95
heterophylla / P / 8-48" / white or cream fls / Cascade to Rockies 229
leucophylla / P / ? / fls pink or lilac, lvs whitish / Colorado 178
linearis / A / to 20" / lvs 3" long; fls viol-wht / B.C.- Alta. & s 25
magellanica / ? / ? / pinkish fls / Peru 64
platycarpa / HH P / to 1½' / fls pinkish-lavender, blue, white / Mex 25
Purshii / A or B / 6-22" / bluish to white fls / c United States 224 / 99
sericea / Bi or P / to 15" / many blue-purple fls / w N. Am. 229 / 25
sericea v. ciliosa / broader lvs & stamens / Idaho, Ore, & Cal. 25
sericea sericea / P/ 4-12"/ hairy, color varies/Pacific nw, not Ore. 142
tanacetifolia / P/ to 4'/ fls blue or lav. / c Calif to Ariz & Mex. 25

PHAEDRANASSA - Amaryllidaceae
Carmioli / B / to 2' / fls scarlet, grn lobes / Costa Rica or Peru 25

PHAGNALON - Compositae
graecum / HH Sh / to 1' / yellowish fls / se Europe 289

PHALACROCARPUM - Compositae
anomalum - see P. oppositifolium 289
oppositifolium / R / 6-12" / white daisy / Spain, Portugal 212 / 289

PHALARIS - Gramineae
arundinacea 'Picta' / P Gr / 2-4' / white-striped lvs / n hemisphere 129
caerulescens / P Gr / 1-5' / purplish or green panicles / Medit. reg. 268

PHARBITIS
purpurea - see IPOMOEA purpurea 116

PHASEOLUS - Leguminosae
caracalla / HH Cl / 20' / fls light yellow to purplish / tropics 28

PHELLODENDRON - Rutaceae
amurense / T / 50' / pinnate lvs, corky bark / Far East 93 / 139
Sachalinense / T / to 45' / bark not corky / China, Korea 25

PHILADELPHUS - Saxifragaceae
californicus / HH Sh / 9' / fls white / California 222
delavayi / Sh / 10' / lvs gray-felted beneath / China, Tibet 139
lewisii / Sh / 6'+ / fls white / Mont., Wash., B. C. 222

lewisii californicus - see P. californicus 222
microphyllus / Sh / to 4' / fls fragrant white / s United States 139
pekinensis / Sh / 6' / creamy fls / n China, Korea 25
X virginalis 'Virginal' - double-flowered hybrid 269 / 25

PHLIESIA - Liliaceae
buxifolia - see P. magellanica 128 / 25
magellanica / Gh Sh / ½-4' / tubular red fls / Chile 194 / 25

PHLEUM - Gramineae
phleoides / P Gr / 4-24" / green or purplish panicles / Eurasia 268

PHLOMIS - Labiatae
alpina / P / 18" / purplish fls / Altai Mts., Mongolia 25
anisodonta / P / 6-20" / rose fls / Iran 204
cashmeriana / P / to 3' / lavender fls / Kashmir 25
chrysophylla / HH Sh / ? / golden-yellow fls / Syria, Lebanon 61
cretica / Sh / to 17" / fls yellow / Greece 288
ferruginea / Sh / to 17" / fls yellow / s Italy 288
fruticosa / HH Sh / 3-4' / fls deep yellow / s Europe 194 / 129
italica / HH Sh / 2' / purple fls / Balearic Isls., Italy 288
lunariifolia - see P. samia 288
lychnitis / Sh / to 24" / fls yellow / sw Europe 288
purpurea / Sh / to 6' / fls purple or pink, rarely white 288
russeliana / P / 3' / fls yellow / Syria 278
samia / P / 3'+ / purple fls / Greece 288
tuberosa / P / 6' / purple or pink fls / c Europe to c Asia 288 / 25
viscosa - see P. russeliana 278

PHLOX - Polemoniaceae
adsurgens / R / 6-8" / pale shell-pink fls / w side of Cascades 404 / 129
alyssifolia / R / to 4" / fls purple to pink / e Rockies, High Pl. 305
amoena / R / 6-10" / pink to purple fls / se United States 129
andicola ssp. parvula / 3" / fls pale lavender to white / Gt. Plains 305
bifida / R / 6-8" / white to pale-violet-purple fls / c US 224 / 129
bifida 'Alba' - fls white 25
buckleyi / P / 6-20" / purple or pink fls / Virginia, West Va. 263
caespitosa / R / to 5" / white to lilac fls / Ore, Mont, s to N.M. 25
carolina / P / 16-40" / fls pink to purple / coastal plain, e US 25
'Chatahoochee' - see P. divaricata ssp. laphamii 'Chaahoochee' 305
condensata - see P. caespitosa 25
diffusa / R / clump / pink to white fls / Sierras, w US 312 / 107
diffusa depressa - perhaps referable to P. multiflora ssp. depressa *
divaricata / R / 9-15" / fls lavender-ble / e North America 224 / 129
divaricata 'Alba' - fls white 25
divaricata ssp. laphamii / with petal blades entire / c US 305
divaricata ssp. laphamii 'Chatahoochee' / deep violet fls / n Florida 305
douglasii diffusa - see P. diffusa 305
drummondii / A / 4-20" / reddish-purple fls / Texas 305
glaberrima ssp. triflora / 2'+ / light purple to white fls / se US 305
hirsuta / HH R / to 8" / fls purple to pink / California 229 / 305
hoodii / R / to 2½" / fls deep lavender to white / Rockies, Plains 305
longifolia / Sh / to 1'+ / fls lilac, pink, or white / nw U. S. 25
multiflora / R / to 6"/decumbent /fls lilac, pink, white / Rockies 229 / 305
nana / R / to 1' / pink, purple, white fls / N. Mex., Texas 25 / 226

nana v. eunana - not a separate of the sp 305
nivalis / R / mats / varii-colored fls / s US, coastal 225 / 107
1 nivalis 'Scarlet Flame' - see P. subulata 'Scarlet Flame'
ovata / P / to 20" / purple to pink fls / Pa., Ind. & s-ward 225 / 25
ozarkiana - see P. pilosa v. ozarkana 259
paniculata / P / to 6' / fls purple, pink, white / e & c US 208 / 305
pilosa / P / 8-18" / purple, pink, red, white fls / c N. Am. 224 / 259
pilosa v. fulgida - fls with hoary pubescence 259
pilosa v. ozarkana - with gland-tipped pubescence 259
procumbens / P / to 1' / fls bright purple / ne North America 25
2 sibirica ssp. pulvinata / R/ loose cushion/ fls many colors/ Rockies
speciosa / Sh / 8-20" / purple to white fls / Rocky Mts. 229 / 305
stolonifera / R / 6-8" / pink shades to white fls / Pa., & s-ward 224 / 107
stolonifera 'Alba' - white form of P. stolonifera
3 stolonifera 'Ariane' - white fls
stolonifera 'Blue Ridge' - fls blue, lvs shiny 25
stolonifera 'Pink Ridge' - fls pink 107
stolonifera X subulata - see P. procumbens
subulata / R / 6", mats / variously colored fls / e US 224 / 25
subulata ssp. Brittonii /R/ mat to 6"/ fls lav-wht/Appalach & Potom. 25
subulata 'Brittonii Rosea' - see P. subulata ssp. Brittonii 25
subulata 'Rosea' / R / mat to 6" / fls pink to red / Appalachians 25
3 subulata 'Scarlet Flame' - near red form of creeping phlox
triovulata / R / 4-16" / fls lilac, pink, white / New Mexico & s 305

PHOENICAULIS - Cruciferae
cheiranthoides / R / 2-8" / fls pink to reddish-purple / Wash., Ida. 229 / 142

PHOENIX - Palmaceae
canariensis / Gh T / 50'+ / graceful palm / Canary Isls. 28

PHORMIUM - Liliaceae
colensoi - see P. cookianum 184 / 36
colensoi 'Tricolor' - see P. cookianum 'Tricolor' 36
colensoi 'Variegatia' - see P. cookianum 'Variegatia' 36
cookianum / HH P / 2-5' / yellow or yellowish-red fls / N. Zealand 36
cookianum 'Tricolor' - lvs white-striped & red-edged 36
cookianum 'Variegatia'- lvs striped green & white 36
tenax / HH P / to 10' / usually dull red fls / New Zealand 129 / 36
tenax 'Purpureum' / 6' / lvs purplish, true-seeding 36
tenax rubrum / P / to 15' / fls dull red; lvs red / New Zealand 25
tenax 'Tricolor' - narrow cream edge 295
tenax 'Variegatum' - tall, lvs striped creamy-yellow 295 / 36

PHOTINIA - Rosaceae
villosa / Sh / to 15' / fls white, frs red / Far East 355 / 25

---

[1] H. Lincoln Foster, Bull. American Rock Garden Society, 25:2

[2] William A. Weber, Rocky Mountain Flora, 1976.

[3] Richard Goher, Quar. Bull. of the Alpine Garden Soc., 39:4.

PHUOPSIS - Rubiaceae
stylosa / R / 6-8" / tiny pink fls / Caucasus 241 / 123

PHYCELLA
bicolor - see HIPPEASTRUM bicolor

PHYGELIUS - Scrophulariaceae
aequalis / HH Sh / to 3' / salmon-pink fls / South Africa 129 / 139
capensis / HH P / to 4' / fls bright red / South Africa 194 / 279
capensis 'Coccineus' - fls crimson-scarlet 129 / 139

PHYLICA - Rhamnaceae
plumosa / Sh / 1-2' / fls covered with stiff buff hairs / S. Africa 61

PHYLLITIS - Polypodiaceae
americana - see P. Scolopendrium v. americana 99
cristata - see P. Scolopendrium 'Cristatum' 159
hybrida / F / lvs ½-1' / sori elliptical / nw Jugoslavia 286
Scolopendrium / F / to 2' / simple-leaved fern / s, w & c Europe 147 / 286
Scolopendrium v. americana - rare North American variety 99
Scolopendrium 'Cristatum" - crimped-leaf form of the sp. 159
Scolopendrium 'Laceratum-Kayes var.' - miniature crests 159
Scolopendrium 'Undulatum' / F / to 1½'/lvs crested, crisper / Eur. 25

PHYLLOCLADUS - Podocarpaceae
asplanifolius / HH T / 60' / Celery-topped Pine / Tasmania 109 / 127

PHYLLOCLADUS - Taxaceae
alpinus / Sh or T / 5-30' / evergreen / New Zealand, in alps 127

PHYLLODOCE - Ericaceae
aleutica / Sh / to 1' / fls light yellowish-green / Alaska & w-ward 273 / 200
X alpine / P. nipponica x aleutica / pink fls 273 / 36
breweri / Sh / 6-12" / deep pink fls / California 312 / 123
caerulea / Sh / to 10" / purplish fls / n hemisphere 148 / 200
caerulea f. yezoensis / Sh / to 6" / fls purple / circumpolar & s 25
drummondii - see P. X intermedia 'Drummond'
empetriformis / Sh / 6-8" / rosy-purple fls / w N. Am. 312 / 107
empetriformis 'Alba' - white fls *
glanduliflora / Sh / 6-8" / pale greenish-yellow fls / w N. Am. 312 / 123
X intermedia / rose-pink fls/ P.empetriformis X glanduliflora / B.C. 132
X intermedia 'Drummond' - purplish-magenta fls 220
jesoensis - see caerulea yezoensis
nipponica / Sh / to 8" / rose to white fls / Japan 273 / 200
nipponica v. amabilis - included in the sp. 200
nipponica v. oblongo-ovata / larger plant / Hokkaido, Honshu 200
tsugifolia / Sh / 6" / fls white / n Japan 25

PHYLLYREA - Oleaceae
latifolia / HH T / 30'+ / white fls / Mediterranean region 288
media - see P. latifolia 288

PHYMATODES - Polypodaceae
scandens / HH F / to 1' / terrestrial or epiphytic / New Zealand 15

PHYSALIS - Solanaceae
alkekengi / P / to 2' / inflated red-orange calyces / Europe 288
alkekengi 'Pygmaea' / 8" / recommended pot-plant 283
franchetti - see P. alkekengi 288
'French Pygmy' - see P. alkekengi 'Pygmaea' *
heterophylla / P / to 3' / yellowish fls; edible frs / e N. Am. 224 / 25
pruinosa / A / 20-32" / yellow fls / e & c United States 263
subglabrata / P / to 4' / fls yellowish, purple throat / c N. Am. 224 / 25

PHYSARIA - Cruciferae
alpestris / R / ? / yellow fls / Washington 229 / 142
australis / P / to 6" / stems numerous; fls yel / Utah, Cal., Ore. 227
didymocarpa / R / 2-3" / pinkish inflated pods / Rocky Mts. 229 / 123
didymocarpa australis / R / to 6" / fls yellow / Utah to Colorado 25
geyeri / R / 4" / yellow fls / w United States 107
vitulifera / P / 4-8" / rectangular pods; fls yel / Cal., Wyo. 229

PHYSOCARPUS - Rosaceae
capitatus / Sh / 3-8' / white fls / Oregon, Calif., B. C. 25
malvaceus / Sh / 6' / white fls / w North America 139
opulifolius / Sh / to 10' / bark thinly exfoliating / Que. to Fla 25
opulifolius aureus - see P. opulifolius 'Luteus' 139
opulifolius 'Luteus' - yellowish lvs 25

PHYSOCHLAINIA - Solanaceae
orientalis / P / 18" / lilac fls / Iberia 279

PHYSOPLEXIS - Campanulaceae
comosa / R / 2-6" / fls pink violet / s Alps 289

PHYSOPTYCHIS - Cruciferae
gnaphalodes / R / to 6" / yellow fls / Caucasus, Iran 6

PHYSOSTEGIA - Labiatae
virginiana / P / 4'+ / rose-purple fls / ne to sc N. Am. 208 / 25
virginiana 'Alba' / 3' / white fls 279
virginiana 'Summer Snow' / 2½-3' / white fls 268

PHYTEUMA - Campanulaceae
austriacum - see P. orbiculare v. austriacum 25
balbisii - see P. cordatum 289
balbisii f. alba / 6" / white fls / Piedmont, Valley Piseo *
betonicifolium / P / to 18" / blue or violet fls / s Europe 210 / 25
canescens - see ASYNEUMA canescens 289
charmelioides - see P. scheuchzeri columnae 289
Charmelii / P / 6-12"+ / dark blue fls / Alps 148 / 25
confusum / R / to 6" / fls dk bl-viol, rarely wht / Eur. to Bulgaria 289
comosum / R / 3-4" / pale lilac-blue fls / Austrian & Italian Alps 148 / 129
cordatum / R / 6-10" / bluish-white fls / Maritime Alps 289
globulariifolium / P / 1/4-4½" / fls dp blue-viol/ Alps & Pyrenees 289
globulariifolium ssp. pedamontanum / 2-5" / lvs acute, 3-dent 289
halleri - see P. ovatum 243 / 289
hemisphaericum / R / 3" / clear blue fls / granitic Alps 148 / 107
humile / R / 2" / violet-blue fls / Switzerland 148 / 123
japonicum / P / 1½-3' / lvs membranous; fls purple / e Asia 200

limonifolium - see ASYNEUMA limonifolia 25
Michelii / P / 1-2' / violet to pale blue fls / s Europe 25
Michelii scorzonerifolium / smaller form / 10-20" 289
nanum - see P. confusum 289
nigrum / R / 1'+ / dense, almost black spike / Bohemia 314 / 123
orbiculare / R / 1' / light blue fls / European Alps 148 / 107
orbiculare v. austriacum -lvs ovate-lanceolate 25
ovatum / P / to 2' / fls dark violet / Alps 25
pauciflorum / R / 3" / fls violet-blue / w Alps, Carpathians 28
pauciflorum ssp. pedemontanum / 2" / dark blue fls / Tyrols 289
pedamontanum - see globulariifolium ssp. pedamontanum 289
persicifolium - see P. Zahlbruckneri 289
scheuchzeri / R / 1' / light blue fls / s Europe 148 / 74
scheuchzeri columnae / fls dp blue; basal lvs cordate / s Europe 289
serratum / R / 1-8" / bracts not longer than petals / mts. Corsica 289
sibirica / R / 2" / globularia-like lvs / Siberia 81
Sieberi / R / 4" / fls dark blue / s Alps & Apennines 148 / 74
Sieberi 'Alba' - white fls *
spicatum / R / 1'+ / cream, white, blue fls / Europe 148 / 123
tenerum / P / 4-20"/ blue-viol fls; lvs finely serrate/ wc & sc Eur. 289
Vagneri / P / to 1'+ / fls dark blue / Hungary 25
villarsii / P / ? / slender stems; fls blue / s Europe 289
Zahlbruckneri / 10-40" / fls deep blue / e Alps to Jugoslavia 289

PHYTOLACCA - Phytolaccaceae
americana / P / to 12' / black-purple frs / Maine to Florida 147 / 25
clavigera / P / 4' / pink fls / China 138 / 279
dioica / HH T / small / frs purplish-black / South America 286

PICEA - Pinaceae
Abies / T / to 150' / needles drk grn; cones 7" / n Hemisphere 25
glauca / T / to 80' / dense, conical habit / N. Am. 338 / 139
Omorika / T / to 100' / needles wht over drk grn;cones 2½"/ s Eur. 25
pungens glauca pendula / T / to 100' / lvs blue; cones 4"/ Wyo. & s 25
sitchensis / T / 150'+ / light brown cones / w N. Am. 126

PIERIS - Ericaceae
floribunda / Sh / 2-6' / fls white / Va. to Ga., in mts. 75 / 107
floribunda compacta / Sh / 3-6' / fls white / se United States 36
formosa / HH Sh / 10'+ / evergreen / e Himalayas 139
formosa v. forrestii / to 8' / fls creamy-white / w China 269 / 36
formosa v. forrestii 'Wakehurst' - shorter, broader lvs 139
X 'Forest Flame' - cross of P. f. v. f. 'Wakehurst' with P. japonica 139
forrestii - see P. formosa v. forrestii 129
japonica / Sh / to 15' / fls creamy-white / Japan 269 / 129
japonica 'Bensai' - very dwarf form 25
japonica f. variegata - lvs creamy-white & pink 222
japonica 'Yakusimanum' / 1½' / smaller lvs 242
nana / Sh / prostrate / white fls / ne Asia 25
phillyreifolia / HH Sh / ? / fls white / Fla., Ga., Ala 25
taiwanensis / HH Sh / to 6' / white fls / Taiwan 129 / 222

PILEANTHUS - Myrtaceae
filifolius / HH Sh / 3-5' / fls pinkish or red / w Australia 38

PILOSELLA
aurantiaca - see HIERACEUM aurantiacum 289

PIMELEA - Thymeleaceae
buxifolia / HH Sh / 3' / white to pink fls / New Zealand, in mts 15
coarctica - see P. prostrata 'Coarctica' 132
ferruginea / HH Sh / 1-3' / rose fls / w Australia 28
ligustrina / HH Sh / 5-6' / white fls / Australia, Tasmania 28
prostrata / Sh / wide mat / white fls; white frs / New Zealand 194 / 25
prostrata 'Coarctica' / prostrate / white fls; white frs 132
sylvestris / HH Sh / 2-3' / pale rose fls / w Australia 46
traversii / HH Sh / 2' / white to pinkish fls / N. Zealand, mts 15

PIMPINELLA - Umbelliferae
major / P / 3'+ / white to deep pink fls / most of Europe 287
saxifraga / P / to 2' / white fls, rarely pinkish / Europe 287

PINELLIA - Araceae
ternata / Tu / 8-16" / green fls, bulbil-bearing / Far East 166 / 200
tripartita / Tu / 8-20" / green & purplish spathes / Japan 166 / 200
tubifera - see P. ternata 200

PINGUICULA - Lentibulariaceae
alpina / R / 3" / white fls / c & n Europe 123
antarctica / R / ? / small pale fls / South America 64
corsica / HH R / to 10" / fls pale blue to pink / Corsica 288
grandiflora / R / 4" / violet-blue fls / w Europe 148 / 123
leptoceras / P / small / fls viol, blue or pink / Alps 288
longifolia / R / rosette / lilac to pale blue fls / s Europe 288
vallisneriifolia / HH R / rosette / large lilac fls / Granada 212 / 96
vulgaris / R / 6" / purple fls / n Hemisphere 148 / 107

PINUS - Pinaceae
albicaulis / T / 30' / green or gray-green needles / w N. America 139
aristata / T / 15-40' / slow-growing conifer / Rocky Mts. & w-ward 127
Ayacahuite / T /to 100'/ ndls in 5's, blu-grn; cones to 15"/C. Amer 25
balfouriana / T / 50'+ / timber-line conifer / California 126
banksiana / T / 25-60' / serotinous cones / n Canada 342 / 127
bungeana /T / to 80' / white, scaling bark / c China 217 / 127
cembra / T / to 75' / slow-growing / Europe, n Asia 314 / 25
cembra v. sibirica - taller tree, larger cones 222
contorta / T / 40'+ / twisted, yellowish-green needles / w N. Am. 73 / 139
elliotii / HH T / 30' / cones armed with prickles / se United States 139
flexilis / T/ to 60'/ needles in 5's, stiff yel-grn /Alta.,s to Tex. 25
griffithii - see P. wallichiana 129
jeffreyi / T / 100'+ / cinnamon-red bark / Sierra Nevadas, Cal. 126
halepensis v. brutia / HH T / 35'+ / orange-brown shoots / Medit. reg 210 / 139
lambertiana / T / 200'+ / cones to 18" long / Ore., Cal. 126
leucodermis /T/to 90'/ brk flaky;ndls in 2's; cones glossy/Balk.Pen. 25
mugo / T / to 30' / broad, bushy habit / Europe, in mts 148 / 127
mugo v. pumilio / shrubby to 6' / c & se Europe 184 / 127

muricata / HH T / 45'+ / long needles, persistent cones / Calif. 139
oocarpa / HH T / 30' / sea-green needles / Central America 139
parviflora / T / 20-50' / white bands on needles / Japan 200
parviflora 'Glauca' - needles more glaucous 127
parviflora nana / T / to 50'+/ needles in 5's, blu-grn, short/ Japan 25
patula / HH T / 30' / bright green needles / Mexico 139
peuce / T / to 60'+/ lvs in 5's, blue-green / Balkans 25
pinaster / HH T / 45' / dull gray needles / w Medit. region 139
pinea / HH T / to 35' / distinct habit / Medit. region 149 / 139
pumila / Sh / to 6' / of interest for the large rock garden / Asia 184 / 200
strobus / T / to 100' / pendant 8" cones / e N. Am. 127 / 139
sylvestris / T / to 100' / reddish-brown bark / Europe 129
thunbergii / T / 80-100' / short, stiff needles / Japan 167 / 36
wallichiana / T / to 150' / long drooping needles / Himalayas 129

PIPTANTHUS - Leguminosae
concolor / HH Sh / 6-8' / yellow fls stained maroon / w China 36
laburnifolius - see P. nepalensis 184 / 25
nepalensis / HH Sh / to 10' / yellow fls / Nepal, Sikkim, Bhutan 138 / 25

PISTACIA - Anacardaceae
chinensis / Sh / 10' / small frs, blue / wc China 139
lentiscis / HH Sh or T / 3-24' / fls yellow, purple / Medit. region 287

PITTOSPORUM - Pittosporaceae
crassifolium / Sh or T / to 35' / bark smooth; fls red-purp / N.Z. 25
heterophyllum / HH Sh / 12' / fragrant pale yellow fls / w China 25
mayi - see P. tenuifolium 36
michiei / Sh / prostrate / fls frag, red outside yel inside / N.Z. 15
phillyraeoides / HH T / to 30' / deep yellow fls / Australia 25
tenuifolium / HH Sh / 12' / chocolate-purple fls / New Zealand 36 / 139
tobira / HH Sh / 10' / creamy fls; glossy lvs / China, Japan 36 / 128
undulatum / HH Sh / 12' / creamy-white fls, fragrant / Australia 194 / 139

PITYOPSIS - Compositae
falcata / P / 4-16" / fls yellow / Cape Cod to s New Jersey 25
graminifolia / P/1-3½'/lvs silvery, to 20"/ N.C. to n Fla to La 25
pinifolia / R / 8-16" / yellow fls / Georgia, N. Carolina 25

PITYROGRAMMA - Polypodiaceae
calomelanus / HH F / 2' / greenhouse fern / S. Am. 93
triangularis / F / 6-12" / gold-backed frond / Oregon, B. C. 73 / 159

PLACEA - Amaryllidaceae
arzae / HH Bb / ? / yellow-cream fls striped purple / Chile 270

PLAGIORREGMA
diphylla - see JEFFERSONIA diphylla *
dubia - see JEFFERSONIA dubia *
mandschuriensis - see JEFFERSONIA dubia *

PLANTAGO - Plantaginaceae
albicans / HH R / ? / silvery spikes & lvs / Spain, s Portugal 212
alpina / R / 4" / prominent yellow stamens; linear lvs / Alps 219
arorescens / Gh Sh / 1' / linear lvs / Canary Isls. 25

argentea - single or few rosetted sp., similar to weedy P. lanceolata 289
asiatica / P / lvs 2-6"; fls 4-20" green-white / Japan 200
asiatica yakusimensis - plant smaller; lvs hairy 200
atrata / R / to 8" / brownish margins to sepals / c & s Eu., in mts 314 / 289
camtschatica / R / to 1' / white-pubescent lvs / Japan 164 / 200
cynops - see P. sempervirens 289
indica / A / 4-14" / glandular-hairy flower heads / Medit. region 211
juncoides - see P. maritima juncoides 289
lanceolata - too common lawn weed 25
lanceolata ssp. lanceolata / P / to 9"+/fls greenish/ Eur. lawn weed 25
lanigera / HH R / short / copious long hairs / New Zealand 15
major 'Atropurpurea' / A / to 1' / bronzy-purple lvs 25
major 'Purpurea' - purple lvs *
major 'Rosularis' - cluster of lvs topping stem 25
major 'Rosularis Rubrifolia' - purplish lvs, stalked rosette 123
major 'Rubrifolia' / 4-5" / purple-stained lvs 123
maritima / R / rosette / whitish fls / Europe 210 / 301
maritima ssp. juncoides / arctic Eur. form / lvs slightly fleshy 289
masonae - see P. triandra 15
maxima / P / to 2' / white feathery spike / Siberia to Hungary 289
media / P / 1 or more rosettes; anther lilac or white / Europe 289
nivalis / HH R / rosette / globular green heads / Spain 212 / 123
ovata / R / to 5" / dense spikes / se Spain 289
purpurea - see P. major 'Purpurea' *
raoulii / HH R / to 6" / pilose lvs / New Zealand 15
reniformis / R / rosette / purplish sepals / Albania, Jugoslavia 289
sempervirens / Sh / to 16" / tiny greenish-yellow fls / sw Europe 289
spathulata / P / ½-4" / spike many flowered / New Zealand 15
tenuiflora / A / rosette / fls inconspicuous / sc & se Europe 289
thalackeri - see P. nivalis 289
triandra / HH R / to 10" / inflores. hidden by stem hairs / N. Z. 15

PLATANTHERA - Orchidaceae
bifolia / P / to 22" / yellowish or greenish fls, vanilla scent / Eu. 148 / 85
chlorantha / P / 1-2' / white fls, often green tinged / s Eu., n Afr. 85
chloroleuca - see P. chlorantha 85
metabifolia / P / 8-20" / white fls / Hokkaido 210 / 200

PLATANUS - Platanaceae
orientalis / T / to 90' / exfoliating bark / w Asia, se Europe 222

PLATYCODON - Campanulaceae
'Cronamere Rose' - see P. grandiflorum 'Cronamere Rose' *
grandiflorum / P / to 3'+ / fls lilac or white / Far East 164 / 200
207 / 25
grandiflorum 'Album' - fls white 25
grandiflorum 'Alpinum' - dwarf selection 25
grandiflorum 'Apoyama' / P / to 2½'/fls blu,lilac,wht,dwarf/ e Asia 25
grandiflorum 'Apoyama' albus - fls white 25
grandiflorum 'Apoyama Leucantha' - paler fls / ? *
grandiflorum 'Autumnalis' - late-flowering 25
grandiflorum 'Azureum' - pale blue fls *
grandiflorum chinensis - the Chinese plant in the sp. 231
grandiflorum 'Cronamere Rose' - selection for pink fls *
grandiflorum 'Maries' / 1' / reddish-purple fls 208 / 147
grandiflorum Plenus / 6-12" / fls blue, wht; semidble / China, Japan 61

grandiflorum 'Pumilum' - dwarf 25
grandiflorum 'Roseum' - pinkish fls 25

PLATYLOBIUM - Leguminosae
obtusangulum / HH Sh / 3-4' / orange-yellow fls / Austr., Tasmania 109 / 61

PLATYSTEMON - Papaveraceae
californicus / A / 1' / cream or yellow fls / w N. Am. 229 / 25

PLECTRANTHUS - Labiatae
umbrosus / P / 2'+ / blue-purple fls / Honshu, in mts 200

PLECTRITIS - Valerianaceae
congesta / P / 4-24" / fls pink or white / s Calif. to B. C. 228

PLEIONE - Orchidaceae
bulbocodioides / Gh / 2'+ / varii-colored fls / Far East 418 / 268
bulbocodioides 'Alba' - the rare white form 268
bulbocodioides 'Blush of Dawn' - pale pink fls 209
bulbocodioides 'Limprichtii' - rich red fls 128
bulbocodioides 'Oriental Splendour' - pink & pale yellow fls 184
bulbocodioides 'Pogonioides' - light purple fls 184
bulbocodioides 'Polar Star' - better of two whites in cult. (244)
bulbocodioides pricei - see P. bulbocodioides 'Oriental Splendour' 184
Delavayi / A / large rose-purple fls / China 61
formosana - see P. bulbocodiodes 129 / 116
formosana 'Alba' - see P. bulbocodiodes 'Alba' 129
formosana pricei - synonym of P. bulbocodioides 268
forrestii / Gh / 3" / yellow to orange fls / Yunnan, Burma 128 / 268
humilis / Gh / 2-3" / white to pale purple fls / India 116 / 268
limprichtii - see P. bulbocodioides 'Limprichtii' 128
pagonioides / A / fls 3", light purple / China 61
'Polar Sun' - a good white for cultivation 184
Pricei - P. bulbocodioides 268
yunnanensis / Gh / 2" / bright magenta-rose fls / China 209 / 268

PLEOPELTIS - Polypodiaceae
ussuriensis / F / to 8" / rhizomes long-creeping / Far East 200

PLEUROSPERMUM - Umbelliferae
camtschaticum / P / 8-16" / large white flowers / Far East 200

PLUCHEA - Compositae
purpurascens / A / to 4½' / pink or purple fls / s & c US 229 / 99

PLUMBAGO - Plumbaginaceae
auriculata / Gh Cl / to 8' / fls blue or white / South Africa 194 / 25
capensis - see P. auriculata 128 / 25

POA - Gramineae
alpina / R Gr / 2-12" / purplish or greenish panicles / n hemis. 148 / 268
alpina v. vivipara - upper spikelet producing new plants 219 / 268
badensis / Gr / 8-12" / spike green or purple / c Europe 290
bulbosa / P Gr / to 2'/ fls dk purp base /n Afr, Eurasia, nat. U.S. 25
chaixii / P Gr / 2-4' / broad-leaved woodland plant / Eurasia 268

colensoi / Gr / to 10" / blue lvs / New Zealand 124
vivipara - see P. alpina v. vivipara 268

PODALYRIA - Leguminosae
argentea / Gh Sh / 1-2' / white fls / South Africa 61
biflora - see P. argentea 61

PODOCARPUS - Podocarpaceae
alpinus / HH Sh / low mound / for protected rock garden / Tasmania 36 / 139
andina / HH T / 20' / evergreen / Chile 28
nivalis / HH Sh / 9' / fleshy red frs / New Zealand 127 / 238

PODOLEPIS - Compositae
acuminata / HH P / 1-2' / yellow daisies / Australia 64
canescens / A / to 1' / lvs linear; fls yellow / Australia 61
jaceoides / A / small / yellow fls / Australia, Tasmania 45
robusta / P / to 2' / stems white, woody; fls yellow / mts Australia 25

PODOPHYLLUM - Berberidaceae
emodi - see P. hexanrdrum 129 / 25
emodi chinensis - see P hexandrum v. chinense 279
edomi 'Majus' - see P. hexandrum 'Majus' 279
hexandrum / P / 1½' / fls white or pinkish / Himalayas 194 / 25
hexandrum v. chinense - fls rose-pink 279
hexandrum 'Majus' - fls larger 279
peltatum / P / 1-1½' / white waxy fls / North America 224 / 25

POGONIA - Orchidaceae
ophioglossoides / P / to 2' / fls rose to white / e N. America 224 / 66

POINIANA
gilliesii - see CAESALOINIA gilliesii 28

POLANISIA - Capparaceae
graveolens / A / 18" / whitish fls; plant foetid / n N. America 99

POLEMONIUM - Polemoniaceae
acutiflorum - see P. caeruleum ssp. villosum (349) / 25
acutiflorum v. laxiflorum / like P. nipponicum / lvs subobtuse 200
acutiflorum laxiflorum album / 16-32" / fls white, deep lobed 200
acutiflorum nipponicum - see P. nipponicum 164
album - white form of P. boreale, carneum or retans 25
Archibaldiae - see P. foliosissimum 95
boreale / R / 8" / pale purple fls / n circumpolar 209
brandegei / R / 6-12" / golden yellow fls / Colorado (548) / 123
caeruleum / P / to 3' / blue fls / Europe 148 / 107
caeruleum 'Albiflorum' - see P. caeruleum v. lacteum
caeruleum v. album - see P. caeruleum v. lacteum 25
caeruleum ssp. amygdalinum - erect inflorescence / B. C., Alaska & s 25
caeruleum 'Blue Pearl' - see P. reptans 'Blue Pearl' 268
caeruleum 'Cashmerianum' - larger pale blue fls (246)
caeruleum grandiflorum album / P / to 3'/fls wht/B.C.-Alta,Siber-Fin 25
caeruleum v. himalayanum - fls lilac blue 25
caeruleum v. lacteum - fls white 25
caeruleum ssp. Van-Bruntiae / showy blue-purple fls / Vt. to W. Va. 313 / 25

caeruleum ssp. villosum / stm lvs smaller / B. C., Alaska, Siberia 25
caeruleum yezoense - see P. yezoense 200
californicum / R / to 1' / fls blue, white, yellow / nw U.S. 25
carneum / R / 6-8" / flesh-pink fls / w N. America 228 / 123
carneum album / P / to 2½' / fls white / Wash to Cal., e to Mont. 25
carneum f. luteum - soft yellow fls 64
cashmirianum - see P. caeruleum 'Cashmerianum' *
caucasicum - see P. caeruleum 288
confertum - see P. viscosum 25
confertum mellitum - see P. mellitum 25
delicatum / R / dwarf / light blue fls / Rocky Mts. 312 / 107
elegans / R / to 6" / blue fls / Washington, B. C. 64 / 228
eximium / HH P / 4-12" / blue & whitish fls / Inyo Co., Calif. 228 / 227
flavum / HH P / 2-3' / fls yellow tinged red / N. Mexico, Arizona 227
foliosissimum / P / ½-2½' / fls blue, viol, cream / Rocky Mts. 95 / 129
foliosissimum var. albiflorum - commonly with white fls 28
grandiflorum / R / 12" / lilac or yellow fls, large / Orizaba, Mex. 96
Haydenii / R / dwarf / blue fls / w United State 107
humile - see P. boreale X P. reptans
lanatum boreale - see P. boreale *
lantanum humile - see P. boreale 25
mellitum / R / dwarf / creamy-white fls / Rocky Mts. 107
mexicanum / P / 1' / fls pale blue / Mexico 81
nipponicum / P / 20-40" / blue fls / alps of Honshu 164
occidentale - see P. caerueum ssp. amygdalinum 228 / 25
pauciflorum / P / 1-2' / yellowish & red fls / Mex., Texas, Ariz. 227 / 25
pulchellum / R / 3-10" / yellow fls / Siberia 209
pulcherrimum / R / to 1' / blue fls / Alaska, Calif., Wyo 229 / 25
pulcherrimum v. calycinum / 12-20" / laxer, more robust / Ida., Mont. 142
pulcherrimum v. pulcherrimum /P / to 1'/ fls blu, yel tube/Alas-Cal. 25
pulcherrimum ssp. tricolor / 12-20"/ laxer, more robust/ Ida., Mont. 276
reptans / P / 6-30" / deep blue fls / e & c U.S. 147 99
reptans album / P/ to 2'/ corolla wht, yel. tube / NH to Ga & w-ward 25
reptans 'Blue Pearl' / 10" / bright blue fls 268
Richardsonii v. pulchellum - see P. pulchellum *
viscosum / R / 4-5" / clear blue fls / Rocky Mts. 312 / 123
yezoense / P / 1-1½' / blue fls / Hokkaido 200

POLLIANTHES - Amaryllidaceae
geminiflora / Tu / 1-2' / red fls, orange-tinged / Mexico 93

POLLIA - Commeliaceae
japonica / P / 12-32" / white fls / Japan 166 / 200

POLYGALA - Polygalaceae
alba / R / to 14" / white fls / w U.S. 229 / 99
amara / R / 2-8" / fls blue, violet, pink, white / ec Eur, in mts. 314 / 287
amarella / R / to 8" / fls variable / much of Eur., except s 287
Balansae / Sh / 5-6' / fls large, deep purple / Morocco 61
calcarea / R / 3-5" / pink to blue fls / England, Europe 212 / 129
Chamaebuxus / Sh / to 1' / fls white to yellow / Europe 133 / 25
Chamaebuxus v. grandiflora / wings purple, petals yellow / Europe 25
Chamaebuxus v purpurea - see P. Chamaebuxus v. grandiflora 184 / 25
Galpinii / Gh Sh / 3-5' / fls pale rosy-lilac / Swaziland 61
paucifolia / R / to 7" / rosy-purple fls / e North America 224 / 25

polygama / B or P / cleistogamous fls / N.J. to Fla. 308 / 99
sanguinea / A / to 16" / fls pink to rose-purple / N. Am. 229 / 99
senega / R / to 1' / fls white or greenish / New Brunswick & w & s 28
vulgaris / Sh / 1'+ / blue, pink or white fls / Europe 287

POLYGONATUM - Lilaceae
biflorum / P / 2-3' / greenish-white fls / e North America 224 / 107
canaliculatum - see P. commutatum 25
commutatum / P / to 6' / yellow-green to green-white fls / N. Am. 192 / 25
falcatum / P / 20-32" / fls greenish-white / Japan, Korea 200
glaberrimum / P / 16-24" / white fls; black frs / Caucasus 4
hookeri / P / stalk short / fls purp or lilac / China, Tibet, Sikkim 61
humile / R / 6-12" / yellowish-green fls / Far East 166 / 200
latifolium / P / 16-20" / frs bluish-black / c Europe, Crimea 4
multiflorum / P / 2-3' / small white fls / Europe 147 / 129
odoratum / P / 1-2' / white fls, toothed green / Eurasia 148 / 93
officinale - see P. odoratum 93
orientale / P / ½-3' / fls scentless, greenish / ec & se Europe 290
pubescens / P / to 3' / fls yellowish-green / e & c N. Am. 224 / 25
racemosum - see SMILACINA racemosa *
roseum / P / to 2' / rose fls; brownish-red frs / c Asia 4
stewartianum / P / 3' / greenish-white fls; red dotted frs / Yunnan 56
verticillatum / P / 1-2' / white fls; violet-red frs / Europe 148 / 4

POLYGONUM - Polygonaceae
affine / R / 4-6" / rosy-red fls / Himalayas 194 / 264
affine 'Darjeeling Red' / 9" / deep pink fls in autumn 184 / 49
affine 'Donald Lowndes' - selected cultivar 49 / 129
affine 'Lownes Var.' - see P. affine 'Donald Lownes' 129
alpinum / P / 12-32" / fls white or pale pink / s, c & e Europe 286
amplexicaule / P / 3-4' / fls bright crimson / Himalayas 129
bistorta / P / to 2' / white fls / Eurasia 314 / 25
bistorta 'Superbum' / P / to 3' / fls light pink 129
bistortoides / P / 2½' / white fls / B.C., Calif., Rocky Mts. 227 / 25
blumei - see P. longisetum 200
capitatum / A / trailing / pink fls / Himalayas 282 / (14)
'Darjeeling Red' - see P. affine 'Darjeeling Red' 49
dentatoalatum / A / to 3' / fr pale green / Japan, e Russia 200
foliosum nikaii / A / 19-24" / fls reddish spikes / Japan, China 200
hayachinense - see P. macrophyllum 25
longisetum / A / erect or ascend / lvs dk grn; fls pale red / Japan 200
macrophyllum / P / to 9", tuft / small pink fls / Himalayas, China 96
millettii / P / 6-18" / deep pink or crimson fls / Himalayas 268
orientale / A / to 6' / rose-pink fls / Asia 224 / 25
scandens - see P. dentatoalatum 200
sphaerostachyon - see P. macrophyllum 268
vacciniifolium / R / 1' / fls rose / Himalayas 48 / 25
viviparum / R / 1' / pale rose or white fls / Eurasia, N. America 229 / 25
Weyrichii / P / to 3' / fls white or greenish / Sakhalin 25

POLYPODIUM - Polypodiaceae
Brownii / F / leathery lvs 6-18" / Australia, New Cal. & Fiji Isls 25
glaucum - probably glaucous form of P. aureum 25
hesperium / F / to 6" / of moist cliffs / Cascade Mts. 142

interjectum / wet-land segregate of P. vulgare / England (424) / 159
Scouleri / F / to 4" / lvs thick & leathery / B. C. to Calif., coast 142 / 25
ussuriensis - see PLEOPELTIS ussuriensis 200
virginianum / F / 10" / on cliffs & rocks / e North America 25
vulgare / F / 3' / fronds evergreen / Eurasia 148 / 25
vulgare 'Cornubiense Trichomanoides' - most dissected form 159
vulgare 'Cristatum' - crested form 159
vulgare v. columbianum - from rock ledges 25
vulgare v. occidentale - from tree trunks, logs & cliffs 25
vulgare v. ramosum - branching repeatedly from the base 159
vulgare trichomanoides - see P. vulgare 'Cornubiense Trichomanoides' 159

POLYSTICHUM - Polypodiaceae
acrostichoides / F / 2' / evergreen fronds / e North America 191 / 25
aculeatum / F / 3' / leathery frond / British Isles 93 / 159
andersonii / F / 1-3' / of deep woods / w North AMerica 142
aristatum 'Variegatum' - see ARANCHNOIDES cristatusm 'Variegatum' 25
braunii / F / to 2' / 2-pinnate frond / Europe 25
cytostegia / HH F / 2-6" / sori large, domed / New Zealand 159
Kruckebergii / F/ 3"/ spinulose tips to lobes / B.C., Wash, & e-ward 142
Lemmonii / F / lvs to 1' long, 2" wide / Alaska to n California 25
Lonchitis / F / 2' / leathery, evergreen fronds / Eurasia, N. Am. 25
mohrioides / F / 4-6" / blunt segments / Calif., Ore., Wash 142
munitum / F / to 3½' / Western Sword-Fern / Alaska s to Montana 25
munitum v. imbricans / F / to 3½' / smaller fronds than P. munitum 25
polyblepharum / F / to 3' / woods-fern / Japan, Korea 272 / 200
richardii / Gh F / indusia white with black dot / New Zealand 193 / 25
rigens / F / 8-16" / coriaceous fronds, lustrous above / Japan 200
setiferum / F / lvs to 2' long, 6" wide and larger / Europe 25
setiferum 'Acutilobum' - narrow fronds with abundant bulbils 159
setiferum 'Congestum' / 6" / recommended dwarf / true from spores 159
setiferum 'Congestum Grandiceps' - magnificent terminal crest 159
setiferum 'Multifidum' - much dissected segments 25
setiferum 'Proliferum' - see P. setiferum 'Acutilobum' 159
tsus-sinense / Gh F / 2' / narrow triangular fronds / Asia 25

POMADERRIS - Rhamnaceae
kumeraho / HH Sh / 6-10' / yellow fls / New Zealand 238

PONCIRUS - Rubiaceae
trifoliata / Sh or T / to 20' / fragrant white fls / n China 129 / 222

PONTEDERIA - Pontederiaceae
cordata / Aq P / to 4' / fls blue / Nova Scotia to Florida & Texas 25

PORLIERIA - Zygophyllaceae
hygrometrica / Gh Sh / 2' / green & white fls / Peru 206

PORTULACA - Portulacaceae
grandiflora / A / 6-12" / varii-colored fls / Brazil 28
smallii / A or P / ? / fls pink or purple / N. Car. to Ga. 250

POTENTILLA - Rosaceae
agrophylla plena / P / 2-3' / lvs palm; fls double yellow / Nepal 61
alba / R / 6" / fls white / c & e Europe 287

alchemilloides / R / to 1' / white fls / Pyrenees 148 / 287
alpestris - see P. crantzii 287
alpinus - name of doubtful application 25
ambigua / R / tufted / yellow fls, solitary / Himalayas 25
andicola / R / caespitose / fls yellow / Colombia 25
anserina / P / procumbent / yellow fls / Europe 287
apennina / R / to 8" / fls white, cream, pink / Balkans, Apennines 287
arbuscula - see P. davurica 110 / 25
argentea / R / 1' / pale yellow fls / n Europe, Alps 148 / 287
argentea v. calabria - see P. calabra 287
arguta / P / to 3' / fls white & crm / N. Bruns. to B.C. 25
argyrophylla / P / 2-3' / fls yellow / Himalayas 25
argyrophylla astrosanguinea - see P. atrosanguinea 28
argyrophylla 'Leucochroa' - silvery lvs 282
atropurpurea - see P. astrosanguinea 64
atrosanguinea / P / 2-3" / dark purple fls / Himalayas (510) / 25
atrosanguinea 'Atropurpurea' - darkest color form *
atrosanguinea 'Gibson's Scarlet' / 1½' / brilliant scarlet fls 268
aurantiaca - see P. aurea ssp. chrysocraspedia 'Aurantiaca' 46
aurea / R / mats / yellow fls / s & c Europe, in mts. 148 / 287
aurea ssp. chrysocraspedia / leaflets always three / Balkans 46 / 287
aurea ssp. chrysocraspedia 'Aurantiaca' - fls soft orange 46
aurea 'Plena' / to 2" / neat tufts, double golden fls 46
aurea rathbonii (rathboneana) - see P. aurea 'Plena' 46
beesii - see P. fruticosa 'Beesii' 132
blaschkiana - see P. gracilis v. glabrata 142
breweri / R / recumbent / yellow fls / Cal. to B. C. 227
buccoana / P / to 2' / fls yellow / nw Anatolia 77
calabra / R / 8", procumbent / yellow fls / Italy, in mts., Balkans 287
californica - see HORKELIA californica 196
canadensis / R / to 6" / fls deep yellow / ne N. Am. 224 / 99
candicans / R / to 6" / fls yellow / Mexico 206
caulescens / R / to 1' / white fls / Alps, s Europe, in Mts. 148 / 287
centigrana / P / decumbent / fls yellow / Japan, Korea, China 200
chinensis / P / to 3' / fls yellow / China, Japan 165 / 258
chrysocraspedia - see P. aurea ssp. chrysocraspedia 287
cinerea / R / mats / yellow fls / c, e, & s Europe 148 / 287
clusiana / R / 2-4" / white fls / Alps, Albania 148 / 287
concolor / P / 12" / hairy; fls yellow, orange / Yunnan 61
coriandrifolia / R / 4-6" / smooth lvs, many; fls white / Himalayas 61
crantzii / R / 8"+ / fls yellow / n Eur., c & e Eur., in mts. 148 / 287
crantzii 'Pygmaea' / 4-6" / yellow fls / circumpolar (248)
crantzii v. tenata - lvs trifoliolate 77
critina / R / 1' / fls golden / Colorado, Utah 229 / 96
cuneata / R / 5" / yellow fls / Nepal 81
davurica / Sh / 2-4' / white fls / China, e Siberia 25
davurica mandschurica / low / gray lvs; white fls 95
davurica 'Veitch' / 3' / fls cream-white 25
delavayi / P / tall / clear yellow fls / China 64
delphinensis / P / 12-20" / yellow fls / sw Alps 287
detommassii / R / to 1' / fls yellow / Balkans, Italy 287
dickinsii / R / 4-8" / fls yellow tinged / Japan 200
divisa - see P. quinquefolia 25
dombeyi / R / to 6" / yellow fls / Peru 25
Durmmondi / P / to 1½' / fls bright yellow / B.C. to Alta & s 25

	erecta / P / 18" / fls yellow / Eurasia		25
	eriocarpa / R / mat / clear yellow fls / Himalayas	46 /	123
[1]	'Firedance' / 2' / almost evergrn; fls red, yel edge		
	fissa / R / to 1' / cream-white fls / S. Dak., Wyo., to N. Mex.		25
	flabellifolia / R / 3-12" / yellow fls / nw North America		228
	foliosa - see P. rupestris		77
	fragariodes / R / stoloniferous / fls yellow / e Asia		258
	fragiformis / R / to 10" / golden-yellow fls / ne Asia	184 /	25
	frigida / R / 1' / yellow fls / Alps, Carpathians		287
	fruticosa / Sh / to 3' / fls yellow / n hemisphere	148 /	287
	fruticosa arbuscula - see P. davurica		25
	fruticosa 'Beesii' / 10" / golden yellow fls / China		132
	fruticosa 'Longacre' / dense, low / fls golden yellow		268
	fruticosa v. mandschurica - see P. davurica mandschurica		95
	fruticosa parvifolia - see P. parvifolia		25
[2]	fruticosa 'Red Ace' / 18-30" / vermilliam red fls, yel underside		
[2]	fruticosa 'Tangerine' / 30" / tangerine-orange fls, turn yellow		
	fulgens / P / to 2' / lvs silvery beneath; fls yellow / Himalayas		25
	gelida - see P. crantzii v. ternata		77
	'Gibson's Scarlet' - see P. astosanguinea 'Gibson's Scarlet'		268
	glabra 'Veitchii' - see P. davurica 'Veitch'		*
	glandulosa / P / 4-28" / fls deep to pale yellow / Calif. & n-ward	229 /	228
	gordonii - see IVESIA gordonii		64
	gracilis / P / 12-32" / yellow fls / Calif., Alaska, Rocky Mts.	229 /	228
	gracilis v. glabrata - lvs green, sparingly hirsute		142
	gracilis ssp. nuttallii / less contrast in lvs / same area		25
	gracilis v. pulcherrima - lvs grayish beneath		142
	gracilis v. rigida - lvs green, sparingly hirsute		99
	grammopetala / R / 2-12" / gr hairy lvs; fls er-yellow / c Alps		287
	grandiflora / R / 4-16" / fls yellow / Alps, Pyrenees	148 /	287
	Hippiana / P / 1-2' / leaflets many; fls yellow / w North America		61
	hirta / P / 4-28" / fls yellow / w. Medit. region		287
	hirta pedata - see P. pedata		287
	hookerana / P / to 16" / 12+ fls yellow / Urals		287
	hyparctica / R / 4" / yellow fls / cirumpolar	209 /	287
	impolita - see P. neglecta		287
	'Longacre' - see P. fruticosa 'Longacre'		268
	matsumurae / R / 4-8" / yellow fls / Hokkaido, Honshu		200
	megalantha / R / 4-12" / fls golden-yellow / Kuriles, Hokkaido		200
	megalantha 'Take' - stands for the author, Takeda		*
	micrantha / R / 2" / fls white, rarely pink / s & c Europe		287
	miyabei / R / to 4" / yellow fls / Hokkaido	165 /	200
	montana / R / to 8 " / white fls / n Iberian Pen., w & s France	212 /	287
	multifida / R / 4" / yellow fls; lvs white-tomentose / Eurasia	148 /	25
	neglecta / P / to 20" / dark to orange-yellow fls / most of Eur.		287
	nepalensis / P / to 18" / fls rose-red / Himalayas		25
	nepalensis 'Miss. Willmott' / to 1' / magenta-rose fls	(249) /	25
	nepalensis 'Roxana' - orange-scarlet fls		129
	nevadensis / R / 6-12" / yellow fls / s Spain		268
	nevadensis condensata - smaller, densely tufted, silky, silvery		268

[1] Wayside Catalog, Fall, 1985.

[2] Wayside Catalog, Spring, 1986.

nitida / R / mats / rosy fls / s Europe, in mts 148 / 107
nitida 'Alba' - fls white 25
nitida 'Compacta' - compact form 25
nitida 'Lissadell' - best pink form 46
nitida 'Rubra' - deeper pink & freer flowering than typical 129
nivea / R / to 8" / fls yellow; lvs white-tomentose/ n Eur. & Asia 148 / 287
norvegica / A / to 1½' / fls yellow / Europe, Asia, N. America 25
nuttallii / see P. gracilis ssp. nuttallii 25
palustris / P / to 18" / fls deep purple / Europe 287
parvifolia / Sh / 3'+ / lvs finely segmented / China 25
pedata / P / to 2' / yellow fls; pinnatifid lvs / se Europe 287
pulcherrima - see P. gracilis v. pulcherrima 142
purpurea / R / tufted / dark crimson fls / Far East 64
pyrenaica / R / 4-16" / fls yellow / n & c Spain 287
quinquefolia / R/ to 8"/few fls yel;lvs wht, tom. beneath/ B.C.-Utah 25
recta / P / to 2' / fls yellow / c, e & s Europe 224 / 287
recta Macrantha' - see P. recta 'Warren' 129
recta 'Warrenii' - fls large, bright yellow 129 / 25
rigoana / R / to 4" / verna group / Monte Polling, s Italy 287
X 'Roxana' - see P. nepalensis 'Roxana' 279
rubricaulis / R / to 8" / yellow fls / Spitzbergen, arctic N. Am. 287
rupestris / P / to 2' / white fls / w & c Eur., Balkans 148 / 287
rupestris ssp. macrocalyx - dwarf, larger fls 148
rupestris 'Nana' - dainty form 25
rupestris v. pygmaea / 1-4" / smaller fls / Corsica, Sardinia 25
sanguisorba / R / 6-12" / fls cream-colored / Siberia 81
speciosa / R / to 1' / white fls / Crete, w & s Balkans 287
ternata - see P. aurea ssp. chrysocraspedia 64
thurberi / P / to 2½' / fls red-purple / Arizona, New Mexico 227 / 25
togasii / P / leaflets 3; petals 5, yellow / Japan 200
tomentilla formosa - see P. tonguei 95
tommasiniana - see P. cinerea 187
tonguei / trails to 1' / fls buff-apricot-yel; red-brn blotch base 95
Tormentilla-formosa - see P. tonguei 95 / 25
tridentata / Sh / 6-10" / white fls / n North America 224 / 107
uniflora / R / to 6" / fls yellow / Tauvia 206
valderia / P / 16" / white fls, gray-tomentose lvs / Maritime Alps 289
verna / R / mat / yellow fls / Great Britain, w & c Europe 123
verna f. nana - in miniature 123
villosa - see P. crantzii 287
villosa v. parviflora / lvs thick, leathery / nw North America 142
warrensii - see P. recta 'Warrenii' *
X wilmottiae - see P. neplensis 'Miss Willmott' 129
yokusiana / R / stoloniferous / fls yellow / Japan 200

POTERIUM
lasiocarpum - see SANGUISORBA minor ssp. lasiocarpa 77
sanguisorba - see SANGUISORBA minor 287
sitchense - see SANGUISORBA sitchensis 142
spinosum - see SARCOPOTERIUM spinosum 77

PRASIUM - Labiatae
majus / HH Sh / to 3' / fls white lilac / Medit. region 140

PRATIA - Campanulaceae
angulata / HH R / carpeter / white fls; purplish lvs / N. Zealand 182 / 107
angulata 'Treadwell' - larger fls, brighter frs (178) / 123
macrodon / HH R / creeping / yellowish to white fls / N. Zealand 15
physaloides / HH P / 3' / pale blue fls; purplish frs / N. Z. 184 / 238
treadwellii - see P. angulata 'Treadwellii' 123

PRENANTHES - Compositae
alba / P / 2-5' / fls dull white / Can. to Ga, & Ill. 28
altissima / P / 3-7' / greenish-yellow lvs / ne North America 224 / 28
Bootii / R / 4-6" / fls straw colored / New Hampshire & New York 99
purpurea / P / to 3' / fls violet & mauve / w Europe, Caucasus 210 / 77

PRIMULA - Primulaceae
abchasica / R / 2½" / fls purple / Caucasus 10
acaulis - see P. vulgaris 123
acaulis caerulea - presumably P. vulgaris cultivar 283
algida / R / to 8" / fls violet / Caucasus, Asia Minor 119 / 107
allionii 'Apple Blossom' / Gh / rosette, light mauve-pink fls 132
alpicola / R / 12" / white, lavender, yellow fls / Tibet 119 / 107
alpicola 'Alba' - white fls, comes true from seed 207
alpicola 'Luna' - pale yellow fls (553) / 107
alpicola v. violacea - violet or purple fls 25
altaica 'Grandiflora' - see P. vulgaris ssp. sibthorpii 136
amoena / R / 5-6" / mauve-pink fls / Caucasus, n Turkey (12) / 123
amoena 'Grandiflora' - selection for larger fls *
angustidens - see P. Wilsonii 119
angustifolia / R / to 1" / solitary fls, purp or wht / Colo.-N.M. 25
X anisiaca - P. vulgaris x P. elatior 28
anisodora / P / 2' / brownish-purple fls / Yunnan 119 / 207
apennina / R / 3" / fls purple or deep pink / Mt. Orsaro 288
X 'Arctotis' / P. auricula x hirsuta / 4-5" / fls lilac-purple 28
'Asthore Hybrids' / 3' / candelabra type, wide range of colors 147
aurantiaca / R / 8-12" / reddish-orange fls / Yunnan (554) / 123
aurantiaca 'Candy Pink' - presumably P. vulgaris 'Candy Pink' *
auricula / R / 4-8" / deep yellow fls / Alps 136 / 129
auricula v. albocincta / lvs densely farinaceous / Monte Baldo 28
auricula 'Alpina' - indicating the wild type 46
auricula 'Arctotis' - see P. X 'Arctotis' 28
auricula v. Balbisii / R /to 8" / fls many colors, frag / Alps 123 / 25
auricula 'Barnhaven Strain' - see P. X polyantha 'Barnhaven Strain' 136
auricula 'Bauhinii' - see P. auricula v. albocincta 96
auricula 'Blue Velvet' - fragrant bluish-purple fls, white eye 136
auricula 'Broadway Gold' - yellow frilled fls 184 / 136
auricula var. ciliata / small, golden-yellow fls / Typol 132
auricula 'Colbury' - white-edged fls 136
auricula 'Decora' / 6" / fls violet-blue 18
auricula 'Dusty Miller' - class of Auriculas classified by color 136
auricula 'Goldlace' - see P. X polyantha 'Goldlace' 136
auricula 'Linnet' - fls mixed green & brown 136
auricula 'Lynn Hall Strain' - recommended strain of Auriculas 135
auricula '(The) Mikado' - large deep maroon fls 136
auricula 'Old Irish Blue' - violet to pale blue fls, frilled 136
auricula 'Old Red Dusty Miller' - wallflower-red fls 136
auricula v. serratifolia - serrate-lvd form from the Banat 96

auriculata / R / 4-14" / rose, violet or lilac fls / Greece to Iran 28
'Barnhaven Strain' - see P. X polyantha 'Barnhaven Strain' 136
'Barrowby Gem' - fragrant clear yellow polyanthus, not true seeding 136
Bayerni / P / lvs glabrous; many fls white and brown / Caucasus 10
beesiana / P / 1½-2' / fls purple-lilac in whorls / w China 129
bellidifolia / R / 4" / red-blue fls / Tibet 119 / 132
X berninae / 2" mat / almost sessile deep pink fls 123
bhutanica / R / to 3" / blue fls, white eye / Himalayas (368) / 25
X biflora / R / to 2" / fls rosy purple, 2 on 2-3" scape / Alps 25
obreio-calliantha / P / 18" / fls rose purple / Yunnan 61 / 119
boveana / Gh / to 10" / yellow fls / Sinai 28
bracteata / R / ? / yellow fls / Yunnan 96
X bullesiana - varii-colored fls: P. beesiana X bulleyana 25
bulleyana / P / 1½-2½' / deep yellow fls / Yunnan (218) / 123
X (Bulleyana X beesiana) - crossed by donor
burmanica / P / to 2' / reddish-purple fls / Upper Burma 241 / 123
calderiana / R / 9" / deep purple fls / Sikkim 100 / 132
'Candelabra Hybrid' / 2' / fls in mixed colors 282
capitata / R / 6-12" / violet fls / Sikkim, se Tibet, Bhutan 194 / 123
capitata ssp. crispata - fls in a flattened head 25
capitata ssp. mooreana - fls deeper in color & more open 93 / 123
capitata ssp. sphaerocephala / funnel-f. fls in globose heads/ China 25
carniolica / R / to 10" /fls rose to lilac, white eye / Alps 148 / 25
cashmiriana - see P. denticulata v. cachemiriana 96
cawdoriana / R / 8" / mauve fls, white-eyed / se Tibet 194 / 25
chionantha / R / to 1' / fragrant white fls / Yunnan 119 / 123
chungensis / P / 2'+ / fls pale orange, fragrant /China,Bhutan,Assam 269 / 25
X chunglenta 'Red Hugh' - vivid orange-red fls, fairly true-seeding 279
chumbiensis / R / ? / wide color range / ? 64
ciliata - now included in P. auricula 288
clarkei / R / 2" / fls rose-red / Kashmir 46
clusiana / R / 4-7" / crimson & white fls / Austrian Alps 148 / 107
Cockburniana / R / 10-14" / orange-scarlet fls / w China 119 / 107
Cockburniana lutea / R / to 1' / fls yellow / Szechwan 61
columnae - see P. veris ssp. columnae 288
commutata - see P. villosa 288
concholoba / R / to 8" / fls bright viol. / Assam-Burma-Tibet border 119
conspersa / R / 9-12" / pale lilac-pink fls / Kansu 96
[1] cordifolia / R / 6-8" / large, clear yellow fls / High Caucasus
cortusoides / R / 6-12" / fls rose-colored / w Siberia 306 / 107
crispa - see P. glomerata 25
cuneifolia / R / to 1' / pink to purplish fls / arctic Asia, Alaska 194 / 209
cuneifolia hakusanensis / P / to 1' / fls red, long corol./c Honshu 25
cuneifolia heterodonta - larger than P. c. hakusanensis / n Honshu 25
cusickiana / P / 2'+ / fls deep violet or white / w United States 268
daonensis / R / to 4" / pink to lilac fls / e Alps 148 / 288
darialica / R / 2'+ / fls rose / Caucasus 107
dealbata - see P. stenocalyx 119
decipiens / 2-6" / fls white / Falkland Isls. (252)
denticulata / R / 6-12" / fls in mauve shades / Himalayas 147 / 129
denticulata 'Alba' - white fls 129 / 207
denticulata v. cachemiriana - deep purple fls, later 123

---

[1] Otto Schwarz, Quar. Bull. Alpine Garden Soc., 43:2.

denticulata 'Grandiflora' - vigorous horticulture form 29
denticulata 'Pritchards Ruby' - glowing ruby-red fls 123
denticulata 'Purpurea' - dark purple fls 28
denticulata 'Rubra' - magenta-red fls 207
deorum / R / to 8" / dark violet fls / sw Bulgaria (550) / 288
X deschmannii - see P. X vochinensis 25
dianae - see P. calderiana 119
X digenea - P. elatior x P. vulgaris 61
edelbergii / Gh / 6-12" / fls yellow / Kabul, Afganistan (252)
edgeworthii / HH R / 3" / variable mauve fls / w Himalayas 129 / 123
edgeworthii 'Alba' - fls white 25
elatior / R / to 1' / fls yellow / s, w, c Europe 148 / 288
elatior ssp. carpathica - rounder lvs., inflated calyx 25
elatior ssp. cordifolia / fls lemon-yellow / high Caucasus 93
elatior ssp. elatior / 8" / fls pale, large throat 154
elatior ssp. intricata / pale yellow fls / s & sc Europe, in mts. 288
elatior ssp. leucophylla / lvs gray-tomentose beneath / Carpathians 288
elatior lofthousei / 4-12" / fls yellow / s Spain 288
elatior ssp. pallasii / fls pale yellow / alp meadows, s Urals 288
elatior ssp. ruprechtii / fls bright yellow / Caucasian alps 288
ellisiae / P / 6" / fls mauve, purple basal blotch / N. America (253) / 154
erosa / R short / lilac fls / Himalayas 96
farinifolia - microform of P. darialica 64
farinosa / P / to 1' / fls lilac & purp, wht rare/ European Alps 148 / 25
farinosa v. denutata - see P. gigantea 10
farinosa v. groenlandica - see P. stricta 10
farinosa laurentiana / P/ to 2½'/ fls lilac / Lab & Que to n Maine 25
fauriei - see P. modesta v. fauriei 200
firmipes / R / to 1' / fls pale yellow, fragrant / Sikkim 64
floribunda / R / to 11" / fls gold to sulphur-yellow / Himalayas 93 / 25
floribunda v. isabellina / 4-8" / fls yellow-buff / w Himalayas 61
floribunda X verticillata - see P. X kewensis 25
florindae / P / to 4' / bright sulphur-yellow fls / se Tibet 129 / 25
forrestii / P / to 3' / fls fragrant, yellow, orange-eyed / China 129 / 25
frondosa / R / 5-7" / pink fls / Balkans (133) / 123
frondosa 'Xanthophylla' - darker form 249
gambeliana / R / to 1' / fls purple-pink to violet-purple / Himalayas (555) / 119
gambeliana gemmifera - see P. gemmifera *
X 'Garryarde Guinevere' / R / 4-6" / pale bluish fls 129 / 268
gaubeana / P / to 20" / yellow fls / Iran 204
gemmifera / R / to 1' / pink to purple fls / China 119 / 25
geraniifolia / R / 6" / deep rose-red fls / Sikkim 132
gigantea / P / 10+ fls, rose-violet, yellow throat / Siberia 10
glabra / R / dwarf to 4" / pinkish-purple, yel. eye / Nepal, Sikkim 119
glaucescens / R / to 4½" / fls pinkish-red to lilac / s Alps 288
glomerata / R / ? / similar to P. deticulata 96
glutinosa / R / 3-4" / fragrant royal-purple fls / Austrian Alps 148 / 96
goebellii - see P. X pubescens 96
'Gold Lace' - see P. X polyantha 'Gold Lace' *
gracilipes / dwarf / bright pink-purp, orange-yellow eye / c Asia 119
gracilis v. pulcherrima - see P. Sieboldii 61
grandis / R / to 1' / fls pale yellow or whitish / Caucasus 10
grandis sredinskya - SREDINSKYA, an alternate genus 10
griffithii / 4-8" / fls dp purp, yel eye / to 18" in fr./ Tibet 119
groenlandica - see P. stricta 10

halleri / R / 3-6" / clear pink to deep rose / European Alps 148 / 123
X heeri - rose-red fls / occurs in Swiss Alps/ rubra X intergrifolia 46
helodoxa / P / 2-3' / fls bright yellow / Yunnan 269 / 129
heucherifolia / R / to 1' / mauve-pink to deep purple fls / Tibet (549) / 25
hidakana Kamuiana / 2-5" / hairy; fls rose, yellow throat / Japan 200
hirsuta / R / 3" / pale lilac to deep purp-red fls / Alps, Pyren. 148 / 288
hirsuta 'Alba' - white fls *
hookeri - see P. algida
hyacinthina / R / to 1' / fls violet, scented / Tibet 119
ianthina / P / to 2' / violet fls / Sikkim 119
inayati / HH R / 1' small fls / Hazara, nw Pakistan (444) / 96
incana / P / 18" / lilac fls, yellow throat / Sask., Col., Utah (551) / 25
'Inschriach Hybrid' - candelabra type, mixed colors 34
integrifolia / R / 2" / reddish-purple to pinkish-lilac fls / Alps 314 / 288
intercedens / R / 4-6" / pink fls, fading white / Great Lakes reg. 25
X intermedia - R / 3-4" / pink fls /occurs in Styria 96
'Inverewe' - candelabra type, brillant brick-red fls (48)
involucrata / R / to 1' / fls white, yellow eye / Himalayas 240 / 25
ioessa / R / 9" / lilac or white fls / Tibet 119 / 24
ioessa 'Alba' - the alternate color 24
ioessa v. subpinnatifida - fls white, or tinged purplish 25
irregulari / R / tiny to 2" / fls bluish purple / Sikkim, Tibet 119
[1] 'Itton Court' / candelabra hybrid / Chinese-red, true from seed
'Jack-In-The-Green' - Polyanthus, calyx a ruffle of green lvs 136
japonica / P / 1½' / fls purplish-red in whorls / Japan 208 / 129
japonica 'Alba' - white flowered 306 / 25
japonica 'Millers Crimson' - dark crimson-purple fls 147 / 207
japonica 'Postford White' - white fls with pink eye 129
japonica 'Red Hugh' - see P. X chuglenta 'Red Hugh' 279
japonica 'Rosea' - rose-colored fls 25
japonica 'Rubra' 25
japonica 'Valley Red' - deep cerise-red fls 282
jesoana / R / 8-16" / lvs rose-purple / Hokkaido, Honshu 164 / 200
jesoana v. pubescens / lvs more pubescent / Hokkaido 200
juliae / R / 3" / fls rose or red / Caucasus 25
X kewensis / P / 1½' / fragrant yellow fls in whorls (418) / 25
kisoana / R / to 8" / fls deep rose or rose-mauve / Japan (378) / 25
kisoana v. shikokiana - see P. kisoana 200
'Lady Greer' - see P. X polyantha 'Lady Greer' 136
latifolia / R / 2-7" / purple to dark violet fls / Alps, Pyrenees 314 / 288
latifolia alba / P / short stem / fls white / European Alps 154
latifolia ssp. cynoglossifolia - smaller lvs 96
latifolia graveolens - see P. latifolia 288
laurentiana / P / to 1½' / lilac fls / Labrador to ne Maine (552) / 25
leucophylla - see P. elatior ssp. leucophylla 288
lichiangensis - see P. polyneura 93
'Linda Pope' - see P. marginata 'Linda Pope' 132
'Lissadell' - hybrids from P. pulverulenta 64
longiflora - see P. halleri 288
longifolia - see P. auriculata 28
lutea - see P. auricula 28
luteola / R / to 1' / fls yellow / e Caucasus 25

---

[1] Goplerud, Far North Gardens Catalog, 1973

'Lynn Hall Strain' - recommended auricula selection 136
macrocarpa / R / small / white fls / Honshu 200
macrophylla / P / 6-10" / fls purple or violet, dk purp eye/w Himal. 194 / (252)
magellanica - see P. farinosa v. magellanica 107
malacoides / A / 4-18" / fls rose to lavender, almost white/ China 25
marginata / R / 3-4" / fls blue-lilac / Maritime & Cottian Alps 148 / 107
marginata 'Alba' - white fls *
marginata 'Linda Pope' - fls rich lavender-blue, white eye (551) / 132
X 'Marven' / R / 6-7" / fls blue-purple, white eye 123
Matsumurae - see P. modesta Matsumurae 200
megaseafolia / R / 2-4½" / fls rose / mts. of Asia Minor 28
melanops / R / 8" / fls rich purple, black eye / sw Szechwan 119 / 123
minima / R / clump / delicate pink fls / se Europe 148 / 107
minima alba / R / very short or rosette / fls white / mts Europe 25
mistassinica / R / 2-4" / pale pink fls / N. Am. 224 / 123
modesta / R / 3-6" / rose-colored fls / Japan, in mts. 273 / 200
modesta 'Alba' - white fls 200
modesta v. fauriei - smaller, from rock cliffs near sea 200
modesta v. fauriei 'Alba' - white selection *
modesta v. fauriei f. leucantha - white fls *
modesta v. Matsumurae - R / 6"+ / fls rose / Hokkaido 200
modesta nemuroensis / 1-5" / fls pinkish purple 61
modesta v. yuparensis 'Alba' - see P. yuparensis 'Alba' 200
mollis / Gr / to 2' / fls rose to crimson / Himalayas 25
'Monacensis' / green lvs, long & narrow / Munich area 96
mooreana - see P. capitata ssp. mooreana 123
muscarioides / R / 8-12" / blue fls / Shensi, n China 119 / 93
nepalensis / R / to 1' / yellow fls / Nepal 119
nipponica / R / to 6" / fls white, yellow eye / Japan 25
nivalis / R / to 1' / rose-violet fls / Siberia 10
nutans / R / to 1' / fls lilac to pink, yellow center / n Russia 132 / 228
obconica v. werringtonenesis / hardier var. / upland China 64
obliqua / R / 12'18" / pale yellow to white fls / Sikkim 119
obtusifolia / P / 1½' / fls bluish-purple / nw Himalayas (556) / 25
oenensis - see P. daonensis 25
officinalis - see P. veris 288
[1] 'Pagoda Hybrid' / 2' / candelabra type, mixed colors
palinuri / HH R / 6-8" / fls deep rich yellow / s Italy 208 / 123
pallasii - see P. elatior ssp. pallasii 93
parryi / R / 6-12" / deep magenta-pink fls / Colorado 312 / 123
patens - see P. sieboldii 28
pedemontana / R / 6" / fls rose or white / sw Alps 148 / 107
pedemontana alba - white form of P. pedemontana
petiolaria / R / 2" / magenta-purple fls / Nepal 119
Poissonii / P / 18" / fls purp-crimson, yel. eye / China, Yunnan 119 / 25
poloninensis - see P. elatior 288
X polyantha / R / 9-12" / fls vaiously-colored 147 / 129
X polyantha 'Barnhaven Strain' - every conceivable color & form 136
X polyantha 'Gold Lace' - bicolored with narrow edging 136
X polyantha 'Lady Greer' - fls cream-colored 136
polyneura / P / 12-18" / varii-colored fls / c China 93 / 107
'Postford White' - see japonica 'Postford White' 129

---

1 Goplerud, Far North Gardens Catalog, 1973.

prenantha / P / fls loose superposed whorls; lg. lvs 119
primulina / R / scape 3" / fls purple, rarely white / Himalayas 119
prolifera / P / 2' / fls golden-yellow / Assam 194 / 268
[1] pruhoniciana 'Wanda'
pseudoclatior - see P. elatior 61
X pubescens / R / 3-5" / fls almost any color 123
X pubescens 'Christine' - miniature old-rose fls 83
X pubescens 'Faldonside' - deep crimson fls 132
X pubescens 'Mrs. J. H. Wilson' / 4-5" / violet fls 90 / 123
X pubescens 'Rufus' - deep terra-cotta red fls 129
X pubescens 'The General' - variety terra-cotta fls 123
pudibunda - see P. sikkimensis v. pudibunda 93
pulverulenta / P / 2-3' / claret-red fls / w China 119 / 129
pulverulenta 'Bartley Strain' - soft pink fls 269 / 129
pusilla - see P. primulina 119
X 'Red Hugh' - see P. X chunglenta 'Red Hugh' 279
reidii / R / 3-4" / white fls / nw Himalayas 194 / 129
reidii v. williamsii - pale bluish-lilac fls (255) / 129
reidii v. williamsii 'Alba' - white form 119
Reinii / R / to 4" / fls rose to purple, yellow eye / c Japan 25
reticulata / P / 7-18" / yellow or white fls / Nepal to Tibet (557) / 93
rosea / R / 3-4" / glowing rose-pink fls / Himalayas 147 / 107
rosea 'Grandiflora' / to 10" / larger lvs 123
rosea 'Micia Visser de Geer' / 6-8" / fls larger & deeper color 123
rotundifolia / R / 6" / fls mauve-pink / Nepal 119 / 132
rubra - see P. hirsuta 288
ruprechtii - see P. elatior ssp. ruprechtii 10
rusby / R / 4-6" / fls bright rose / N. Mex., Arizona 227 / 28
sandemaniana / R / to 6" / fls bluish-purple / s Tibet 61
saxatilis / R / to 1' / fls rose-lilac / ne Asia 119 / 25
scandinavica / R / 6" / fls dark purple / Norway, Sweden 288
scotica / R / 2" / fls deep purple / n Scotland 314 / 123
secundiflora / R / to 1' / fls dark ruby-purple / sw China 119 / 107
serratifolia / P / ? / pale yellow fls / sw China 28
sibirica - see P. nutans 288
sibthorpii - see P. vulgaris sibthorpii 288 / 25
Sieboldii / R / 6-9" / fls dark pink to white / Japan 147 / 129
Sieboldii alba - white form of P. Sieboldii 25
sikkimensis / P / 2' / yellow fls / Sikkim 194 / 107
sikkimensis v. hopeana - fls white; slender plant 61
sikkimensis v. pudibunda - alpine form, smaller fls 93
sinensis / Gh / ? / yellow fls / Ethiopia 28
sinensis 'Stellata' / Gh / any of several colors, tiered fls 25
sinoplantaginea / R / 6" / violet-purple fls / Yunnan (558) / 132
sinopurpurea / R / 4-16" / purple fls / w China 119 / 25
smithiana / P / to 2' / fls pale yellow / Himalayas 119 / 25
'Snow Cushion' / 2" / juliae hybrid / white fls 18
sonchifolia / P/to 1'/fls blue, with yel-wht margined eye/Himalayas 25
sorachiana - small-flowered, possibly a variant of P. yuparensis 200
spectabilis / R / to 6" / pinkish-red to lilac fls / s Alps 148 / 288
specuicola / R / 6" / pink fls / Rocky Mts. 123
stenocalyx / R / to 6"/ bluish-lavender to pink fls / Tibet, w China 119

[1] Hillier's Hundred Catalog, 83-84.

stricta / R / 1' / fls violet to lilac / Scandinavia, Iceland, B. C. 194 / 288
strumosa / R / 1'+ / yellow fls / Bhutan, Nepal 100 / 119
stuartii / R / 8-12" / golden-yellow fls / nw Himalayas, Nepal 119
suffrutescens / R / 4" / deep rose fls / California 312 / 123
takedana / R / to 6" / white fls / Hokkaido, in mts, 200
tanneri - ssp. of P. griffithii 119
X tommasinii - P. veris x vulgaris 288
tosaensis / R / 4-6" / fls rose purple / Japan 200
tschukschorum / R / to 10" / violet fls, lavender eye / n Arctic reg. 13 / 209
turkestanica - fls viol-black; lvs violet spotted, wht waxy margin 10
tyrolensis / R / 1/4-2" / fls bright purple-pink / Italian Dolomites 288
uralensis - see P. veris ssp. macrocalyx 25
'Valley Red' - see P. japonica 'Valley Red' 282
X variabilis - see X tommasinii 288
veitchi - see P. polyneura 25
X venusta - 3-4" / crimson to purple fls 123
veris / R / 4-8" / fragrant yellow fls / Europe 148 / 107
veris algida - see P. algida 107
veris ansiaca - see P. X anisiaca 28
veris ssp. canescens / gray-tomentose / lowlands, c Europe 288
veris 'Coccinea' - fls crimson 25
veris ssp. columnae / white-hairy plant / mts of s Europe 288
veris ssp. macrocalyx - long calyx, orange-yellow fls 25
veris suavolens - see P. veris ssp. columnae 288
veris uralensis - see P. veris var. macrocalyx 10
veris X vulgaris - natural hybrid found in most Europe 288
verticillata / Gh / to 2' / yellow fls / s Arabia 25
verticillata boveana - see P. boveana 28
verticillata sinensis - see P. sinensis 28
vialii / P or B / 1-1½' / fls bluish-violet / nw China 147 / 129
villosa / R / 4" / fls rose or lilac / Alps 148 / 107
villosa comuttata / R / to 6" / fls rose to lilac, wht eye/ se Aus. 25
violacia / P / short / 1-15 fls, violet / Switzerland, Pyrenees 61
viscosa - see P. latifolia 184 / 288
X vochinensis - dwarf / deep red fls / P. minima x wulfeniana 25
vulgaris / R / 4-5" / fls cream-yellow / Europe 147 / 136
vulgaris 'Heterochroma' - smaller form in wide range of colors 64
vulgaris ssp. ingwerseniana / white fls / Mt. Olympus, Greece 64
vulgaris ssp. sibthorpii / fls usually red or purple / Balkans 288
waltonii / P / 2'+ / fls purple to red / Tibet, Bhutan 119 / 93
warschenewskiana / R / mat / bright pink fls / e Afganistan 119 / 19
werringtonensis - see P. obconica v. werringtonensis 25
X wettsteinii - 5" / blue fls / P. minoma x clusiana (256)
wilsonii / P / to 3' / fls purple / China 119 / 25
wulfeniana / R / 2" / fls rose-colored / Austrian Alps 148 / 107
yargonensis / P / 1' / fls pink or purple, white eye / Himalayas 119 / 25
yuparensis / R / 9" / pale rose-purple fls / Hokkaido 119 / 200
yuparensis 'Alba' / R / 2"/ white fls (unnoted) / Hokkaido 200

PRINSEPIA - Roseaceae
sinensis / Sh / 6' / yellow fls / Manchuria 184 / 25
uniflora / Sh / to 4' / white fls; red frs / nw China 28

PROBOSCIDEA - Martyniaceae
jussieuii - P. louisianica 93

louisianica / A / to 12" / whitish to purplish fls / sc US 25

PROSARTES
oregana - see DISPORUM hookeri v. oreganum 160

PROSTANTHERA -Labiatae
cuneata / Gh Sh / to 4' / fls pale mauve / Australia 45
nivea / Sh / 3-6' / fls white, blue-tinged / se Australia & Tasmania 25

PROTEA - Proteaceae
cynaroides / Sh / to 6' / bracts pink or nearly red / South Africa 25
repens / Gh Sh / 6" / evergreen lvs / Cape of Good Hope (559) / 207
scolymocephala / Gh Sh / 3' / pale green bracts / South Africa 25

PRUNELLA - Labiatae
asiatica - see P. vulgaris v. lilacina 200
asiatica albiflora - see P. vulgaris f. albiflora 99
X bicolor - P. grandiflora x vulgaris v. lilacina (257)
grandiflora / R / 4" / purple-violet fls / Europe 148 / 123
grandiflora 'Alba' white fls 107
grandiflora 'Loveliness' - usually des. by color: 'Lilac', ' Pink' 129 / 49
grandiflora 'Pink Loveliness' - pink fls 123
grandiflora 'Rosea' - darker pink fls 25
grandiflora 'Rubra' - selection for red fls 25
grandiflora 'White Loveliness' - white selection 49
incisa - see P. vulgaris v. pinnatifida 81
laciniata / R / to 1' / fls yellowish-white / s, w & c Europe 210 / 288
laciniata 'Alba' - selection for white fls *
vulgaris - naturalized European weed 308 / 107
vulgaris 'Alba' - see P. vulgaris v. lanceolatum f. candida 259
vulgaris f. albiflora - white fls on typical P. vulgaris 99
vulgaris asiatica - see P. vulgaris v. lilacina 200
vulgaris atropurpurea - included in typical color range 9
vulgaris laciniata - see P. laciniata 81
vulgaris v. lanceolata f. candida - narrow lvs, white fls 259
vulgaris v. lanceolata f. rhodantha - narrow lvs, pink fls 99
vulgaris v. lilacina / blue-purple fls / Japan 200
vulgaris v. pinnatifida / cut leaved, purple fls / s Europe 81
vulgaris rosea - see P. vulgaris v. lanceolata f. rhodantha 99
webbiana / possibly a horticultural hybrid / fls bright purple 147 / 25
webbiana 'Lilac Loveliness' - belongs with 'Loveliness' group above *

PRUNUS - Roseaceae
emarginata / Sh / 3-10' / fls tinged green; frs black / B. C. & s 28
grandulosa / Sh / 3-5' / fls white to pink / Far East 110 / 25
incisa / Sh or T / 15-30' / fls red & pink / Japan 36 / 25
laurocerasus /Sh or T/ 20'/fls white; frs dk purp./ se Eur, sw Asia 38 / 25
Mume / HH T / to 30' / fls white to dark red / Japan 129 / 25
nipponica v. kurilensis / T / to 25' / fls wht; fr blk / Kurile Is. 35
padus / T / 30' / white fls; black bitter frs / Eurasia 274 / 139
padus 'Colorata' - lvs purplish 36
pensylvanica / T / to 35' / white fls; light red frs / e & c N. Am. (342) / 25
prostrata / HH Sh / 18" / fls lt. rose; reddish frs / Levent 132
pumila / Sh / to 8'/ fls wht; fr purp-black /sandy shores, Gt Lakes 25
pumila v. depressa / Sh/ prostrate/ edible frs/ N. Brunswick - Mass. 222

sargentii / T / 20' / single pink fls / Japan 129
serotina / T / 80'+ / white fls in racemes / e & c N. Am. (342) / 25
subhirtella / T / 25' / fls pink to white / Japan 269 / 25
tenella / Sh / to 4½' / bright pink fls / e & ec Europe 292 / 286
virginiana / Sh or T / to 30' / white fls / N. Am. 28
virginiana demissa prostrata /Sh or T/droops/fl wht; frs red/w N.Am. 38 / 25

PSEUDOGALTORIA - Liliaceae
1 clavata / HH B / to 4' / fls green / Natal, ne Cape

PSEUDOLARIX - Pinaceae
amarilis - see PSEUDOLARIX Kaempferi 25
Kaempferi / T / to 130' / lvs brt green, turn yellow / China 25

PSEUDOMUSCARI - Liliaceae
azureum / Bb / 4-6" / pale blue fls, darker stripe / Caucasus 185
azureum f. album / 3" / white fls, weak grower 17
azureum v. amphibolus / 12" / pale blue fls / Turkey 17
chalusicum / Bb / 4" / pale china-blue fls / Iran 185

PSEUDORLAYA - Umbelliferae
pumila / R / to 8" / fls white or purple / Medit., s Portugal 287

PSEUDOTAENIDIA - Umbelliferae
montana / P / 2½' / yellow fls / Allegheny Mts. 313

PSEUDOTSUGA - Pinaceae
Menziesii / T / to 300' / needles blue-grn, lvs to 4½"/B.C.-Mex. 25

PSIDIUM - Myrtaceae
cattleianum / HH Sh or T / 20' / Strawberry Guava / Brazil 28

PSORALEA - Leguminosea
pinnata / Gr Sh / 12' / fls blue & white / South Africa 61
subacaulis / R / 4-8" / lvs 5-7 sg; fls blue-purple / Tennessee 225

PSYLLIOSTACHYS - Plumbaginaceae
suworowii / R / 12"+ / rose-pink fls / Caucasus, Iran 25

PTELEA - Rutaceae
angustifolia - see P. baldwinii 222
baldwinii / Sh / 10'+ / small lvs; large fls / sw US, Mexico 139
trifoliata / T / to 24' / frs winged wafers / e N. Am. (342) / 222

PTERIDOPHYLLUS - Papaveraceae
racemosum / R / 6-10" / fls white / Japan 200

PTEROCEPHALUS - Dipsacaceae
papposus / A / to 2' / fls pink or purplish / e Medit. region 289
parnassi - see P. perennnis 184 / 289
perennis / R / to 5" / fls pink to pale purplish / s & e Greece 138 / 289
plumosus - see P. papposus 289

---

1 Sima Elivson, Wild Flowers of Southern Africa, 1980.

PTEROSPORA - Pyrolaceae
andromeda / P / 3' / red, white fls; parasitic / w N. America 228

PTEROSTYLIS - Orchidaceae
barbata / Gr 6" / green fls / temperate Australia 45
nutans / HH R / 6" / greenish-yellow fls / Australia 45 / 206
oliveri / Gr / ? / green & white fls / New Zealand 184

PTEROSTYRAX - Styracaceae
hispida / T / 50' / creamy white fls / China, Japan 28

PTILOSTEMON -Compositae
hispanicus / HH P / 3'+ / purple fls / s Spain 289

PTILOTRICHUM - Cruciferae
laperousianum / Sh / to 1' / white fls / e Pyrenees, e Spain 286
macrocarpum / HH Sh / 8" / white fls / s France 286
purpureum / R / low tufts / rosy-purple fls / s & se Spain 212
pyrenaicum / Sh / 20" / white fls / e Pyrenees 314 / 286
reverchonii / HH Sh / 20" / white fls / se Spain 286
spinosum / Sh / 2' / white or purplish fls / e & s Spain 90 / 286
spinosum 'Coccineum' - reddish fls *
spinosum 'Purpureum' - purplish fls *
spinosum 'Roseum' - rare pink form (133) / 132

PTYCHOTIS - Umbelliferae
saxifraga / B / 12-28" / whitish fls / w Medit. region 287

PUERARIA - Leguminosae
lobata / HH Cl / to 60' / red-purp fls / rampant in s U.S., Japan 61 / 25

PULMONARIA - Borginaceae
angustifolia / R / 6-12" / bright blue fls / Europe 288
angustifolia 'Munstead Blue' / 9" / bright blue fls 268
angustifolia 'Salmon Glow' - fls soft rose (258)
azurea - see P. angustifolia 288
filarszkyana / P / lvs to 12"; fls red / Carpathians 288
longifolia / R / 1'+ / fls violet to violet-blue / w Europe 210 / 288
officinalis / R / to 1' / fls rose-violet to bl; lvs wht spots/ Eur. 314 / 25
'Salmon Glow' - see angustifolia 'Salmon Glow' *

PULSATILLA - Ranunculaceae
alba / P / 2' / fls white / c Europe, in mts (259) / 286
albana / R / 5" / fls in pale colors / e Caucasus to India 96
albana 'Albo-cyanea' - white & blue fls *
albana 'Albo-cyanea Lutea' - se P. albana 'Lutea' *
albana ssp. armena / violet-blue fls / Turkey, Iran 77
albana 'Lutea' - included in the sp. 5
albana 'Violacea' - see P. albana ssp. armena 77
albo-cyanea - see P. albana 'Albo-cyanea' *
albo-violacea - see P. albana ssp. armena 64
alpina / P / 1½' / white fls / European Alps 129 / 148, 286
alpina alpicola - see P. alba
alpina ssp. apiifolia - fls pale yellow 284 / 286

alpina ssp. austriaca - see P. alba 286
alpina 'Burseriana' / 16" / fls white, flushed blue or red / w Alps 61
alpina 'Millefoliata' - smaller plant, lvs finely cut 96
alpina ssp. sulphurea - see P. alpina ssp. apiifolia 129 / 286
ambigua / R / to 1' / fls blue-violet, rarely pink / s Siberia 5
amoena - see P. vulgaris 96
apiifolia - see P. alpina ssp. apiifolia 286
armena - see P. albana ssp. armena 77
aurea / R / to 14" / golden yellow fls / Caucasus 5
'Budapest' - see P. vulgaris ssp. grandis 268 / 286
bungeana / P / 6-20" / lvs before blue fls / w Siberia 5
burseriana - see P. alpina 'Burseriana' 61
campanella / R / to 10" / fls violet-blue / w Siberia, w Asia 5
caucasica - see P. albana (61)
cernua / R / 4-16" / red-brown or purple fls / Far East 165 / 200
chinensis / R / 8-16" / fls violet / China 258
daurica / P / big plant / bright yellow fls / Dahuria 64
flavescens / R / 6"+ / yellowish fls / Siberia 5
georgica / R / to 1' / fls pale blue / Caucasus 5
globosa - see ANEMONE globosa 64
grandis - see P. vulgaris ssp. grandis 286
[1] grandis 'Alba' - see P. vulgaris ssp. grandis 'Alba'
halleri / R / 6-12" / fls pale mauve-violet / e Alps 148 / 129
halleri 'Alba' - white fls 96
halleri 'Budapest' - see P. vulgaris ssp. grandis 268
halleri ssp. grandis - see P. vulgaris ssp. grandis 286
halleri ssp. halleri / 1-3" / fls dark violet 286
halleri ssp. rhodopea - basil lvs petioled 286
halleri ssp. slavica - basal lvs divided into 3 / w Carpathians 19 / 286
halleri ssp. slavica 'Rubra' - reddish fls *
halleri ssp. styriaca - basal lvs divided into 5 / se Austria 286
halleri ssp. taurica - divisions of basal lvs sessile / Crimea 286
ludoviciana - see P. patens ssp. multifida 276
lutea - see P. aurea 5
montana / R / 4" / fls bluish to dark violet / w Europe 148 / 286
'Mrs. Van der Elst' - see P. vulgaris 'Mrs. Van der Elst' 123
multifida - see P. patens ssp. multifida *
nigricans - see P. pratensis ssp. nigricans 286
nipponica / R / to 1' / pale yellow & white fls / Hokkaido, Honshu 200
nuttalliana - see P. patens ssp. multifida *
occidentalis / P / to 2' / white fls / w N. Am. 107 / (261)
patens / R / 5" / bluish-violet fls / e & ec Europe 286 / (51)
patens 'Alba' - white fls *
patens flavescens - see P. flavescens 5
patens multifida - Siberian & North American form 286
patens slavica - see P. halleri ssp. slavica *
pratensis / R / 4" / fls variously-colored / c & e Europe 314 / 286
pratensis ssp. hungarica / fls muddy yellow / se range 286
pratensis ssp. nigricans / dark purple fls / n range 286 / (51)
ranunculoides - see ANEMONE ranunculoides *
'Red Cloak' - see P. vulgaris 'Red Cloak'
regeliana - see P. ambigua 5

[1] Joseph Starek, Bull. Am. Rock Garden Soc., 24:1.

rhodopea - see P. halleri ssp. rhodopea 286
rubra / R / 4" / dark reddish fls / France, Spain 212 / 286
slavica - see halleri ssp. slavica 286
styriaca - see P. halleri ssp. styriaca 286
turczaninovii / R / 2-14" / blue-violet fls / Manchuria, e Mongolia 5
vernalis / R / 4-8" / white fls 129
violacea / P / lvs narrow; fls lilac or white / Caucasus 5
vulgaris / R / 4-8" / fls violet-purple / Europe 212 / 129
vulgaris 'Alba' - white fls 129
vulgaris 'Budapest' - see P. vulgaris ssp. grandis 268
vulgaris 'Coccinea' - bright red fls (61)
vulgaris ssp. grandis - lvs appearing after the fls 286
vulgaris ssp. grandis 'Alba' - white fls (51)
vulgaris grandis rubra / P / fls red, before lvs / c Europe 286
vulgaris ssp. gotlandica / like ssp. grandis / from Gotland 286
vulgaris 'Mrs. Van der Elst' - pale shell-pink fls 123
vulgaris 'Red Cloak' - dwarf with red fls 46
vulgaris 'Rubra' - deep red fls 129 / 123
vulgaris slavica - see P. halleri ssp. slavica 286

PUNICA - Punicaceae
granatum / Gr Sh / to 10' / red, yellow, white fls / s Asia 194 / 177
granatum 'Nana' / Gr Sh / to 2' / fls orange-scarlet 128 / 129

PUSCHKINIA - Liliaceae
libanotica - see P. scilloides 267
libanotica 'Alba' - see P. scilloides 'Alba' 267
scilloides / Bb / to 6" / fls pale blue / Near East 267 / 129
scilloides 'Alba' - white fls 25

PUTORIA - Rubiaceae
calabrica / HH Sh / to 1' / rosy-red fls; foetid lvs / Medit. region 212 / 211

PUYA - Bromeliaceae
alpestris - in cult. is P. berteroniana 129 / 25
berteroniana / Gh / 10-12" / blue-green fls / c Chile (262) / 268
caerulea / Gh / 3-4' / blue fls / Chile 28
chilensis / Gh / to 15" / yellow fls / Chile (110) / 25

PYCNANTHEMUM - Labiatae
viginianum / P / to 3' / pinkish fls / e U.S. 224 / 25

PYCNOSTACHYS - Labiatae
urticifolia / HH P / 5-7' / bright blue fls / Tropics, S. Africa 28

PYGMAEA - Scrophulariaceae
pulvinaris / R / dense cushion / pale lav. fls / New Zealand 238

PYRACANTHA - Rosaceae
angustifolia / Sh / to 12' / frs orange-yellow / w China 139
coccinea / Sh / 10'+ / rich red frs / s Europe, Asia Minor 139
coccinea v. 'Kasan' / Sh / 6-15' / fls wht; fr scarlet /se Eur, Asia 25
crenato serrata / Sh / to 18' / fls downy; fr persistant / China 38
Rogersiana 'Flava' / HH Sh / 15' / bright yellow frs 129 / 139
x watereri / to 8' / a vigorous dense shrub / fls white 36

PYROLA - Pyrolaceae
asarifolia / R / to 1' / pink fls / N. Am., Asia 224 / 107
asarifolia incarnata - see P. japonica 200
asarifolia v. purpurea - despite name separated by leaf characters 99
chlorantha / R / to 1' / yellowish-green fls / Europe 288
dentata / R / to 1' / fls cream to greenish-white / Calif., Montana 142
elliptica / R / to 1' / fls milk-white or creamy / n N. Am. 229 / 99
grandiflora / R / to 8" / creamy-white fls / circumpolar 13 / 25
japonica / R / 6-12" / fls 5-12, white / Japan 200
media / R / 6-12" / fls white to lilac-pink / n & c Europe 288
minor / R / 8" / fls whitish to lilac-pink / Europe 148 / 288
picta / R / to 1' / purplish fls / Pacific Slope of N. America 107
picta v. dentata - see P. dentata 142
promiscua secunda - see P. secunda *
rotundifolia / R / to 1' / white fls / Europe, N. Am. 148 / 123
secunda / R / 5" / fls greenish-white / boreal n hemisphere 164 / 200
secunda v. obtusata / fls creamy-white / Asia, N. Am. 25
uniflora - see MONESES uniflora 25
virens - see P. chlorantha 288

PYRUS - Rosaceae
salicifolia / T / to 30',thorny / fls wht; fr yel /se Eur, Cauc. 25

PYXIDANTHERA - Diapensiaceae
barbulata / Sh / creeping / white or pinkish fls / New Jersey 224 / 25

QUAMOCLIT
pennata - see IPOMOEA quamoclit 225 / 25

QUERCUS - Fagaceae
Sadleriana / Sh / 3-5' / serrate lvs / n Calif., sw Oregon 139
vaccinifolia / HH Sh / to 10' / evergreen gray-gr beneath / sw U.S. 139

QUILLAJA - Rosaceae
saponaria / HH T / to 60' / white fls / Chile 25

RAMONDA - Gesneriaceae
1 'Mont Serrat' - see R. myconi 'Mont Serrat'
myconi / R / to 8" / fls pale mauve / n Pyrenees 148 / 129
myconi 'Alba' - pure white fls 123
myconi f. alborosea /R/2"/fls lilac to pink-wht/ cultivar R. myconi 25
1 myconi 'Mont Serrat' - large rosettes, correspondingly fine fls
myconi 'Pyrenaica' - see R. myconi
myconi 'Rosea' - pale clear pink fls 232 / 123
Nathaliae / R / 4-6" / lavender-blue fls / Balkans 149 / 107
pyrenaica - see R. myconii 129
serbica / R / 3-4" / mauve fls / Balkans 93 / 123

---

1 Carl Worth in correspondence.

RANUNCULUS - Ranunculaceae

abnormis / R / to 8" / yellow fls / w & c Spain, Portugal	212 /	286
acetosellaefolius / R / to 8" / white fls / Spain	212 /	286
aconitifolius / P / to 20" / sepals reddish to purplish / c Europe		286
acris / P / 20-40" / common Buttercup, weedy in US / Eurasia		220
acris v. nipponicus / 6-32" / alpine plain of Japan & n Korea	273 /	200
adoneus / R / 4-8" / yellow fls / Col., Mont., Utah in mts.		107
aduncus / of montanus group / pubescent lvs / sw Alps, Spain		286
alpestris / R / to 5" / white fls / Europe, in mts.	148 /	286
alpestris traunfellneri - see R. traunfellneri		286
amplexicaulis / R / 6-8" / white fls / Pyrenees	148 /	129
amplexicaulis v. grandiflorus - larger fls		184
anemoneus / R / ? / pure white fls / Austrian alps		(263)
arvensis / A / 20-24" / fls pale greenish yellow / w & c Europe		286
asiaticus / Tu / 6-12" / varii-colored fls / Asia Minor		129
auricomus / P / variable heights / yellow fls / Europe	1314 /	286
bilobus / R / 4" / yellow fls / s Tyrols	148 /	25
brevifolius / R / 8" / yellow fls / most of Europe		286
brevifolius pindicus / basal lvs 4-10 / Greece, Kriti		286
buchanani / HH R / to 1' / white fls / subalpine New Zealand	(220) /	15
bulbosus / P / to 20" / yellow fls / most of Europe		286
bullatus / P / fls frag, grn, pubescent, hon-lvs yel. / Medit.		286
cadmicus / HH R / 4" / yellow fls / Cyprus		77
calandrinioides / HH R / to 1' / pink or white fls / North Africa	132 /	123
californicus / P / to 2' / fls yellow / Oregon & California		25
cardiophyllus / P / 6-18" / yellow fls / Sask. to Colorado	229 /	58
carinthiacus / R / ? / yellow fls / Alps, Balkans	314 /	286
carpaticus / P / 6-16" / basal lvs 1-3, pubescent / e Carpathians		286
cassubicus / P / 20" / yellow fls / c & e Europe		286
chordorhizos / P / to 3" / fls 2", pale yellow / New Zealand		15
constantinopolitanus / P / 18" / yellow fls / se Europe		286
cortusifolius / HH P / 3'+ / yellow fls; coriaceous lvs / Azores		286
crenatus / R / 4" / white fls / e Alps, Iceland		286
creticus / P / to 1' / fls golden yellow / Greek Isls.		25
crithmifolius / P / glab to 3" / petals 5, yellow / New Zealand		15
demissus / R / to 8" / yellow fls / s Europe, in mts.	212 /	286
demissus var. hispanicus / 4" / golden fls / Sierra Nev., Europe		96
dissectifolius / conventional buttercup / Australian alps		109
ensyii / HH R / to 1' / fls bright to pale yellow / New Zealand		15
Enschscholtzii / to 1' / yellow fls / nw North America	13 /	142
fascicularis / P / to 1' / fls yellow / Ontario to Texas		25
Ficaria / R / to 1' / fls yellow / Europe		286
Ficaria alba / P / to 1' / fls white / Europe, w Asia, nat. N. Amer.		25
flabellatus / see R. paludosus		286
flammula / P / to 32" / yellow fls / Europe, mostly n		286
geraniifolius - see R. montanus		286
glaberrimus / R / 2-6" / yellow fls, aging white / n Sierra Nevada		228
glacialis / R / 3-6" / white fls tinged purplish / Eur., Greenland	148 /	25
gouanii / R / 1' / fls yellow / Pyrenees	314 /	286
gramineus / R / 4-8" / bright yellow fls / s Europe	212 /	129
		25
graminiifolius - see R. gramineus		
graniticola / P / stems lg, sprawling; fls yellow / New South Wales		109
gunniianus / HH R / dwarf / yellow, pink, white fls / Australia	109 /	64
Haastii / R / 3-6" / golden fls / alps of New Zealand	193 /	96

hirtus / P / to 24" / hairy; fls yellow / New Zealand		15
hornschuschii - see R. oreophilus		286
hyperboreus / P/ creeps or floats/ sol. fls, beak short/ arctic Eur.		286
illyricus / P / 20" / pale yellow fls / c & se Europe		286
insignis / HH P / 3' / waxy yellow fls / New Zealand	194 /	238
kitadakeanus / R / to 16" / yellow fls / alps of Honshu		200
lappaceus / HH R / 1'+ / yellow fls / Australia	45 /	25
Lingua / P / very short, creeping/ fls bright yel / Europe, Siberia		25
Lingua v. grandiflora / aquatic / 3'+ / golden fls / England		96
lobulatus / HH P / to 20" / bright yellow fls / New Zealand		15
Lyallii / HH P / to 4' / white fls / New Zealand	194 /	238
macranthus - see R. orthorhyncus var. platyphyllus		25
macrophyllus / P / 1-2' / yellow fls / w Medit. region		286
magellensis - see R. bilobus		286
micranthus / R / 6" / fls pale yellow / e & c United States		99
millefoliatus / R / 6" / yellow fls / s & ec Europe		286
'Molten Gold' - see R. montanus 'Molten Gold'		(265)
monspeliaceae / P / 20" / yellow fls; lvs whitish / w Medit. region		286
montanus / R / variable heights / yellow fls / Alps	148 /	96
montanus carinthiacus - see R. carinthiacus		286
montanus 'Molten Gold' - free-flowering cultivar	48 /	(265)
montoi / P / tufted / 2-8 fls, yellow / New Zealand		15
muelleri / R / rosette / yellow fls / Austrian alps		64
nemerous / P / 3'+ / fls golden-yellow / Europe		286
niphophilus / P / ½-1" / golden fls / New South Wales, Australia		109
nivicolus / HH P / to 3' / golden yellow fls / New Zealand		15
novae-zealandiae / R / rosette / yellow fls / New Zealand		64
obesus / P / 18" / golden fls / n Armenia		77
occidentalis / P / to 1½' / fls greenish yellow / B.C. to Ore.		25
oreophilus / P / to 20" / fls yellow / mts. of c & s Europe		286
orthorhyncus var. platyphyllus / to 4" / yellow fls / Alaska		196
pachyrrhius / P / mat to 2" / fls bright yellow / New Zealand		15
paludosus / P / to 20" / yellow fls / w Europe, Medit. region		286
parnassifolius / R / 4"+ / fls white or reddish / European Alps	210 /	286
pascuinus / P / ½-2' / hairy; fls tall, gold / Tasmania, Australia		109
pedatifidus / R / 1' / yellow fls / Spitzbergen		286
platanifolius / P / to 4' / lvs 5-7 lobed, glabrous above / c&s Eur.		286
pseudomontanus / as P. montanus except achene short-beaked / Balkans		286
pygmaeus / R / to 6" / pale yellow fls / arctic regions	148 /	209
pyraneus / R / to 1' / white fls / Alps, Pyrenees, Corsica	286 /	25
rupestris / HH R / to 12" / yellow fls / s Portugal, s Spain		286
Seguieri / like R. glacialis/ sepals glabr, recept. hairy / Alps		25
sericophyllus / R / 10" / fls golden yellow / New Zealand, in mts.	238 /	15
spruneranus / P / 10-16" / yellow fls / Balkans, Syria		77
Thora / R / to 1' / yellow fls / Europe, in mts.	148 /	286
Thora carpaticus - from Carpathians		286
traunfellneri / R / 4" / white fls / se Alps	314 /	286
trichophyllus / Aq / White Water-Crowfoot / n hemisphere		209
villarsii - see R. aduncus		286
weyleri / HH R / to 8" / yellow fls / Mallorca		286

RANZANIA - Berberidaceae

japonica / R / 6-16" / fls before lvs, pale lavender-violet / Japan	165 /	268

Entry	Ref.	Page
RAOULIA - Compositae		
australis / R / carpet / fls pale sulfur-yellow / New Zealand	182 /	123
Buchananii / R / cushion / dark crimson fls / Mt. Alta, N. Zealand	184 /	15
eximia / R / cushion / scarlet fls, rarely seen / New Zealand	(62) /	268
glabra / carpet / green lvs; white fls / New Zealand		107
grandiflora / R / cushion / silvery lvs / New Zealand	238 /	15
Haastii / R / cushion 2-3' across / bracts creamy-wht to dark/N.Z.		25
Hookeri / R / mat / yellow fls / New Zealand	(177) /	238
Hookeri v. albo-sericea - compact habit, snow-white fls		15
Hookeri v. apice-nigra - inner phyllodes black-tipped		15
lutescens - see R. australis	129 /	15
mammilaris / R / woody cushion / white fls / New Zealand		25
Monroi / R/ prostrate / yellow-green fls / New Zealand		15
parkii / R/ prostrate / bracts white-tipped / Mt. Alta, N. Z.		15
petriensis / R / tuft / silvery lvs, showy frs / New Zealand		64
rubra / R / 6" cushion / crimson to purplish fls / New Zealand		15
subsericea / R / compact / silvery-lvs; white fls / N. Zealand		64
subulata / R / tufted / glabrous lvs / New Zealand		15
tenuicaulis / R / carpet / silvery lvs; yellow fls / N. Zealand		107
Youngii / R / carpet / silvery lvs; white daisies / N. Zealand	193 /	64
RAPHIOLEPIS - Rosaceae		
umbellata / HH Sh / to 6' / white fls; frs bronzy-black / Far East	36 /	139
RATIBIDA - Compositae		
columnaris - see R. columnifera		25
columnifera / P / 1-3½' / yellow fls / c North America	308 /	25
pinnata / P / to 4' / yellow fls / e & c North America	257 /	25
tagetes / P / 18" / yellow or purple fls / Kan., Col. to N. Mex.	227 /	25
REICHSTEINERA		
leucotricha - see SINNINGIA leucotricha	128 /	25
REHMANNIA - Scrophulariaceae		
angulata - see R. elata	184 /	25
elata / P / to 6' / fls rose-purple & yellow / China		25
glutinosa / R / to 10" / fls purple-brown & creamy-yellow / China	194 /	25
RESEDA - Resedaceae		
alba / A or P / 12-32" / white fls / s Europe, Jugoslavia	211 /	286
odorata / A / decumbent / fls yel-wht, fragrant / n Africa	286 /	25
RESTIO - Restionaceae		
tetraphyllus / HH P / 3' / apetalous fls; rush related / Australia		206
RETAMA - Leguminosea		
raetam / Gh Sh / ? / white fls / Near East deserts		270
RHAMNUS - Rhamnaceae		
Alaternus / HH Sh / to 20' / bluish-black frs / s Europe		25
alnifolia / Sh / 3' / black frs / North America		25
cathartica / Sh or T / 15'+ / frs shining black / Europe	210 /	139
frangula / Sh / 18' / frs red changing to black / Europe		139
pumilus / Sh / to 8" / black frs / Alps, s Eur. in mts		139
purshiana / T / 30' / large lvs / w North America		139
saxatilis / Sh / 2-3' / spiny branches, black frs / c & s Europe		222

RHAPHITHAMNUS - Verbenaceae
cyanocarpus / T/ 25'/ evergreen; lilac fls; blue frs / Chile, Argen. 93 / 95

RHAPONTIUM
scariosum - see LEUZEA rhapontica 289

RHAZYA —
orientalis / P / to 3' / fls blue to lilac / Greece to Asia, Minoa 25

RHEUM - Polygonaceae
alexandrae / P / 3-4' / straw-colored bracts / China, Tibet 207
emodi / HH P / 6-10' / purple plumey fls / Himalayas 207
nobile / HH P / to 4' / straw-colored bracts / Himalayas 194 / 25
palmatum / P / to 6' / lvs deeply, palmately lobed / ne Asia 129 / 25
palmatum 'Bowles Red' / crimson lvs with blood-red veins / ne Asia 184
palmatum tanguticum - more robust, coarsely-cut lvs 95

RHEXIA - Melastomaceae
alifanus / P / to 40" / pink fls / Fla., La., N. C. 105
lutea / HH R / 1' / fls yellow / N. Carolina to Florida 28
mariana / P / to 2' / pale purple fls / N. J., Ky., Fla. 225 / 107
mariana v. purpurea - crimson fls 43
nashii - see P. mariana v. purpurea 43
virginica / P / 12-18" / rosy-purple fls / Maine to Fla. & Mo. 224 / 107

RHINANTHUS - Scrophulariaceae
alectrolophus / P / 8-32" / yellow fls / Alps, Pyrenees, Apennines 148
alpinus / Bb / to 20" / pale yellow fls / se & ec Europe, in mts 288
aristatus / P / to 20" / yellow fls / sc Europe 288
crista-galli / A / 2'+ / yellow fls / circumboreal 224 / 25
minor / P / to 20" / yellow fls / most of Europe 138 / 288

RHINOPETALUM - Lilaceae
bucharicum / Bb / to 1' / fls green, white, light violet / c Asia 4
karelinii / Bb / 4-8" / fls rosy violet / c Asia 4
stenantherum / Bb / 6" / light rosy-violet fls / c Asia (444) / 4

RHIPOGONUM - SEE RIPOGANUM 25

RHIZOBOTRYA - Cruciferae
alpina / R / caespitose / white fls / s Alps 138

RHODIOLA - Crassulaceae
articum - see SEDUM articum 25
dumulosa / R / to 8" / white fls / China - see SEDUM 64
himalalense - see SEDUM himalense genus ref. 25
himalalense v. stephanii - dwarfer; longer fls; narrower lvs 96
integrifolia - see SEDUM integrifolium 276
integrifolia v. rosea - see SEDUM roseum 25
rosea - see SEDUM roseum 25

RHODODENDRON - Ericaceae
aberconwayi / HH Sh / to 8' / white or pinkish fls / e Yunnan 25
aechmophyllum / Sh / to 6' / corolla white-rose / Szechwan 25

alabamense / Sh / low stoloniferous / fls white, frag. / Alaska 25
albiflorum / Sh / to 6' / white fls / B. C. to Col. & Ore. (562) / 25
albrechtii / Sh / 5' / deep rose-pink fls / c & n Japan 232 / 129
amagianum / HH Sh / to 12' / orange-red fls / Japan 25
ambiguum / Sh / to 15' / pale yellow fls / Szechwan 194 / 36
anthopogon / HH Sh / 2' / white or pink fls / Tibetan Himalayas (569) / 25
arborescens / Sh / 9' / white or pinkish fls / Pa. to Ala. 222 / (563)
arboreum / HH Sh / to 20' / white to blood-red fls / Himalayas 139
artosquameum / Sh / to 8" / rose to purple fls / se Tibet 25
atlanticum / Sh / 3'+ / white fls, occ. pinkish / Del. - S. Car. 138 / 25
augustinii / HH Sh / to 10' / pinkish to mauve fls / w China 232 / 25
augustinii v. chasmanthum / 10' / lavender fls / Yunnan 55A
aureum - see R. xanthostephanum 25
auriculatum / Sh / 15' / white fls / China 139
auritum / Sh / to 10'/ white to rose-pink fls / Tibet 25
austrinum / Sh / to 10' / fls frag, yel-orange / Fla. & Ga. 25
baileyi / Sh / to 6' / fls deep red-purple / s Tibet 25
barbatum / HH Sh / 18' / crimson-scarlet fls / Nepal, Sikkim 139
'Blue Diamond' (augustinii X 'Intrifast')/bush to 3'/ lav-blue fls 139
'Blue Peter' / vigorous hybrid / fls cobalt-blue 139
'Boule de Neige' / compact / white fls 95
'Bow Bells' / compact / pink fls 139
brachyanthum / Sh / to 5' / pale yellow-green fls / Yunnan, Japan 139 / 25
brachyanthum v. hypolepidatum / 3' / yellow fls / China, Tibet 132
brachycarpum / Sh / to 9' / fls white or pinkish, grn spots / Japan 200
brachycarpum v. roseum / Sh / to 10' / fls pink / Japan 25
'Britannia' / compact / pinkish-scarlet fls 129 / 36
Bureavii / Sh / to 6' / fls rose or reddish / Yunnan 25
'Buzzard' / straw-yellow / Knap Hill azalea 139
caesium / Sh / to 4' / greenish-yellow fls / w Yunnan 25
calendulaceum / Sh / 12' / fls yellow to scarlet / e U.S. 110 / 139
californicum / Sh / 8'+ / fls rosy-purple / California & n-wards 28
callimorphum / HH Sh / to 10' / rose fls / w Yunnan 25
calophytum / Sh / 10'+ / fls white or pink / Szechwan 139
calostrotum / Sh / 4' / fls in rose shades / Upper Burma (570) / 36
calostrotum 'Gigha' - fls deep claret-red; gray green lvs 139
campylocarpum / Sh / 5'+ / clear yellow fls / Nepal 139
campylogynum / Sh / 12-18" / brownish-red fls / China 123
campylogynum v. cremastum / deep rose-purple / Yunnan 139
campylogynum v. myrtilloides / fls plum-purple / Upper Burma (564) / 139
camtschaticum / Sh / to 1' / red fls / Japan, Alaska 13 / 200
canadense / Sh / 3'+ / fls rose-purple / ne North America 22 / 139
canadense 'Albiflorum' - white fls 25
canescens / Sh / 10' / fls white flushed pink / se United States 139
'Caracticus' / 10'+ / dark purple fls 264
'Carmen' / dwarf or prostrate / dark crimson fls 133 / 139
carolinianum / Sh / to 6' / fls pale rose-purple / e U.S. (560) / 25
carolinianum 'Album' - white fls 25
catawbiense / Sh / to 10' / lilac-purple fls / se U.S. 93 / 25
caucasicum / HH Sh / 3' / pink or yellowish-white fls / Caucasus 25
cephalanthum / HH Sh / to 4' / white or pink tubular fls / s China 25
cephalanthum v. crebeflorum / 1-2' / pink fls / Assam 139
chaetomallum / Sh / 6-10' / waxy red fls / se Tibet 139
chameunum / Sh / dwarf / rosy-purple fls / w & sw China 139
Chapmanii / Sh / to 6' / pink fls / w Florida 139

charitopes / HH Sh / 1½-3' / pink fls / Burma		123
charitostreptum - see R. brachyanthum v. hypolepidotum		54
chartophyllum / HH Sh / 8'+ / pale purple fls / Yunnan	(565) /	25
chasmanthum - see R. augustini v. chasmanthum		139
chengshieanum - see R. ambiguum		
chrysanthum / Sh / prostrate/ yellow fls / n Asia	132 /	(266)
chryseum / Sh / 1½-2' / fls bright yellow / China	133 /	123
ciliatum / Sh / 3-4' / pink fls aging white / Nepal to Sikkim	129 /	36
X Cilpinense / 3' / fls white, flushed pink		139
cinnabarinum /Sh / to 6'/ corolla brick-cinnabar red /Sikkim, Himal.		25
commodum - see R. sulfureum		36
concatenans / HH Sh / 6' / apricot fls, tinged purplish / se Tibet		25
concinnum / HH Sh / to 7' / purple fls / Szechwan		25
concinnum v. pseudoyanthinum / deep ruby-red or purple-red fls		139
'Conewago' - lavender-pink fls		62
[1] 'Cream Crest' / 3' / yellow fls		
crebreflorum - see R. cephalanthum v. crebreflorum		139
cremastum - see R. campylogynum v. cremastum		139
cuneatum / Sh / 6'+ / rose-lilac fls / Yunnan		139
'Cunningham's White' / 3' / white fls / caucasicum x ponicum album		139
cyclium - see R. callimorphum		139
dalhousiae / Gh Sh / straggling / tubular white fls / ?		25
dasypetalum / Sh / 1' / fls rose-purple / Yunnan		132
dauricum / Sh / 6' / fls rosy-purple / Far East		36
dauricum v. albiflorum / fls white / Siberia to Japan		10
davidsonianum / HH Sh / to 10' / pink fls, red spotted / w Szechwan		25
'Day Dream' / Exbury Hybrid azalea / crimson buds, pink fls		139
decorum / Sh / to 20'/ fls frag, wht to soft rose / Szechwan,Yunnan		25
degronianum - see R. Metternichii v. pentamerum		200
diaprepes / Sh/ to 25'/ fls frag, wht or soft rose/Yunnan, ne Burma		25
dichroanthum / HH Sh / to 5' / yellowish or purplish fls / Yunnan	(565) /	25
dichroanthum v. scyphocalyx / coppery-orange fls / Burma		139
didymum - see R. sanguineum v. didymum		139
drumonium / Sh / to 2' / mauve fls / Yunnan		25
edgarianum / Sh / to 3' / blue-purple fls / w China		139
elaeagnoides - see R. lepidotum		139
'Elizabeth' / dwarf / dark red fls	129 /	36
'Exbury Isabella' (auriculatum v. griffithianum) / S or T/ wht fls		139
eximeum / T / to 30' / fls rose or pink / Bhutan		25
X 'Fabia' / domed bush / fls scarlet shaded orange		139
falconeri / HH Sh / 15' / creamy-yellow fls / Nepal, Bhutan		139
fargesii / Sh / 10'+ / frs rosy-lilac / w China		139
fastigiatum / Sh / to 3' / light purple fls / Yunnan		25
fauriei - see R. brachycarpum		200
ferrugineum / Sh / 3½' / rosy-crimson fls / European Alps	148 /	129
ferrugineum album - white fls		139
ferrugineum X hirsutum / Sh / 3-4' / fls usually rose / Eur. Alps		36
fimbriatum / Sh / 1-3' / fls purplish / Szechwan		139
flammeum / Sh / to 6' / fls orange to scarlet / S. Car. & Ga.		25
flavidum / Sh / to 3' / primrose-yellow fls / w China		139
flavidum ' Album' / 3' / lax habit / large white fls		139
Fletcherianum / Sh / to 3' / fls pale green-yellow / Yunnan		25

---

[1] Greer Gardens Catalog, 1973.

floccigerum / Sh / 3-5' / fls red or yellow / se Tibet 139
forrestii / Sh / 6" / red to crimson fls / China, Tibet 123
forrestii v. repens - lvs pale green beneath 139
fortunei / Sh / to 12' / fls lilac to pink / e China 25
glaucophyllum / Sh / 3-6' / lilac-rose fls / e Nepal to Sikkim 36 / 139
glaucophyllum v. luteiflorum / lemon-yellow fls / Burma 139
[1] glaucophyllum v. tubiforme - Ludlow & Sheriff coll.
glomerulatum / Sh / to 3' / fls purple-mauve / Yunnan 139
grande / HH Sh / 15' / ivory-white fls / Nepal, Sikkim 139
griersonianum / Sh / to 6' / strong red fls / w. Yunnan 129
gymnocarpum / Sh / to 3' / crimson fls / Yunnan 139
haematodes / Sh / 3'+ / scarlet-crimson fls / Yunnan 139
X halense - intermediate between parents: ferrugineum x hirsutum 36
hanceanun 'Nanum' / 1'+ / fls yellowish / sw Szechwan 18 / 139
heliolepis / Sh / to 10' / fls red to rose / Yunnan 25
hemitrichotum / Sh / 3'+ / white or pink fls / Yunnan 139
hippophaeoides / Sh / 3'+ / lavender-rose fls / Yunnan (571) / 139
hirsutum / Sh / 1-2' / rosy fls / c Europe 148 / 123
hirsutum aureovariagatum - varigated lf form of R. hirsutum 36
Hodgsonii / Sh or T / to 20' / lvs glossy above, cor. magenta/Himal. 25
hormophorum / HH Sh / to 3' / rose fls, brown marked / sw Szechwan 25
hypolepidotum - variety of R. brachyanthum 25
impeditum / Sh / 6-18" / mauve fls / China (267) / 123
imperator / Sh / 6" / pinkish-purple fls / Burma 123
insigne / Sh / to 12' / fls wht-pink, crimson spots/ sw Szechwan 25
X intermedium - see R. X halense 36
intricatum / Sh / 6-12" / lavender-blue fls / China 123
irroratum / Sh or T/to 25'/fls wht, tinged rose, crims. spots/Yunnan 25
japonicum / Sh / to 6' / vermillion to yellow fls / Japan 110 / 200
japonicum aureum / Sh / to 6' / fls yellow / Japan 222
Kaempferi / Sh / 3-9' / fls in red shades / Japan 139 / 25
Kaempferi album / Sh / to 3' / fls white / Japan 222
keiskei / Sh / 3'+ / pale greenish-yellow fls / Japan (268) / 200
keiskei cordifolia / Sh / to 3' / fls pale yellow-green / Honshu 200
keiskei 'Yaku Fairy' / Sh / prostrate / fls 1½", dull yellow/ Japan 36
keleticum / Sh / 6-12" / clear pink fls / se Tibet 36 / 123
kiusianum / Sh / to 3' / lilac-purple fls / Kyushu 110 / 139
kiusianum 'Album' - white fls 25
kotschyi / Sh / to 2' / rose or white fls / c Europe, in mts. (561) / 25
X Lady Chamberlain 'Salmon Trout' - salmon-pink fls 139
lanatum /Sh or T/to 15'/fls yel, spot. crims-purp/Assam,Sikkim,Tibet 25
lapponicum / Sh / to 20" / purple fls / n Europe, n North America 288
ledoides - see R. trichostomum 139
lepidostylum / Sh / to 1' / fls pale yellow / w Yunnan 25
lepidotum / Sh / to 3' / pink, purple, yellow fls / Szechwan, Tibet 139
lepidotum 'Album' - white fls 139
lepidotum 'Eleagnoides' - included in the sp. 36
leucaspis / Sh / 1-2' / solid white fls / Tibet 132 / 123
Lindleyi /Sh/often epiphytic to 15'/fls & frs. wht/Burma,Sik.,Tibet 25
linearifolium / Sh / to 4' / fls rose-lilac / Japan 25
Loderi / Sh / ? / fls frag, white or pink / parents from Himalayas 25

[1] Gibson, Jour. Royal Horticultural Soc., August 1967.

lowndesii / Sh / 6" / pale yellow fls / Nepal			132
ludlowii / Sh / dwarf / yellow fls / Tibet	100	/	139
lutescens / HH Sh / 10'+ / yellow fls, green spotted / w Szechwan	110		25
luteum / Sh / to 12' / fls fragrant yellow / Caucasus & Europe			25
macrophyllum / Sh / to 12' / fls rose to rose-purple / B.C. to Cal.	248	/	25
macrosepalum - variety of R. linearifolium			25
Maddenii / Gh Sh / white fls, faint rose flush / Himalayas			25
mageratum / Sh / to 2' / fls yellow or cream / Burma & e Himalayas			36
Makinoi / Sh / to 7' / fls pink, often spotted / w Yunnan, n Burma			25
maximum / Sh / 10'+ / fls purple-rose to white / e U.S.			139
mekongense / Sh / lvs more hairy than R. trichocladum			36
Metternichii / Sh / to 12' / rose-colored fls / Japan			200
Metternichii pentamerum / Sh / 6' / fls rose, 5-merous/mts. Honshu	200	/	36
micranthum / Sh / to 6' / white bell fls / China			139
microleucum / Sh / 18" / white fls / Yunnan			268
minus / Sh / to 12' / fls magenta-pink / Tenn., N. Carolina	(566)	/	25
molle - see R. japonicum			139
mollicomum / Sh / to 6' / fls rose or crimson / Yunnan			25
moupinense / HH Sh / 3' / white fls, spotted / w China	129	/	222
moupinense rubra - red form			139
mucronatum / Sh / to 6' / white fls / Japan	(567)	/	25
mucronulatum / Sh / 3-6' / fls rose-purple / Far East			200
myrtifolium / Sh / 20" / fls clear pink / Carpathians			288
X myrtifolium / R. hirsutum x minus / fls lilac-pink			25
myrtilloides - see R. campylogynum v. myrtilloides			25
nakaharai / Sh / to 3' / brick-red fls / Taiwan			(242)
neriiflorum / Sh / 6'+ / fls rose to crimson / Yunnan			139
nipponicum / Sh / 3-6' / white fls / Honshu, in mts.			200
nitens / Sh / to 3' / deep purple fls / ne Burma			139
nudiflorum - see R. periclymenoides	(568)	/	25
Nuttallii / Sh? / to 30', sometimes epiphytic / fls wht-yel / Bhutan			25
obtusum / HH Sh / to 3' / fls rose-red to purple / Japan	208	/	25
obtusum v. japonicum / Sh / 15" / rosy-purple fls			132
occidentale / Sh / 6'+ / white to pink fls / w North America	36	/	139
oleifolium / Sh / to 6' / fls pink or white / Yunnan			25
oreotrephes / HH Sh / 10'+ / mauvish fls / se Tibet, Yunnan			25
orthocladum / Sh / 3' / blue-purple fls / w China			139
parvifolium / Sh / to 3' / fls rose-purple / Japan, Korea, Siberia			200
patulum / Sh / flat / fls pink, sometimes spotted / Burma			36
pemakoense / Sh / 6-12" / pale purple fls / se Tibet	134	/	129
pentamerum - see R. Metternichii pentamerum			36
pentaphyllum v. nikoense / 9' / fls rose / Japan			200
periclymenoides / Sh / 9' / white, pink, purplish fls / e U.S.			25
'Pink Drift'/ dwarf / lavender-rose fls			139
'Pioneer' / 4' / lavender-rose fls			62
X 'P.J.M.' / 4' / lavender-pink fls			120
ponticum / Sh / 10'+ / mauve to lilac-pink fls / Asia Minor			139
poukhanense - see R. yedoense v. poukhanense			200
prinophyllum / Sh / to 8' / white to deep pink fls / e & c N. Am.			25
prostratum / Sh / 4" / pinkish-purple fls / China			123
prunifolium / Sh / to 10' fls apricot to orange / Ga. & Ala.			25
pseudochrysanthemum / Sh / dome / bell fls pink or wht, drk lines			139
pubescens / Sh / to 4' / rose or lighter fls / w China			25
'Puck' / 3' / pink fls			120

pumilum / Sh / prostrate / fls pink or rose / Sikkim 100 / 139
quinquefolium / HH Sh / to 25' / nodding white fls / Japan 129 / 25
racemosum / Sh / 5-6' / fls white or pink / China 269 / 129
racemosum 'Album' - white fls 132
racemosum 'Forrests Dwarf' - bright pink fls 139
X 'Racil' - blue-pink fls 139 / 268
radicans / Sh / 3-4" / solitary purple fls / Tibet 25
ravum / Sh / to 6' / rose to purple fls / Yunnan 25
repens - see R. forrestii v. repens 123
reticulatum / Sh / to 24' / fls rose-purple to magenta / Japan 222
roseum - see R. prinophyllum 25
X 'Rosy Bell' / to 6' / bell-shaped rose-pink fls 139
rupicola / Sh / 1-2' / fls dark purple-crimson / Yunnan 25
russatum / Sh / 2-4' / fls blue-purple / China 25
russatum v. cantabile / 2' / blue-purplish fls / Yunnan 132
saluenense / Sh / 12"+ / deep purple fls / Yunnan 132
sanguineum / Sh / 3'+ / bright crimson fls / Yunnan, se Tibet 93 / 123
sanguineum v. didymum / 2' / dk purple-red fls / Tibet 139
sargentianum / Sh / 1' / fls white or yellow / Szechwan 129
scabrifolium / Sh / small-med. / white to deep rose / Yunnan 139
schlippenbachii / Sh / 9'+ / fls rose-pink / Korea (568) / 139
scintillans / Sh / 1-3' / fls lavender-blue / Yunnan 139
'Scintillation' / Dexter Hybrid / deep pink fls
scyphocalyx - see R. dichroanthum v. scyphocalyx 139
searsiae / Sh / 5-8' / white, rose, mauve fls / Szechwan 139
setosum / Sh / to 4'/ cor. br. purp-pink; widely funnel form/ Himal. 25
smirnowii / Sh / 10' / rose-pink fls / Caucasus 139
souliei / HH Sh / to 15' / white fls, tinged pink / w China, Tibet 25
sperabile / Sh / to 6' / bell-shaped fleshy fls, red / upper Burma 139
spiciferum / HH Sh / 3' / pink fls / Yunnan 25
spinuliferum / Sh / to 8' / fls crimson to brick red / Yunnan 25
'Spinulosum' - bell-shaped, deep pink fls 139
sulfureum / Sh / 2-4'/fls clusters 4-8, br yellow/ nw Yunnan, Burma 36
tashiroi / Sh / fls 2-3, rose / Kyushu, Ryukyus & Formosa 200
telopeum /Sh / 5' / yellow fls / Yunnan, Tibet 139
tephropeplum / Sh / 4' / fls pink, carmine-rose / Tibet 139
Thomsonii /Sh or T/to 20'/lvs wht below;cor. red/Sik. Bhutan, Tibet 25
trichanthum / Sh / 10'+ / violet-purple fls / Szechwan 139
trichocladum / Sh / to 5' / fls yel, grn tinge / nw Yunnan, n Burma 36 / 25
trichostomum / Sh / 3-5' / fls white, pink, rose / Yunnan 139
trichostomum v. radinum - densely scaly corolla 132
triflorum v. mahoganii / fls mahogany blotch or suffused/ se Tibet 139
tsangpoense / Sh / 3' / fls pink, crimson, or violet / Burma, Tibet 139 / 25
Tschonoskii / Sh / 3' / white fls / Japan, Korea 200
Tschonoskii v. trinerve - slightly larger in all parts 200
tsusiophyllum / Sh / prostrate / fls white umbellate/ mts Honshu 200
uniflorum / Sh / to 1' / purple fls / s Tibet 25
vaseyi / Sh / 9' / rose-pink to white fls / North Carolina 269 / 139
venator / Sh / 6'+ / scarlet, red-orange fls / se Tibet 139
villosum - see R. trichanthum 139
virgatum / Sh / 4' / lilac-purple fls / Bhutan 139
viridescens / Sh / to 4' / fls pale yel-grn, spotted grn / Tibet 133 25
viscosum / Sh / 5' / fragrant white fls / e N. Am. 269 / 139
wadanum / Sh / 6-12' / fls rose-purple / Honshu, in mts. 200
wardii / Sh / 8' / clear yellow fls / Szechwan 129

wasonii / Sh / sm-med / fls wht-pink, crimson spots / w Szechwan 139
williamsianum / Sh / 4-5' / soft rosy-red fls / Szechwan 129 / 36
Weyrichii / Sh/ to 15'/ fls before lvs, brick-red/Japan, Cheju Is. 25
'Windbeam' - apricot fls fading white 62
'Winsome' - wavy-edged deep pink fls 139
'Wyanoke' - white fls 62
xanthostephanum / Sh / straggly / fls yellow / China 62
yakusimanum / Sh / 3' / white fls / Island of Yakushima 129
yakusimanum 'Exbury' / to 5'/ all white fls 36
yedoense / Sh / 3-5' / rosy-purple fls / from cult., Japan, Korea 139
yedoense v. poukhanense / 5' / purplish fls / Far East 200
yunnanense / Sh / to 12' / fls wht-pink, spotted red / Yunnan 25

RHODHYPOXIS - Amaryllidceae
baurii / HH Bb / 1-3" / rose-red fls / South Africa 185 / 123
baurii 'Margaret Rose' - bright pink fls 132
baurii v. platypetala - white fls (572) / 93
baurii v. rubella - see R. rubella 64
platypetala - see R. baurii v. platypetala 93
rubella / HH Bb / to 3" / fls bright pink / Drakensberg Mts., S. Afr. 185

RHODOPHIALA - Amaryllidaceae
andicola / HH Bb / to 10" / purple fls / Andes, Patagonia (271) / (270)
bifida / HH Bb / 12" / dark red fls / Argentina (139)
elwesii - see R. mendocina (270)
mendocina / HH Bb / to 1' / fls yellow / Argentina (271) / (270)
rhodolirion / HH Bb / to 8" / fls carmine-pink / Chile, Patagonia (270)
rosea - see HIPPEASTRUM roseum 25

RHODOTHAMNUS - Ericaceae
chamaecistus / Sh / 1' / pure pink fls / European Alps 148 / 107

RHODOTYPOS - Rosaceae
kerriodes - see R. scandens 278
scandens / Sh / 6' / fls white; shiny black frs / Far East 36 / 287
tetrapetala - see R. scandens 222

RHOPALOSTYLIS - Palmaceae
sapida / HH T / 25' / Feather-Duster Palm / New Zealand (573) / 25

RHUS - Anacardiaceae
aromatica / Sh / 3-5' / fls yellowish / e & s N. Am. 75 / 139
copallina / Sh / 15'+ / red frs & autumn lvs / e & c N. Am. 139
cotinus - see COTINUS coggyria 25
cotinus atropurpurea - see COTINUS coggyria f. purpureus 139
glabra / Sh / to 10' / scarlet frs / e North America 139
sylvestris / HH T / to 30' / frs brownish-yellow / Far East 222
trichocarpa / HH T / 25'+ / whitish frs / e Asia 167 / 25
trilobata / Sh / to 6' / fls grn, appear before lvs; fr red / Calif. 25
typhina / Sh or T / to 30' / crimson frs / e North America 184 / 129
typhina 'Laciniata' - cut-leaved form, tenderer & smaller 28

RHYNCHELYTRUM - Gramineae
repens / A Gr / 3' / rose-purple panicle / South Africa 25

RIBES - Saxifragaceae (Grossulariaceae)
aureum / Sh / to 6' / fls yellow; fr. black / Wash-Mont, s to Cal. 25
bracteosum / Sh / to 9' / greenish or purplish fls / Alaska, Cal. 222
cereum / Sh / 3' / fls wht or grn; frs bright red/ B.C., Mont., Cal. 131 / 25
lacustre / Sh / 3'+ / pale yellow to white fls / w North America 139
Lobbii / HH Sh / to 6' / fls purple-red; purple frs / B. C. to Cal. 222
odoratum / Sh / to 6' / fls frag, large, yellow / c-sw N. America 25
petraeum / Sh / to 6' / fls red or pink; fr dark red / mts Europe 25
punctatum / HH Sh / 3' / greenish-yellow fls / Chile 206
sanguineum / Sh / 12' / fls red; frs bluish-black / B.C. to Calif. 194 / 25

RICHEA - Epacridaceae
continensis / Sh / alpine / rosettes; fls creamy / N. S. Wales 184 / (123)
dracophylla / Gh Sh / 6-12" / white fls / Tasmania, in mts 194 / 268
scoparia / Sh / to 5' / fls white-pink; full branched / Tasmania 61

RICINUS - Euphorbiaceae
communis / A / 3-15' / great variety in leaf / tropical Africa 212 / 25

RICOTIA - Cruciferae
davisiana / P / lvs fleshy; fls lilac-pink / Turkey 77

RIGIDELLA - Iridaceae
flammea / HH Bb / 2' / bright orange-red fls / Mexico 279
orthantha / HH Bb / 2' / crimson fls / Mexico 206 / (243)

RIPOGONUM - Liliaceae
scandens / Sh / much branched; sm fls green-white / New Zealand 25

RIVINA - Phytolaccaceae
humilis / Gh / 6-24" / fls white; frs crimson / Florida, S. America 128

ROBINIA - Leguminosae
X ambigua 'Decaisnea' / T / pink-fls / Black Locust hybrid 25

RODGERSIA - Saxifragaceae
aesculifolia / P / 2½-6' / white fls; compound lvs / China 28
pinnata / P / 3' / rose fls / Yunnan 147 / 93
pinnata 'Alba' / 3-4' / creamy-white fl'd form / China 207
pinnata 'Superba' / to 4' / fls delicate rose; frs dark red 28
podophylla / P / to 5' / yellowish-white fls / Far East 147 / 25
sambucifolia / P / to 3' / white fls / Yunnan 93
tabularis / P / 3' / creamy-white fls; large lvs / n China, Korea 147 / 279

ROHDEA - Lilaceae
japonica / HH R / 4-8" / pale yellow fls; red frs / s Japan, China 166 / 200

ROMANZOFFIA - Hydrophyllaceae
californica / R / to 6" / creamy-white fls / nw N. Am. 123
sitchensis / R / to 10" / white fls / on wet cliffs, Ore., & Calif. 229 / 142
sitchensis suksdorfii - included in the sp. 142
Suksdorfii - see R. californica, if tuberous-rooted 142
Tracyi / R / to 4" / wht fls; glandular lvs / coastal bluffs, nw US 142
unalaschensis - see R. Tracyi 276

ROMNEYA - Papaveraceae
coulteri / HH Sh / to 8' / satiny white fls / California 269 / 129
trichocalyx - differs only slightly from R. coulteri 175

ROMULEA - Iridaceae
atroviolacea - see R. requienii 290
bulbocodium / C / 6" / bright violet fls / s Europe 212 / 267
bulbocodium 'Album' - white fls 267
bulbocodium v. leichtliniana - white or creamy fls 149
bulbocodium 'Pylia' - white fls marked purple 61
citrina / HH C / 5" / bright yellow, starry fls / Namaqualand 267
clusiana / C / 6" / violet-mauve fls / Spain 212 / 267
columnae / C / 6" / whitish or pale mauve fls / c Portugal (245) / 212
columnae ssp. rollii / lvs slender, filaments hairy/ e from France 290
crocea / C / 6" / bright golden-yellow fls / Asia Minor 267
cruciata v. parviflora / HH C / pale violet, streaked fls / S. Afr. 121
duthiae / HH C / 6" / fls white flushed lavender / Stellenbosch 121
flaveola - see R. columnae ssp. rollii 290
gaditana / C / 6" / violet-purple fls / sw Spain 212
grandiscapa / HH C / ? / purple & orange fls / n Africa 64
leipoldtii / C / 12-18" / orange-yellow with wht / South Africa 268
linaresii / C / 6" / violet & purple fls / France to Turkey 211
linaresii ssp. graeca / perianth segments acute / Greece, Aegean 290
longifolia / HH C / 6" / rose fls, yellow cup / S. Africa 185 / 61
longituba / HH C / 6" / yellow fls / South Africa 121
longituba v. alticola / 2' in flower / yellow fls / e S. Africa (243) / 185
lutea nigra - see R. longituba v. alticola 185
macowanii / HH C / ? / golden-yellow fls / mts. of e Cape Province 267
nivalis / HH C / 6" / yellow, white, lilac fls / Lebanon, Israel (272) / 185
parviflora - see R. cruciata v. parviflora (574) / 121
ramiflora / C / 6" / pale to deep lilac fls / w Medit. region 185
ramiflora gaditana / lvs slender; perianth lilac-pink/ Iberian Pen. 290
requienii / C / 5" / deep purple-violet fls / Corsica 267
rosea / HH C / 5" / deep pink fls / Cape Districts, S. Africa 267
rosea 'Tabularis' / to 1' / rose-violet fls / Cape Peninsula 121
sabulosa / HH C / 3" / bright cherry-red fls / Natal (444) / 111
saldanhensis / HH C / 4" / brilliant yellow fls / sw Cape 185
speciosa / HH C / 5" / deep carmine-pink fls / Cape of Good Hope 267
tabularis - see R. rosea 'Tabularis' 121
tempskyana / HH C / 2-4" / bluish-lilac fls / Medit. region 185
thodeli / HH C / 6" / deep violet fls / Drakensberg Mts., S. Africa 185
tortuosa / C / fls 2-3, small, bright yellow / Sutherland 121
zahnii / C / ? / white & gold & violet fls / Greece 64

ROSA - Rosaceae
acicularis / Sh / to 3' / dense, prickly; fls rose-pink / n hemis. 61 / 139
arkansana / Sh / 1½' / pink fls / c U.S. 222
arvensis / Sh / dense mounds or drapes / scentless white fls / Europe 139
blanda / Sh / 6' / pink fls; usually thornless / e N. Am. 25
X cantabrigiensis / 5' / fls pale yellow 129 / 139
carolina / Sh / 3-5' / fragrant rose-pink fls / e N. Am. 71 / 139
centifolia 'Cristata' / 3' / pink fls, crested petals 25
chinensis / HH Sh / 5'+ / crimson or pink fls / c China 139
chinensis 'Minima' / Sh / 1½' / rose-red fls 25

cristata - probably R. centifolia 'Cristata' *
Ecae / 3-5' / buttercup yellow fls / Afghanistan 139
Eglanteria / Sh / 3-6' / bright pink fls / Eurasia 25
Elegantula / Sh / to 5' / fls pink / w China 25
Farreri / Sh / 6' / pale pink or white fls / s Kansu, China 139
Farreri f. persetosa - see R. Elegantula 110 / 2
'Frau Dagmar Hastrop' - see R. rugosa 'Frau Dagmar Hastrop'
gallica / Sh / 5' / deep pink fls / c & s Europe 222
glauca - see R. rubrifolia 139
gymnocarpa / Sh / to 8' / rose fls, usually solitary / nw N. Am. 2
hirtula / Sh / large / pale pinkish fls / Honshu 167 / 20
holodenta - see R. moyesii f. rosea 221
horrida / Sh / low / one fl, white, 1" across / s Eur, Asia Minor 25
Hugonis / Sh / to 8' / bright yellow fls, solitary / China 208 / 2
Moschata / Sh / 10' / creamy-white fls / w Asia 208 / 13
Moyesii / Sh / 10' / blood-crimson fls / w China 269 / 139
Moyesii 'Geranium' / 8' / lighter green lvs; orange hips 129
Moyesii f. rosea - fls light pink 222
multiflora / Sh / 10' / clusters of wht fls; red frs / Japan, Korea 208 / 25
nitida / Sh / 1½' / fragrant pink fls; drk red frs / Nfld. to Conn. 25
nutkana / Sh / 8' / bright pink fls / w North America 139
omeiensis v. pteracantha / 10" / red prickles, white fls / China 28
palustris / Sh / 6' / pink fls / e North America 222
pendulina / Sh / 4' / magenta-pink fls / c & s Europe, in mts. 148 / 139
pendulina pyrenaica - included in the sp. 287
persetosa farreri - see R. Elegantula 139
pimpinellifolia - syn. R. spinossima 25
pisocarpa / Sh / 6' / pink fls / B. C. to Idaho 222
primula / Sh / to 6' / pale yellow fls / Asia 184 / 25
pteracantha - see R. omeiensis v. pteracantha 28
Roxburghii / Sh / 8' / bright pink fls / China, Japan 25
rubiginosa - see R. eglanteria 194 / 25
rubrifolia / Sh / to 6' / fls clear pink / c & n Europe 129 / 139
rugosa / Sh / 6' / pinkish-rose fls / ne Asia 208 / 139
rugosa 'Alba' - white fls 139
rugosa 'Frau Dagmar Hastrop' / 5' / fls pale rose-pink 129 / 139
rugosa 'Hansa' / 5' / double, deep crimson-purple fls 281
rugosa 'Rubra' - fls wine-crimson 184 / 139
rugosa 'Scabrosa' - lg fr; excellent folliage; viol-crimson fls 139
sericea / HH Sh / to 8' / white fls / Himalayas (575) / 25
sericea pteracantha - see R. omeiensis v. pteracantha 25
setigera / Sh / 6' / fls deep rose / c North America 28
spinosissima / Sh / to 3' / fls cream-white / Eurasia 184 / 25
spinosissima v. altaica / few prickles / white fls / Asia 269 / 25
spinosissima 'William III' / 2' / fls semi-double, lilac-crimson 278
Sweginzowii / Sh / 10'+ / rose-pink fls / nw China 139
ultramontana / Sh / 9-10' / fls small, rose-colored / e Ore., e Wash. 142
villosa / Sh / 4½' / pink fls / c & s Europe 287
virginiana / Sh / 4' / bright pink fls / e North America 110 / 139
Wichuraiana / Sh / trailing / white fls; small red frs / Japan 139
Woodsii ultramontana - see R. ultramontana 131
Woodsii v. fendleri / 6' / fls lilac-pink / w North America 129

ROSCOEA - Liliaceae

alpina / R / to 6"/ dk purple & white fls, wht tube/ Himalayas-Burma 182 / 25

capitata / R / to 16" / fls blue or purplish / Himalayas 25
cautleoides / R / to 1' / pale lemon-yellow fls / w China 93 / 129
cautleoides Beesii / P / 10-12: / lvs glossy; fls pale yel / China 61
cautleoides 'Grandiflora' - larger yellow fls 95
Humeana / R / to 8" / large violet-purple fls / w China 129 / 25
Humeana lutea / R / to 8" / fls yellow / China 61
procera / R / 8-12" / white fls, purple tipped / Himalayas 185
purpurea / R / 1'+ / purple fls / Kuamon, Sikkim 184 / 123
purpurea v. procera - see R. procera 185

ROSMARINUS - Labiatae
lavandulaceus / HH Sh / prostrate / blue fls / ? 139
officinalis / HH Sh / 2-4' / pale blue fls / circum-Medit. region 211 / 25
officinalis 'Prostratus' / HH Sh / 2" / pale violet fls 123
prostratus - see R. lavandulaceus 139

ROSULARIA - Crassulaceae
aizoon / R / rosettes / fls pale yellow / Armenia, Iran 77
chrysantha - see R. pallida 156
pallida / R / rosettes, white or yellow, red striped / Asia Minor 25
sempervivoides - see SEDUM sempervivoides 77
Sempervivum / R / flat rosette / deep pink fls / Caucasus, Iran 268
serrata / HH R / rosettes / purplish fls / Greek Isls., Asia Minor 77

RUBUS - Rosaceae
arcticus / R / 3-15" / fls pink; frs red / arctic Eurasia 210 / 25
calycinoides / HH Sh / creeping / white fls / Taiwan 139
illecebrosus / Short / white petals, red fr / Japan, nat. n Eur. 287
lasiocarpus - see R. niveus 222
leucodermis / Sh / to 6' / fls white; purple-black frs / nw US 222
niveus / HH Sh / to 6' / fls rosy-purple / India, w China 222
odoratus / Sh / 7' / fls purple-rose / e North America 139
parviflorus / Sh / 3½-7½' / white fls / w North America 139
parvus / HH Sh / 2' / white fls / New Zealand 139
pedatus / R / trailing, mat-forming / white fls / Alaska to Ore. 25 / 142
phoenicolasius / Sh / to 9' / clustered pale pink fls / Far East 139
saxatilis / P / 4-20" / petals small, white; stamens white/ Europe 287
spectabilis / Sh / 3-4' / bright magenta-rose fls / w N. America 139
X Tridel / Sh / to 9' / glistening white fls 139
ursinus / Sh / trailing or clambering / fls wht, pale pink/ nw N. A. 142

RUDBECKIA - Compositae
californica / P / to 5' / fl disc green, rays yel / Sierra Nev, Cal. 25
californica v. glauca / P / to 5'/ yellow fls / coastal Ore., Calif. 25
fulgida / P / to 3' / orange-yellow fls / e & se U.S. 225 / 25
fulgida 'Gloriosa Daisy' - see R. hirta 'Gloriosa Daisy' 268
fulgida 'Golden Flame' - see R. hirta 'Golden Flame' 268
fulgida v. speciosa - basal lvs entire 268 / 25
fulgida v. speciosa 'Goldsturm' / P / 2' / orange-yellow fls 300 / 268
fulgida v. sullivantii - uppermost lvs reduced to bracts 25
hirta / A or B / 1-3' / gold-yel fls, 'Black-eyed Susan' / N. Am. 224 / 25
hirta 'Gloriosa Daisy' / A / 3-3½' / gold & bronze shades 268
hirta 'Golden Flame' / A / 2' / golden-yellow fls, dark centers 268 / 282
laciniata / P / 1½-9' / yellow fls / e & c North America 99
maxima / P / 3-9' / ray fls yellow, 2" long / Mo.- La., Texas 25

speciosa - see R. fulgida v. speciosa 268
sullivantii 'Goldsturm' - see R. fulgida v. speciosa 'Goldsturm' *
triloba / B or P / to 4½' / yellow fls / e & c U.S. 224 / 25

RUELLIA - Acanthaceae
carolinensis / P / to 3' / fls / lavender to lilac-blue / se US 99
ciliosa - see R. carolinensis 99
humilis / P / to 2½' / fls lavender to bluish / se U.S. 224 / 99
humilis v. longiflora - fls about twice length of the type 99
pedunculata / P / to 30" / fls violet or purple / c U.S. 224 / 99
Purshiana / P / to 2' / fls violet or purple / se US 99
strepens / P / 3' / blue fls / sc U.S. 229 / 96

RULINGIA - Sterculiaceae
hermannifolia / Sh / sprawling / clusters tiny wht fls / N.S. Wales 109 / 61

RUPICAPNOS - Papaveraceae
africana / HH R / decumbent / white fls / sw Spain, nw Africa 273 / 286

RUMEX - Polygonaceae
angiocarpus / P / basal lobes, lvs divided; fls erect / Europe 286
lunaria / Sh / inflores. compound panicle / nat. Sicilia, Sardegna 286
salicifolius ssp. salicifolius - see R. angiocarpus 286
scutatus / P / 10-20" / helmet-shaped lvs / c & s Europe, in mts. 148

RUSCUS - Liliaceae
aculeatus / HH Sh / to 3' / green flattened stems; red frs / s Eu. 129 / 139
hypophyllum / Sh / to 1' / fls 5 orb, white; fr red / Madeira-Cauc. 25

RUTA - Rutaceae
bracteosa - see R. chalepensis 287
chalepensis / HH P / 8-24" / fringed yellow fls / s Europe 211 / 287
corsica / P / to 18" / pale yellow fls / Corsica, Sardinia, in mts 287
graveolens / P / 18" / yellow fls / Balkans 274 / 287
graveolens 'Jackmans Blue' / compact / glaucous blue lvs 129 / 139
macrophylla / Gh Sh / 3' / pinnate lvs / n Africa 81
montana / P / 6-28" / leaf-segments linear / sw Europe 287
patavina - see HAPLOPHYLLUM patavina 287

RYDBERGIA
grandiflora - see ACTINELLA grandiflora 123

SABATIA - Gentianaceae
angularis / A / to 3' / rose-pink fls, greenish eye / e U.S. 224 / 25
campestris / A / to 15" / fls solitary, lilac / Mo. & Kans,s to Tex. 25
kennedyana / R / 8-12" / fls 1-24, frag., roseate / Mass., R.I. 99
stellaris / A / to 2' / pink fls, yellow & red eye / coastal e U.S. 224 / 25

SABINA
chenensis - see JUNIPERUS chenensis 222
chenensis v. procumbens - see JUNIPERUS procumbens *

SAGINA - Caryophyllaceae
caespitosa / R / cushion / white fls / Iceland, Scandinavia, in mts. 194 / 286
glabra 'Aurea' / tiny creeping plant / golden lvs 184 / 123
nodosa / P / 1-6" / fls small, white / Lab. to Hudson Bay 99
pilifera aurea / P/creeps, to 2"/lvs stiff; fls yel /Corsica, Sard. 61
subulata / R / mat / white fls / w & c Europe 286

SAGITTARIA - Alismataceae
latifolia / Aq / 4' / white fls / North America 308 / 25
sagittifolia / Aq / 18" / European Arrowhead / Europe 25 / 301

SALICORNIA - Chenopodiaceae
australis - see S. quinquefolia
quinqueflora /P/?/no lvs;tiny fls;jointed stems/salt marsh,sw Austrl 109

SALIX - Salicaceae
arbuscula / Sh / to 6' / catkins in length 3X width / n Europe 286
arctica / Sh / 4", creeping / large catkins / arctic N. Am. 312 / 25
glauca / Sh / to 3' / catkins with the lvs / temperate zone 25
herbacea / Sh / glossy green lvs / arctic region, s in mts. 148 / 25
lanata / Sh / T / 1½-3' / yellow-gray woolly catkins / n Eurasia 139
lanata stuartii /prob. hyb of S. l. & S. lapponum/silver hair on lvs 36
mackenzieana / T / 15-30' / branches pale yellow-green / w N. Am. 139
myrsinites / Sh / procumbent / lvs shiny / n Europe 286
myrtilloides / Sh / to 20" / rounded lvs / n & c Europe 286
nivalis / Sh / tufted, creeping / dark green lvs / nw N. Am. (576) / 25
pedicillaris / Sh / to 3' / young lvs purplish / n Am. 99
reinii / Sh / thick brown branches; catkins slender, 1-2" / Japan 200
repens / Sh / creeping / small gray-green lvs / Eurasia 23 / 139
reticulata / Sh / dwarf / catkins slender / Arctic & Antarctic 148 / 25
retusa / Sh / low / cylindrical catkins / Europe, in mts. 184 / 25
sachalinensis / Sh or T / 12'+ / large catkins / ne Asia 139
saximontanta - see S. nivalis 25
uva-ursi / Sh / prostrate / catkins w. lvs / Alaska-Lab., s in mts. 25

SALPICHROA - Solanaceae
origanifolia /A or P/gr.cover/fls & frs wht-yel/ Argen.,esc. Cal.,Tx. 25

SALPIGLOSSIS - Solanaceae
sinuata / A / 1-3' / varii-colored fls / Chile 129

SALVIA - Labiatae
acetabulosa - see S. multicaulis 268
aethiopis / B or P / 1-3' / white fls / s & se Europe 212 / 288
amplexicaulis / P / to 32" / fls violet / Balkan Peninsula 288
argentea / B or P / to 3' / white fls / e Medit. region 210 / 268
azurea v. grandiflora / P / to 6' / large deep bl. fls / Minn.-Texas 25
beckeri / P / 1½' / blue & violet fls / Caucasus 207
blepharophylla / HH Sh / 1-1½' / fls crimson-red / Mexico 268
bulleyana / P / to 3' / yellow to purple fls / China 268
caespitosa / Sh / 3-6" / fls violet or blue / Anatolia (275) / 268
canariensis / Gh Sh / 3-6' / purplish fls / Canary Isls. 25
carnosa / Sh / 2' / blue fls / nw North America 268
chamaedryoides / HH R / 1' / blue fls / Mexico 206
chamaelaegna / Gh Sh / 3' / pale bluish-purple fls / South Africa 286

clevelandii / Gh Sh / 3' / blue fls / lower California 227 / 25
coccinea / P gr, as A / 2' / fls deep scarlet / Brazil 308 / 268
columbariae / A / 4-20" / blue fls / California, Utah, Arizona 229 / 25
[1] compacta / R / to 1' / pinkish-lavender fls / Turkey
Dorrii / Sh / 1-2½' / much branched/ fls blue / e Wash, Ida, s-Cal. 25
farinacea / P / 2-3' / fls violet-blue / New Mexico, Texas 226 / 25
forskaohlei / P / 10-40" / fls violet-blue / se Balkans 288
fruticosa / Sh / 2-4½' / fls lilac, pink, rare wht / Sicily-Syria 25
gesneriiflora / Gr Sh / 2-6' / fls scarlet, showy / Columbia 61
glutinosa / P / 3' / yellow fls / Europe to sw Asia 148 / 268
grahamii / HH Sh / 3-4½' / ruby-red fls / Mexico 129 / 268
grandiflora / Sh or P / 3'+ / lilac, pink, violet-blue fls / Balkans 288
greggii / Gh Sh / 3' / red or purplish-red fls / Texas, Mexico 308 / 25
haematodes - see S. pratensis 129 / 288
hians / P / 2-3' / sapphire-blue fls / Kashmir 194 / 25
hierosolymitana / HH P / 2'+ / blue-violet fls / Israel, Syria 25
hispanica / A / 2' / pale blue fls / Mexico to Peru 25
horminium / A / 2' / fls blue, white, purple / s Eu., sw Asia 268
indica / HH P / 4' / bluish fls / Syria, Iraq, Lebanon 268
japonica / P / 2' / dark lilac fls / Japan 200 / 268
judaica / P / to 3½' / fls violet / Lebanon, Israel 25
jurisicii / P / 12-16" / fls violet-blue to pink / Serbia 268
lavendulifolia / HH P / to 20" / bl. to violet-blue fls / Spain, Fr. 210 / 25
lemmonii / HH P / 10-30" / rose to crimson fls / s Arizona, Mexico 227
lyrata / P / 8-24" / fls violet / e United States 99 / 25
microphylla v. neurepia / HH P / to 4' / cherry-red fls / Mexico 268
moorcroftiana / P / 2' / violet fls / Himalayas 268
multicaulis / Sh / 1' / violet fls / sw Asia 268
nemerosa / P / 1-2' / violet-blue fls / c, se & s Europe 93 / 288
nemerosa superba - sterile hybrid 25
neurepia - see S. microphylla v. neurepia 268
nutans / P / 5'+ / violet fls / c Hungary, sc Russia 288
officinalis / Sh / 2' / violet, blue, pink, white fls / s & se Eu. 201 / 288
officinalis tricolour /Sh/ to 2'/fls viol-wht;lvs varieg/Spain-Asia 25
patens / HH P / 30" / fls bright, deep blue / Mexico 93 / 268
pitcheri - see S. azurea v. grandiflora 300 / 25
pratensis / P / to 3' / blue fls / Europe, including Britain 148 / 268
pratensis 'Rosea' - rosy-purple fls 28
ringens / P / 2' / blue or blue-violet fls / s & e Balkans 25
roemeriana / HH P / 1-2' / deep scarlet fls / Texas / Mexico 28
sclarea / B or P / to 4' / fls pinkish / Europe 229 / 268
sclarea v. turkestanica - white fls tinged pink 129
sonomemsis / P / 4-16, creeping / fls blue-violet / California 25
splendens / P gr. as A / to 3' / scarlet fls / Brazil 268
taraxifolia / HH P / 18" / fls pale pink / Morocco 268
tricolor / P gr as A / 2' / white & salmon-red fls / unknown origin 25
triloba - see S. fruticosa 25
verbascufolia / P / to 2' / white fls; lvs white-woolly / Caucasus 268
verticillata / P / to 3' / fls lilac or lilac-blue / Europe 268
verticillata 'Alba' - white fls *
virgata / P / to 40" / fls violet-blue / Italy, s & e Balkans 288

[1] Carl Worth in correspondence.

SAMBUCUS - Caprifoliaceae
caerulea / Sh or T / to 50' / fls yellowish-white / B. C. to Mont. 28
callicarpa / Sh / 10'+ / scarlet frs / Alaska to California 28
canadensis / Sh / to 8' / edible purple-black frs / e N. Am. 308 / 25
canadensis 'Acutiloba' / 10' / lvs much dissected 222
canadensis 'Laciniata' - see S. canadensis 'Acutiloba' 222
Ebulus / P / to 4' / fls white; fr small & black / Eur., n Afr., Asia 25
glauca - see S. caerulea 28
pubens / Sh / to 15' / inedible scarlet frs / e N. Am. 25
racemosa / Sh / to 10' / bright scarlet frs / Eu., w Asia 148 / 139
racemosa v. arborescens - see S. callicarpa *

SANDERSONIA - Liliaceae
aurantiaca / HH Tu / 2' / fls pale orange / Natal 186 / 267

SANGUINARIA - Papavaraceae
canadensis/ R / 6-9" / fls white / e North America 224 / 107

SANGUISORBA - Roseaceae
albiflora / P / 1-2½' / fls wht, often reddish / Japan 200
canadensis / P / 4-5' / whitish fls / North America 224 / 129
hakusanensis / P / 16-32" / deep rose-purple fls / Japan 200
minor / P / to 2½' / sepals green, pink margins / Europe 138 / 287
minor ssp. lasiocarpa / frs hairy / Turkey 77
obtusa / P / 12-20" / fls white / Honshu (520) / 200
officinalis / P / 2-4' / small red-brown heads / n. hemisphere 148 / 287
sitchensis / P / 1-3' / sepals greenish or pinkish / nw N. Am. 229 / 142
stipulata / see S. sitchensis 99

SANICULA - Umbelliferae
arctopoides / R / flat leaf-rosette; yellow fls / Calif. to Ore. 25
marilandica / P / 1½-4' / fls greenish-white / n North America 28

SANTOLINA - Compositae
chamaecyparissus / Sh / to 3½' / silver-gray lvs / s Europe 129
chamaecyparissus 'Nana' / dense Sh / under 1' / whitish lvs 276 / 123
elegans / Sh / 2-8" / gray tomentose; florets yel / s Spain 289
ericoides / HH Sh / 2' / creamy fls, gray-green lvs / s Europe 99
incana 'Nana' - see chamaecyparissus 'Nana' 123
neapolitana / Sh / dwarf / fls bright lemon-yellow / Italy (277) / 139
virens / Sh / 2' / lvs dark green, glabrous / Medit. region 110 / 25

SANVITALIA - Compositae
procumbens / A / 6" / yellow fls / Mexico 208 / 129

SAPINDUS - Sapindaceae
saponaria / HH T / to 30' / orange-brown frs / s Fla., W. Indies 28

SAPONARIA - Caryophyllaceae
bellidifolia / R / 8-16" / yellow fls / Balkans 148 / 286
X boissieri / mat / clear pink fls 61
'Bressingham Hybrid' / R / 3-4" / clear pink fls 123
caespitosa / R / dense / fls purplish / Pyrenees 148 / 286
calabrica / A / 6-12" / lvs lively rose-red / Italy, Greece 211 / 93
calabrica 'Compacta' - a dwarf form 93

glutinosa / A or Bi / 10-20" / fls purp / se Eur, e Spain 286
lutea / R / to 4" / fls yellow / s, w & c Alps 148 / 286
nana - see S. pumila 25
ocymoides / R / procument / fls pale purplish / s Europe 148 / 286
ocymoides 'Alba' - pure white fls 154
ocymoides 'Rubra Compacta' / dwarf / deep pink fls 129
ocymoides 'Splendens' - fls larger & darker 300 / 25
officinalis / P / 1-3' / fls flesh-pink / Eu., naturalized in U.S. 257 / 286
X Olivana - R / mat / bright pink fls 123
pamphylica / R / 6-12" / fls pink to purple / Turkey 77
pulvinaris - see S. pumilio 286
pumila / R / to 3" / fls rose or white / e Alps, se Carpathians 25
pumilio / R / densely caespitose / crimson or purple fls / w Turkey 148 / 25
X wiemannii / to 6" / pale rose fls 93

SARCOCAPNOS - Papaveraceae
baetica / R / mat / fls wht to yel; lvs varied / s Spain 286
crassifolia / R / tufted / pinkish fls / s & se Spain 212 / 286
enneaphylla / R / to 1' / fls white to yellow / Spain 212 / 286

SARCOCOCCA - Buxaceae
confusa / Sh / 3' / fragrant fls; black frs / origin unknown 184 / 139
hookerana / Sh / to 6' / fls tiny; fr black or purp 25
hookerana v. digyna / frs purplish-black / China 110 / 25
hookerana v. humilis / under 2' / bright pink anthers / China 25
humilis - see S. hookerana v. humilis 184 / 25
ruscifolia / Sh / 3' / dark red frs / c China 139

SARCOPOTERIUM - Rosaceae
spinosum / Sh / 2½' / green sepals, white rimmed / e Medit. region 149 / 77

SAROTHAMNUS
Scoparius - see CYTISUS scoparius

SARRACENIA - Sarraceniaceae
alabamensis / c Alabama pl / of the S. rubra group 242
alata / HH P / 10-40" / greenish-yellow fls / Ala. to Texas 242 / 116
drummondii - see S. leucophylla 242
flava / HH P / to 4' / yellowish-green fls / Va. to Fla. 225 / 25
jonesii - see S. rubra ssp. jonesii 242
leucophylla / HH P / to 1' / red-purple fls / Ga., Fla., Miss. 242 / 25
minor / HH P / 8-24" / pale yellow fls / N. Carolina to Florida 28
psittacina / HH P / to 1' /lvs decumbent; fls red-purp./ Ga., Miss. 242 / 25
purpurea / P / 12" / red-purple fls / e North America 128 / 107
purpurea v. venosa / P / to 2' / fls yellow / fla., Ala., n to Va. 99
rubra / HH P / 6-20" / red-purple fls / N. Carolina to Florida 28
rubra v. jonesii - 2'+ / bright red fls / Carolina mt. form 242

SATUREJA - Labiatae
alpina - see ACINOS alpina 28
arkansana / R / to 16" / lavender or purplish fls / e & c U.S. 99
calamintha - see CALAMINTHA officinalis 25
coerulea - see S. montana 'Coerulea' 129
Douglasii / R / trailing / fls white or purplish 228
gloryiana / Sh / to 2' / fls pink to lavender / N.C. s to Fla. 25

hortensis (Summer Savory) / A / 12-18"/ fls pale lav. to wht /Medit. 25
montana / R / 6-12" / white or pink fls / Medit. region 240 / 25
montana 'Coerulea' - fls deeper shade of lilac 129
montana ssp. illyrica / stems nearly glabrous / Albania, w Jugoslavia 288
rupestris - see MICROMERIA thymifolia 288
thumbra / Sh / 8-12"/ lvs gray; fls pink or purp / Agean, s Sardinia 288
thymifolia - see MICROMERIA thymifolia 288
vulgaris / P / 1-2' / fls red-purple / Europe 99
vulgaris var. neogaea / fls whitish to lilac-pink / North America 99

SAUSSUREA - Compositae
alpina / R / 1'+ / violet-blue fls / Alps 148 / 219
alpina ssp. depressa / R / to 4" / compact corymb / Alps 289
chionophylla / R / ½-4" / cobwebby lvs; fls purp, tiny / Japan 200
controversa / P / 10-32"/ lvs wht-tom. beneath; fls bl-viol/e Russia 289
depressa - see S. alpina ssp. depressa 289
discolor / R / 4" / purplish-blue fls / w Alps 148
gossipiphora / R / 9" / white shaggy lvs; fls hidden / Himalayas 25
obvallata / R / rosette / large pale yellow fls / Turkestan 64
pygmaea / R / to 8" / blue-violet fls / c Europe, in mts. 25
riederi v. insularis / 20-34" / purplish fls / Hokkaido 200
reideri yezoensis / P / stems leafy / lvs long; fls purplish / Japan 200
stella / R / 2" rosette / blue-purple fls / Tibet (279) / 123
yanigisawae / R / to 1' / purplish cobwebby fls / Hokkaido 200

[1] SAXIFRAGA - Saxifragaceae

1. Micranthes. Northern. Small horicultural importance.
2. Hirculus. Bog plants. Limited horticultural importance.
3. Robertsoniana. The "London Pride" group.
4. Miscopetalum. Small-flowered. Negligible horticultural value.
5. Cymbalaria. Annuals. Limited horticultural useage.
6. Tridactylites. Annuals, mostly. Little horticultural value.
7. Nephrophyllum. Meadow Saxifrages group. Of limited interest.
8. Dactyloides. Mossy Saxifrages group. Of considerable interest.
9. Trachyphyllum. Mat-forming. Of no great importance.
10. Xanthizoon. One northern sp. of negligible interest.
11. Euaizoon. Encrusted group. Of major interest.
12. Kabschia. Includes "Englerias." Of considerable interest.
13. Porphyrion. Mat-forming. Of definite importance.
14. Tetrameridium. Asian ssp. Mostly of unknown value.
15. Diptera. From Japan or China. Doubtful hardiness.

6 adscendens / B / to 10" / white fls / Europe, in mts. 286 / 286
aemula - see S. X borisii 'Aemula' (280)
10 aizoides / R / to 10" / yellow, orange, red fls / Europe 286 / 148
aizoides 'Atrorubens' /R/to 6"/ fls yel-red/ Eur,Asia,arc N.A. 219 25
† aizoides aurantia / R / 2-12" / loose mat / fls orng / Austria
aizoon - see S. paniculata 286

[1] Sections - left margin - as in Harding's Saxifrages, A. G. S. 1970.

† Fritz Kohlein, Saxifrages and Related Genera, 1980. Trans. David Winstanley, 1985.

	aizoon 'Alba' - see S. paniculata v. 'Alba'		*
	aizoon balcana - see S. paniculata v. balcana		96
	aizoon baldensis - see S. paniculata 'Minutifolia'		126
	aizoon brevifolia - see S. paniculata v. brevifolia		96
	aizoon correvoneana - see S. paniculata 'Correvoneana'		96
	aizoon eriophylla - see S. paniculata 'Eriophylla'		
	aizoon icelandica - see S. paniculata 'Icelandica'		
	aizoon lutea - see S. paniculata v. lutea		132
	aizoon minima - see S. paniculata v. minima		96
	aizoon minor - see S. paniculata v. minor		132
	aizoon 'Minutifolila see S. paniculata 'Minutifolia'		126
	aizoon notata - see S. paniculata 'Notata'		96
	aizoon pardoxa - see S. X paradoxa		96
	aizoon portae - see S. paniculata 'Portae'		*
	aizoon 'Rex' - see S. paniculata 'Rex'		132
	aizoon rosea - see S. paniculata v. rosea		132
	aizoon rosularia - see S. paniculata 'Rosularis'		96
	aizoon sturmiana - see S. paniculata 'Sturmiana'		96
	aizoon venetia - see S. paniculata 'Venetia'		*
	albertii - see S. Callosa 'Alertii'		*
12	X 'Amitie' / 2" / lilac fls / S lilacina x dalmatica		(280)
8	amoena / R / short stems / pink fls / Europe		132
12	andersonii / R / 2" / pink fls / Sikkim		132
3	X andrewsii / R / 6" / pink fls, spotted purple	182 /	96
8	androsacea / R / 3" / white fls / Europe, in mts	148 /	286
†	androsacea tridentata - very slight dif. from S. androsacea		
	appennina - see S. oppositifolia ssp. appinina		126
12	X apiculata / R / 2-4" / primrose-yellow fls	147 /	129
	X apiculata 'Alba' - white variant		123
8	'Apple Blossom' / 4-8" / pink fls		107
8	aquatica / P / 2' / wht. fls / e & c Pyrenees on stream margins		288
7	arachnoidea / R / decumbent / fls straw-colored / Alps		286
8	X arendsii / to 4" / fls rose & blood-red	93 /	25
	aretiastrum - see S. X boydii 'Aretiastrum'		(280)
12	aretioides / R / to 2½" / bright yellow fls / Pyrenees	286 /	148
	Arguta - see S. Nelsoniana		
	ascendens - see S. adscendens		*
9	aspera / R / to 8" / fls white or cream / Alps, Pyrenees		288
	aspera bryoides - see S. bryoides		268
	assimilis - see S. X salmonica 'Assimilis'		(280)
	autumnalis - see S. aizoides		286
	baldensis - see S. panniculata 'Minutifolia'		126
†	'Banshee' - white fls		
12	'Bellisant' / 2" / fls light rose		132
12	X 'Bertolonii' / mounds / short-stalked red fls		107
12	X 'Biasolettii' / small rosettes / blood-red nodding fls	107 /	(579)
1	X Biasolettii 'Crystalie' / 7" / crimson-red fls		
13	biflora / R / small / reddish-purple fls / Alps		126
7	biternata / HH R / procumbent / large white fls / s Spain	212 /	126

---

† Fritz Kohlein, Saxifrages and Related Genera, 1980. Trans. David Winstanley, 1985.

1 Harding, Quar. Bull. Alpine Garden Society, 37:2

	X 'Borisii' / S. Ferdinandi-Coburgii x S. marginata/to 3"/yel.		25
	boryi - see S. marginata		288
12	X boydii 'Aretiastrum' - pale yellow fls		(280)
†	X boydii 'Cherry Tress' /4-5" diam/ sev.lem-yel fls, 2-3" tall		
	X boydii 'Faldonside' - yellow fls		(280)
	X boydii 'Hinhead Seedling' - very light yellow fls		(280)
7	bracteata / R / 4"+ / white fls / arctic Asia & N. Am.	165 /	209
12	'Bridget' / 6" / fls pale pink, red calyces		24
9	bronchialis / R / 6" / pale yellow fls / n Eurasia & N. Am.	312 /	286
	bronchialis ssp. austromontana - entire lvs	(580) /	(281)
	bronchialis rebunshirensis - see S. cherleroides v. rebunsh.		200
	bronchialis 'Stelleriana' - denser pl, bronzed lvs in winter		96
2	Brunoniana / R / 4" / yellow fls; crimson runners/ Himalays	126 /	25
9	bryoides / R / 3" / fls yellow / European Alps	148 /	268
	bucklandii - see S. cuneifolia v. bucklandii		96
11	X burnatii / 6" / white fls		18
12	Burserana / R / 2" / white fls / e Alps	148 /	286
	Burserana 'Crenata' - compact, crenate petals	123 /	25
	Burserana 'Falstaff' - white fls		(280)
	Burserana 'Gloria' - large-flowered selection	129 /	126
	Burserana 'Grandiflora'/ R / to 4"/1 wht fl, lgr than S.B./Eur.		25
	Burserana 'Lutea' - refer. to S. Burserana 'Major Lutea'?		*
	Burserana 'Magna' / R / to 4" / petals white, large / Eur.		25
	Burserana 'Major' - fls larger	(581) /	25
	Burserana 'Major Lutea' - sulphur-yellow fls on red stems	123 /	126
	Burserana 'Minima' - diminutive fls		28
	Burserana 'Minor Crenata' - see S. Buserana 'Crenata'		123
	Burserana 'Sulphurea' - soft yellow fls		123
	Burserana 'Tridentina' - see S. Buserana 'Major'	(582) /	25
12	caesia / R / 4" / white fls / c & s Europe	148 /	286
	caespitosa / R / to 6" / fls tiny, white / N. A., n Eur., Asia	227	25
	caespitosa decipiens - see S. rosacea		288
	calabrica 'Tumbling Waters' - see S. X 'Tumbling Waters'		25
11	callosa / R / rosettes / fls white / w Medit. region	182 /	286
	callosa 'Albertii' / 12" / fls white (as S. lingulata cv.)		18
	callosa 'Albidia' - see S. callosa 'Albertii'		96
†	callosa v. australis / P / lvs linear / fl 10"/ c&s Italy		126
†	callosa ssp. bellardii / P/rosettes ineg/lvs 4"/fls wht/s Alps		
	callosa ssp. catalaunica / infl. glandular / ne Spain		286
	callosa 'Kathleen Pinsent' / 9" / rose-pink fls; silvery lvs		126
11	callosa v. lantaoscana / to 8" / dense white fls		132
	callosa 'Superba' / 12" / fls white / Maritime Alps form		18
8	camposii / HH R / 6" / white fls / se Spain		286
	canaliculata / P / 3-10" / color ?; lvs viscid / n Spain		286
11	X canis-dalmatica / 12" / sprays of white fls, spotted purple		123
2	cardiophylla / P / 2' / orange-yellow fls / Yunnan, Szechwan	126 /	64
	Careyana / R / 9-12" / fls white, yellow spots / Allegheny Mts.		99
	carinthiaca / a form of S. aizoon / small, neat cream fls		61
8	'Carnival' / to 6" / crimson-rose fls		126
1	caroliniana / R / 12"+ / white fls, yellow spots / Va., Tenn.		99
7	carpathica / R/ to 6" / white fls, tinged pink / sw Bulgaria		286

† Fritz Kohlein, Saxifrages and Related Genera, 1980. Trans. David Winstanley, 1985.

	cartilaginea - see S. paniculata ssp. cartilaginea		77
	cartilaginea kolenatiana - see S. kolenatiana		77
	caterhamensis - see S. cotyledon v. caterhamensis		132
8	cebennesis / R / 4" / fls white or cream / s France	123 /	286
	'Cecil Davis' / 6" / flat rosettes, white fls		132
7	cernua / R / to 10" / white fls / n N. Am., n Eurasia	148 /	99
	cervicornis - see S. pedemontana ssp. cervicornis		286
8	cespitosa / R / 3" / fls dull white / arctic Europe		286
	cespitosa 'Palmata' / 3-6" / fls white, variable habit		61
	cespitosa v. uniflora / 2½" / gr.-white fls, Jura Mts., Norway		148
9	cherlerioides / R / 2½" / creamy fls / Kamtchatka, Siberia		(281)
	cherlerioides var. rebunshirensis - 4" / whitish fls / Japan		200
2	chrysanta / R / 2" / yellow fls / Colorado, New Mexico, in mts.	(577) /	25
	chrysosplenifolia / very similar to S. rotundifolia		126
	X chrystalie - see S. X Biasolettii 'Chrystalie'		126
11	X Churchillii / 9" / fls white / e Alps		132
	clusii / R / to 1' / fls white / sw Europe		286
4	cochlearis / R / to 8" / fls white, red spotted / Maritime Alps	148 /	286
	cochlearis 'Major' - larger plant		132
	cochlearis 'Minor' - smaller plant	126 /	132
	cochlearis probynii - see S. cochlearis 'Minor'		132
8	conifera / R / 4" / white fls / n Spain	212 /	286
8	continentalis / R / 4-12" / fls white / n Spain, s France		212
8	corbariensis / HH R / 5-10" / pure white fls / e Pyrenees		126
	coriifolia - see S. rotundifolia v. coriifolia		77
†	X 'Corona' - form of S. Kabschia		
	correvoniana - see S. paniculata 'Correvoniana'		61
	cortusifolia / P / to 1½' / fls white / e Asia		25
†	cortusifolia v. fortunei 'Rubrifolia'/lvs & stems red;fls wht		
	corybifera - see S. luteoviridis		
	corymbosa - see S. luteoviridis	(192) /	286
	corymbosa luteoviridis - see S. luteoviridis		268
11	cotyledon / R / to 20" / white fls / Alps, Pyrenees, Iceland	148 /	286
	cotyledon v. caterhamensis / fls heavily spotted red / Norway		132
	cotyledon 'Icelandica' / to 4' / bronzy, leather lvs		25
	cotyledon 'Montefoniensis' - stocky, red-stemmed plant		96
	cotyledon v. norvegica / 18" / red stems / Norway	(583) /	132
	cotyledon 'Pyramidalis' - broadly conical inflorescence	(584) /	126
	cotyledon 'Southside Seedling' / 18" / white fls spotted red	(419) /	18
12	'Cranbourne' / 1½" / large rose fls	(585) /	123
11	crustata / R / 1' / fls white / e Alps, Jugoslavia	148 /	286
	crustata vochinensis - included in the sp.		25
8	cuneata / R / 6" / fls white / w Pyrenees, n Spain	212 /	286
3	cuneifolia / R / 6" / fls white / s & sc Europe, in mts.	148 /	286
	cuneifolia v. bucklandii - larger plant		96
	cuneifolia v. capillipes / smaller plant / fls speckled purple		126
	cuneifolia v. subintegra / larger / bluntly toothed lvs		126
5	cymbalaria / A / decumbent / fls yellow / sw Asia, n Africa		286
	cymbalaria ssp. huetiana - smaller fls		286
	dalmatica - see Farrer's reference to S. dalmatica		96
1	davurica / R / 7" / small white fls / Siberia, N. Am.		126

---

† Fritz Kohlein, Saxifrages and Related Genera, 1980. Trans. David Winstanley, 1985.

	decipiens - see S. rosacea		286
	decipiens alba - see S. rosacea		286
	decipiens 'James Bremner' - see S. X 'James Bremner'		126
	decipiens sternberii - see S. rosacea 'Sternbergii'		96
	demissa - see S. mutata ssp. demissa		286
8	demnatensis / R / 6" / white fls / Atlas Mts.	64 /	77
12	diapensioides / R / deep, rigid cushion / white fls / sw Alps	148 /	286
	diapensioides 'Lutea' - yellow fls		25
	diapensioides primulina - see S. diapensioides f. lutea		132
	diapensioides primuloidea - see S. diapensioides f. lutea		132
2	diversifolia / R / 10" / golden-yellow fls / Burma, China	182 /	126
	dolomitica - see S. hostii		286
11	'Dr. Ramsey' / 9" / white fls; lvs spoon shaped		126
	'Dulcimer' - see S. petraschii 'Dulcimer'		(280)
†	'Elegantissima' - dense low-growing cushions		
	X 'Elizabethae' / R / blue-grn mat/ fl stem to 3"; petals yel.		25
12	X Elizabethae 'Jason' - pale yellow fls		(280)
†	X Elizabethae 'L.C. Godseff' / round bl-grn hummocks/yel fls 2"		
	X Elizabethae 'Millstream Cream' - fls cream to near yellow		(280)
11	X engleri / R / 12" / white fls / Carpathian & Venetian Alps		132
8	erioblasta / R / compact cushion / white fls / se Spain in mts.		286
11	'Esther' / 6" / light yellow fls		132
8	exarata / R / 4" / fls white to pale cream / Alps, Balkans	148 /	286
7	exilis / R / tuft / large white fls; ivy lvs / Siberia		64
	'Eudoxiana' / 2" / fls light yellow		123
8	'Fairy' / 3" / white fls, late-flowering		126
	'Faldonside' - see S. X Boydii 'Faldonside'	(586) /	(280)
	'Falstaff' - see S. Burserana 'Falstaff'		(280)
12	Ferdinandi-Coburgii / R / 2-3" / fls bright yellow / Bulgaria	129 /	126
	Ferdinandi-Coburgii v. Pravislawii - superior f., Rhodope Mts.		126
	Ferdinandi-Coburgii v. Radoslavovii - fls somewhat darker		93
†	Ferdinandi-Coburgii v. rhodopaea - darker lvs than type		
	ferruginea / R / 4-18" / fls white / Calif. to Alaska, Rockies		288
2	flagellaris / R / 3-6" / yellow fls / Europe, N. America	312 /	25
	flagellaris ssp. komarovii / Himalayan form S. flagellaris		126
	flagellaris platypetala - see S. platypetala		286
†	X fleischeri Bucholzii /R / 2" / yel flowered Engleria hyb.		
11	florulenta / P / 18" / flesh fls / Maritime Alps	148 /	126
	'Flowers of Sulphur' / 4" / creamy-yellow fls		46
	forsteri / R / loose mats, stems 2" / fls yel / n Tyrol		61
15	fortunei / R / 9" / cream-white fls / Far East	194 /	46
	fortunei v. incisilobata / cut-lobed / predominant in Japan		200
11	'Francis Cade' / 9" / pure white fls on reddish stems		126
†	X franzii - see S. X paulinae 'Franzii'		
	frederiki-augusti - see S. porophylla		25
	fusca v. kikubuki /P/to 12"/fls red, grn-brn; lvs pubes./Japan		25
	gaetneri - see S. mariae-theresiae 'Gaetneri'		(280)
	X gaudinii / R / 1' / white fls / spotted red		96
	'G. F. Wilson' / mossy hybrid / brilliant carmine fls		6
	gemmiferae - see S. hypnoides 'Kingii'		28
8	geranioides / R / 8" / white, scented fls / e Pyrenees	148 /	286

† Fritz Kohlein, Saxifrages and Related Genera, 1980. Trans. David Winstanley, 1985.

	geranioides corbariensis - see S. corbariensis		*
3	X Geum - intermediate of parents: S. hirsuta x umbrosa		25
	X Geum dentata - see S. hirsuta		28
	X Geum hirsuta - see S. hirsuta		286
	glaucescens - see S. granulata		286
8	globulifera / HH R / to 5" / small white fls / s Spain	212 /	126
8	'Gnome' / to 4" / crimson fls, late flowering		126
12	'Gold Dust' - yellow fls		48
7	granulata / P / to 20" / white fls / n, c & w Europe	148 /	286
	granulata 'Flore Pleno' - fls double		25
	'Guildford Seedling' / 3-4" / fls deep red		61
12	Grisebachii / R / 9" / pinkish fls / Greece, Albania	129 /	286
	Grisebachii 'Alba' - albino form		*
	Grisebachii 'Wisley Var.' / larger rosettes / stem lvs red	132 /	24
	groenlandica - see S. cespitosa v. uniflora		148
12	X Gusmussii / R / 4" / reddish-orange fls		132
12	X Haagii / R / 2-3" / golden-yellow fls	123 /	126
8	hartii / R / to 2" / pure white fls; sepals reddish / nw Eire		126
1	hieracifolia / R / 4-16" / gr-purple fls / arctic Eu., N. Am.	148 /	286
	'Hinhead Seedling' - see S. Boydii 'Hinhead Seedling'		(280)
2	hirculus / R / 8"+ / bright yellow fls / n, c & e Europe	312 /	286
11	hirsuta / R / caespitose / fls white / Pyrenees		286
	hirsuta ssp. hirsuta / P/lvs hairy;fls pale pnk/Pyren., sw Ire.		286
	hirsuta ssp. paucicrenata / dwarf / on limestone / Pyrenees		286
11	Hostii / R / to 1' / white fls, may be spotted red / s Alps	148 /	126
	Hostii v. altissima - fl stems to 18"		126
	Hostii ssp. dolomatica / P / 10-24"/fls pale, red spots /e Alps		286
	Hostii ssp. rhaetica / narrow lvs / Italian Alps		126
	huetiana - see S. cymbalaria ssp. huetiana		286
8	hypnoides / R / to 6" / fls white / Europe	148 /	25
	hypnoides 'Densa' / 4-5" / fls creamy-white, dense growth		61
	hypnoides 'Kingii' / dwarfed / red-tinted lvs in fall		126
	hypnoides 'Whitlayi Compacta' - carpet of white fls		46
	incrustata - see S. crustata		
	X intermedia / red fls / like S. griesbachii, one parent		29
	intricata - see S. nervosa		
12	'Iris Pritchard' / 2" / buff-apricot fls	(585) /	126
7	irrigua / R / to 1' / fls white / Crimea, in mts.		286
12	X Irvingii / 1½" / pale pink fls	132 /	123
	X Irvingii 'Jenkinsiae' - fls light rose		(280)
12	X jaeggeana / 2-3" / white fls		96
8	X 'James Bremner' / 8-10" / huge white fls		126
	'Jason' - see S. X Elizabethae 'Jason'		(280)
	X Jenkinsae - see S. Irvingii 'Jenkinsae'	129 /	(280)
12	juniperifolia / R / caespitose / yellow fls / Caucasus		77
	juniperifolia ssp. sancta /R/to 3"/fls yel,nearly glab/Bulgaria		25
	'Kathleen Pinsent' - see S. callosa 'Kathleen Pinsent'		126
	X kingii - see S. hypnoides 'Kingii'		126
†	'Kingscote White' / handsome cush / lg wht fls, 4" high		
11	Kolenatiana / R / 6-9" / purple fls / Caucasus		77
	Kolenatiana 'Major' - see S. paniculata 'Major'	(549) /	25

---

† Fritz Kohlein, Saxifrages and Related Genera, 1980. Trans. David Winstanley, 1985.

12	X Kyrillii / 4-5" / fls clear citron-yellow		96
	laciniata / R / to 4" / white fls / Hokkaido, Korea		200
12	laevis / R / short shoots / yellow fls / Caucasus		96
	lantoscana - see S. callosa		138
7	latepetiolata / B / 1' / white fls / e Spain	212 /	126
	latina - see S. oppositifolia ssp. latini		
	leucanthemifolia - see S. michauxii		99
12	lilacina / R / flat mat / lilac fls / Himalayas	306 /	126
	lingulata - see S. callosa		286
	lingulata albertii - see S. callosa 'Albertii'	(549) /	18
	lingulata albida - see S. callosa 'Albertii'		96
	lingulata catalaunica - see S. callosa ssp. catalaunica		286
	lingulata lantoscana - included in S. callosa		286
	lingulata superba - see S. callosa 'Superba'		18
†	loeflingi (syn. S. grata loeflingi) - dp yel fls on 2" scapes		
8	X 'Lloyd Edwards' - mossy hybrid, fine foliage		249
11	longifolia / R / 18" / starry white fls / Pyrenees	148 /	129
	longifolia 'Francis Cade' - silvery rosettes		123
	longifolia 'Tumbling Waters' - see S. X 'Tumbling Waters'		126
12	luteoviridis / R / 4" / fls pale greenish-yellow / Bulgaria		286
11	X Macnabiana / 9" / white fls, heavily spotted	182 /	18
8	maderensis / Gh / 6" / white fls / Madeira		126
1	manshuriensis / R / 1'+ / soft pink fls / Far East	126 /	93
12	marginata / R / 3½" / fls white or pale pink / Italy, Balkans	50 /	286
	marginata v. coriophylla - column of lvs		286
	marginata v. coriophylla minor - see S. marginata 'Minor'		(280)
†	marginata 'Intermedia' / orig. unknown/ sm cushions, wht fls		
	marginata v. Karadzicensis / 1" / shiny white fls		132
	marginata 'Millstream'/ a Kabschia hyb.; Lincoln Foster, dev.		249
	marginata 'Minor' - white fls		(280)
†	marginata 'Purpur' - white fls; reddish buds		
	marginata v. Rocheliana - flat rosettes	126 /	286
	marginata v. Rocheliana 'Lutea' - pale yellow fls		132
12	X mariae-theresiae 'Gaertneri' - rose fls		(280)
12	media / R / to 6" / fls purplish-pink / Pyrenees	148 /	286
12	X 'Megasaeflora' - rose-pink fls, large petals	(418) /	126
	melaleuca / R / 6" / white or purplish fls / China		65
	Merkii / R / 2½" / white or creamy fls / e Asia		209
1	Mertensiana / R / ½-2" / white fls / Calif., Alas., Rocky Mts.	229 /	228
	Mertensiana bulbifera - see S. cernua		286
1	Michauxii / R / to 20" / white fls / Va., Ga., Tenn.	60 /	99
	'Millstream Cream' - see S. Elizabethae 'Millstream Cream'		(280)
	minima - see S. paniculata v. minima		96
	minutifolia - see S. paniculata 'Minutifolia'		*
	montavoniensis - see S. cotyledon 'Montafoniensis'		96
	montenegrina - see S. Grisebachii	28 /	286
	montenegrina alba - see S. Grisebachii 'Alba'		*
8	moschata / R / to 4" / varii-colored fls / s & c Europe	148 /	286
	moschata f. acaulis - densely tufted		61
	moschata 'Alba' - selection for white fls		*
	moschata f. compacta - densely caespitose		77

† Fritz Kohlein, Saxifrages and Related Genera, 1980. Trans. David Winstanley, 1985.

	moschata f. intermedia - dense-growing		61
	moschata v. laxa - loose mats		126
	moschata f. pygmaea - selection for compactness		126
	moschata 'Rhei' / 6" / large rose-colored fls		28
	moschata 'Rosea' - selection for pink fls		*
8	'Mrs. E. Piper' /R /mossy hybrid /red fls; 2-4" stems/Cultivar	107 /	126
	murithiana - see S. oppositifolia ssp. glanduliera		126
8	muscoides / R / mats / fls white or yellow / Alps		126
	muscoides rhei - see S. moschata 'Rhei'		28
11	mutata / R / to 20" / orange fls / Alps	148 /	286
	mutata ssp. demissa / 5" / spreading inflorescence		126
12	X 'Myra' - deep pink fls; raised by R. Farrer		126
-	nathorstii / R / to 2" / fls pale rose to purp. / e Greenland		209
1	Nelsoniana / P / to 18" / sm. wht - pale pnk fls / n hemisphere		25
	neogana - see S. paniculata 'Neugana'		249
	nervosa / R / 1-4" / fls wht; lvs scented / c Pyrenees		126
8	nevadensis / HH R / 3" / white fls, occ. red-veined / s Spain		126
	nipponica / similar to S. rufescens / Japan		126
	nishidae - differs slightly from S. bronchialis		126
1	nivalis / R / to 8" / fls white or pink / arctic Eu. & N. Am.	148 /	286
	norvegica - see S. cotyledon v. norvegica		132
12	X Obristii / 4" / ivory-white fls		25
1	occidentalis v. rufidula / R / to 1' / purplish fls, nw N. Am.		142
12	X ochroleuca - pale yellow fls		64
13	oppositifolia / R / mat / fls pink to purple / Europe, N. Am.	148 /	286
	oppositifolia v. alba - small starry white fls		132
	oppositifolia apennina / red fls / compact, s Italy		126
	oppositifolia ssp. arcto-alpina - tight mat		219
	oppositifolia ssp. glandulifera - purple, starry fls		126
	oppositifolia 'Grandiflora' - large-flowered selection		160
	oppositifolia ssp. latina / clear rose-pink fls / Apennines	123 /	126
	oppositifolia Nathorstii / R / many stems/fls sm purp to wht		61
	oppositifolia ssp. pyrenaica - large crimson fls		28
	oppositifolia ssp. rudolphiana / rose-purple fls /compact /Alps		126
	oppositifolia 'Ruth Draper' - large flowered	(418) /	(79)
	oppositifolia 'Splendens' - large purplish-crimson fls	129 /	25
	oppositifolia 'Wetterhorn' / rich crimson fls / Bernese Oberl.		132
1	oregana / P / 1½-3' / white fls / w North America	229 /	142
	pallida / R / basal rosette / to 3" / fls wht / Himalayas	126 /	61
11	paniculata / R / to 1' / fls white to pale cream / c & s Europe	212 /	286
	paniculata aizoon - see S. paniculata		
	paniculata 'Alba' - white fls		25
	paniculata 'Balcana' / 5" / fls heavily dotted crimson		96
	paniculata 'Baldensis' - ashy-gray lvs, short & thick		25
	paniculata 'Brevifolia' - white fls		25
	paniculata ssp. Cartilaginea / acuminate lvs / Caucasus, Iran	286 /	77
	paniculata ssp. Cartilaginea v. kolenatiana-see S. kolenatiana		77
	paniculata 'Correvoniana' / sm. rosettes/ 3"/ sm. dull wht fls		61
	paniculata 'Cristata' /lvs narrow, silvery; petals cream / N.A.	119 /	25
	paniculata v. hirtella / 6"/ cream fls, few red spots / Corsica		132
	paniculata 'Icelandica' - see S. cotyledon 'Icelandica'		25
	paniculata ssp. kolenatiana - see S. Kolenatiana		77
	paniculata 'Lagraveana' / 6" / waxy fls, creamy		25
	paniculata 'Lutea' - fls yellow		25
	paniculata 'Major' - oblong-linear lvs		25

	paniculata 'Minima' - small white fls	25
	paniculata 'Minor' - see S. paniculata 'Minima'	25
	paniculata 'Minutifolia' / 2" / creamy-white fls / Mt. Baldo	126
	paniculata 'Neugana' - fls rose-colored	249
	paniculata notata - see S. cotyledon	
†	paniculata 'Orientalis' /8-10"/med rosettes,cr-wht fls,red spots	
†	paniculata orientalis 'Orjen' - local form of S. paniculata 'O'	
	paniculata portae - see S. X portae	132
	paniculata 'Rex' - silvery lvs, red stems, creamy-white fls	123
	paniculata 'Rosea' - fls bright pink	25
	paniculata 'Rosularis' - fls white; lvs incurved	25
†	paniculata sturmiana / R /sm rosette/ser. lvs;fls wht / c Alps	
	paniculata 'Venetia' / flat mat / pure white fls on 2" stems	126
9	paradoxa / R / to 6" / pale green fls / se Alps	286
2	paradanthina / R / to 1' / deep orange fls / sw China	126
†	X patens / nat hyb. of S. aizoides x S. caesia / Alps	
†	X paulinae 'Franzii' - up to 6 pale yel fls per 2" scape	
8	'Pearly King' / mossy hybrid / white fls	107
11	X pectinata / 6" / creamy fls / natural hyb. in Carniola	126
8	pedemontana / R / to 6" / white fls / sw Alps	<u>148</u> / 93
	pedemontana ssp. cervicornis /lvs tiny, not glandular / Corsica	286
	pedemontana cymosa / R / 2-4"/ lvs thin; fls white / e Carpath.	126
	pedemontana ssp. prostii / rosette lvs not incurved / s France	286
	pensylvanica / P / 2' / yellowish-white fls / Maine to Minn.	280 / 99
	pentadactylis / R/ to 3"/ fls starry, wht;lvs aromatic / e Pyr.	126 / 25
	perdurans - see S. Wahlenbergii	(578) / 286
8	'Peter Pan' / mossy hybrid / 2-4" / crimson fls	<u>129</u> / 126
	petraea / R / 4-8" / white fls / e Alps	286
	X Petraschii / S. marginata x tombeanensis / white fls	107
12	X Petraschii 'Dulcimer' - white fls	280
1	petrophila / R / 3" / by inference, yellow fls / Himalayas	64
8	'Pixie' / mossy hybrid / red fls	107
	platypetala - European plant of S. flagellaris	286
	platysepala - see S. flagellaris	
	Polodiae - listed name of no botanical standing	25
12	porophylla / R / 3" / purplish-pink fls / c & s Italy, Greece	<u>148</u> / 286
	porophylla montenegrina / P / 4-6" / fls purp-pk / Balkan Pen.	286
	porophylla sibthorpiana - see S. sempervivum	(192) / 286
	porophylla thessalica - see S. sempervivum	(192) / 126
11	X portae / 6" / creamy fls	132
8	pratermissa / R / to 5" / white fls / Pyrenees	286
12	pravislavia / R / 2" / deep yellow fls / Macedonia	132
10	X primulaize / 3" / red, salmon or carmine fls	123
	primuloides - see S. X urbium 'Primuloides'	129
	primuloides 'Elliotts Var.' - see S. X urbium 'Clarence Elliot'	123
12	probynii - see S. cochlearis 'Minor'	132
	X Prosenii / 2-3" / fls reddish-orange	123
	prostii - see S. pedemontana ssp. prostii	286
	'P.S. Borisii' - see S. marginata	286

---

† Fritz Kohlein, Saxifrages and Related Genera, 1980. Trans. David Winstanley, 1985.

3X11	X pseudo-forsteri / 3" / green rosettes, white fls			61
8	pubescens / R / 1-4" / white fls / Pyrenees			25
	pubescens ssp. iratiana / 2"+ / petals veined red / Pyrenees			286
	pulchella - see S. hypnoides			64
	punctata - see S. Nelsoniana	228	/	25
12	X Pungens / 2" / yellow fls			96
	pyrenaica - see S. oppositifolia ssp. pyrenaica			28
	radiata / R / 2-16" / white fls / arctic Asia, N. Am.			209
	radoslowowii - see S. Ferdinandi-Coburgii v. Radislavovii			93
1	reflexa / R / 3-18" / spotted white fls / nw North America			209
13	retusa / R / mat / purplish-red fls / Pyrenees to Bulgaria	148	/	286
1	rhomboidea / R / 4-8" / white fls / nw U.S.	229	/	286
8	rigor / R / 2½" / fls white, campanulate / se Spain			286
	'Riverslea' (S. lilacina & S. porophylla)/R/small; fls wine red			126
7	rivularis / R / 3" / white or purplish fls / circumboreal	22	/	25
	rocheliana - see S. marginata v. rocheliana			286
8	rosacea / R / to 10" / white to red fls / nw & c Europe			25
	rosacea 'Alba' - white fls			*
	rosacea ssp. sponhemica - leaf segments sharply tipped			126
	rosacea 'Sternbergii' / solid massed plant / cream-white fls			96
	'Rosaleen' - see S. X salmonica 'Rosaleen'			(280)
†	'Rosemarie' / magnificent cush. / 1-2 lg rose-pink fls			
	'Rosinae' / R / cush.forming /2" /fls, wht / cult. orig unknown			61
4	rotundifolia / R / to 16" / white fls, spotted / c & s Europe	148	/	286
	rotundifolia v. coriifolia - leathery lvs			77
	rudolphiana - see S. oppositifolis ssp. rudolphiana			126
1	sachalinensis / R / 4-16" / white fls / Hokkaido, Kuriles	165	/	200
12	X salmonica 'Assimilis' - white fls			(280)
	X salmonica 'Rosaleen' - white fls			(280)
†	X salmonica 'Salomonii' - form of Kabschia			
†	X salmonica 'Schreineri' /lax cush to 6" diam /wht fls 2" high			
12	sancta / R / 1½" / yellow fls / Greece, Turkey	93	/	77
8	X 'Sanguinea Superba' / 9" / deep crimson fls			64
	sarmentosa - see S. stolonifera			286
12	scardica / R / to 4" / white or rose fls / Medit. region	126	/	25
	scardica 'Erythrantha' - pink or red form			126
	scardica v. obtusa / 3-4½" / pure white fls			126
	scardica v. spruneri - see S. spruneri			286
8	sedoides / R/ mat / greenish-yellow fls / e Alps, Balkans	148	/	126
12	sempervivum / R / to 5½" / fls purplish-pink / Balkans	149	/	286
†	sempervivum f. sempervivum / R / darkest of all Saxifrages			
†	sempervivum f. stenophylla -close set lvs;cluster blood-red fls			
12	X Schotii / flat rosettes / small yellow fls			126
	sibirica / R / to 8" / fls few, white / se Europe			25
5	Sibthorpii / A / trailing / fls yellow / Greece			77
8	'Sir Douglas Haig' - crimson-flowered mossy hybrid			64
	'Southside Seedling' - see S. cotyledon 'Southside Seedling'	(302)	/	18
3	spathularis / R / cespitose / fls white, red spots / Portugal			286
12	Spruneri / R / 3" / fls white / Balkan Peninsula	126	/	286
12	squarrosa / R / to 4" / white fls / se Alps	148	/	286
	stansfieldii / garden f. of S. decipiens / fls wht			61

† Fritz Kohlein, Saxifrages and Related Genera, 1980. Trans. David Winstanley, 1985.

1	stellaris / R / 8" / white fls, yellow spots / Eu., in mts.	148 /	286
	stelleriana - see S. cherleriodes		160
	Sterngergii - see S. roscaea		25
†	stolitzkae / R /cushion/ lvs dk & pale grn; 4-6 wht fls/Nepal		
15	stolonifera / HH P / 9-24" / fls white / China, Japan	165 /	128
12	Stribrnyi / R / caespitose / fls purplish-pink / Bulgaria	91 /	286
	Stribrnyi v. Zollikoferi - fls more prominent		132
1	strigosa / HH R / 8" / white fls / Himalayas		96
12	X Stuartii / 4" / brick-red fls		126
12	X Suendermannii / 2" / large white fls		132
	X Suendermannii 'Major' / 2½" / pale blush-pink fls		126
	Symons-Jeunii - see S. X calabrica 'Tumbling Waters'		126
	taygetea / R / 6" / fls white / Albania, Greece		286
8	tenella / R / to 6" / fls creamy-white / se Alps	148 /	286
	tenuis / R / taller than S. nivalis / white / Scand, Ice., Rus.		126
	thessalica - see S. sempervivum		126
12	X tirolensis / to 3½" / white fls / Alps		25
1	Tolmiei / R / mat / white fls / California to Alaska		286
12	tombeanensis / R / 3" / fls white / Italian Alps		286
9	tricuspidata / R / 2-9" / fls wht or cream, spotted / n N. Am.	(566) /	25
6	tridactylites / A / to 8" / small white fls / Europe	210 /	126
	tridens - see S. androsacea		
8	trifurcata / R / 4-12" / fls 3/4" across,white fls / n Pyranees	286 /	25
	X tyrolensis - see S. X tirolensis		25
11	X 'Tumbling Waters' /R /silvery rosettes/ wht fls, stalk to 2'		126
3	umbrosa / R / caespitose / white fls, red spots / Pyrenees	148 /	286
	umbrosa 'Aurea-Variegata' - see S. X urbium 'Variegata Aurea'		126
	umbrosa 'Melvillei' - see S. X urbium 'Melvillei'		96
	umbrosa 'Ogilvieana' - see S. X urbium 'Ogilvieana'		*
	umbrosa 'Primuloides' - see S. X urbium 'Primuloides'		126
	umbrosa 'Primuloides Elliot's Form' - see S. X urbium 'P. C. E'		126
	umbrosa 'Variegata' - see S. X urbium 'Variegata Aurea'		*
	X urbium / P / to 18" / large pink or white fls		25
3	X urbium 'London Pride' / 6-18" / white fls, red spots		126
	X urbium 'Melvillei' - lvs nearly round		96
	X urbium 'Ogilvienana' - starry pink fls, darker spots		96
	X urbium 'Primuloides' / 8" / deep pink fls		129
	X urbium 'Primuloides Clarence Elliot' / 5-6" / rose-pink fls		126
	X urbium primuloides (Elliot's var.) 'Hartside Pink'-sel. form		126
	X urbium 'Primuloides Ingwersens Form' - see below		126
	X urbium 'Primuloides Walter Ingwersen' / smaller pl./ rose-pink		126
	X urbium 'Variegata Aurea' - lvs gold-blotched		126
	vahlii / R / rosettes/ acaulescent white fls / ?		107
11	valdensis / R / to 5" / white fls / sw Alps	148 /	286
	valdensis 'Pygmaea' - smaller plant		*
†	'Valerie Finnis' - see S. X boydii 'Aretiastrum'		
12	'Valerie Keevil' / S. X anglica group / deep rose fls		(280)
8	Vayredana / R / to 4" / fls white / ne Spain		286
	veitchiana / R / to 4" / fls white / w China		126
	venetia - see S. paniculata 'Venetia'		126

---

† Fritz Kohlein, Saxifrages and Related Genera, 1980. Trans. David Winstanley, 1985.

1 9 vespertina / R / 4" / fls white, yellow spots / Wash., Oregon 107
1 virginiensis / R / 3-12" / fls white / e North America 224 / 286
8 wahlenbergii / R / to 3" / white fls / w Carpathians
† X wallacei - see S. camposii
† wendelboi / 3-4 wht fls per scape / Iran
8 'White Pixie' / 2" / white fls 129
11 'Whitehill' / 5-6" / white fls, blue-gray lvs 126
whitlavei compacta - see S. hypoides 'Whitelavei' 126
12 'Winifred' / 2" / large deep rose fls 126 / 123
8 X 'Winston S. Churchill' / 4" / clear pink fls 46
11 X zimmeteri / 4" / starry white fls / Austria 132

SCABIOSA - Dipsacaceae
alpina - see CEPHALARIA alpina 25
alpina 'Nana' - see CEPHALARIA alpina 'Nana'
argentea / B or P / 6-24" / fls white or cream / s Eu, Near East 77
atropurpurea / A / 3' / fls in various colors / s Europe 283
atropurpurea 'Nana' / A / dwarf strain of Sweet Scabious 25
caucasica / P / 18" / lavender-blue fls / Russia, Iran 147 / 77
caucasica 'Clive Greaves' - large mauve fls 168
caucasica 'Miss Willmott' - cream fls 207
cinerea ssp. hladnikiana - bluish-violet fls; lvs greenish 289
columbaria / P / 3' / lilac-blue fls / Europe 289
columbaria 'Alpina' / 2-5" / dwarf forms of European Alps 123
columbaria ssp. ochroleuca - see S. ochroleuca 289
cretica / R / 4-12" / often leafless, corolla lilac / s Medit. 289
graminifolia / R / 6-8" / pinkish-lavender fls / s Europe 49 / 123
hladnikiana - see S. cinera ssp. hladnikiana 289
japonica / P / 12-32" / fls blue / Japan, in mts 164 / 200
japonica v. acutiloba / lvs more acute at apex / Hokkaido, Honshu 200
japonica v. alpina / dwarfed / larger fls / Honshu, Shikoku 200
lucida / P / to 2' / rose-lilac fls / c Europe 148 / 25
lucida 'Rosea' / P / to 2' / fls rose-pink / c Europe 25
maritima - see S. atropurpurea 28
micrantha / A / to 2' / reddish fls / c Balkans, Crimea 289
ochroleuca / B or P / 2½' / fls pale yellow / Europe, w Asia 279 / 25
olgae / P / 10-18" / grayish-blue fls / Caucausus 12
parnassi - see PTEROCEPHALUS parnassi 123
parnassifolia - see PTEROCEPHALUS parnassi (284)
pterocephalus - see PTEROCEPHALUS parnassi 123
pyrenaica / R / 8-16" / fls clear blue-lilac / Pyrenees, sw Alps 148
rhodopensis / like S. graminifolia/ corolla pale yellow / Rodopi 189
rumelica - see KNAUTIA macedonia 279
scabra - see CEPHALARIA scabra 25
silenifolia / R / 6-12" / fls blue-violet / sc Europe (284) / 25
speciosa / P / 2' / fls lavender to mauve / Himalayas 25
succisa - see SUCCISA pratensis 268
sylvatica - see KNAUTIA sylvatica 29
turolensis / A or P / upper lvs pubes., wht / cor. red-purp/s&c Spain 289

---

[1] Calder & Savile, Brittonia, 11:4.

† Fritz Kohlein, Saxifrages and Related Genera, 1980. Trans. David Winstanley, 1985.

ukranica - see S. argentea 77
variifolia / HH Sh / 3' / fls pinkish / Rhodes, Anatolia 77
vestita - habit of S. graminifolia 61

SCHEFFLERA - Araliaceae
digitata / Gh T / 25' / globose purple frs / N. Zealand 184 / 25

SCHINUS - Anacardaceae
latifolius / HH T / ? / fls white / Chile 29
molle / HH T / 15'+ / evergreen lvs; rosy-red frs / S. America 139
terebinthifolius / Gh T / 20' / frs bright red / Brazil 25

SCHISANDRA - Schisandraceae
chinensis / Cl / white or pinkish fls; red frs / e Asia 25
grandiflora v. rubriflora / HH Cl / fls deep crimson / w China (418) / 139
repanda / Cl / small / creamy-white lvs; blue-black frs / Japan 200
rubriflora - see S. grandiflora v. rubriflora 184 / 139

SCHIVERECKIA - Cruciferae
bornmulleri - see S. doerfleri 77
doerfleri / P / 4-6" / fls yel, ½" across / Europe 95
podolica / R / to 10" / fls white / Ukraine, Roumania 286
scoparium / P Gr / to 5' / Prairie Beard Grass / United States 25

SCHIZANTHUS - Solanaceae
grahamii - see S. retusus 'Grahamii' 25
hookeri / A / 2' / fls pale rose / Chile 81
retusus 'Grahamii' / A / lavender & yellow fls 25

SCHIZOCODON - Diapensiaceae
illicifolius - see S. soldanelloides v. illicifolious 200 / 107
macrophyllus - see SHORTIA soldanelloides 200
soldanelloides v. illicifolius / R / 4-8" / fls rose / Japan 200 / 107
soldanelloides minor - see SHORTIA soldanelloides minima

SCHIZOPETALON - Cruciferae
walkeri / A / 12"+ / fragrant, white fringed fls / Chile 25

SCHIZOPHRAGMA - Saxifragaceae
hydrangeoides / Cl / to 36' / cream fls & bracts / Japan 167 / 139

SCHIZOSTYLIS - Iridaceae
coccinea / HH Tu / 2½' / scarlet to pink fls / South Africa 267 / 129
coccinea 'Grandiflora' - probably not distinct from 'Major' *
coccinea 'Major' - larger fls, lvs 267
coccinea 'Mrs. Hegarty' - rose-red fls 267 / 111
'Mrs. Hegarty' - see S. coccinea 'Mrs. Hegarty' 111

SCHOENOLIRION - Liliaceae
album / P / 1-5' / fls white or greenish / Oregon, California 228

SCIADOPITYS - Taxodiaceae
verticillata / T / to 100' / 3-5" linear lvs / Japan 127 / 25

SCILLA - Liliaceae

Adlamii / Bb / 5-9" / mauve-purple / Natal			61
amethystina - see BRIMEURA amethystina	(588)	/	25
amoena / Bb / 6" / starry fls, deep indigo-mauve / c Europe	182	/	267
autumnalis / Bb / 8" / fls blue-lilac to purple / Europe	267	/	(79)
bifolia / Bb / 3-6" / varii-colored fls / s Europe	148	/	129
bifolia 'Rosea' - fls rose			25
bithynica / Bb / 4-9" / fls pale blue / e Bulgaria			290
campanulata - see ENDYMION hispanica			267
campanulata rosea - see ENDYMION hispanica 'Rosea'			*
chinensis - see S. scilloides			185
festalis - see ENDYMION non-scripta			28
hispanica - see ENDYMION hispanica			268
hispanica 'Alba' - see ENDYMION hispanica 'Alba'			25
hispanica 'Excelsior' - see ENDYMION hispanica 'Excelsior'			
hispanica 'Rosea' - see ENDYMION hispanica 'Rosea'			25
hohenhackeri / Bb / to 10" / pale to mid-lilac-blue fls / Iran			185
Hughii / HH Bb / to 10" / deep violet fls / Marettimo			290
hyacinthoides / Bb / 18" / pale lilac-blue fls / Medit. region	211	/	267
italica - see ENDYMION italicus			25
latifolia / Bb / 1-1½' / 30-60 fls in raceme, lilac / Canary Ils.			61
lilio-hyacinthus / Bb / 1½' / fls pale lilac-blue / Medit. region	148	/	267
Litardierei / Bb / to 10" / blue campanulate fls / Jugoslavia			25
Litardierei 'Amethystina' - vigorous form			25
Litardierei 'Robusta' - same as above ?			*
messeniaca / Bb / 6-8" / mid-blue fls / s Greece	184	/	185
monophyllus / Bb / 8" / starry blue fls / Spain, Portugal	212	/	267
natalensis / Gh Bb / to 3' / blue fls / South Africa	186	/	25
nivalis - see S. bifolia			
non-scripta - see ENDYMION non-scripta			268
nutans - see ENDYMION nonscriptus			25
obtusifolia / Bb / 4-12" / fls pink to lilac/ ne Spain, isles w Med.			290
paucifolia / Gh Bb / ? / greenish fls / South Africa			61
persica / Bb / 1' / pale blue fls / w Iran			(285)
peruviana / HH Bb / 6-10" / lilac-blue fls / Spain, c Medit. reg.	210	/	129
peruviana 'Alba' - white fls			267
pratensis - see S. Litiardierei	129	/	25
pratensis v. amethystina - brighter blue fls			267
pushkinioides / Bb / 4-8" / pale azure fls / c Asia			4
ramburei / HH Bb / 4-12" / bright mid-blue fls / s Spain & Portugal	212	/	185
reverchonii / Bb / 4-10" / fls deep blue / se Spain			290
rosenii / Bb / 1' / fls pale azure to whitish / Asia Minor	(256)	/	4
scilloides / Bb / 6" / pink fls in autumn / China, Korea	(210)	/	185
sibirica / Bb / to 8" / gentian-blue fls / Caucasus	267	/	129
sibirica 'Alba' - white form			267
sibirica v. atrocoerulea - earlier & stronger			267
sibirica 'Spring Beauty' - see S. sibirica v. atrocoerulea			267
sibirica v. taurica - paler blue, dark tipped fls			267
tubergeniana / Bb / 6" / fls pale blue, nearly white / nw Iran	267	/	129
verna / Bb / 6" / fls pale blue-mauve / n Europe	148	/	267
vicentina - see HYACINTHOIDES vicentina			
violacea - see LEDEBOURIA socialis			185

SCIRPUS - Cyperaceae
cyperinus / semi-Aq / 1' / reddish-brown involucre / e & c U.S. 53 / 99

SCLERANTHUS - Caryophyllaceae
biflorus / R / mat to 6" / white fls / Australia, Tasmania 46 / 45
Brockiei / R / mat / green lvs / New Zealand 15
perennis / R / to 8" / may be procumbent / most of Europe 286
perennis ssp. dichotomus / 8" / large frs / Alps 286
uniflorus / B or P / moss-mats / yellowish fls / New Zealand 238

SCLEROCACTUS - Cactaceae
Whipplei / R / 8" / yellow, pink, purple, white fls / Col., Utah (306) / 25

SCOLIOPUS - Lilaceae
Bigelovii / Bb / 8" / green & purple fls / California 19 / 228

SCOPIOLA - Solanaceae
carniolica / P / 8-24" / brownish-violet fls / c & se Europe 148 / 288

SCORZONERA - Compositae
austriaca / P / to 20" / pale yellow fls / c Europe 148 / 289
glabra - see S. austriaca 289
hispanica / P / to 3' / fls yellow / Europe 212 / 25
purpurea / P / 2'+ / pale lilace fls / s Europe 149 / 289
purpurea ssp. rosea / P / 2' / pale purplish fls / ec Europe 289
rosea - see S. purpurea ssp. rosea 289

SCROPHULARIA - Scrophulariaceae
canina ssp. hoppii / P / 8-24" / fls dark purplish-red/ Jura, s Alps 288
chrysanta / to 2' / yellow fls / Caucasus 25
hoppii - see S. canina ssp. hoppii 288
ningpoensis / P / 3' / fls purplish-brown / se China 258
nodosa / P / 12-32" / fls green & brownish / Europe 288
peregrina / P / to 3' / dark red to purplish fls / Medit. region 288
scopolii / P/ to 3'/fls greenish with purple/ se&ec Eur. s to Italy 288
vernalis / B or P / 3'+ / yellow fls / Europe, mostly s 288

SCUTELLARIA - Labiatae
alpina / R / 10" / purple & white fls / Europe 148 / 107
alpina 'Alba' - all white fls *
alpina 'Rosea' - pink fls 28
amoena - see S. baicalensis
angustifolia / R / to 1' / fls deep blue-violet, large /B.C., Calif. 228 / 142
antirrhinoides / R / to 1' / fls deep blue-violet / Idaho, Ore. 229 / 142
baicalensis / R / to 15" / bluish-purple fls / e Asia 25
baicalensis 'Coelestina' - fls bright blue 25
balearica / P / to 3" / lvs purp. beneath; fls purp./ Balearic Isles 288
Brittonii / R / 6-10" / deep violet-blue fls / Wyo., Col., N. Mexico 25
coelestina - see S. baicalensis 'Coelestina' 25
galericulata / P / 1-2' / fls violet blue / global in temp. regions 25
hastifolia / P / 6-20" / fls violet-blue / Europe 288
incana / P / to 3' / blue fls / New York, Alabama, Kansas 99
indica / R / 1' / bluish fls / Far East 25
indica 'Alba' / P / to 1' / fls white / China, Japan 25
indica v. japonica - see indica v. parvifolia 25

indica c. parvifolia / 4-7" / fls lilac to blue 138 / 25
integrifolia / P / to 2' / fls blue-purple & white / United States 99
laeteviolacea / R / 6" / purplish fls / Japan 200
laterifolia / P / 1-2' / fls blue to whitish / U.S. 229 / 224
maekawae / P / 3-8" / lvs purple beneath; fls purple / Japan 200
orientalis / R / 3" / yellow fls / Altai Mts. 210 / 107
ovata / P / 2'+ / blue & white fls / s United States 99
ovata v. rugosa / R / sprawling, blue & white fls / Ark., W. Va. 99
parvifolia - see S. indica parvifolia
pontica / R / creeping / magenta fls / ? 64
resinosa / R / 9" / violet-blue fls / Kansas to Texas 257 / 25
rubicunde / P/ 2-16"/lvs glab.; fls purp-blue, rare wht/Italy,Balk. 288
scordifolia / R / 6" / pure, deep blue fls / Korea 123
serrata / P / 2' / blue fls / s N.Y. to Tenn. 224 / 99
supina - see S. supina alpina
supina alpina / P / 2-12" / lvs glabrous; fls yel / U.S.S.R. 288
tuberosa / HH R / mats to 5" / violet-blue fls / Ore. to S. Calif. 227 / 25

SEBAEA - Gentianaceae
exacoides / A / ? / yellow fls / Cape of Good Hope 223

SECURIAGEA - Leguminoseae
securidacia / A / to 1' / fls yellow / s Europe 61

SEDUM - Crassulaceae
acre / R / 1-3" / bright yellow fls / Europe, Asia Minor 148 / 129
adenotrichum / R / rosettes fleshy/ fls white / Himalayas 61
aizoon / P / 1½' / fls yellow to orange / Japan, Siberia 49 / 25
albo-roseum - see S. erythrostictum 25
album / R / to 7" / fls white / Europe 286
alpestre / R/ creeping to 3"/ fls grn-yel./ mts Eur., Asia Minor 25
altissimum / see S. sediforme 286
amplexicaule - see S. tenuifolium 25
anacampseros / R / to 10" / fls lilac or dull red / Pyrenees 148 / 286
annuum / A or B / spotted or streaked / 1-3" / fls yel / n Eur. 286
arcticum / R / lvs crowded / fls yel / n Russia 25
atropurpurea - see S. Telephium ssp. maximum 'Atropurpureum' 207
bellum / R / 3-6" / fls white / w Mexico 25
beyrichianum - see S. Nevii 25
brevifolium / R / creeping / white fls / sw Europe, nw Africa 25
Brownii - see S. kamtschaticum
bupleruoides / R / to 1' / fls 5-merous, red-purp / Himalays 25
caeruleum / A / 4-8" / fls blue or white / Medit. reg. 194 / 63
caespitosum / A / 2"/ white fls, tinged pink / s & sc Europe 286
camtschaticum - see S. kamtschaticum *
carpaticum - see S. Telephium 25
caucasicum / P / 1'+ / shell lvs with ears / Caucasus 77
cauticola / A / 3" / rose-purple fls / Japan 129
cepaea / A or Bi/ 6-12"/ weak stems / fls pink, red rim/s S.C., Eur. 286
Cocherellii / R / to 8" / fls white / s California & n Mexico 25
crassipes stephani - see S. stephani 96
cyaneum / R / creeping to 4" / fls lilac-pink / e Siberia 25
dasyphyllum / R / to 3" / fls white & pink / s Europe 148 / 286
debile / R / to 5" / fls yellow / Nevada to Wyoming 25
divergens / R / to 6" / yellow fls / Calif. to B. C. 228 / 25

	Douglasii - see S. stenopetalum		63
	dumulosum / R / 4-8" / white / n China		25
	Ellacombianum - see S. kamtschaticum v. ellacombianum	94 /	25
	erythrostictum / A / 1-2' / fls white, pistils pink/ Orient?		25
	Ewersii / R / 4-8" / pink or mauve fls / c Asia, Himalayas	147 /	286
	Ewersii v. homophyllum / very dwarf / glaucous lvs		64
	floriferum / A / 6" / yellow fls / China		25
	Forsteranum / R / 1' / ascending / yellow fls / w Europe		286
	furusei / P / fls sessile, greenish wht / Japan		200
	glaucophyllum / R / 4" / white fls / W. Va. & Va.	224 /	99
1	globosum / R / 4" / whitish fls / ?		
	gracile / R / to 4" / fls white, greenish midvein / Caucasus, n Iran		77
	grisebachii - see S. annuum		
	himalense / A / 6-12" / fls purple / Himalayas		25
	hirsutum / R / to 6" / fls white / sw Europe	212 /	25
	hispanicum / A or B / to 7" / fls white / se Europe, sw Asia		25
	hispanicum v. minus / R / 2" / fls white with red nerve	184 /	94
	humifusum / HH R / 1-2" / bright yellow fls / Mexico		123
	hybridum / R / to 1' / golden-yellow fls / e Eur., Siberia,Mongolia	286	25
	hyperaizoon / P / 3' / yellow fls / e Siberia		25
	integrifolium - see S. rosea ssp. integrifolium	227 /	25
	involucratum / R / mats / white starry fls		96
	kamtschaticum / R / 4" / yellow fls / Asia		123
	kamtschaticum ssp. ellacombianum / light green lvs / n Japan	260 /	25
	kamtschaticum ssp. Middendorffianum / 10" / golden fls / e Asia		63
	kamtschaticum ssp. Middendorffianum 'Diffusum' / 4" / gold-yel fls		46
	kamtschaticum 'Nanum' - dwarfed form		25
	kamtschaticum 'Variegatum' - white-margined lvs	(82)	123
	Kirilkowii / R / to 1' / fls brownish-red / Himalayas, China		25
	lanceolatum / R / to 8" / fls yellow / w North America	229 /	25
	laxum / R / 8-16" / pink or white fls / Calif., Ore.		228
	laxum ssp. Heckneri / R / 1'+ / pink or white fls / Ore., Calif.		63
	Leibergii / R / 3-5" / short thick lvs; yellow fls / Ida., Wash.		229
	lidakense / like S. cauticola/purp. lvs; carmine-rd fls/ unkn. orig.		46
	linearifolium / R / 4-5" / large white fls / India		96
2	litorale / P / 1'+ / golden fls / Popov Is., Peter the Great Bay		
	lydakense - see S. lidakense		*
	lydium / R / 2-5" / white fls / Asia Minor	94 /	25
	magellense / R / 3-6" / white fls / Greece, Asia Minor		25
	Maximowiczii - see S. aizoon		25
	maximum - see S. Telephium ssp. maximum		286
	maximum 'Atropurpureum' - see S. Telephium ssp. m. 'Atropurpureum'		*
	Middendorffianum - see S. kamtschaticum ssp. Middendorffianum		63
	Middendorffianum 'Difussum' - see S. kamtschaticum ssp. M. 'D.'		46
	monregalense / R / 6" / white fls / sw Alps		286
	montanum - see ochroleucum montanum		
	Nevii / R / 4" / white fls / se U.S.	194 /	107
	oaxacanum / P / decumbent / fls yellow / Mexico		25
	obtusatum / P / 4-10" / yellow fls / mts. California	229 /	25
	ochroleucum / R / creeping mats / white fls / Eurasia		25

---

[1] Allen, Bull. Am. Rock Garden Soc., 19:3.

[2] Yuzepchuk, Flora U.S.S.R., Vol. IX, 1971 (trans).

	ochroleucum montanum - fls bright yellow		286
	oppositifolium - see S. spurium		286
	oreganum / R / 2-3" / yellow fls / w North America	156 /	123
	oregonense / R / 2-11" / yellow to whitish fls / n Calif., s Ore.		228
1	pallescens / P / 1-2' / whitish or pinkish fls / e Siberia, Dauria		
	palmeri / HH R / 6" / deep orange fls / Mexico		28
	pilosum / R / 2-3" / rose-pink fls / Asia Minor	(248)	123
	pluricaule / R / caespitose / fls purple / Sakhalin Isls.	48	25
	pluricaule 'Rosy Carpet' / 3" / pink fls		200
	pluricaule v. yezoensis / P / 2-3" / fls rose-purple / Japan, China		200
	populifolium / Sh / 18" / fls pale pink or white / Siberia	94	107
	primuloides pachyclados / Sh / 2-3" / stout stem, fls wht / Yunnan		61
	pruinatum - see S. rupestre		25
	pulchellum / B / 4-12" / rosy-purple fls / e United States		25
	Purdyi - see S. spathhulifolium ssp. Purdyi		63
	purpurea-viride - see S. Telephium		25
	purpureum - see S. Telephium		286
	reflexum / R / 6-12" / bright yellow fls / c Europe	148 /	286
	reflexum 'Cristatum' - crested form		25
	rhodanthum / P / 18" / rose-colored fls / Rocky Mts.		107
	Rhodiola / 6" / fls yel; lvs thick blue-grey / mts.		96
	Rhodiola rosea - see S. rosea		99
	rosea / R / 6-16" / fls yellowish or purplish / circumboreal	312 /	25
	rosea ssp. integrifolium / 5" / purple fls / nw N. Am.		25
	rubens / A / 2-5" / white or pink fls; lvs reddish / s & w Europe		286
	X rubrotinctum / Gh Sh / 1' / yellow fls in winter	187 /	25
2	'Ruby Glow' - spectabile X cauticolum		
	rupestre / R / 6-12" / fls golden-yellow / s Europe	147 /	25
	rupestre v. Forsteranum - see S. Forsteranum		286
	'Schorbusser Blut' - see S. spuriumm 'Schorbusser Blut'		129
	sediforme / P / to 2' / fls straw-colored / Medit. region	212 /	286
	Selskianum / A / 1-1½' / fls yellow, many stemmed / Manchuria		25
	Semenowii / 20" / white & pink fls / Turkestan		93
	sempervivoides / B/ 6-10" 2nd season/ crimson fls/ Asia Minor, Cauc.	156 /	25
	sichotense / R / 4-8"/ fls greenish yellow / Primorsk, USSR		156
	Sieboldii / R / 6-8" / pink fls / Japan		129
	'Silver Moon' - yellow fls: S. laxum ssp. heckneri x spathulifolium		63
	spathulifolium / R / 4"+ / yellow fls / w N. Am.	147 /	107
	spathulifolium atropurpureum - see S. cepaea		286
	spathulifolium 'Aureum' / 3" / golden variegated lvs		61
	spathulifolium 'Capa Blanca' - silvery-gray lvs		129
	spathulifolium ssp. Purdyi / lvs gr., closely compacted / Ore., Cal.		63
	spathulifolium 'Purpureum' - lvs purple	269 /	25
	spathulifolium 'Rosum' - tinted lvs?		25
	spectabile / A / 12-22" / fls pink / China, Korea		25
	spinosum - see OROSTACHYS spinosa		286
	spurium / R / 9" / fls pink to purple / Caucasus	94 /	25
	spurium 'Album' - fls white		25
	spurium coccineum / P / to 9" / deep pink / Nfld to N.Y. & Pa.		99
	spurium 'Dragons Blood' - see S. spurium 'Schorbusser Blut'		241

---

1 Yuzepchuk, Flora U.S.S.R., Vol. IX, 1971 (trans.)

2 Hillier's Hundred Catalog, Index, p. 55

spurium 'Roseum' - rosy fls 25
spurium 'Schorbusser Blut' - fls deep carmine-crimson, decumbent 129
stellatum / A / 6" / sensile fls, purplish / Crete to Portugal 25
stenopelatum / R / 12" / yellow fls / nw North America 312 / 63
stephani / dwarfer S. Rhodiola / lg fls, nar. lvs 96
stoloniferum / R / 6-12" / pink fls / Caucasus, Asia Minor 25
tatarinowii / R / 4-8" / fls pinkish-white / China 94 / 25
telephiodes / P / 6-32" / pale pink fls / N. Y. to Ill. 224 / 99
Telephium / P / 12-18" / fls pink or white / Europe 286
Telephium Borderi / A/1-2'/lvs deeply toothed, fls red-purp / France 25
Telephium ssp. maximum / to 32" / fls whitish 286
Telephium ssp. maximum 'Atropurpureum' - vivid purple lvs 207
Telephium 'Roseo-Variegatum' / 1-2' / lvs pink in spring 279
tenellum / R / to 4" / reddish fls / Caucausus, Iran, Armenia 25
tenuifolium / 1-3" / yel fls 6-10 / s Eur., N. Afr., Asia Minor 25
ternatum / R / 3-6" / white fls / e United States 25 / 107
Treleasii / HH R / to 12" / bright yellow fls / Mexico 128
tsugarense / P / to 20" / fls pale green / Honshu 200
villosum / R / 6" / lilac or pale pink fls / w & c Europe 286
viviparum / R / 8" / fls white to greenish / Manchuria 156
Watsonii - see S. oregonense 63

SELINUM - Umbelliferae
tenuifolium / P / 3-5' / white fls; black anthers / Himalayas 184 / 279

SELLIERA - Goodeniaceae
radicans / HH R / creeping / white fls / Australia, N. Z., Chile 28

SEMIAQUILEGIA - Ranunculaceae
adoxoides / R / to 1' / cream, chocolate, red-brown sepals/ Far East 25
ecalcarata - see G. adoziodes 25
simulatrix / R / 12" / dark reddish-violet fls / China 24

SEMPERVIVILLA - Crassolaceae
alba / R / carpeting 1-2" / fls greenish white / Himalayas 95

SEMPERVIVUM - Crassulaceae
acuminatum - see S. tectorum v. glaucum 25
'Ambergreen' - glaucous green lvs, shaded rose-violet 241
andreanum / R / 4½" / metallic-pink fls / n Spain (591) 268
arachnoideum / R / 2-6" / pink to crimson fls / Europe, in mts. 148 / 129
arachnoideum 'Album' - white fls *
arachnoideum v. doellianum - glabrescent 93
arachnoideum v. glabrescens - rosettes ovoid, web rather lax 268
arachnoideum grandiflorum / R / fls 8-10 paired; rose-red hyb. 61
arachnoideum 'Minus' - rosettes small *
arachnoideum v. oligotrichum - webbing sparse or disappearing 156
arachnoideum 'Rubrum' - lvs reddish *
arachnoideum 'Stansfieldii' - red fls 241 / 18
arachnoideum v. tomentosum - flat rosettes, abundant webbing 25
arachnoideum 'Webbianum' - see S. arachnoideum v. tomentosum 93
arboreum - see AEONIUM arboreum 286
arenarium - ros. to 3/4"; fl stems 3-5"; fls pale yel / e Alps 25
arvernense - see S. tectorum 286
assimile - see S. marmoreum 25

atlanticum / R / to 1' / white fls, red banded / Morocco	(592)	/	25
atropurpureum - see S. tectorum 'Atropurpurea'			25
ballsii / R / 1½" rosette / fls dull rose / nw Greece			268
blandum - see S. marmoreum			25
X barbulatum - variable between S. arachnoideum & montanum	(589)	/	25
Braunii - see S. montanum v. Braunii			25
X calcaratum / 12-16" / fls dull red-purple			268
calcareum / R / 1'+ / pale pink fls / French Alps	(289)	/	286
calcareum 'Mrs. Guiseppi' / 2" rosette / light green, flame tipped	(245)	/	18
cantabricum / R / open rosettes / fls pink & green / Spain	(212)	/	132
cantalicum - see S. tectorum			25
carpathicum - see S. montanum ssp. carpathicum			156
ciliosum / R / 3-5" / pale yellow fls / Balkans	156	/	268
ciliosum f. borisii - hairs clustered brush-like at the tip			156
Clusianum - possible syn. for S. tectorum			
'Commander Hay' / 15" / pink fls / somewhat tender			129
Doellianum - see S. arachnoideum v. Doellianum			93
dolomiticum / R / 1½" rosette / deep rose fls / e Alps	148	/	268
erythraeum / R / 6-8" / fls deep reddish-purple / Bulgaria	(593)	/	286
X Fauconnettii - small dense rosettes: S. arachnoideum x tectorum	(590)	/	25
X fimbriatum - is either X barbulatum or X roseum			25
X Funckii / 8" / purple-rose fls	148	/	268
giganteum - see S. tectorum 'Giganteum'			18
guiseppii - see S. calcareum 'Mrs. Guiseppi'	(594)	/	18
glaucum - see S. tectorum v. glaucum			25
grandiflorum / R / 4-8" / fls lemon-yellow / Alps, s Switz.	148	/	286
Hausmanni - see S. Fauconnetti			25
Heuffelii / see JOVIBARBA Heuffelii	(590)	/	268
hirtum - see JOVIBARBA hirta			268
'Jubilee' / 3" / bronze lvs			18
Kindingeri / R / 8-10" / yellow & red fls / Balkans			132
Kochii - see S. arenarium			25
kopaonikense - see JOVIBARBA kopaonikense			93
kosaninii / R / 6-8" / rose-purple fls / Jugoslavia	(595)	/	268
leucanthum / R / 4-5" / greenish-white fls / Bulgaria	(596)	/	268
macedonicum / R / long stolons / dull rose fls / sw Jugoslavia	(597)	/	268
marmoratum - see S. marmoreum			268
marmoratum 'Rubrifolium Ornatum' - see S. marmoreum			268
marmoreum / R / 8" / fls red, white margins / Balkans	(598)	/	286
marmoreum ssp. blandum - glabrous lvs			156
marmoreum 'Brunneifolium' - lvs brown, red in winter			25
marmoreum 'Ornatum' - lvs reddish, green tipped	147	/	268
marmoreum f. rubricundum - lvs much reddened			156
marmoreum rubrifolium - see S. marmoreum f. rubricundum			156
Mettenianum / R / 1½" rosette / rose fls / c Europe			28
montanum / R / 8-12" / pinkish-purple fls / Europe, in mts	148	/	132
montanum v. Braunii - fls white or yellowish-white			25
montanum v. burnatii - robust form from Piedmont & Maritime Alps	(599)	/	156
montanum ssp. carpathicum - ros.lg. closed; fl. 6-8; lvs short-taper			156
montanum v. stiriacum / to 7" / lvs red tipped / e Alps	(600)	/	25
X Morelianum - possibly derived from S. arachnoideum			25
nevadense / R / 3-5" / clear rose fls / Sierra Nevada, Spain	(601)	/	268
nevadense v. hirtellum - larger rosettes			268
octopodes / R / small rosettes / yellow fls / sw Macedonia	(445)	/	268
ornatum - see S. marmoreum			268

ossetiense / R / dense rosette / fls white & purple / Caucasus (602) / 156
patens - see JOVIBARBA Hueffelii *
pernkoffii - see S. X rupicolum v. pernkoffii 156
X pilosella - see S. X Fauconnettii 25
pittonii / R / to 6" / pale yellow fls / e Alps 148 / 286
X pomellii - see S. X Fauconnettii 25
pumilum / R / 2-3" / deep cherry-red fls / Caucasus (600) / 268
pyrenaicum / Pyrenees form of S. tectorum / light pink fls 96
'Queen Amalie' - see S. Regina-Amaliae *
'Rauhreif' - see SENECIO cineraria 'Rauhreif' *
Reginae-Amalie - see S. marmoreum (590) / 25
X roseum - S. arachnoideum hybrid 96
rubicundum - see S. marmoreum f. rubicundum 156
rubrifolium ornatum - see S. marmoreum 'Ornatum' *
X rupicolum v. pernkofferi / fls yel-red; lvs less gland.-hairy 156
ruthenicum / R / 1' / greenish-yellow fls / e Europe (603) / 268
sabanum / R / 1-2" rosettes / light green lvs / ? 241
Schlehanii - see S. marmoreum 25
X Schottii / purple fls / S. montanum x tectorum 268
1 'Silverine' - silvery-green lvs
simonkianum - see JOVIBARBA simonkiana 64
soboliferum - see JOVIBARBA sobolifera 268
speciosum - see S. tectorum 25
staintonii / R / to 5" / fls yellow; lvs purplish / Turkey 77
Stansfieldii - see S. arachnoideum 'Stansfieldii' 18
tectorum / R / 1'+ / dull rose fls / Europe, in mts. 148 / 268
tectorum ssp. alpinum - fls pink; lvs reddish-brown 156
tectorum ssp. atlanticum - see S. atlanticum 156
tectorum 'Atropurpurea' - darker lvs 96
tectorum 'Atroviolaceum' - colored leaf form 25
tectorum ssp. boutignyanum - Pyrenees form of ssp. alpinum 25
tectorum ssp. boutignyanum f. pallescens / paler fls / High Pyrenees 156
tectorum calcareum / P / 8-18" / fls purp-red, brn-pur tip lvs/ Eur. 286 / 25
tectorum ssp. calcareum 'Mrs. Guiseppi' - see S. calcareum 'M. G.' *
tectorum 'Densum' - garden form 96
tectorum 'Giganteum' - huge green rosettes 18
tectorum Guiseppiae - see S. calcareum 'Mrs. Guiseppi' *
tectorum v. glaucum - lvs glaucous, whitish at base 25
tectorum 'Lamotte' - fl stems to 15" 96
tectorum 'Mahogany' - lvs tinged dark red-brown 48
tectorum 'Nigrum' - lvs tipped reddish-brown 123
tectorum ssp. pyrenaicum - see S. t. ssp. boutignyanum f. pallescens 156
tectorum 'Triste' - lvs dark purplish-brown; fls pinkish 147 / 156
tectorum X versicolor - S. zeleborii x S. marmoreum 286
tectorum 'Violaceum' - leaf-color form 96
X Thomayeri - see S. X Fauconnettii 25
thompsonianum / R / ? / fls many colors, spherical pl / Greece (604) / 156
X thompsonii - see arachnoideum X marmoreum 268
X vaccari / reddish fls / S. arachnoideum X grandiflorum 156
vincentei / R / to 1' / lilac-banded fls / n Spain 268
violaceum - see S. tectorum 'Violaceum' 28
Webbianum - see S. arachnoideum v. tomentosum 25

---

1 George Schenk, The Wild Garden, n.d.

Wulfenii / R / 6-8" / yellow fls; glabrous lvs / Alps 148 / 268
Zeleborii - see S. ruthenicum 25

SENECIO - Compositae
abrotanifolius / R / 10-12" / orange-yellow fls / c Europe 148 / 123
abrotanifolius ssp. abrotanifolius /4-12"/with brn stripes/Alps,Jugo 289
abrotanifolius carpathicus /capitula solit;upper lvs sm/Carpath,Balk 289
abrotanifolius v. tiroliensis - plant smaller, larger ragged fls 107
adonidifolius / P / 18" / orange-yellow fls / mts. s Europe 93 / 25
alpinus / P / 3' / orange-gold fls / Alps 93
antennariifolius / R / 12" / yellow fls / Virginia, West Virginia 25
argyreus / HH Sh / 12"+ / yellow fls / Chile 64
aurantiacus - see S. integrifolius ssp. aurantiacus 289
aureus / P / to 30" / yellow fls / Nfld. to Fla. & Texas 224 / 25
bellidioides / R / creeping / yellow fls on 1" stems / New Zealand 193 / 15
bicolor / Sh / 10-20" / fls light yellow / Medit. 289
bicolor ssp. nebrodensis /invol. subglab. or wht-tom.n. /n Sicily 289
bidwillii / HH Sh / to 3' / white to buff tomentum of pl / N. Z. 15
Bolanderi v. Harfordii / P / to 2' / yellow fls; lvs thin / Oregon 142
brunonis / T / 6-12' / lvs sticky; fls dark gold / sw Australia 109
campestris - see S. integrifolius 209
candidus - see S. bicolor ssp. nebrodensis 289
canus / P / 18" / yellow fls / w North America 229 / 25
capitatus / R / to 12" / fls yellow or orange / c Europe, in mts. 148 / 25
carniolicus - see S. incanus v. carniolicus 123
carthamoides / R / to 1' / yellow fls; plant leafy / Col., Wyo. 235
chilensis / Gh Sh / 2'+ / yellow daisies / Andes Mts. 64
cineraria / HH P / 2½' / white-woolly lvs / Medit. region 28
cineraria 'Rauhreif' - selection for whiteness of leaf 93
congestus / R / 3-8" / yellow fls, woolly infloresence / cirumpolar 209
cruentus / Gh / 2-3' / purple fls / Canary Isls. (418) / 25
doria / P / 3-4' / yellow fls / Europe 289
doronicum / P / 2-3' / large bright yellow fls / c Europe 148 / 207
elaeaganifolius / Sh / to 10' / fls yel; lvs grn over gray / N.Z. 25
elegans / A / 2' / purple fls / South Africa 184 / 25
fendleri / P / 4-24" / yellow fls / Wyoming, Utah 229
Flettii / R / to 12" / fls yellow / Washington, in mts. 228 / 25
Fremontii / R / 4" / yellow fls / California to Canada 228
fuchsii - see S. nemorensis 200
glaberrimus / sim. to S. doronicum /glabrous or subglabrous / e Eur. 289
glastifolius / Gh Sh / to 5' / purple to pink fls / South Africa 25
Greyi / HH Sh / 6' / yellow fls / New Zealand 110 / 238
Greyii laxifolius - see S. laxifolius 238
haastii / HH R / to 14" / yellow fls / New Zealand 15
halleri / R / to 4" / orange-yellow fls / sw & sc Alps 289
Hartfordii - see S. Bolanderi v. Hartfordii 25
harveianus 'MacOwen' - see S. inaequidens 289
heritieri / Gh Sh / 3'+ / fls white & crimson & purple / Canary Isl. 139
Herreianus / P / suc. 1-2' / hanging basket plant / sw Africa 25
huntii / HH Sh / to 18' / yellow fls / Chatham Isls. 238
inaequidens / Gh Sh / 8-20" / gld-yellow fls / S. Afr., escape Italy 289
incanus / R / 2-4" / golden buttons; silvery lvs / Alps 148 / 123
incanus v. carniolicus - larger flower-heads 219 / 123
integrifolius / P / stock short / 1-15 fls. corymb / Eur. 289
integrifolius ssp. aurantiacus /to 2'/fls orng-red;bracts purp/ec Eur 289

[1] integrifolius ssp. capitatus / 12", orange fls
kawakamii / R / 12" / decumbent / yellow fls / alps of Japan 200
lautus / A / small / yellow fls / Australia 45
laxifolius / HH Sh / to 5' / yellow fls / New Zealand 269 / 129
leucophyllus / R / to 8" / yellow fls / France, in mts. 148 / 25
leucostachys / similar to S. cineraria / whiter lvs / Argentina 93
longilobus / Sh / to 5' / yellow fls / Europe 129
lyallii / HH P / to 20" / light yellow fls / New Zealand 193 / 238
macroglossus 'Variegatus' /Gh/ trailing/ yel fls, varieg, lvs/S. Af. 25
macrophyllus / P / 3-4' / yellow fls / Europe 28
martinensis / HH R / 6" / rayless fls / Andes 64
Millefolium / P / 6-21" / fls yellow / Virginia to South Carolina 99
Monroi / HH Sh / 3-5' / yellow fls / New Zealand 139
nemorensis / P / 3'+ / fls yellow / Europe, Siberia, Far East 164 / 200
nemorensis v. fuchsii - included in the sp. 200
obovatus / P / to 2' / fls yellow / Massachusetts to Florida 25
palustris / see S. congestus 209
pauperculus / P / 2-24" / deep yellow fls / n N. Am. 225
pectinatus / HH R / creeper / orange-yellow or paler fls / Tasmania 64
primuleifolius / P / ? / daisy-like fls / s Tasmania 109
Przewalskii - see LIGULARIA Przewalskii 268
pseudoarnica / P / 18" / yellow fls / seashores, e Asia, N. Am. 13 / 25
repens / Gh / to 1' / fls white / South Africa 93
resedifolius / R / 4" / yellow daisies / n North America 25
revolutus / HH Sh / to 20" / yellow fls / New Zealand 15
robbinsii / P / 3' / yellow fls / ne North America 99
rodriguezii / HH Sh / dwarf / purple fls, tinged yellow / Minorca 212
scandens / HH Cl / 30' / bright yellow fls / Far East 164 / 222
scorzoneroides / HH P / to 20" / white fls / New Zealand 25
smithii / P / 2-3' / invol. dk grn or blk, ligules wht /nat.Scotland 289
soldanella / R / 6" / yellow fls / Colorado, in high mts. 229 / 25
subalpinus / P / 1-2' / rich yellow fls / Carpathians, Balkans 219
subdiscoideus / HH R / 6" / orange fls / South America 64
thapsoides / P/8-28"/stems woody; dense wht-tom, achenes glab/Balk. 289
tomentosus / P / 8-30" / bright yellow fls / N.J., Fla., Ark. 224 / 225
tyrolensis - see S. abrotanifolius v. tirolensis 107
uniflorus / see S. halleri 148 / 289
Websteri / R / short / ragged yellow fls / ne Asia, Wash. 258 / 25
werneriifolius / P / to 6" / fls yellow / mts. w North America 25

SERAPIAS - Orchidaceae
neglecta / R / to 1' / brick-red fls / Medit. region 210 / 85
oblia / R / short / fls pale violet to purple 290
vomeracea / R / fls in spikes, 2-10, pale red / s Europe 290

SERIOCARPUS - Compositae
asteroides / P / 6-24" / white fls / Maine, Michigan, Florida 224

SERRATULA - Compositae
centauroides / P / to 32" / purple fls / ne Asia 258
coronata / P / to 5' / fls wine-red / e Europe & Siberia 25
lycopifolia / P / to 3' / florets purplish / ec Eur, s U.S.S.R. 289

[1] Ruffier-Lanche, Bull. American Rock Garden Society, 21:2.

seoanei / R /smaller than S. tinctoria, very nar. lvs / sw Eur. 289
Shawii - see S. seoanei 46 / (290)
tinctoria / P / 3' / fls red-purple / Eurasia 210 / 25
wolfii / P / to 5' / purple fls / s & c USSR, Roumania 289

SESAMUM - Pedaliaceae
indicum / A / to 3' / fls pale rose or white; oil seeds / Tropics 184 / 25

SESELI - Umbelliferae
caespitosum - see OLYMPOSCIADUM caespitosum 77
libanotis / B or P / to 4' / fls white or pink / Europe 184 / 287

SETARIA - Gramineae
glauca / A G / to 3' / fls loosely spiral up / Eurasia, nat. U.S. 99
italica / A / to 5' / Foxtail Millet / warm-temperate Asia 210 / 268
pumila - see S. glauca 99

SETIECHINOPSIS - Cactaceae
mirabilis / small cactus / to 6" / fls white / n Argentina 25

SEVERINA - Rutaceae
buxifolia / Gh T / small / fls white; frs black / s China 25

SHORTIA - Diapensaceae
galacifolia / R / 6-8" / white fls / s Appalachians 225 / 107
soldanelloides / R / to 9" / fls deep rose to white / Japan (606) / 25
soldanelloides 'Alba' - the white form *
soldanelloides v. illicifolia - leaf-margins coarser 25
soldanelloides v. magna - lvs larger, many small teeth 25
soldanelloides minima / R / very dwarf form from high mountains 200
soldenelloides uniflora - crossed by donor?
uniflora / R / 3-6" / shell-pink or white fls / Japan 164 / 129
uniflora 'Grandiflora' - larger fls (249) / 25
uniflora 'Rosea' - deep rose-pink fls 25

SIBARA - Cruciferae
virginica / A / 4-16" / fls white or pink-tinged / e & c U.S. 99

SIBBALDIA - Rosaceae
macrophylla / R / 4" / small yellowish fls / c Asia 7
procumbens / P / procumbent / small yellow fls / N. Am., Eurasia 99 / 25
procumbens 'Grandiflora' - larger fls *

SIBIRAEA - Rosaceae
laevigata / Sh / to 5'/ bark scaly; lvs bl-grn; fls gr-wht /Siberia 25

SIDALCEA - Malvaceae
campestris / P / 2-7' / fls white or pale rose / Oregon 288
candida / P / 2-3' / white fls / Wyo., Col., Utah 229 / 25
'Malvaeflora Hybrids' / P / 4' / fls in shades of pink 279
oregans / P / 1-6' / fls pink / Idaho to Washington 25
ranunculaceae - see S. spicata v. ranunculacea 227
spicata v. ranunculacea / under 3' / pink fls / Sierra Nevada 227
'Stark Hybrids' / P / 3' / white, pink, purplish fls 135

SIDERITIS - Labiatae
candicans / Gh Sh / to 3' / yellow fls / Canary Isls. 25
euboea - see S. syriaca 288
glacialis / R / to 6" / yel fls, brn. markings / s Spain, in mts. 288
hyssopifolia / P / 16" / yellow fls / sw Europe 148 / 288
incana / P / to 24" / 6-flowered, yellow and pink 288
macrostachys / Sh/ to 3'/ lvs br-grn above, wht beneath /Canary Ils. 25
mertensiana / R / to 1' / fls wht, anthers red / Rockies, N.Amer. 126
scordioides / R / to 1' / yellow fls / s Europe 212 / 288
syriaca / P / 4-20" / yellow fls / s Europe 288
taurica / see S. syriaca 288

SIEVERSIA
ciliata - see GEUM triflorum 299
pentapetala - see GEUM pentapetalum 200
reptans - see GEUM reptans 25

SILENE - Caryophyllaceae
acaulis / R / 1-4" / deep pink fls /arctic & alpine in n hemisphere 148 / 286
acaulis 'Alba' - fls white 25
acaulis v. elongata - fl stems to 4" 25
acaulis v. exscapa - fl stems to 3/16" 25
acaulis ssp. longiscapa - flowering stems extended 286
acaulis pannonica / P / mat-forming / fls deep pink / arctic Europe 286
acaulis 'Pedunculata' - see S. acaulis v. elongata 25
acaulis v. subacaulescens - longer peduncles 213
alba / A or P / to 32" / fls usually white / most of Europe 286
alpina - see S. vulgaris ssp. prostrata 286
alpestris / R / to 1' / usually white fls / Alps, n Balkans 148 / 286
alpestris 'Flore Pleno'- densely rosetted fls 123
altaica / P / to 20" / fls white / e Russia 286
apetala / A / 4-14" / petals may be absent / s Europe 286
armeria / A or B / 16" / fls usually pink / c & n Europe 228 / 286
armeria 'Lady Pink' - garden selection, clear pink fls (291)
asterias / P / 3'+ / deep purple fls / Balkans, in mts. 286
boryi / R / mat / fls pink above, red below / Spain 212 / 286
californica / R / flopping / brillant scarlet fls / w U.S. 227 / 107
caroliniana / R / tuft / deep pink to white fls / e U.S. 224 / 107
caroliniana ssp. pennsylvanica - calyx glandular 25
caroliniana Wherryi /calyx without glands, broad/dry,sand, N.H.-Ala. 25
caryophylloides Echinus /P/to 16"/petals pink;calyx, no glands/Turk. 25
caucasica / P/ to 10"/ wht or pink; petals lobed/ mts. Cauc.-Anarct. (105) / 25
ciliata / P/to 12"/petals wht or pink, gr-red below/Port. to Greece 25
ciliata v. graefferi / fls wht-pnk, grn-red beneath;lvs large/Greece 286
coeli-rosea /?/ 8-20" / pink fls / se Europe 286
colorata / A / branched / fls pink or white / s Europe, n Africa 194 / 25
compacta / A or B / 15" / pink fls / se Europe 77 / 286
compacta pendula - see S. pendula 'Compacta' 28
conica / A / 6-20" / pink fls, rarely white / c & s Europe 286
cretica / A / to 12" / petals pink / n Africa, s Europe 25
cucubalus - see S. vulgaris 286
Delavayi / R / 3-8" / fls deep purple / Yunnan 61
dianthoides / P / tufted / fls white / Asia Minor 61
dinarica / R / mat forming / petals pink / mts. Romania 25

dioica / P / 30" / red fls / Europe	148 /	286
dioica 'Rose Queen' - garden selection		*
Douglasii / P / 16"+ / fls white or pink / Calif. to B. C.	228 /	229
Douglasii var. macounii - high mt. form		28
Douglasii v. monatha - dwarfer, scarcely hairy		142
echinata / P / ½-2' / fls pink or white / c & s Italy		286
Elizabethae / R / 5" / fls dark red to reddish-purple / Alps	148 /	286
flavescens / P / to 12" / densely hairy; fls yellow / mts Hungary	289 /	25
foetida / R / to 8" / pink fls / n Portugal, nw Spain	212 /	286
fortunei / P / 1½-3' / fls rose or white / China		28
furcata / R / to 1' / whitish fls / arctic Europe		286
Frivaldszkyana / P / 1'+ / fls whitish / Balkan Peninsula	77 /	286
gallica / A / to 1½' / fls white or pink / s & c Europe	211 /	286
gallica v. quinquevulnera - fls crimson-spotted	165 /	286
glauca - see S. secundiflora		286
gracillima / P / to 3' / fls white / mts c Japan		25
hallii - see S. scouleri ssp. hallii		
heuffellii / B / to 32" / white fls / n Balkans		286
hifacensis / P/ 8-12"/ heavy stem;fls red or purp/Baleric Ils, Spain		286
Hookeri / R / 6" / pink, purp., violet, white fls / s Ore., n Calif.		25
Hookeri v. Bolanderi / pure white / laciniated fls / Siskiyous		(221)
hyponica / A or B / to 16" / fls usually pink / Ukraine		286
Ingramii - included in S. Hookeri; in hort. a cherry-red segregate		(292)
Ingramii 'Alba' - white fls		*
integripetala / HH R / 4-16" / pink fls / s Greece		286
Keiskei / R / 6-8" / fls in shades of pink / Japan		107
Keiskei alba - white form		25
Keiskei minor /3"/lvs, ped., calyx soft-hairy; br pk fls;lvs brn-grn	123 /	25
laciniata / P / to 3' / crimson fls / Calif., N.Mex., in mts.	227 /	25
Lerchenfeldiana / R / 10" / fls reddish / Carpathians		286
linicola / A / ? / pink fls / c Europe; weed in flax fields		286
Macounii - see S. Douglasii var. Macounii		28
macropoda / R / to 10" / white fls / Bulgaria, in mts.		25
maritima - see S. vulgaris ssp. maritima	184 /	286
maritima islandica - see S. vulgaris var. islandica		286
[1] maritima 'Robin White Breast' - silv-gray fol. cush.;wht dble bell fls		
maritima 'Rosea' - see S. vulgaris ssp. maritima 'Rosea'		123
maritima 'White Bells' - see S. vulgaris ssp. maritima 'White Bells'		*
marshallii / P / ½-1' / fls white / sw Asia		286
noctiflora /A/to 3'/ fls wht or pink, frag /open night/Eur, sw Asia		25
nodulosa / P / 16" / fls white above, red below / Corsica		286
odontopetala / R / to 1' / fls whitish to pinkish / Near East		77
orientalis - see S. compacta		61
nutans / P / to 2' / fls variable in color / Europe	210 /	286
paradoxa / P / 2'+ / fls cream or yellow / Europe		286
pauciflora - see S. nodulosa		286
pendula / A / to 2' / pink fls / s Medit. region		286
pendula 'Compacta' - cushion to 4"		25
petraea - see S. saxifraga		286
pseudovelutina / HH P / 2' / whitish fls / s Spain	212 /	286
pumilio - see SAPONARIA pumilio		286

[1] Thompson & Morgan Catalog, 1986.

pusilla / R / 6" / fls white, rarely pink or lilac / s & c Europe	148 /	286
pulsilla ssp. albanica / P / to 6" / fls wht; rare pnk-lilac/Albania		286
pygmaea / R / to 8" / fls rose / Caucasian alps		25
quadridentata - see S. pusilla		286
quadrifida / see S. pusilla		286
quadrifida villosa - see S. alpestris		
regia / P / fls crimson / Ohio to Mo., s to Ga.		25
Regis-Ferdinandii / R / 4" / solitary white fls / Bulgaria	64 /	286
Reichenbachii / R / 6-12" / white fls / w Jugoslavia		286
repens / P / to 2' / stoloniferous fls white / Rus., Alaska, Yahu		25
repens latifolia / P / 4-12"/ densely hairy; fls pl rose-wht / Japan		200
'Robin White Breast' - see S. vulgaris ssp. maritima		203
'Rose Queen' - see S. dioica 'Rose Queen'		*
rubella / A / 4-20" / pink fls / w Medit. region		286
rupestris / R / 10" / fls white or pink / w & c Europe	148 /	286
ruprechtii - see S. saxatilis		77
sachalinensis / P / mats to 4" / white fls / e Siberia		25
saxatilis / P / to 20" /fls greenish-yellowish-whitish/ Turkey, Iran		25
saxifraga / R / 8" / fls whitish or greenish / s Europe	148 /	286
saxifraga v. parnassica / P/to 8"/single fls, wht or grn/se Eur,mts.		286
schafta / R / 3-6" / fls pink or purple / Caucasus	93 /	25
scouleri ssp. hallii / P/ tall / fls wht, drk veins/ Rockies, slopes [1]		
secundiflora / A / 4-20" / fls pink or white / s & e Spain		286
sicula / P / ½-3' / fls red or purple / s Italy, Sicily, Greece		286
stellata / P / to 3' / fringy white fls / Mass. to Texas	224 /	225
stricta / P / 6-16" / fls pink, red or green veined / Medit., Port.		286
succulenta / HH P / procumbent / fls white / Isls. of Medit. Sea		286
succulenta ssp. corsica / procumbent / white fls / Corsica		286
uniflora / R / 6" / fls white / arctic Russia		245
uralensis ssp. apetala - see S. Wahlbergella		276
vallesia / R / 6" / fls pale pink above, red below / w Alps	148 /	286
vallesia graminea / P/mats/lvs linear; fls pink-red/Alps Ital, Balk.		286
virginica / P / 2'/ fls crimson to scarlet / e U.S.	224 /	107
viscaria - probably LYNCHIS viscaria		*
viscariopsis / R / 8-12" / deep purple fls / s Macedonica		286
vulgaris / P / 2' / fls whitish / all of Europe	148 /	286
vulgaris var. islandica / subcaespitose / fls whitish		286
vulgaris ssp. maritima / 10" / fls whitish / coasts of Europe	210 /	286
vulgaris ssp. maritima 'Flore Pleno' - double fls, more compact	147 /	123
vulgaris ssp. maritima 'Rosea' - pink fls		123
vulgaris ssp. maritima 'White Bells' - white selection		*
vulgaris ssp. prostrata / P/ mat to 10"/ fls white / Alps	286 /	25
vulgaris 'Rosea' see S. vulgaris ssp. maritima 'Rosea'		*
Wahlbergella / R / to 8" / fls dull reddish-purple / arctic Europe		286
Waldsteinii / R / 6" / white fls / mts. of Balkan Penisula		286
Wherryi - see S caroliniana ssp. Wherryi		107
'White Bells' - see S. vulgaris ssp. maritima 'White Bells'		*
Zawadskii / P/ to 1'/ lvs to 5"; calyx hairy/ e Carpath. Rom. & Ukr.		25

---

[1] William A. Weber, Rocky Mountain Flora, 1976.

SILPHIUM - Compositae
perfoliatum / P / to 8" / yellow fls / c North America 257 / 25
terebinthinaceum / P / 3-9' / fls yel-grn / prairies S. Dak.-Mo. 99

SILYBUM - Compositae
marianum / B / 1-3' / fls pink to purple / Medit. region, Iran 129 / 77

SINNINGIA - Gesnariaceae
cardinalis / Gh / under 1' / scarlet fls / Brazil 25
leucotricha / Gh / 10" / salmon-red or rose fls / Brazil 25

SINOMENIUM - Menispermaceae
acutum / Cl / to 36' / yellowish fls; black frs / c & w China 139

SISYMBRIUM - Cruciferae
luteum / P / to 4' / fls yellow / Far East 200

SISYRINCHIUM - Iridaceae
albidum / C / 6-18" / fls white or pale violet / e North America 25
anceps - see S. angustifolium 25
angustifolium / R / 8-10" / pale to deep blue fls / e & c N. Am. 229 / 99
angustifolium 'Album' - white fls 25
arenicola / P / to 20" / blue-violet fls / e N. Am. 99
arizonicum / HH R / 8" / orange fls / New Mexico, Arizona 227
atlanticum / P / to 28" / blue-violet fls / e & c N. Am. 225 / 99
bellum / HH P / 4-20" / amethyst-violet fls, veined purple / Calif. 229 / 25
bellum 'Album' - white fls 25
bermudiana / HH P / 1-2' / violet blue fls / Bermuda 129 / 25
bermudiana 'Album' - white fls *
birameum / P / 12-20" / dark blue fls, yellow eye / sw Washington 25
boreale - see S. californicum 25
brachypus - see S. californicum 25
californicum / HH R / 6-16" / bright yellow fls / California 229 / 25
californicum album /B /lvs glaucous; 1 wht fl/Ore, Cal, nat. Ireland 290 / 228
campestre / P / 4-20" / fls blue, violet, white / c U.S. 224
'Chilean Andes' / HH Bb / 6-12" / fls purple, yel at base / Brazil 61
coleste - listed name of no botanical standing 25
convolutum / HH R / 6-12" / yellow fls, veined br. / Mex. to Ecuador 25
cuspidatum / HH P / 1-2' / yellow fls / Chile 25
depauperatum / Gh / short / pink fls / Chile, possibly 18
Douglasii / R / 6-10" / satiny purple fls / w N. Am. 24 / 107
Douglasii 'Album' - white fls 28
Douglasii v. inflatum - base of column inflated 73
filifolium / R / 6-8" / white fls, veined red / Falklands (293) / 123
graminifolium / HH R / 6-8" / fls yellow / Chile 93
graminifolium v. maculatum / 9"/ yellow fls, red-spotted/ Valparaiso 121
graminoides - see S. angustifolium 25
grandiflorum - see S. Douglasii (479) / 25
grandiflorum 'Album' - see S. Douglasii 'Album' 107
idahoense / P / 8-20" / fls dark violet-blue / Idaho, Mont. 229 / 25
idahoense v. Macounii - purple fls 276
inflatum - see S. Douglasii v. inflatum 229 / (293)
iridifolium - possibly S. convolutum 25
junceum - see S. juncifolium 25
junceum 'Roseum' - see S. juncifolium *

juncifolium / HH P / 18-28" / fls rose-colored / Chile 25
littorale / R / 15" / violet-blue fls / B.C. to Alaska 25
Macouni - see S. idahoense v. Macouni 276
Macouni 'Album' - see S. angustifolium 'Album' *
micranthum / A / 3-8" / white fls, suffused yellow / South America 25
montanum / P / 1½' / blue-violet fls / c U.S. to B.C. 224 / 227
montanum v. crebrum - lvs dark green or tinged purplish 25
mucronatum / R / 1' / blue fls / Mass., Va., Mich. 93
mucronatum 'Album' - fls white 25
pachyrhizum / Gh / 2' / yellow fls / Brazil, Argentina 25
sarmentosum - see S. angustifolium 142
striatum / HH P / 15-30" / greenish-yellow fls, brown veins / Chile 129 / 25
tenuifolium / HH R / tufted / greenish-yellow fls / Mexico 64

SKIMMIA - Rutaceae
X Formannii / Gh Sh / 20" / scarlet frs 25
japonica / HH Sh / 3' / white fls; red frs; lvs evergreen / Japan 110 / 139
japonica 'Foremannii' - see S. X Foremannii 25
japonica v. intermedia f. repens/low shrub/ flat lvs/ Japan, in mts. 200
japonica repens - see S. japonica v. intermedia f. repens 167 / 200
Reevesiana / HH Sh / 2½' / crimson-red frs / China 133 / 139

SMELOWSKIA - Cruciferae
calycina / R / to 6" /white, cream, purpl. fls / arctic N. Am., Asia 13 / 25
calycina v. americana - included in the sp. 96

SMILACINA - Liliaceae
amplexicaulis / P / 3' / plumey white fls / w North America 107
japonica / P / 8-24" / fls whitish / Far East 200
oleracea / P / to 4' / fls wht, rose tinged; fr purple / Sikkim 61
racemosa / P / 2-3' / creamy-white fls / North America 22 / 129
sessilifolia / P / 2' / white fls; purple frs / Pacific States 107
stellata / P / 18" / starry white fls / U.S. & Canada 224 / 107

SMILAX - Liliaceae
aspera / Gh Cl / high / pale green fls; red frs / s Europe 211 / 128
herbacea / Cl / to 6' / bluish-black frs / ne U.S., s to Ala. 224 / 25
rotundifolia / woody Cl / rampant / blue-black frs / e U.S., Canada 192 / 25

SMYRNIUM - Umbelliferae
perfoliatum / B / to 5' / frs brownish-black / s Europe, Czechosl. 210 / 268

SOLANUM - Solanaceae
aculeatissimum / P or Sh / 1-2' / frs orange, yellow / tropics 28
aviculare / Gh Sh / 10' / fls violet-blue / Australia, N. Z. 45 / 25
capsicastrum / HH Sh / 1-2' / frs orange-red / Brazil 28
carolinense / P / 8-28" / violet or white fls / e N. Am. 250
ciliatum - see S. aculeatissimum 28
crispum / HH Cl / to 18' / purple-blue fragrant fls / Chile 208 / 139
crispum autumnale - see S. crispum 'Glasnevin' 184 / 139
crispum 'Glasnevin' - flowering season extended 110 / 139
eleagnifolium / P / 2' / silvery lvs; purplish fls / c U.S. 229
incanum / HH Sh / 3' / gray-woolly lvs; violet fls / tropics 211
laciniatum - see S. aviculare 25
pseudocapsicum / Gh / pot plant / Jerusalem Cherry / tropics 93 / 25

seaforthianum / Gh / trailing / pink or lilac fls / S. America 128

SOLDANELLA - Primulaceae
alpina / R / 2-3" / fls pale to deep mauve / Alps 148 / 129
austriaca / R / 4" / pale violet fls / Austrian Alps 288
carpatica / R / to 6" / violet fls / w Carpathians 288
carpatica 'Alba' - white fls *
cyanaster / R / 6" / pure blue fls / Bulgaria (12)
hungarica / R / 4" / violet fls / ec Europe, in mts. 288
hungarica major / P / 8-10" / fls violet / ec Eur, Balkan Pen. 288
minima / R / 4" / fls pale bluish-purple / s Alps 148 / 25
minima 'Alba' - white form of S. minima 288
minima X austriaca - crossed by donor? 288
montana / R / 3-6" / fls pale or deep mauve / Pyrenees 148 / 129
montana 'Alba' - white form of S. montana 288
montana ssp. hungarica - see S. hungarica 288
pindicola / R / to 3½" / fls rose-lilac / Albania, Greece 149 / 25
pusilla / R / 2-3" / pale lavender fls / European Alps 123 / 148
villosa / R / 4-5" / blue-amethyst fls / Pyrenees 18 / 123

SOLENOMELUS - Iridaceae
chilensis / HH P / 12-18" / yellow fls / Chile 25
pedunculatus - see S. chilensis 61
sisyrinchium / HH P / ? / blue fls / Chile 93

SOLIDAGO - Compositae
algida - see S. multiradiata 25
alpestris - see S. Virgaurea v. alpestris 97
altissima / P / to 6½' / dense pyramidal panicle / e & c N. Am. 308 / 25
bicolor / P / to 32" / fls cream & yellow / e N. Am. 99
brachystachys - see S. Virgaurea 25
caesia / P / 2-3' / yellow fls / e N. Am. 224 / 207
californica / P / to 4' / fls yel. in dense nar. thyrse/sw Ore-Baja 25
canadensis / P / 1-4' / fls yellow-green / ne North America 99
canadensis v. scabra - see S. altissima
Cutleri / R / to 1' / yellow fls / Maine to N. Y., in mts. 22 / 107
gigantea / P / 3-7' / yellow fls / North America 210 / 99
glomerata / P / 1-4' / yellow fls / N. Carolina, Tenn. 60 / 225
graminifolia / P / 1½-4' / lvs narrow lanc; 12-15 yellow fls/ne N.A. 99
lapponica - see S. Virgaurea 28
'Lemore'
minuta - see S. Virgaurea v. minuta 123
minutissima - see S. Virgaurea v. minutissima 200
missouriensis / P / to 2 1/4' / fls yel heads in panicles/w N.A. 25
mollis / P / 6-20" / bright yellow fls / c United States 99
multiradiata / P / 18" / yellow fls, dense corymbs / w N. Am. 229 / 25
multiradiata v. parviceps / P / stems slender/1-16"/fls dense/Nfld. 99
multiradiata scopulorum - a synonym 25
nemoralis / P / to 4' / fls pale yellow / e & c N. Am. 224 / 99
nemoralis v. decemflora - larger flower head 259
nemoralis v. longipetiolata - see S. nemoralis v. decemflora 259
odora / P / to 3' / fls yellowish; lvs anise-scented / e & c U.S. 224 / 99
puberula / P / to 3' / fls orang-yellow / e North America 99
roanensis / P / 16-32' / deep yellow fls / W. Va., Tenn., Ga. 225 / 99

rugosa / P / to 6' / fls yellow, in large panicles / ne N.A. 99 / 25
sempervirens / P / to 8' / succulent lvs; fls lg panicles/Nfld-N.J. 99 / 25
spathulata / P / to 2' / yellow fls; aromatic lvs / s coastal dunes 228 / 142
spathulata f. nana / yellow fls / 6-12" / Rocky Mts. 228
spathulata v. neomexicana / 2½'/infl elongate, not arom./w N. Amer. 25
squarrosa / P / 3'+ / yellow fls / e North America 313 / 99
tenuifolia / P / 10-32" / bright yellow fls / ne North America 224 / 99
Virgaurea / P / to 3' / yellow fls / Europe 210 / 25
Virgaurea v. alpestris - see S. Virgaurea ssp. minuta 289
Virgaurea v. asiatica / 6-28" / yellow fls / Japan 200
Virgaurea ssp. leicarpa /P/to 3'/fls yel, dense term.thyrs/Eur, mts 95
Virgaurea ssp. minuta/to 8" / rather coriaceous lvs/ arc. & alp. Eu. 289
Virgaurea v. minutissima / under 4" / high mts. of Kyushu 200

SOLLYA - Pittosporaceae
fusiformis / Gh Cl / to 6' / fls bright blue / Australia 164 / 128

SONCHUS - Compositae
congestus / P / to 2' / fls brn-yel, 2-3" across / Canary Islands 95
heterophylla - see S. fusiformis 194

SOPHORA - Leguminosae
japonica / T / to 30' / fls yellowish-white / China, Korea 167 / 25
macrocarpa / HH Sh / 5-8' / yellow fls / Chile (418) / 139
microphylla / HH T / to 30' / yellow fls / New Zealand 238
prostrata / Sh / to 6' / fls orange to yellow / New Zealand 193 / 25
secundiflora / HH T / 35' / fls violet-blue / Texas, N. Mexico 28
tetraptera 'Grandiflora' - slightly larger fls 139
tetraptera v. microphylla - see S. microphylla 238
viciifolia / HH Sh / 3-7' / bluish white fls / China 139

SORBARIA - Rosaceae
arborea / Sh / to 18' / white fls / China 25
grandiflora / Sh / 3' / white fls / e Siberia 222
sorbifolia / Sh / to 15' / creamy-white fls / Japan 300 / 139

SORBUS - Rosaceae
alnifolia / T / to 60' / small, bright red frs / Far East 300 / 139
alnifolia v. submollis - lvs broader, pubescent beneath 139
americana / Sh or T / to 30' / bright red frs / e N. Am. (609) / 25
aria 'Decaisneana' / T / 35' / large crimson frs 139
aria 'Majestica' - see S. aria 'Decaisneana' 139
aucuparia / T / 20-40' / frs bright red / Europe to w Asia 148 / 139
aucuparia 'Fastigiata' / habit narrow / frs large (610) / 25
aucuparia lanuginosa / more hairy / lfts coarsely toothed / s Eur. 36
aucuparia xanthocarpa / hyb. from European sp / frs amber yellow 139
cashmiriana / T / 15-30' / soft pink fls; white frs / Kashmir 129 / 139
chamaemespilus / Sh / 4½' / pink fls; scarlet frs / c & s Europe 287
commixta / T / 20'+ / scarlet frs / e Asia 167 / 25
cuspidata / T / 35'+ / lvs whitish beneath / Himalayas 139
decora / T / to 30' / red frs / Labrador, N. Y., to Minn. 25
discolor / T / 15-30' / creamy-yellow frs / n China 139
'Embley' / T / sm-med / heavy shiny red lvs; orange-red fr/Cultivar 139
esserteauana / T / to 30' / small scarlet frs / w China 139
hedlundii - of cult. is S. cuspidata 139

hupehensis / T / 15'+ / lvs bluish green; frs white / w China 139
hupehensis v. obtusa / T / 15'+ / pink frs 139
[1] 'Joseph Rock' / T / 25'+ / yellow frs / origin unknown 129 / 139
lancastriensis / T / small / crimson frs / Britain 287
lanuginosa - see S. X thuringiaca 'Decumens' 36
matsumurana / T / small / white fls; red frs / Japan, in mts. 110 / 200
minima / Sh / to 9' / lvs gray-tomentose beneath; fls wht / Wales 287
occidentalis / Sh / to 9' / frs red, bloomy / B. C. to Oregon 222
pohuashanensis / T / 30'+ / heavy frs, orange-red / n China 139
poteriifolia / T / 15'+ / deep rose-pink frs / China 139
prattii / Sh / 10' / pearly-white frs / China 129 / 139
prattii v. subarachnoidea - white frs; cobwebby lvs 139
reducta /Sh/1-2'/13-15 dk grn lflts; sm fr wht-rose/n Burma, w China 129 / 139
rufo-ferruginea / T / 15' / reddish frs / Japan 139
sambucifolia / Sh / to 7' / red frs / Far East 273 / 222
sargentiana / T / 25' / small scarlet frs / China 139
scalaris / T / 30' / orange-scarlet frs / China 129
scopulina / Sh / 3-12' / frs orange to scarlet / Rocky Mts. & w (611) / 25
serotina / T / 15' / small, bright orange-red frs / w China 139
sitchensis / Sh / 4-15' / fls 5" across; br red frs/Alas, B.C., Id. 299 / 25
sitchensis v. grayii / Sh / to 4' / lvs nearly entire/ B. C. to Calif. 142
X thuringiaca 'Decarrens' / 5-7 pr. lflts / frs not freely born 36
vilmorinii / Sh or T / 10-18' / round red frs / China 129 / 25
Zahlbruckneri / T / to 20' / lvs grn over gray; fls wht; fr red/China 61

SORGHASTRUM - Gramineae
nutans / P Gr / 2' ornamental yellowish fls / e & c U.S. 73 / 268

SORGHUM - Gramineae
nigrum / A Gr / 4' / former conservatory plant / India 206

SPARAXIS - Iridaceae
bulbifera / HH C / 9" / violet fls / Cape of Good Hope 206
elegans / HH C / 6-8" / white, orange, maroon fls / sw Cape Prov. 185
grandiflora ssp. acutiloba / Bb / 4-14" / yellow fls / S. Africa 185
X streptanthera - bi-colored hybrids 267
tricolor / HH C / 15" / orange-red fls / Natal 180 / 111

SPARTIUM - Leguminosae
junceum / Sh / to 10' / deep yellow fls / s Europe 212 / 129

SPECULARIA
speculum - see LEGOUSIA speculum-veneris 208
sepculum-veneris grandiflorum - see LEGOUSIA s-v. 'G.' 283

SPERGULA - Caryophyllaceae
arvensis / A / to 2' / white fls, weedy / temperate zone 25

SPERGULARIA - Caryophyllaceae
media / R / 2-16" / fls white or pink / coasts of Europe, N. Z. 193 / 286
rubra / R, sometimes A / 10" / pink fls / Europe 210 / 286
rupica / R / 2-14" / uniform pink fls / Atlantic coasts of Europe 210 / 286

[1] Hilliers Hundred Catalog, p. 35.

SPHAERALCEA - Malvaceae
ambigua / P / to 36" / fls orange-red / Utah to Mexico 229 / 25
coccinea / R / decumbent / fls orange to red / c N. Am. 229 / 25
Munroana / P / 3' / orange fls / B. C., Utah, Nevada 229 / 25

SPIGELIA - Loganiaceae
marilandica / P / 1-2' / tubular red fls / N. J. to Wisc. 226 / 25

SPILANTHES - Compositae
oleracea / A / creeping / edible lvs / tropics (612) / 25

SPILOXENE - Amaryllidaceae
capensis / Gh Bb / 1' / white or yellow fls / Cape of Good Hope 184 / 223

SPIRAEA - Rosaceae
alba / Sh / 1-4' / small white fls / nc North America 224 / 25
X arguta (S. X multifora x S. Thunbergii) / 3-6'/ wht fls/n. hemis. 25
bella / HH Sh / 3' / fls white to pink / Himalayas 253 / 25
betulifolia / Sh / 2-3' / fls white / ne Asia, Japan 25
betulifolia aemiliana / Sh / sm thorn. sp./ fls wht / Japan, Siberia 200
betulifolia v. lucida / Sh / 1-2' / white fls / B. C. to Oregon 142
X Billardii / Sh / to 6'/ like S. Douglassii / fls pink / B.C.-Calif 25
bullata / Sh / to 1½' / deep rose fls / Japan, ex cult. 123 / 25
X Bumalda / Sh / to 2' / fls white to deep pink 110 / 25
X Bumalda 'Nana' / prostrate to 6" / bright rose fls 123
decumbens / Sh / almost prostrate to 10" / fls wht / s Europe 25
densiflora / Sh / 1-3' / small rose fls / Calif. to B.C., in mts. 247 / 25
densiflora lucida - see S. betulifolia v. lucida 142
digitata 'Nana' - see FILIPENDULA palmata 'Nana' *
Douglasii / Sh / 3'+ / fls purplish-rose / w N. Am. 107
[1] humilis / P / 2' / rose-colored fls / Caucasus
hypericifolia v. acuta / Sh / 3'+ / yellowish-white fls / Eurasia 222
japonica / Sh / 4' / pink fls / Far East 139
japonica 'Alpina' / Sh / to 1' / rose-pink fls 300 / 139
X Lemoinei 'Alpestris' / Sh / 3' / fls rose pink / n hemisphere 25
thunbergii / Sh / 5' / white fls / Far East 110 / 25
tomentosa / Sh / 3'+ / deep rose or rose-purple fls / n N. Am. 224 / 222
trichocarpa / Sh / 6' / white fls / Korea (454) / 222

SPIRANTHES - Orhidaceae
cernua / P / 6-25" / fls white or yellowish / e North America 107
gracilis / P / 8-18" / fragrant white fls / e North America 28
Romanzoffiana / P / to 2' / white fls / U.S. & Canada 229 / 25
sinensis / P / 4-28" / fls red, pink or white / Far East 258

SPRAGUEA
multiceps - see CALYPTRIDIUM umbellatum 25
umbellata - see CALYPTRIDIUM umbellatum 25

---

[1] Robert M. Senior, Quar. Bull. Alpine Garden Society, 40:2.

SPREKELIA - Amaryllidaceae
formosissima / HH Bb / to 1' / fls bright crimson / Mexico 129 / 25

STACHELINA - Compositae
dubia / P / 8-16" / purple fls / sw Europe 212 / 289

STACHYS - Labiatae
Alopecurus / P / 8-24" / fls pale yellow / c & s Europe 288
alpina / P / to 3' / fls dull purple / w, c & s Europe, in mts 148 / 288
betonica - see S. officinalis 288
byzantina / P / 6-32" / small purple fls; lvs woolly / Turkey 300 / 288
candida / HH R / 4-8" / fls white with purple spots / s Greece 149 / 288
citrina / Sh / lvs 1½" long; fls yellow / Greece 61
coccinea / P / 1-2' / fls scarlet-red / w Texas, Arizona 28
cretica / P / 8-32" / lvs grn; fls pink & purple / s Eur, s France-e 288
densiflora - see S. monieri 288
discolor / R / 1' / rose or yellowish white fls / Caucasus 28
germanica / P / to 3' / pink or purple fls / w, s & c Europe 210 / 288
germanica v. boisseri - see S. cretica 288
grandiflora / P / 18" / violet fls / Caucasus 25
grandiflora 'Superba' - cultivar 25
hyssopifolia / P / 8-40" / fls pink spotted purple & white / c U.S. 99
iva / R / 6-16" / wht. lanate pl; fls yel. / s Jugoslavia, n Greece 288
lanata / see S. byzantina 129 / 25
lavandulifolia / R / 6-8" / red-purple fls / Caucasus to Iran 93
macrantha - see S. grandiflora 129 / 25
monieri / R / 4-16" / fls pink / Alps, Pyrenees (295) / 288
nivea / HH R / 6-12" / white fls / Syria 25
officinalis / P / to 3'/ fls bright reddish-purple / most of Europe 288
officinalis alba / P / to 3' / fls white / Europe, Asia 25
recta / P / 6-40" / pale yellow fls / c Europe 210 / 288
saxicola / HH R / prostrate / white-woolly lvs; cream fls / Morocco 64
spinosa / HH R / to 12" / pale pink fls / s Aegean region 288
sylvatica / P / 1-4' / fls dull reddish-purple & white / most of Eu. 288

STACHYURUS - Stachuruaceae
praecox / HH Sh / 12' / greenish-yellow fls / Japan 110 / 25

STACKELINA
dubia / SH / 8-16" / fls purple, bracts red & grn / sw Europe 289
uniflorsculosa / Sh / to 20" / lvs dk grn; fls purple / Balkan Pen. 289

STACKHOUSIA - Stackhousiaceae
monogyna / HH P / to 2' / fls cream / Australia 45 / 28

STAPELIA - Asclepiadaceae
hirsuta / Gh / 8" / purple-brown fls / South Africa 25

STAPHYLEA - Staphyleaceae
holocarpa / Sh or T / 25' / fls white; fr 1½' long / China 25
holocarpa rosea - sim. to S. holocarpa, with pink fls 25
pinnata / Sh / to 15' / inflated fruit capsules / Eurasia 28
trifolia / Sh / 6-15' / white fls / e & c North America 28

STATICE
bellidifolia - see LIMONIUM bellidifolium 288
cosyriensis - see LIMONIUM cosyriensis 123
dumosa - name of no botanical standing 25
minutum - see LIMONIUM minutum *
perezii - see LIMONIUM perezii 28
peticulum - see LIMONIUM pectinatuna 61

STAUNTONIA - Lardizabalaceae
hexaphylla / Cl / to 40' /fls wht, frag, viol. tinge / Japan, China 25

STEIRODISCUS
tagetes - see GAMOLEPIS Tagetes 25

STEIRONEMA - Primulaceae
ciliata / P / 1-4' / yellow fls / United States 28

STELLARIA - Carophyllaceae
alaskana / R / to 5" / white fls / Alaska, Yukon 20
graminea / R / 6" / white fls / Europe 28
pubera / R / to 1' / fls white / New Jersey - Illinois & s-ward 99 / 263
ruscifolia / R / to 6" / fls few, solitary / alpine c & n Japan 25

STEMMATUM
narcissoides - see LEUCOCORYNE narcissoides (191)

STENANTHIUM - Liliaceae
gramineum / R / 8" / wht to grn to purplish fls / e U.S. s to Texas 224 / 99
gramineum v. robustum - se Pa., Md. and D.C. to Ind.
occidentale / P / to 2' / fls purplish-green / nw North America 229 / 25
robustum - see S. gramineum v. robustum 99

STENOLOBUM
stans - see TECOMA stans 28

STENOPHYLLUS
capillaris - see BULBOSTYLIS capillaris 99

STEPHANANDRA - Rosaceae
incisa / Sh / to 8' / whitish fls / Japan, Korea 25

STERNBERGIA - Amaryllidaceae
1 candida / Bb / to 10" / fls white; fr / Turkey
clusiana / Bb / 1' /fls bright golden-yellow / A. Minor, w Iran 185 / 267
colchiciflora / HH Bb / 3" / pale yellow fls / se Europe 185 / 111
Fischerana / Bb / 6" / bright yellow fls / Caucasus, Iran (613) / 185
lutea / Bb / 6" / bright yellow fls / Medit. region to Russia 210 / 185
lutea sicula / Bb / 2-4" / lvs wider; fls yellow / s Italy, Sicily 290
sicula - see S. lutea sicula 290

STEVIA - Compositae
ivifolia / Gh / white fls / Mexico 206
satureifolia / Gh Sh / to 18" / fls white / Mexico 61

---

1 Brian Mathew and Turhan Baytop, The Bulbous Plants of Turkey, 1984.

STEWARTIA - Theaceae

koreana / T / 10'+ / white fls / Korea	(433) /	139
Malacodendron / Sh / 12' / purple eyed white fls / se U.S.	134 /	25
monadelpha / HH T / to 80' / white fls / c & s Japan	167 /	139
ovata / Sh / 10' / white fls / se U.S.	184 /	139
Pseudocamellia / T / 30'+ / white fls / Japan	300 /	139
pteropetiolata / HH Sh / 10'+ / white fls / Yunnan		139
sinensis / T / to 30' / fls white, 2" across / China		25

STILBOCARPA - Araliaceae

polaris / HH P / to 4' / yellowish fls / New Zealand		29

STIPA - Gramineae

avenaceae / P Gr / to 3'+ / Black Oat Grass / e & c U.S.		99
barbata / P Gr / 1-2' /narrow panicle, few spikelets/ w Medit. reg.		290
calamagrostis / P Gr / 3' / violet awns / s Europe		129
capillata / P Gr / to 3' / glaucous lvs, stiffly erect / Eurasia		25
gigantea / P Gr / to 7' / yellow spikelets / Spain, Portugal		184
Joannis / P Gr / 20-30" / Feather Grass / Altai, w Siberia		266
pennata / P Gr / 2½' / bearded awns / Europe, Siberia	148 /	129
pulcherrima / P Gr / 3'+ /pinnatte Feather Grass / se Europe	292 /	93
sibirica / Siberian Needle-grass / hardy at North Dakota		303
spartea / P Gr / 2-3' / Porcupine Grass / prairies, Pa. to N. Mex.		53
tenacissima / Gr / long leaf blades, awns to 2½" long / w Medit.		25

STOKESIA - Compositae

'BLue Moon' - see S. laevis 'Blue Moon'		135
caerulea - see S. laevis 'Caerulea'		29
cyanea - see S. laevis		25
laevis / P / 1-1½' / fls lavender-blue / North America	(296) /	129
laevis 'Alba' - white fls		207
laevis 'Blue Danube' / 12-15" / deep blue fls	300 /	135
laevis 'Blue Moon' - large silvery blue to lilac fls		135
laevis 'Caerulea' - light blue fls		29

STRANGWEIA - Liliaceae

spicata / Bb / 2½" / blue fls / Greece		185

STREPTANTHERA - Iridaceae

cuprea / C / 9" / fls coppery-yellow / South Africa		267

STREPTOCARPUS - Gesneriaceae

Fanniniae / Gh / creeping / fls pale blue to whitish / Natal		25
gardenii / Gh / 4-12" / pale violet fls / South Africa		25
grandis / Gh / 2'+ / blue or lavender fls / Natal, Zululand		93
X multiflorus / Gh / to 1' / large bluish-purple fls		28
polyanthus / P / large lvs; fls pale violet / Africa		25
primulifloris / Gh / 10" / blue-violet fls / S. Africa	194 /	25
rexii / Gh / 8" / white, bluish, mauve / South Africa		25
sylvaticus / Gh / 6" / pale violet fls / South Africa		25
woodii / Gh / 9" / fls bluish-mauve, purple spots / Natal		61

STREPTOPUS - Liliaceae

amplexifolius / P/ 3' / greenish-white fls / Eurasia	148 /	107

amplexifolius v. americanus - the North American variety 229 / 25
amplexifolius v. chalazatus - glabrous stems; entire lvs 142
roseus / P / 1-2' / fls purple or rose / e U.S. 22 / 107
streptopoides / R / to 1' / fls in light shades / w N. Am., Far East 200

STROBILANTHES
alternatus - see HEMIGRAPHS alternata 93

STROPHOSTYLES - Leguminosae
helvola / A / to 6', twining / pink or purple fls / e N. Am. 224 / 99
umbellata / Cl / low / fls purplish-pink / se & sc U.S. 225

STRUTHIOPTERIS
spicant - see BLECHNUM spicant 28

STYLIDIUM - Stylidiaceae
bulbiferum macrocarpum / P / to 6" / fls grn-purple / Australia 61
graminifolium / HH R / 6-12" / pink fls / Tasmania, Australia (152) / 123

STYLOMECON - Papaveraceae
heterophylla / A / to 2' / fls brick red, purple center / Cal & Baja 25

STYLOPHORUM - Papaveraceae
diphyllum / P / to 20" / yellow fls / e United States 129 / 25
lasiocorpum / P / to 18" / fls 4-5 on stem, yellow / c & e China 61

STYPANDRA - Lilaceae
caespitosa / Gh / 1' / dark blue fls / Australia 45
glauca / P / 1-3' / lvs 3-4"; fls blue / Australia 61

STYRAX - Styraxaceae
americanus / HH Sh / to 10' / whte fls / Va. to Fla. (318) / 25
Hemsleyana / HH T / 15'+ / white fls / c & w China 139
japonicus / Sh / 15'+ / fragrant white fls / Far East (418) / 25
obassia / Sh / 18' / fragrant white fls / Japan 167 / 139
officinalis v. fulvescens / large shrub / white fls / s California 294
Wilsonii / HH Sh / 6-10" / white fls / w China 139

SUCCISA - Dipsaceae
pratensis / P / 6-32" / fls lilac to dark violet / Caucasus 148 / 77

SULLIVANTIA - Saxifragaceae
oregana / R / 5" / white fls / gorge of the Columbia River 228
sullivantii / R / to 1' / white fls / Ohio, Kentucky, Indiana 99

SUTERA - Caryophyllaceae
microphylla / Gh Sh / to 1' / purple fls / South Africa 25

SUTHERLANDIA - Leguminoseae
frutescens / HH Sh / to 10' / terra-cotta fls / South Africa 139

SUTTONIA
nummularia - see MYRSINE nummularia 15

SWERTIA - Gentianaceae
alpestris / P / 6-24" / dark violet fls / Europe 288
bimaculata / A or B / 20-32" / creamy-white fls / Japan, in mts. 200
carolinensis / B / 3-4' / greenish-yellow fls, purp. dots / e N. Am. 99
columbiana - see FRASERA albicaulis v. columbiana 142
japonica / A or B / sm. / fls wht., purp. nerves / Far East, in mts. 200
[1] kilimandscharica / HH P / to 3' / white fls / Mt. Kilimanjaro
kingii / P / 2' / fls greenish-white / Himalayas 184
longifolia / P / 2' / yellow fls / Iran 204
multicaulis / P / to 8" / fls blue, fringed / Himalayas 25
perennis / R / 6-12" / dull purple fls / n hemisphere 148 / 123
perennis v. cuspidata / to 12"/ dark blue fls/ high mts. of Japan 25
perfoliata / P / to 4' / white fls, blotched dark blue / Himalayas 25
petiolata /P/to 4'/fls wht, dk blue blotch; seed not winged/Himal. 25
radiata - see FRASERA speciosa 226 / 25
speciosa - see S. perfoliata 25
thomsonii / A / ? / blue-gray fls / Himalayas 61

SYMPHORICARPOS - Caprifoliaceae
albus / Sh / 4' / white frs / e North America 139
albus v. laevigatus / larger frs / 6' / Alaska, Calif., to Colorado 142
[2] X chenaultii 'Hancock'
mollis v. hesperius / Sh / 18" / white frs / Washington, Idaho 142

SYMPHOSTEMON - Iradaceae
narcissoides / P / 1-1½' / lf very narrow; fls yel stripe / s Amer. 61

SYMPHYANDRA - Campanulaceae
armena / R / 1' / blue-violet fls / Caucasus 182 / 107
cretica / P / to 20" / fls blue / Crete, Greece 194 / 25
Hoffmannii / R / 1' / fls white / Bosnia 30 / 123
langezura - see S. zanzegura *
orientale / P / 3' / white fls / Turkey 206
pendula / R / to 1' / creamy-yellow fls / Caucasus 30 / 107
pendula 'Alba' / P / to 2'/ fls white / Caucasus 25
Wanneri / R / 6" / tyrian-purple fls / Roumania 93 / 123
zanzeguri / R / to 1' / large violet-blue fls / Caucasus (606) / 12

SYMPHYTUM - Boraginaceae
grandiflorum / P / to 16" / fls pale yellow / Caucasus 25
officinale / P / 3' / Comfrey; fls variied-colored / Eurasia 229 / 25

SYMPLOCOS - Stryracaceae
paniculata / Sh or T / to 40' / white fls; blue frs / Far East 110 / 25

SYNEILESIS - Compositae
palmata / P / 2½-3'+ / fls whitish / Japan 200

SYNNOTIA - Iridaceae
bicolor - see S. villosa 25

---

[1] David Mabberley, Quar. Bull. Alpine Garden Soc., 39:3.

[2] Hillier's Hundred Catalog, p. 35.

Metelerkampiae - see S. variegata v. Metelerkampiae 185
variegata v. Meterlerkampiae /6-8"/purp. fls, spotted orange/S. Afr. 185
villosa / HH C / to 18" / yellow fls, tinged violet / S. Africa 25

SYNTHYRIS - Scrophulariaceae
bullii - see WULFENIA bullii 99
lanuginosa - see S. pinnatifida v. lanuginosa
missurica / R / to 16" / bright bl.-purple fls/ Wash., Calif., Idaho 229 / 25
missurica 'Alba' - white fls *
missurica stellata - see S. stellata *
pinnatifida v. canescens / R/ 6"/ violet-blue fls/ sw Mont., c Idaho 142
pinnatifida v. lanuginosa - pl permanently white-tomentose (616) / 142
platycarpa / R / to 1' / violet-blue fls / Idaho 184 / 25
reniformis / R / to 1' / blue or purple fls / Oregon, Washington 306 / 107
reniformis 'Alba' - white fls 25
rotundifolia - see S. reniformis 25
schizantha / R / 4-12" / purplish fls / Oregon, Washington 228
stellata / R / to 1' / blue-purple fls / Columbia River gorge 64 / 107
stellata 'Alba' - white fls 25

SYRINGA - Oleaceae
amurensis / Sh / to 18' / white fls / Manchuria, Korea 139
amurensis japonica - see S. reticulata 139
emodi / Sh / to 10' / pale lilac fls / Himalayas 139
X Henryi / Sh / tall / pale viol-purp. fls / S. josikea x villosa 25 / 139
josikea / Sh / to 10' / deep violet-mauve fls / c & e Europe 139
josikea 'Pallida' - fls blue-violet fading white 93
Meyeri 'Palibin' - compact selection, violet fls (393)
microphylla / Sh / to 6' / fls rosy-lilac / n & w China 139
microphylla 'Superba' / Sh / "small" / fls pink / China 25
oblata v. dilitata / Sh / to 12' / fls lilac; lvs long / China 25
oblata v. giraldii / 10'/ purplish-violet fls / n China 139
Palibiniana - see S. Meyeri 'Palibin' (393)
pekinensis / Sh / to 15' / fls yellowish-white / n China 222
reflexa / Sh / 10' / purplish-pink & white fls / c China 138
reticulata / Sh / 10'+ / creamy-white fls / Japan 167 / 139
Sweginzowii / Sh / to 10' / lilac fl. panicles to 8" / China 25
Sweginzowii 'Albida' - fls paler 25
velutina / Sh / 5-6' / fls pale to pinkish lilac / Korea 139
vulgaris 'Nana' - dwarf or slow-growing clone 309
Wolfii / Sh / 8' / pale violet-purple fls / Korea 139
yunnanensis / Sh / to 10' / pinkish fls / Yunnan 139

SYRINGODEA
luteo-nigra - see ROMULEA longituba v. alticola 185

TACITUS - Crassulaceae
bellis / P / lvs rosette; 1-10 fls, deep pink / w Mexico 25

TAENIDIA - Umbelliferae
integerrima / P / to 3' / fls yellow / ec North America 99 / 25

TAGETES - Compositae
minuta / A / to 3' / fls light yellow / Peru, Brazil, nat U.S., Afr. 99 / 25

TALINUM - Portulacaceae
calycinum / R / 8" / cherry-red fls / Arkansas, Missouri 224 / 107
crassifolium - see T. triangulare 160
Mengesii / R / 1'+ / pink fls / Tennessee to Alabama 25
okanoganense / R / 1½" / white fls / nw N. Am. (617) / 107
patens / Gh Sh / grown as A / 1-2' / fls carmine / South America 28
paniculatum / HH P / 1-5' / carmine fls / W. Indies, S. Am. 250
parviflorum / R / 10" / rose or lavender fls / Texas to Minn. 229 / 227
rugospermum / R / to 8" / roseate fls / Ind. to Minn. 224 / 99
spinescens / R / to 1'/ fls magenta-purple / Washington 268
teretifolium / R / to 1' / rose-purple fls / Ga. to Pa. 224 / 225
triangulare / A / to 20" / pink fls / South America (365)

TAMARIX - Tamaricaceae
africana / HH Sh / 15' / white or pink fls / Mediterranean coast 212

TAMUS - Dioscoreaceae
communis / Tu / vine to 12' / fls green-yellow; fr red / sw Europe 290

TANACETUM - Compositae
boreale - see CHRYSANTHEMUM vulgare 200
cineariifolium / P / 6-18" / wht. fls; lvs silvery-gr. / w Jugoslav. 289
clusii - see T. corymbosum ssp. clusii 289
corymbosum ssp. clusii / P / 1-4' / fls white / Carpathians, Alps 289
densum / HH Sh / 10" / yell. fls; lvs white-tomentose / Anti-Lebonan 77
densum amanum - see T. haradjanii *
1 gossypinum / R / 1-2" / silver foliage; yel fls / Nepal
haradjani / R / 9-15" / fls yellow; lvs light gray / Anatolia 77 / 268
Herderi / P / to 1' / fls bright yellow / Turkestan 25
huronense / P / 16-32" / yellow fls / Michigan 224 / 99
huronense v. terra-novae / R / to 8' / wht lvs; yellow fls / w Nfld. 99
pallidum ssp. spathulifolium / fls white or yel./ high mts. of Spain 212
vulgare / P / 2-3' / yellow fls / Europe, adventive in U.S. 28

TANAKAEA - Saxifragaceae
radicans / R / 4-12" / fls white / wet shaded rocks / Honshu 200

TARAXACUM - Compositae
albidum / R / 1' / white fls / Honshu, Shikoku, Kyushu 200
alpicola / R / 1' / orange-yellow fls / Honshu 200
alpinum / P / to 8" / fls golden yellow / Pyranees 25
arcticum / R / 4" / variable in fl color / cirumpolar 209
ceratophorum / R / 4-10" / yellow fls, dark stripped / Alps 289
officinale album - see T. albidum 200
pulustre / P / 1-8" / petioles purp, fls yellow / n Europe 289
retzii / T. Fulvum group/ lvs bright grn, ligules pale yel/ nwc Eur. 289
rubifolium /T.spectabile grp/lvs dull grn,lig dp yel,red-purp stripe 289
trigonolobum / P / bracts dk green, fls orange-yel. / Hokkaido, Jap. 200
yatsugatakense / R / 10" / orange-yellow fls / alps of Honshu 200

---

1 Ron McBeath, Journal Scottish Royal Garden Club, Vol. 18, pt. 2, No. 71.

TAXUS - Taxaceae
baccata / T / to 60' / scarlet fleshy frs / Eu., N. Africa, w Asia 222
baccata 'Adpressa' / T / small / needles glossy on top / Europe 139 / 25
baccata repandens / T / nearly prostrate / lvs bluish-grn/ n Hemis. 139 / 25
baccata 'Standishii' - columnar, yellow lvs 25
brevifolia / T / 15-45' / lvs dark yel-green / Mont to B.C. & Cal. 36 / 25
canadensis / Sh / 4', often prostrate / red frs / North America 139
cuspidata / Short / to 50' / lvs to 1"; dull grn above / Jap., China 139 / 25
cuspidata capitata - the typical form of T. cuspidata 25
floridana / HH T / to 25' / lvs narrow / Florida 126
media 'Columnaris' / T / broadly columnar / United States 25
media Hatfieldii - conical, branches ascending 139 / 25
media 'Hicksii' / T / columnar form, upright 139 / 25
media 'Kelseyii' - upright, dense, fruits heavily 25

TCHIHATCHEWIA - Cruciferae
isatidea / R / 6-9" / fls bright rose red, vanilla scent / Armenia 95

TECOMA - Bignoniaceae
stans / HH Sh / 18'+ / yellow fls / South America 28

TECOMARIA - Bignoniaceae
capensis / Gh Sh / 6-8' / vermilion fls / South Africa 128

TECOPHILAEA - Amaryllidaceae
cyanocrocus / HH C / 4" / deep blue fls, white eye / Chile 123 / 185

TEESDALIOPSIS - Cruciferae
conferta / R / to 6" / fls white / nw Spain, n Portugal, in mts. 286

TELEKIA - Compositae
speciosa / P / 3-4½' / yellow fls / Europe, w Asia 292 / 77

TELEPHIUM - Caryophyllaceae
imperiati / Sh / 6-20" / white fls / s Europe 148 / 286

TELESONIX - Saxifragaceae
jamesii / R / 8" / fls reddish-purple / nc N. America 229 / 25
jamesii v. heucheriformis / cherry-red fls. / Big Horn Mts. 362

TELINE - Leguminosae
monspessulana / HH Sh / 3-5' / yellow fls / Medit. region 212 / 287

TELLIMA - Saxifragaceae
grandiflora / P / 18" / greenish fls, aging red / Calif. to Alaska 229 / 107
grandiflora 'Purpurea' / 1' / lvs red, fls yellow 61

TELOPEA - Proteaceae
orneades / HH T / 30-40' / fls crimson / Australia 61
seciosissima / Gh Sh / to 8' / red fls / New South Wales 128 / 25
truncata / Gh Sh / to 4' / red fls / Tasmania 45 / 25

TEMPLETONIA - Leguminosae
retusa / Sh / to 10"/ fls red, rarely yellow / s & w Australia 25

TEPHROSIA - Leguminosae
virginiana / P / to 2' / yellowish & pink-purple fls / Me. to Fla. 224 / 25

TETRACLINIS - Cupressaceae
articulata / HH T / shrubby in cult. / round cones / Medit. region 139

TETRAGONOLOBUS - Leguminosae
maritimus / P / 4-20" / pale yellow fls / c & s Europe 148 / 287
purpureus / A / 4-16" / crimson fls / s Europe, Ukraine 212 / 287

TETRANEMA - Scrophulariaceae
mexicanum - see T. roseum 184 / 25
roseum / Gh / 4-8" / purple fls / Mexico, Guatemala 25

TETRAPATHAEA - Passifloraceae
tetrandra / Cl / high / fls greenish yellow / N. Zealand 238

TEUCRIDIUM - Verbenaceae
parviflorum / HH Sh / 4'+ / fls blue or white / New Zealand 238

TEUCRIUM - Labiatae
ackermannii / Sh / 4-6" / soft rose-crimson fls / Asia Minor 640 / 123
arduini / Sh / 1' / fls whitish / Jugoslavia, Albania 288
aroanium / R / procumbent / fls blue or lilac / s Greece, in mts. 194 / 288
asiaticum / Sh / fls pink to purple / Balearic Isls. 288
botrys / A / to 1' / purplish-pink fls / s, w & c Europe 288
canadense / P / to 3' / pinkish-purple fls / North America 229 / 25
Chamaedrys / Sh / 1-2' / fls pale to deep purple / Eurasia 210 / 25
flavum / P / to 2' / fls yellow; lvs leathery / Medit. 25
fruticans / Gh Sh / to 4' / blue or lilac fls / w Medit. region 149 / 25
fruticans album / Sh/ 7-8'/ stems square, felty; fls white / Morocco 36
hircanicum / P / 2' / fls dark lilac / Iran 279
laciniatum / P / 6-8" / lvs lacinate; fls wht, fragrant/Tex & N.Mex. 227 / 229
lucidum / Sh / to 2' / fls pale to deep purple / sw Alps 288
marum / HH Sh / to 20" / purplish fls / Isls. of Mediterranean Sea 288
massiliense / HH Sh / to 20" / pink fls / w Medit. region 288
montanum / Sh / 4-10" / cream fls / s & c Europe 148 / 288
orientale / P / 1' / violet or blue fls / Caucasus 28
Polium / Sh / 5-6" / yellow or white fls / Europe, w Asia 211 / 123
Polium ssp. capitatum / Sh / 2-20" / fls red or white, compound 288
pyrenaicum / R / 4-6" / fls mauve & white / s Europe 148 / 129
Scorodonia / Sh / to 20" / pale yellow fls / Eur., nat. Ont.& Ohio 288 / 25
subspinosum / R / to 8" / bright pink fls / Asia Minor 212 / 123
webbianum / Sh / thinly grey tomentose / fls purp; rare wht/se Spain 288

THALICTRUM - Ranunculaceae
acteaefolium / P / 12-28" / white fls / Japan 200
adiantifolium minus - see T. minus 25
alpinum / R / 2-8" / fls purplish-green / arctic Europe & N. Am. 148 / 286
amurense - see T. flavum
aquilegifolium / P / 2-3' / fls purplish-pink / Europe, N. Am. 148 / 129
aquilegifolium 'Purpureum' / P / to 2' / fls purple / Cultivar 25
baicalense / P / 20-32" / fls creamy-white / Far East 200
Chelidonii / P / to 5' / lilac fls / Himalayas 279 / (618)
contortum - see T. aquilegifolium 200

coreanum / R / 6" / rose-pink fls / Korea 123
Delavayi / P / 2-3' / lilac fls / w China 279 / (639)
diffusiflorum / R / 12"+ / mauve to violet fls / se Tibet 123
dioicum / P / 8-30" / purplish fls / e & c N. Am. 224 / 99
dipterocarpum / P / to 3' / fls rose-violet / w China 147 / 25
dipterocarpum 'Album' - white fls 135
fendleri / P / tall and leafy / fls wht / foothills of Rockies 229
filamentosum / P / 8-28" / white fls / Amur reg., Manchuria, Korea 200
filamentosum v. yakusimense / P/ small / fls waxy wht/ Korea, Manch. 200
flavum / P / to 4' / whitish fls / Eurasia 210 / 25
flavum ssp. glaucum / yellow fls / Spain, Portugal 212 / 286
flavum 'Illuminator' / to 5'/ sulphur-yellow fls / yellow lvs 268
foetidum / P / 4-16" / fls yellow / e, c & sw Europe 286
glaucum - see T. flavum ssp. glaucum 286
kiusianum / R / 4" / mauve-pink fls / s Japan 107 / 123
koreanum - see T. coreanum *
minus / P / to 5' / yellow lvs / Europe 300 / 207
minus adiantifolium - synonym of the sp. 207
minus ssp. saxatile / 6-12" / coriaceous lvs / c & e Europe 286
occidentale / P / 16-40" / fls whitish or purplish / nw N. Am. 229 / 228
petaloideum / P / to 20" / white fls / e Asia 25
polycarpum / P / 3' / aromatic lvs / California 28
polygamum / P / 3-8' / fls whitish / e N. Am. 22 / 25
Rochebrunianum / P / 2-3' / fls pale purple / Honshu 165 / 200
rubescens - see T. aquilegifolium
simplex / P / 8-48" / yellowish fls / most of Europe 286
sparsiflorum / HH P / 12-40" / fls whitish / s California 228
speciosissimum / P / 2-5' / yellow fls / w Medit. region 25
strictum - see T. simplex 286
tuberosum / P / 8-16" / fls yellowish-white / Pyrenees 212 / 286

THASPIUM - Umbelliferae
trifoliatum / P / 2-5' / yellow fls / Rocky Mts. 99 / 65

THELESPERMA - Compositae
filifolium / P / 10-26" / lvs narrow; fls red-brn / Tex., N.Mex. 227

THELYMITRA - Orchidaceae
grandiflora / Gh / 9" / fls in blue shades / Australia 61
ixiodes / Gh / 1'+ / fls blue / Australia 28
longifolia / P / to 16" / fls white to violet / New Zealand 238
pauciflora - see T. longifolia 28

THELYPTERIS - Polypodiaceae
hexagonoptera / F / to 20" / Broad Beech Fern / e North America 159
phegopteris / F / 9" / Long Beech Fern / temperate n hemisphere 148 / 25
palustris v. pubescens / F / 20" / Marsh Fern / e & C North America 25

THERMOPSIS - Leguminosae
caroliniana / P / 5' / bright yellow fls / North Carolina 300 / 107
chinensis / P / 2' / fls yellow / China 258

[1] divaricarpa / P / fls ?; fr spreading / foothills of Rockies
lanceolata - see T. lupinoides 25
lupinoides / R / to 1' / yellow fls / Siberia, Alaska 25
mollis / P / to 5' / yellow fls / mts Va. to Ga. 99
montana / P / 3' / yellow fls / w North America 147 / 107
villosa / P / 3' / yellow fls / Tennessee, Alabama 225

THERORHODIAN
camtschaticum - see RHODODENDRON camtschaticum 222

THEVETIA - Apocynaceae
thevetioides / Gh T / to 15' / orange fls / Mexico 25

THLASPI - Cruciferae
alpestre / P / 4-20" / fls white or purplish / Europe, in mts. 286
alpinum / R / 4-6" / fls white / Alps 148 / 286
arvense / A / to 32" / Penny-cress pods, weedy / Europe 229 / 25
bellidifolium / R / densely caespitose / fls dark purple / Albania 286
brevistylum / R / 4" / white fls / Corsica, Sardinia 286
bulbosum / R / 4" / fls dark lilac to violet / Greece 286
cepiifolium - see T. rotundifolium ssp. 25
cilicum / R / 6-10" / white fls / Turkey 77
dacicum / R / to 10" / white fls / e & n Carpathians 286
densiflorum / B / to 4" / fls white / Turkey 25
fendleri / P / to 12" / fl petals 4-11; white / Alaska to Cal. 142
goesingense / R / to 16" / white fls / Balkans 286
Jankae / R / 4-8" / white fls / Czecho-Slovakia, Hungary 286
limosellifolium - see T. rotundifolium (619) / 286
montanum / R / 4-12" / fls white / c Europe, w N. Am. 229 / 286
nevadense / R / to 4" / fls white / Spain 286
ochroleucum / R / to 10" / fls white, anthers yel. / Balkan pen. 286
praecox / R / 4-8" / white fls / s Austria, Balkans 286
rotundifolium / R / 2-4" / fls pale slaty-purple / Alps 148 / 129
rotundifolium ssp. cepiifolium / basal lvs smaller, notched 25
stylosum / R / 1" / purplish fls / c & n Apennies 148 / 286
violascens / A / 8" / fls violet / Turkey 77

THOMASIA - Sterculiaceae
macrocarpa / Gh Sh / 3' / purple fls / w Australia 45

THRYPTOMENE - Myrtaceae
payneii - see T. 'Paynes Hybrid' 45
'Paynes Hybrid' / HH Sh / to 4' / pink fls / s Australia 45

THUJA - Cupressaceae
gigantea - see T. plicata 28
occidentalis 'Holmstrupensis' / dwarf / conical, deep green lvs 304
occidentalis 'Recurva Nana' / dwarf / branchlets recurved 25
plicata / T / to 200' / bright green lvs / Alaska to Montana 28

THYMBRA - Labiatae
spicata / Sh / to 1' / pink fls / Turkey, Greece (620) / 288

---

[1] William Weber, Rocky Mountain Flora, 1976.

THYMOPHYLLA
belenidium - see DYSSODIA pentachaeta v. belenidium. 39
tenuifolia - see DYSSOIDA tenuifolia 25
tenuiloba - see DYSSODIA tenuiloba 25

THYMUS - Labiatae
caespititius / HH R / prostrate / purplish-pink fls / se Europe 133 / 25
camphoratus / Sh / 6-12" / bracts purple; fls rose / s Portugal 25
cilicicus / R / 4-6" / pale pink fls / Asia Minor (302) / 123
dzevanowskyi - see T. pannonicus 288
Doerfleri / R / 1-2½" / pale rose-purple fls / Albania 194 / 268
Drucei - see T. praecox ssp. arcticus 288 / 25
hackeliana - see T. serpyllum v. lanuginosus *
Herba-barona / R / to 4" / pale purple fls / Corsica 288
hirsutus / R / mat / deep pink fls / Crimea, Balkans 123
hirsutus Doefleri - see T. Doerfleri 288
integer / sub sp. / mat forming / bracts purp; fls rose / Cyprus 61
latifolius - see T. pannonicus 288
leucotrichus / R / creeping / purplish bracts / c & s Greece in mts. 288
Marschallianus - see T. pannonicus 288
Mastichina / HH Sh / 8-20" / white fls / Spain, Portugal 212 / 288
membranaceus / HH R / dwarf / white fls / Spain 107
nitens - sim. to T. Herba-barona; lvs & fls longer 288
nitidus - see T. richardii ssp. nitidus (621) / 288
pannonicus / R / to 8" / pink-toned fls / ec & e Europe 288
praecox / R / 4" / purple fls / Europe 25
praecox 'Albus' - white form 25
praecox ssp. arcticus / corolla rose-purple 25
praecox 'Coccineus' - fls crimson 25
praecox ssp. polytrichus / purple fls / s & c Europe, in mts. 148 / 288
przewalskii - see T. quinquecostatus 200
pseudolanuginosus/creeping mat/ hairy lvs, pale pink fls/ orig. unkn. 25
quinquecostatus / Sh / prostrate / rose-purp. fls / Far East in mts. 200
quinquecostatus albus / Sh / prostrate to 4" / fls white /Mong.-Jap. 25
Richardii ssp. nitidus / HH R / 4" / purple fls / w Sicily 286
Serpyllum / R / to 4" / purple fls / nw Europe 210 / 25
Serpyllum 'Albus' - see praecox 'Albus' (577) / 25
Serpyllum 'Aureus' - golden variegated lvs 28
Serpyllum 'Coccineus' - see T. praecox 'Coccineus' 129 / 25
Serpyllum 'Lanuginosus' - see T. pseudolanuginosus 25
thracicus / R / procumbent /purp. fls/ c Balkan Pen., n Aegean reg. 25
vulgaris / R / 4-12" / fls whitish to pale purple / w Medit. reg. 201 / 286

THYSANOTIS - Lilaceae
multiflorus / Gh / 9" / purple fls / w Australia 45
tuberosus / Gh / 2' / fls purple /Australia 45

TIARELLA - Saxifragaceae
cordifolia / R / 10" / white starry fls / e North America 49 / 123
cordifolia collina - see T. wherryi 25
Oliveri / T / to 60' / small fls; lvs whitish beneath / c China 222
polyphylla / P / 18" / white fls / Asia 129 / 107
trifoliata / R / 10" / white fls / w North America 229 / 107
trifoliata v. unifoliata - see unifoliata 276
unifoliata / lvs usually simple / Alaska to Mont. & California 247 / 107
Wherryi / R / 10" / clumpy plant / e North America 129 / 107

TIGRIDIA - Iridaceae
chiapensis / HH C / 12" / fls white & yellow, purple spots / s Mex. 185
Pavonia / HH C / 1-1½' / varii-colored fls / Mexico 128 / 129
Pavonia 'Alba Immaculata' - fls pure white 25
Pavonia 'Conchiflora' - bright yellow fls 28
violacea / HH C / 18" / fls light purple-blue / Mexico 185

TILINGIA - Umbelliferae
tachiroei / R / 2-6" / fls white / Hokkaido, Japan 200

TILLIA - Tiliaceae
tomentosa / T/ 60-100'/fls dull wht, 5-10 fl raceme/Balk.,Hung.,Rus 139 / 36

TIPULARIA - Orchidaceae
discolor / P / to 2' / greenish to purplish fls / N. J. to S. Car. 224 / 25
japonica / B / 8-10' / fls yellow-green / Honshu, Japan 200

TITHONIA - Compositae
rotundifolia / A / to 6' / fls orange & scarlet / Mex., & c America 25
[1] rotundifolia 'Torch' / A / tall / vivid orng-scarlet 3" fls/Cultivar

TODEA - Osmundaceae
supurba / F / tall / Tree Fern; fronds to 4' / New Zealand 28

TOFIELDIA - Lilaceae
alpina - see T. palustris 129
calyculata / R / 6" / yellowish fls / Europe 148 / 93
coccinea / R / to 4" / purplish fls / Siberia to e Greenland 209
glutinosa / R / 3-20" / fls whitish / e & nc N. America 224 / 99
glutinosa v. brevistyla - western form, Alaska, B. C., to Ore. 142
nuda / R / 14" / fls white / Japan 166 / 25
nutans coccinea - see T. coccinea 200 / 25
okuboi / P / fls green-white / Hokkaido, Honshu 200
palustris / R / 2-4" / whitish fls / arctic Europe 93
pusilla / R / to 12" / whitish or greenish fls / n N. America 209 / 99

TOLMIEA - Saxifragaceae
Menziesii / P / 1½-2' / green & chocolate fls / nw N. Am. 182 / 207

TONESTUS
pygmaeus - see HAPPLOPAPPUS pygmaeus 142

TOONA
sinensis - see CEDRELA sinensis 222

TOWNSENDIA - Compositae
eximia / P / to 20" / fls purplish to bluish / s Colorado, in mts. 227 / 229
exscapa / R / low mound / fls white or purplish / Rocky Mts. 229 / 107
exscapa v. wilcoxiana - included in the sp. 25
florifera / A or B / 10" / fls white to pink / nw U.S. 229 / 25
formosa / R / 3" / lilac fls / New Mexico 228 / 123

---

[1] Thompson & Morgan Catalog, 1986.

glabella / R / to 3" / white or pale lavender fls / nw Colorado		229
grandiflora / B / 2-3" / wht. fls, pink-striped beneath / S.D., N.M.	182	25
Hookeri / R / mat to 3" / white fls; pinkish below / B.C., Alta-Utah	169	25
incana / R / 2" / lilac or white fls; silvery lvs / Wyo. to Ariz.	229	235
mensana / R / 1½" / fls whitish to dull red / n Utah, se Nevada		229
montana / R / caespitose / blue or violet fls / Mont., Ida., Wash.	229	142
Parryi / R / 8" / purple fls / Wyo., Idaho & n-ward	229	107
Rothrockii / R / 2-3" / blue-purple fls / Colorado	312	25
sericea / see T. exscapa		123
spathulata / R / spat. shaped / fls pink, orng, lav. / c Wyoming		229
strigosa / R / to 2½" / pappus of rays reduced / Wyo. to N. Mex.	229	235
Wilcoxiana - see T. exscapa		25

TRACHELIUM - Campanulaceae

asperuloides / HH R / flat mat / fls purple to pale pink / Greece	132	64
caeruleum / HH Sh / 2' / fls in shades of blue / Medit. region	212	74
jacquinii / R / to 6" / bluish-lilac fls / ne Greece		289
jacquinii ssp. rumelianum / 6-12" / fls bluish / Bulgaria, Greece		289
myrtifolium / R / 2" / tubular whitish fls / Caucasus		74
rumelianum - see T. jacquinii ssp. rumelianum		289

TRACHYMENE - Umbelliferae

humilis / P/mats /sm lvs; stalkless pnk-wht ½ bells/Tas,NSW, Austrl		109

TRADESCANTIA - Commelinaceae

hirsuta / Gh / 1½" / fls bright purple / South America		29
hirsuticaulis / R / to 1' / fls in several colors / sc U.S.	225	25
longipes / P / lvs 6-10"; fls pink, purp., or blue / Missouri		99
'Pauline' - see T. virginiana 'Pauline'		135
pinetorum / P / lvs narrow; fls. blue, rose, viol / Ariz, N.Mex.		227
rosea / P / 8-16" / fls pink or roseate / Florida to N. Carolina		99
rosea v. graminea - see CUTHBERTIA graminea		25
virginiana / P / to 3' / fls violet-purple / North America	129	25
virginiana 'Pauline' / P / 18-24" / rose-mauve fls	300	135
[1] virginiana 'Zwanenburg Blue' - royal blue fls		

TRAGOPOGON - Compositae

dubius / B / fls pale yellow / sw Ukraine		206
major - see T. dubius	224	289
porrifolius / B / to 4' / purple fls; salsify edible root / s Europe	229	25
pratensis / B or P / 3' / fls yellow / Europe, weedy in U.S.		28

TRAPA - Trapaceae

natans / P / Aq / fls white / Orient		25

TRAUTVETTERIA - Rannunculaceae

carolinensis / P / 2-3' / fls white / Pa., s & w-ward		28
japonica / P / to 2' / small white fls / n Japan		200

TRICHOCEREUS - Cactaceae

litoralis / Gh / 3' / white fls / Chile		25

---

[1] Wayside Catalog, Spring 1986.

TRICHOCLINE - Compositae
sinuata - South American genus of large short-rayed fls 64

TRICHOMANES
cristata - see ASPLENIUM trichomanes 'Cristata' 25

TRICHOPETALUM - Liliaceae
1 pulmosum / HH P / 3' / greenish fls / Chile 25

TRICHOSANTHES - Cucurbitaceae
cucumeroides / A Cl / to 15' / Snake Gourd; wht-fringed fls / Japan 164 / 25

TRICHOSTEMA - Labiatae
arizonicum / Sh / 1-2' / white & blue fls / New Mexico 227
dichotomum / A / 4-32" / blue or pink fls / Mass. to Texas 28
lineare - see T. setaceum 99
setaceum / P / 4-12" / fls blue, slender / Fla. to La. and n-ward 99

TRICUSPIDARIA
dependens - see CRINODENDRON patagona 93
hookerianum - see CRINODENDRON hookerianum 139
lanceolata - see CRINODENDRON hookerianum 139

TRICYRTIS - Liliaceae
affinis / P / 2'+ / fls white, spotted purple / Japan 25
Bakeri / P / 2-3' / fls yellow, spotted red / Japan 61
flava / P / to 20" / fls yellow, purp spotted inside / Japan 200 / 25
formosana / R / 1' / fls whitish-purple / Taiwan 61
hirta / P / 1-2' / white fls, spotted dark purple / Japan 166 / 200
hirta 'Alba' - unspotted white fls (306)
latifolia / P / 16-32" / yellow fls, purple spots / Japan 166 / 200
macrantha / P / 16-32" / yellow fls / Shikoku 200
macrantha ssp. macranthopsis - see T. macranthopsis 200
macranthopsis / P / to 2½' / fls yellow, spotted choc. inside/Japan 200
macropoda / P / 1½-2½' / greenish-yellow fls / China, Japan 166 / 207
nana/much smaller than T. hirta/fls yel, purp spots/Honshu, Jap. 200
ohsumiensis / P / 8-20" / yellow fls / Kyushu 166 / 200
ohsumiensis 'Takokuma Hototugisu' / 6" / yellow fls 96
perfoliata / P / 20-28" / yellow fls / Kyushu 200
stolonifera formosa - see T. formosana 61

TRIENTALIS - Primulaceae
americana - see T. borealis 99
borealis / R / 6" / starry white fls / e North America 22 / 99
europea / R / to 10" / white fls / Eurasia 148 / 25
latifolia / P / to 8" / fls rose pink or white / California 25

TRIFOLIUM - Leguminosae
agrarium / A / 22" / yellow fls, Hop Clover / Eur., weedy in U.S. 99
alpinum / P / 8-20" / fls pink, purple, cream / Alps 148 / 287
arvense / A or B / 1½-16" / whitish or pink fls / Europe 287
badium / R / 4-10" / golden yellow fls / c & s Europe 287

---

[1] Hossain, Notes from the Roy. Bot. Garden, Edinburgh, 23:3.

[1] badium ssp. rhytidoseminum / 1'+ / creamy-white fls / Iran, Turkey
dasyphyllum / P/mat, tap root/fls lil-purp, age brn/Mont.,Col.,Utah 142
fragiferum / P / creeping 1' long / fls pink to white / Medit. 25
incarnatum / A / 8-20" / red, pink, cream, white fls / e Europe 148 / 287
lupinaster / P / 6-20" / red or white fls / e Europe 287
macrocephalum / R / 2-3" / deep rose-pink fls / nw N. Am. 229 / 123
microcephalum / A / weak-stemmed / fls white to pinkish / nw N. Am. 142
montanum / P / to 2' / fls white or yellowish / Europe 287
nanum / R / flat growth / reddish-purple fls / w U.S. 229 / 107
Parryi / P / low / fls purple / Wyo., Colo, Utah 25
productum / P / plant hairy / fls whit, purp. tip / Ore.-Sierra Cal. 142
repans 'Purpurascens' / R / carpet / bronze-purple lvs 184
[1] rhytidoseminum - see T. badium ssp. rhytidoseminum
spadiceum / P / 8-16" / yellow fls, aging brown / Pyrenees, Alps 148
stellatum / A / 8" / mostly pink fls / Medit. region 287
[2] stenolobum / bicolor alpine clover, endemic to Rocky Mountains
thalii / R / to 6" / fls white or red / Alps, Apennies 211 / 287
Thompsonii / P / 8-20" / fls red-lavender to deep orchid / c Wash. 229
virginicum / P/ 4" / white fls / Md., W. Va., Va., Pa., in mts. 99
wormskjoldii / P/ leafy stem,4-24"/ fls red-purp, wht tip/B.C.-Mex. 142 / 229

TRIFURCIA - Iridaceae
lahue ssp. amoena - Gh C / to 1' / violet fls / n Argentina, Uruguay (23)
lahue ssp. caerulea - HH C/to 8"/fls blue-purple/Texas, Louisiana (23)

TRIGLOCHIN - Juncaginaceae
maritimum / P / 2½' / inconspicuous bloom / n hemisphere 28
palustris / P / to 28" / rush-like fleshy lvs / N. Am. 99

TRIGONELLA - Leguminosae
Foenum-graecum / A / to 2' / fls whitish / s Eur., Asia 25

TRILLIUM - Liliaceae
albidum / Tu / 1-2' / white fls / California to Washington (308)
camtschaticum - see T. kamtschaticum 200
Catesbaei / Tu / 1½' / fls pink, rose, white / N. Car., Ala., Ga. 225 / 25
cernuum / Tu / 1½' / white or pinkish fls / e N. Am. 224 / 267
chloropetalum / Tu / 1½' / greenish-white fls / nw N. Am. 279 / 129
chloropetalum v. gigateum - red fls (129)
chloropetalum rubrum - see T. chloropetalum v. giganteum (308)
cuneatum / Tu / 10" / dark colored fls / N. Car. & s-ward 224 / 25
[2] decumbens / P / mat forming / fls brn-purple / Ala, Ga.
discolor / P / 4-12" / yellow petals / N. Car. to Ga. 225
erectum / Tu / 1' / varii-colored fls / e North America 224 / 267
erectum f. albiflorum - white fls 25
erectum 'Album' - see T. erectum f. albiflorum 25
erectum v. blandum / creamy fls / not ill-scented / Tenn. 99
erectuf X flexipes picotee hyb. - crossed by donor?
erectum f. luteum - clear yellow fls 99
flexipes / Tu / to 2' / white fls / e & c U.S. 229 / 99

---

[1] Hossain, Notes from the Roy. Bot. Garden, Edinburgh, 23:3.

[2] Alpines of the Americas. Report of the First Inter. Rock Garden Plant Conference, 1976.

flexipes f. walpolei / - fls maroon or purple 99
foetidum - see T. erectum
giganteum - see T. chloropetalum v. giganteum
govanianum / Tu / ? / greenish fls / Himalayas 64
grandiflorum / Tu / 1-1½' / white fls / e North America 147 / 129
grandiflorum 'Roseum' - pink fls 207
hibbersonii - see T. ovatum f. hibbersonii (550) / 276
Hugeri - see T. cuneatum (433) / 25
kamtschaticum / Tu / tiny / green & white fls / Far East 64
ludvicianum / P / 6-19" / red or purp-brn fls / coast pl Fla-La. 225
luteum - see T. viride v. luteum 224 / 25
nivale / Tu / 6" / white fls / Pa., Minn., Mo. 224 / 107
ovatum / Tu / to 20" / white fls, fading rose / Calif. to B.C. 224 / 25
ovatum f. hibbersonii / 2-3' / pink fls / Vancouver Island (309)
ozarkanum / see T. pusillum v. ozarkanum 259
pusillum / Tu / to 1' / fls white, aging purplish / se U.S. 25
pusillum v. ozarkanum / larger fls / Mo. & Ark. 25
pusillum v. pusillum / P / to 1' / fls white / S.C., Mo. & Ark. 25
pusillum v. virginianum / fls sessile; petals tiny / se Va. 99
recurvatum / P / to 1½' / fls brown-purple / c & sc U.S. 224 / 25
rivale / Tu / 5" / white fls with purple base / Oregon 19 / 267
sessile / Tu / 1' / maroon or greenish-yellow fls / North America 129 / 25
sessile 'Album' - white fls *
sessile californicum - see T. chloropetalum 25
sessile californicum v. rubrum - see T. chloropetalum v. giganteum (308)
sessile chloropetalum - see T. chloropetalum 129
sessile luteum - see T. sessile f. viridiflorum 300 / (308)
sessile rubrium - see T. chloropetalum v. giganteum (308)
sessile f. viridiflorum - yellowish-green fls (308)
simile - see T. erectum v. blandum 225
Smallii / Tu / 8-16" / petals, if present, purplish / Japan 200
stamineum / P / 4-12" / lvs blotched; fls purp / Ga. to Miss. 225
stylosum - see T. Catesbaei 25
Tschonoski / P / to 1½' / fls white / Japan & Korea 25
underwoodi / shorter stalks & stamens than T. cuneatum / N.Car-Ala. 225
undulatum / P / 20" / white fls, purple veined / Que. to Ga. 225 / 25
vaseyi / Tu / 10-20" / fls dark purple / Tennessee, Georgia 99
viride / Tu / 15" / greenish fls, purple base / Ill. to Ark. 229 / 259
viride v. luteum / fls yellow, lemon scented / Ky. to Ga. 25

TRIMEZIA - Iridaceae
martinicensis / HH C / 16" / yellow fls, brown base / West Indies (327) / 185

TRIODANIS - Campanulaceae
perfoliata / P / to 3' / fls deep purple to lav. / New England & s 224 / 25

TRIOSTEUM - Caprifoliaceae
aurantiacum / P / 2-4' / red-purple fls; red frs / c N. Am. 224 / 225
pinnatifidum / P / to 3' / fls pale green, fr white / China 200

TRIPETALEIA - Ericaceae
bracteata / Sh / 4'+ / green & white & pink fls / Japan 139
paniculata / Sh /7' / fls white to roseate / Japan, in mts. 167 / 200

TRIPLEUROSPERMUM - Compositae
maritimum / P / prostrate to 1' / disc. fls yel, ray fls white/Eur. 25
Tchihatchewii / R / mats / white daisies on 1" stems / Asia Minor 25

TRIPTEROSPERMUM - Ericaceae
japonicum / P / 16-32" / fls light blue-purple / Far East 164 / 200

TRISTAGMA - Liliaceae
nivalis / Bb / little / fragrant green or purplish fls / Chile 64

TRITELEIA - Liliaceae
Bridgesii / C / 18" / deep blue-purple fls / n. Calif., s Oregon 185
crocea / C / to 1' / bright yellow fls / nw Calif., sw Oregon 267
grandiflora / C / to 2' / fls bright blue to white / w N. Am. 25
Hendersonii / P / to 1' / fls yel, dk purple midvein / Oregon 25
Hendersonii v. Leachiae - fls white, often suffused blue
hyacinthina / C /1-2' / fls blue or white / B. C. to Calif. 185
hyacinthina 'Alba' - selection for white fls *
ixiodes / C / 8-30" / fls cream to golden yellow / Calif., Ore. 186 / 185
ixiodes v. scabra - fls straw, cream or yellow 25
ixiodes ' Splendens' - see T. ixiodes v. scabra 25
laxa / C / 2-3' / deep blue fls / n Calif., s Oregon 185
peduncularis / HH C/ to 3'/ wht, lilac-tinged fls / coast mts., Cal. 25
X tubergenii / 9"+ / rich blue-lilac fls 185
uniflora - see IPHEION uniflorum 267

TRITICUM - Gramineae
aegilops - possible ancestor of wheat / s Europe 143

TRITOMA
uvaria - see KNIPHOFIA uvaria 28

TRITONIA - Iridaceae
crocata / HH C / 1-1½' / fls bright tawny-yellow / South Africa 180 / 25
crocata aurantiaca - included in the sp. 25
hyalina / HH C / 15" / fls salmon-pink / Natal (622) / 111
hyalina 'Roseline' / 2-3' / deep pink fls 111
hyalina 'Salmon Queen' - salmon-orange fls 111
rosea - see T. rubrolucens 25
rubrolucens / HH C / to 2' / rose fls / South Africa 25
viridis - see ACIDANTHERA viridis 61

TROLLIUS - Ranunculaceae
acaulis / R / 6" / lemon-yellow fls / Himalayas (623) / 107
albiflorus - see T. laxus v. albiflorus (428) / 142
altaicus / P / 1-2' / yellow or pale orange fls / Siberia 207
asiaticus / P / to 2' / orange fls / ne Russia 194 / 286
aurantiacus - see T. X cultorum 96 / 268
chinensis / R / 1' / golden fls / n China 93 / 207
X cultorum / P / 2½'+ / yellow to orange fls 268
X cultorum 'Aurantiacus' / 2-3' / hybrid 96 / 268
X cultorum 'Golden Queen' - orange-yellow fls 268

1 X cultorum 'Orange Princess' / 12" / brt orange 3" fls; 24" stem		
X 'Goldquelle' / 2½' / rich yellow-orange fls	49 /	268
europeus / P / 4-28" / lemon-yellow fls / Europe	148 /	268
europeus 'Lemon Queen' / 2' / very pale yellow fls		207
europeus 'Nana' - dwarf form		207
europeus ssp. transilvanicus / R / 4-8"/ lemon-yel fls / Carpathians		268
hybridus - listed as T. X cultorum		268
laxus / P / to 2' / fls pale greenish-yellow / n U.S.	229 /	107
laxus ssp. albiflorus / 8-32"/ fls grn-white to white / w N. Am.		142
Ledebourii - in cult., mostly T. chinesis	300 /	25
Ledebourii 'Golden Queen' - see T. X cultorum 'Golden Queen'		129
'Lemon Queen' - see T. europeus 'Lemon Queen'		207
patulus - see T. ranunculinus		25
pulcher / P / 8-24" / fls orange-yellow / Hokkaido, Japan		200
pumilus / R / to 12" / golden fls / n India, w China	50 /	123
pumilus yunnanensis - see T. yunnanensis		64
ranunculinus / P / 3-12" / fls golden yel. / Caucasus, Armenia		25
ranunculoides - resembling T. pumilus / China		64
riederianus / R / to 1' / yellow fls / Kuriles, Kamtchatka	165 /	200
riederianus v. japonicus / lvs more deeply cut / Hokkaido, Honshu		200
stenopetalus / P / 2½' / light yellow fls / China		279
yunnanensis / P / 2' / flat buttercup fls / China	279 /	207
yunnanensis stenopetalus - see T. stenopetalus		279

TROPAEOLEUM - Tropaeolaceae

azureum / Cl / lvs 5-lobed; fls blue / Chile		61
minus / A / dwarf scrambling / fls dk central spot / S. Amer., Andes		25
pentaphylla / HH Cl / slender / scarlet fls / South America		93
peregrinum / A Cl / to 12' / canary-yellow fls / Peru	194 /	93
polyphyllum / HH R / trailing / fls orange or yellow / S. Am.	129 /	25
speciosum / HH Cl / high / vermilion-red fls / Chile	129 /	25
tricolorum / Gh Cl / 3' / lemon-yellow fls / Chile, Bolivia	(312) /	128
tuberosum / HH Tu / to 9' / orange-red fls / Peru		129

TSUGA - Pinaceae

canadensis / T / 60-100' / small cones / e N. Am.		127
canadensis 'Curtis Spreader' / 6' / pendulous horizontal branches		79
canadensis 'Fremdii' - slow-growing, conical		36
canadensis 'Macrophylla' - larger needles		25
canadensis 'Pendula' - Sargent's Weeping Hemlock	(550) /	25
caroliniana / T / 70' / larger cones / se U.S.	79 /	127

TSUSIOPHYLLUM - Ericaceae

tanakae / Sh / 10" / white fls / Japan	132 /	123

TUBERARIA - Cistaceae

globularifolia / P / to 16" / fls yel; lvs bract-like/nw Port, Spain		287
guttata / A/ 1' / yellow fls / s & w Europe		287
lignosa / P / to 16" / yellow fls, unspotted / w Medit. region	212 /	25
praecox / A / to 12" / fls yellow, unspotted / c Medit.		25

[1] Wayside Catalog, Spring 1986.

TULBAGHIA - Liliaceae

cepacea / Gh Tu / 9" / brownish fls / Cape of Good Hope 206
fragrans / Gh Tu / 1½' / bright lilac fls, sweet-scented / S. Africa 184 / 25
natalensis / Gh Tu / 1'+ / fragrant white fls / South Africa 61
pulchella / Gh Tu / 1½' / violet fls / Transvaal 61
violacea / Gh Tu / 2½' / bright lilac fls / South Africa 180 / 25

TULIPA - Liliaceae

acuminata / Bb / 18" / fls light yellow & red / Pyrenees 306 / 111
Aitchisonii / Bb / 3-4" /wht fls, flushed crimson/ Chittral, Kashmir 267
Aitchisonii ssp. cashmeriana - totally yellow 267
[1] armena / Bb/lvs gray-grn; sm fls sol, brn-red, blue dot/nw Iran,Cauc.
Aucheriana / Bb / 3" / pink fls / n Iran 267 / 111
aurpestanica / Bb / fls 1 or 2 / varying colors / c Kashmir 61
australis / Bb / to 10" / fls frag, yel inside, red outside/Medit. 25
Bakeri / HH Bb / 4" / mauve-pink fls / Crete (125) / 185
Batalinii / Bb / to 6" / pale yellow fls / Usebek USSR 267 / 25
Batalinii 'Bronze Charm' - fls apricot & orange 111
biebersteiniana / Bb / 6-12" / yellow fls, violet tinged / c Asia 4
biflora / Bb / 5" / white fls, stained red & green / s Russia 267 / 25
biflora turkestanica - see T. turkestanica 267
bithynica / Bb / 10" / dull red & scarlet fls / Asia Minor 96
cashmeriana -see T. sylvestris ssp. australis 267
celsiana - see T. sylvestris ssp. australis 290
chrysantha - see T. clusiana v. chrysantha 185
Clusiana / Bb / 12-16" / white & pinkish-crimson fls / Iran 267 / 129
Clusiana v. chrysantha - fls all yellow inside, stained red outside 185
Clusiana v. stellata / yellow blotch inside / Afghanistan, nw India 25
cretica / Bb / to 8" / white fls flushed red / Crete (180) / 267
dasystemon - see T. tarda 129
Didieri / Bb / to 1' / fls crimson with blotch / s Europe 25
Eichleri / Bb / 12-18" / brilliant scarlet fls / Turkestan 129
ferganica / Bb / 4-16" / yellow fls violet tinged / c Asia 4
florenskyi / Bb / 4-8" / red or yellow fls / Caucasus 4
Fosterana / Bb / 12-18" / vermilion-scarlet fls / c Asia (624) / 267
Gesnerana /Bb/to 2'/ fls crimson or olive, dk blotch/e Eur,Asia Min. 25
Greigii /B/to 8"/fls orange-scarlet, yel marg, blk blotch/Turkestan 25
Greigii 'Aurea' / 6-8", golden yellow fls 17
[2] Greigii 'Tororito' / 6" / small bronze anthers; red fls
Hageri / Bb / to 1' / fls coppery to scarlet / Greece, Asia Minor 267 / 25
hissarica / small / yel inside, viol tinge outside / c Asia 125
Hoogiana / Bb/ to 1'/ fls scarlet, olive /blk basal blotch / c Asia 25
humilis / Bb / to 4" / pale purple fls, reddish-green outside / Iran 267 / 25
iliensis / Bb / 4-12" / yellow fls, tinged violet / c Asia 4
ingens / Bb / to 10" / glossy vermilion fls / c Asia 25
julia / Bb / 6" / deep scarlet fls, margined yellow / Transcaucasia 125
Kaufmanniana / Bb /to 8" /white, pink, scarlet, yel fls / Turkestan 267 / 25
Kolpakowskiana / Bb / 1' / yellow or pink fls / e Turkestan 267 / 129
kuschkensis / Bb / 8-20" / red fls / c Asia 4
linifolia / Bb / 6-12" / scarlet fls /Turkestan 267 / 129

---

1 Brian Mathew and Turhan Baytop, Bulbous Plants of Turkey, 1984.

2 McClure and Zimmerman Catalog, 1985.

Marjolettii / Bb / 2' / yellow fls aging white / Savoy, se France (624) / 25
Maximowiczii / Bb / to 1' / scarlet fls / Bokhara 267
Micheliana / Bb / to 1' / fls vermilion-scarlet / Turkmen, Iran (624) / 25
mogoltavica / indistinguishable form T. Greigii / c Asia 125
montana / Bb/ 9-12" / cherry-red fls / n Iran 267
oculis-solis / Bb / 12-16" / scarlet, green & brown fls / s Europe 93
Orphanidea / Bb / 15" / fls variable in color / Greece 267
Orphanidea v. pontica - Bulgarian form of a variable tulip 262
Ostrowskiana / B /to 8" / scarlet fls, olive blotch / e Turkestan 267 / 25
patens / Bb / to 1' / whitish or yellowish fls / Siberia 25
persica - see T. patens (624) / 25
platystigma - see T. schrenkii 290
polychroma / Gr Bb / 6" / starry white fls / Near East 306 / 111
praecox / Bb / to 20" / fls dull scarlet, olive blotch / n Italy 25
praestans / Bb / to 1' / brick-red fls / c Asia 129 / 25
primulina / Bb / 1' / fragrant primrose-yellow fls / e Algeria 267 / 111
pulchella / Bb / 4" / pinkish-crimson fls / Asia Minor 267
pulchella humilis - see T. humilis 185
pulchella 'Violacea' - see T. violacea 267
rhodopea / Bb / 8"+ / fls dull brownish-scarlet / Bulgaria 61
saxatilis / Bb / to 1' / pinkish lilac-magenta fls / Crete 267
schrenkii / Bb / 6-16" / red, yellow or white fls / Caucasus 4
sharonensis / Gh Bb / 6" / dark scarlet fls / Israel (318) 25
Sprengeri / Bb / to 1' / fls light brownish-crimson / Asia Mnor (442) / 25
stellata - see T. clusiana v. stellata (624) / 25
stellata chrysantha - see T. clusiana v. chrysantha 129 / 185
sylvestris / Bb / to 1' / yellow fls, outside gr.-tinted / Eu., Iran 267 / 185
sylvestris ssp. australis /tinged pink, crims outside /range of sp. 290
sylvestris 'Tabriz' - free-flowering form from n Iran (624) / 185
tarda / Bb / to 8" / creamy, star-like fls / Turkestan 267 /129
thracica - see T. orphanidea v. pontica 125
turkestanica / Bb / to 8" /ivory-white fls, orange blotch /Turkestan 129 / 25
uniflora / Bb / dwarf / yel outside, stained green / c Asia 125
urumiensis / Bb / 2" / fls golden yel, red, olive inside / nw Iran 25
violacea / Bb / 4-8" / cerise-violet fls / Iran, Kurdistan 267 / 129
Vvedenskyi / Bb / to 8" / fls scarlet to orange / c Asia 268
Whitallii / Bb / to 1' / fls bright orange / Izmir, w Turkey 129 / 25

TUNICA
nanteulii - see PETRORHAGIA nanteuilii 286
prolifera - see PETRORHAGIA prolifera 286
rhodopea - see PETRORHAGIA illyrica ssp. haynaldiana 25
saxifraga - see PETRORHAGIA saxifraga 286

TURNERA - Turneraceae
ulmifolia / Gh Sh / to 2' / yellow fls / tropical America (540) / 25

TUSSILAGO - Compositae
Farfara / R / 6" / lt. yel. fls, early; pl weedy / Eu., U.S. escape 148 / 25

TWEEDIA
caerulea - see OXYPETALUM caeruleum 25

TYMBRA
spicata - see THYMBRA spicata *

TYPHA - Typgaceae
minima / Aq / 2½' / male & female fls separated / Europe 25

UGNI - Myrtaceae
molinae / Gh Sh / to 6' / pink fls; edible berries / Chile, Bolivia 25

ULEX - Leguminoisae
minor / Sh / dwarf / golden-yellow fls / sw Europe 139

UMBELLULARIA - Myrtaceae
californica / HH T / 20-30' / dark purple frs / Oregon 139

UMBILICUS - Cupressaceae
aizoon - see ROSULARIA azoon 77
pendulinus - see U. rupestris 286
rupestris / P / 8-20" / straw-colored fls / s & w Europe 182 / 25

UNGANDIA
speciosa - error for UNGNADIA speciosa

UNGERNIA - Amryllidaceae
victoris / Bb / 8-10" / yellowish-rose fls / c Asia 4

UNGINEA - Liliaceae
maritima / Gh Bb / 1-3' / white fls / circum-Mediterranean 211

UNGNADIA - Sapindaceae
speciosa / HH Sh / 5-10' / rose-col. fls / Texas 28

UNIOLA
latifolia - see CHASMANTHE latifolium 141 / 25

URGINEA - Liliaceae
ferganica / Bb / 4-8" / pale ochreous fls, purple tipped / c Asia 4
maritima / Gh Bb / 4-5' / fls white to rosy-yellow / Medit. reg. 211 / 25

UROSPERMUM - Compositae
Dalechampii / P / 12-20" / yellow fls / Medit. region 210 / 93

URSINIA - Compositae
anethoides / A / 1-2' / orange-yellow fls / South Africa 93 / 129
calenduliflora / A / 14" / deep yellow fls / South Africa 25
chrysanthemoides geyeri - see U. geyeri 268
geyeri / Gh / dwarf / bright red & purple fls / Cape Province 268

URTICA - Urticaceae
pilulifera / A / to 3' / stinging nettle / Medit. region 211

UTRICULARIA - Lentibulariaceae
bifida / Gh Aq / to 6" / yellow fls / Far East 200
capensis / Gh Aq / minute yellow & lavender fls / South Africa 223
yakusimensis / Gh Aq / 6" / pale purple fls / Japan 200

UVULARIA - Liliaceae
floridana / B / 4-16" / narrow petals grn-yel / Ga., Fla., Ala   225
grandiflora / P / 18" / pale yellow fls / e North America   224 / 123
perfoliata / P / 15" / paler yellow fls / e North America   224 / 107
sessilifolia / R / 10" / pale greenish-yellow fls / e N. Am.   224 / 107
sessilifolia 'Variegata' - lvs marked white   *

VACCINIUM - Ericaceae
angustifolium / Sh / 6-18" / light blue frs / e N. Am.   107
angustifolium v. laevifolium / Sh / 2' / narrow lvs / ne N. Am.   25
Arctostaphylos / HH Sh / 9' / frs purplish-black / Turkey   288
corymbosum / Sh / 15' / Highbush Blueberry / Maine to Florida   129 / 25
Delavayi / HH Sh / small / frs dark crimson / w China   222
fracteatum / Sh / to 6' / fls white; fr red / Orient   139
floribundum / Sh / small / fls racemes pink;fr red, edible / Ecuador   139
glauco-album / HH Sh / 4' / frs dark blue / Sikkim   129
hirtum / Sh / to 3' / frs red / Japan   200
japonicum / Sh / 2-3' / decid; sol fls, pink; fr br red /Jap.,Korea   36
macrocarpon / Aq Sh / creeping / Cranberry / North America   184 / 25
membranaceum / Sh / to 4½' / black frs / Mich, to Ore. & B.C.   222
moupinense / Sh / dwarf / frs purplish-black / w China   139
myrtilloides / Sh / 1-2' / edible blue frs / n N. Am.   25
Myrtillus / Sh/ to 2'/lvs br. grn; fls grn-pink; fr blk/ Eur, n Asia   25
Nummularia / HH Sh / small / fls rose-red to pinkish / Himalayas   222
Oldhamii / Sh / to 12' / fls red-pink; fr black / Korea, Japan   25
ovalifolium var. coriaceum / 1-3' / frs blue-black / n hemisphere   200
ovatum / HH Sh / 2-8' / frs black / California   139 / (313)
Oxycoccos / Sh / creeping to 1'/ fls pink; Cranberry / n hemisphere   229 / 25
parvifolium / Sh / 6-12" / frs red / Alaska & s-ward   299 / 139
praestans / Sh / 6" / white fls, tinged pink / ne Asia   (626) / 25
retusum / HH Sh / to 3' / pink fls / e Himalayas   139
scoparium / Sh / to 16" / sts yel-grn; fls pink; fr red/ Alta-Calif.   25
uliginosum / Sh / 2' / pink fls; dark blue frs / n hemisphere   243 / 25
uliginosum alpinum /to 2'/few fls wht-pink tinge;fr blue-black,sweet   99
vacillans / Sh / to 3' / dark blue edible frs / e N. Am.   71 / 25
Vitis-idaea / Sh /to 1'/ fls pink or wht; frs red / Eur, n Asia   148 / 25
Vitis-idaea 'Minor' - see V. Vitis-idaea v. minus   107
Vitis-idaea v. minus /dwarf/dense mat/fls pink to red/Ma. to Alaska   107 / 25
Vitis-idaea 'Variegata' - creamy wht margined lvs, not very constant   139

VALERIANA - Valerianaceae
alliarifolia / P / to 3' / pink fls / e Greece   289
alpestris / P / to 2' / white fls / Caucasus   77
arizonica / R / 12" / whitish or pinkish fls / Col., Utah & s-ward   227 / 25
celtica / R / 4-5" / brownish-yellow fls, aromatic / Alps of Eu.   96
columbiana / R / 1+ / creamy fls / Wenatchee Mts.   229 / 142
fonkii / HH R / mats / pale pink or lilac fls / Argentina   64
montana / R / 12" / fls pink to white / Europe   148 / 25
moyanoi / HH R / flat rosettes / yellow fls / Argentina   64
officionalis / P/ 3½-5'/fls wht-pink-lav, frag/ Eur, w Asia, n N.A.   99 / 25
phu aurea / P / to 3' / glabrous, young growth br yel /Caucasus   61
scouleri / P / 1-2' / white fls / California to Alaska   229

sitchensis / P / to 4' / fragrant pink fls / nw N. Am., in mts. 229 / 25
supina / R / 4-6" / fragrant pink fls / c Europe 148 / 25
tripteris / R / 4-16" / pink or white fls / s Europe 210 / 289

VALLOTA - Amaryllidaceae
purpurea - see V. speciosa 128
speciosa / Gh Sh / 2' / fls scarlet / South Africa 129 / 25

VANCOUVERIA - Berberidaceae
chrysantha / R / 8-12" / soft yellow fls / Oregon, California 228 / 123
hexandra / R / 10" / white fls / Wash. to Calif., B. C. 228 / 107
planipetala / R / 9" / white fls / Oregon, California 107

VAUQUELINIA - Rosaceae
californica / HH Sh / 5-15' / evergreen lvs / California 294

VELLA - Cruciferae
spinosa / HH Sh / low / yellow fls / s & se Spain, limestone mts. 212 / 286

VELTHEIMIA - Liliaceae
bracteata / Gh Bb / 18" / soft pink fls / South Africa 186 / 279
capensis / Gh Bb / 18" / pink fls tipped cream / South Africa (314) / 279
glauca / Gh Bb / 15" / white fls / Cape Province 111
viridiflora - see V. capensis 25
viridifolia - see V. bracteata 128 / 279

VERATRUM - Liliaceae
album / P / 3-4' / yellowish-white fls / Europe, n Africa 148 / 207
album lobelianum - see V. lobelianum 4
californicum / P / 3-6' / greenish bell-shaped fls / w U.S. 207
Eschscholtzii - see V. viride 25
grandiflorum / P / 3-4½' / fls greenish-white / Japan 200
lobelianum / P / 2-5' / yellowish-green fls / Eurasia 4
Maackii / P / to 2' / fls dark purp / e Asia, Japan 25
nigrum / P / 3' / fls blackish-purple / s Europe, Asia 148 / 93
stamineum / P / 20-40" / white fls / Hokkaido, Honshu 200
viride / P / 2-5' / green fls / North America 229 / 207

VERBASCUM - Scrophulariaceae
acaule / R / 4" / bright yellow fls / Greece, in high mts. 288
acaule 'Alba' - white fls *
arcturus / HH P / 1-2' / yellow fls / Crete 149 / 288
banaticum / B / 20-40" / yellow fls / se Europe 288
Blattaria / B / 1-4' / yellow fls / Europe, U.S. escape 224 / 288
Blattaria f. albiflorum / white fls / rare in nature 207
Blattaria virgatum - see V. virgatum 99
bombyciferum / P / 4-6' / golden fls; felted lvs / Asia Minor (145) / 207
bugulifolium / P / 8-30" / yellowish fls / Turkey, Bulgaria 288
Chaixii / P / 20-40" / yellow fls / s, c & e Europe 212 / 288
Chaixii 'Album' - white fls 207
'Cotswold Beauty' / P / 3-4' / biscuit-color fls 268
'Cotswold Gem' / P / 3-4' / soft rosy-amber fls 268
creticum / HH B / to 5' / yellowish fls / w Mediterranean region 288
dumulosum / R / 6-10' / fls bright yellow / Asia Minor 194 / 129
georgicum / B / to 5' / yellow fls / Caucasus, Turkey 288

graecum / B / to 5' / yellow fls; whitish lvs / s Balkan Penin. 288
heldreichii - see V. banaticum 288
lychnitis 'Alba' / B / to 5' / white fls / Europe 288
nigrum / P / 20-40" / fls yellow / Europe 148 / 288
nigrum 'Album' - white-flowered form 135
olympicum / B / 5-6' / golden fls / s Europe 147 / 208
pestalozzae / R / to 8" / yellow fls / Asia Minor 93
phoeniceum / P / 1-3' / violet fls / se & ec Europe 194 / 288
phoeniceum album - white form 25
phlomoides / B / to 4' / fls yellow / c & s Eur., nat. U.S. 25
roripifolium / B / to 5' / metallic-bronze fls / s Bulgaria 288
spinosum / HH Sh / low / axillary yellow fls / Crete 129 / 211
Thapsus / B / 1-6' / yellow fls / Eu., common weed in U.S. 224 / 288
undulatum /B or P/1-4'/var. wht, gray, yel, tomentose/s&w Balk.Pen. 288
vernale - see V. Chaixii 61
virgatum - roadside adventive in U.S. from Eu., similar to V. b. 99
Wiedemannianum / B / 3' / indigo-blue to violet fls / Asia Minor 93

VERBENA - Verbenaceae
bipinnatifida / P / 6-18" / fls lilac or purple / S. Dak. to Mexico 229 / 107
bonariensis / P gr. as A / 4'+ / fls reddish-purple / Brazil 207 / 228
canadensis / P / 6-18" / rosy-purple fls / Okla., Kan., Col. 224 / 229
hastata / P / 3-5' / violet fls / North America 229 / 207
peruviana / P gr as A / 4-6" / scarlet to pink fls / South America 286
pumula / R / 6" / fls white / Texas, Oklahoma 226
rigida / HH P / 1-2' / purp. or magenta fls / Argentina, escape N.C. 129 / 25
stricta / P / to 1½' / fls purple or blue / n & w United States 25
venosa / see V. rigida 129

VERBESINA - Compositae
enceliodes / A / 3' / deep yellow fls / w U.S., Mexico 229 / 25

VERNONIA - Compositae
altissima / P / 5-10' / purple fls / Pa. - Fla. 28
noveboracensis / P / 6' / fls purple / Mass. to Miss. 224 / 25

VERONICA - Scrophulariaceae
Allionii / P / 4-20" / deep blue fls / Alps, in pastures 148 / 288
alpina / R / 6" / fls blue / Eurasia 148 / 107
alpina 'Alba' / 6-8" / white form 135
aphylla / R / procumbent / fls deep blue / c & n Europe 148 / 288
armena / R / to 4" / fls deep or violet-blue / Armenia 306 / 25
austriaca / P / 10-20" / bright blue fls / e, ec & se Europe 288
austriaca ssp. austriaca /P/10-20"/calyx & cap hairy, glab/ec&s Eur. 288
austriaca ssp. bipinnatifida / P / 10-20" / bright blue fls / Europe 288
austriaca v. jaquini - see V. austriaca v. austriaca 288
austriaca ssp. teucrium - P / 1-3' / blue fls / most of Europe 288
Balfouriana - see HEBE balfouriana 15
baumgartenii / R / 2-5" / cor. blue; lvs serrate-subentire/ Carpath. 288
bellidioides / R / to 8" / fls lilac to violet-blue / Pyrenees 148 / 288
Bidwillii - see HEBE decora 132
bipinnatifida - see V. austriaca ssp. bipinnatifida 262
Bombycina / HH R / neat tuft / reddish fls; wooly lvs / Syria (315) / 96
bonarota - see PAEDEROTA bonarota 288
cana v. miqueliana / R / 6-12" / racemes 3-10fls, pale rose / Japan 200

catarractae - see PARAHEBE catarractae 139
caucasica / R / to 6" / fls loose racemes, wht, veined viol / Cauc. 25
cinerea / P / shruby, mat form / 2-8" / fls blue / Turkey 77
Cusickii / R / 9" / blue or violet fls / Oregon, California 229 / 25
densiflora / R / to 6" / large pale blue fls / Altai Mts. 81
derwentia / P / woody, 2-3'/ fls blue or wht, fragrant / Austrl,Tas. 61
epacridae - see HEBE epacridea 15
exaltata - see V. longifolia 129 / 25
Forrestii / R / 6-15" / fls reddish / Yunnan 28
fruticans / R / 6" / clear blue fls / nw Europe 148 / 107
fruticulosa / R / 6" / veined pink fls / Europe, in mts. 148 / 123
fruticulosa 'Alba' - all white fls *
gentianoides / P / 6-24" / fls pale blue, darker veins / Caucasus 129 / 25
gentianoides 'Nana' / R / dwarf 123
glauca / R / to 8" / deep blue fls / s Balkan Peninsula 288
grandiflora / R / to 4"/ blue-purple fls / w Aleutian Isls. 13
grandis v. holophylla / P / 2' /glossy leathery lvs; blue fls / Japan 25
X Guthrieana - see HEBE X guthrieana *
Hendersonii - see HEBE X andersonii 28
holophylla - see V. grandis v. holophylla 25
hookeriana - see PARAHEBE hookeriana *
1 'Icicle' - white fls on 15" stems
incana / P / to 2' / blue fls / n Asia, Russia 147 / 25
incana 'Nana' - dwarfer 25
incana spicata - see V. spicata ssp. incana 288
incana 'Wendy' - lavender-blue fls 268
Kellereri / P / ? / fls blue / Morocco 25
latifolia - see V. austriaca ssp. teucrium 288
1 latifolia 'Crater Lake Blue' / 12" / gentian blue star fls, wht eye
longifolia / P / 2' / lilac-blue fls / n, e & c Europe 269 / 135
longifolia 'Alba' - all white fls 25
2 longifolia 'Blue Giantess' / 2-4" / densly packed dp purp blue fls
longifolia v. subsessilis - fls larger, deep blue 25
lutea - see PAEDEROTA lutea 288
lyallii - see PARAHEBE lyallii 139
lycopodioides - see HEBE lycopodioides 139
macrantha - see HEBE macrantha 139
minuta - see V. telephiifolia 77
nipponica / R / to 4" / fls pale blue-purple / Honshu 164 / 200
nivea - see PARAHEBE hookeriana *
Nummularia / R / procumbent / fls blue or pink / Pyenees 148 / 288
officinalis / R / prostrate / fls pale blue / Eurasia, N. Am. 224 / 25
orchidea - S. spicata v. orchidea 25
orientalis / R/ 4-10"/ corolla blue, rarely pink/ Krym, Asia Minor 288 / 77
orientalis ssp. orientalis /lvs sparsely puberulent,often near marg. 77
perfoliata - see PARAHEBE perfoliata 279
pirolaeformis / R / short / fls rose or pale violet / China 28
prostrata / R / 2-8" / fls deep blue / Eurasia 129 / 135
prostrata 'Heavenly Blue' - sapphire-blue fls 135
prostrata nana / P / prostrate / fls to 10", dp blue / Europe 25

1 Wayside Catalog, Spring 1986.

2 Thompson and Morgan Catalog, 1986.

prostrata 'Rosea' - fls purplish-pink 93
prostrata 'Spode Blue' - clear pale China-blue fls 123
pubescens - see HEBE pubescens 15
pyrolaeformis - see V. pirolaeformis 28
repens / R / 4" / fls pink / Corsica, Sardinia 208 / 288
rupestris - see V. prostrata 129
saturejoides / R / flat creeper / clear blue fls / Balkans 107
saturejoides 'Alba' - white form 154
saxatilis - see V. fruticans 25
Schmidtiana / R/ 4-10"/ fls pale blue-purp, dp striations / n Japan 200
Schmidtiana bandiana / lf lobes shallow, subacute / Honshu, alp reg 200
serpyllifolia / R / to 8" / fls white or pale blue / Europe 288
'Shirley Blue' / 9" / bright blue fls 268
spicata / P / to 1½' / fls dense, blue or pink / n Eur., Asia 148 / 25
spicata 'Caerulea' - light blue fls 25
spicata 'Corymbosa' / R / to 2½' / fls blue or pink 25
spicata ssp. incana - dense gray-wht, tomentose and eglandular 288
spicata 'Nana' / to 1½' / fls blue, lav, pink 107 / 25
spicata v. orchidea - corolla convolute in bud 25
spicata 'Rosea' - fls purplish-pink 93
spicata 'Wendy' - see V. incana 'Wendy' 268
stelleri - see V. Wormskjoldii 276
subsessilis / P / 20-40" / fls blue-purple / Honshu 200
surculosa / R / 3" / white fls, purplish-eyed / Asia Minor 93
telephiifolia / R / 1" mat / mid-blue fls / Armenia 24
teucrium - see V. austriaca ssp. Teucrium (627) / 288
teucrium 'Blue Tit' - see V. austriaca Teucrium 154
teucrium 'Rosea' - the rarely rose form 28
teucrium 'Trehane' / 9" / light blue with gold-grn lvs 129
thymifolia / HH R / decumbent /fls lilac-blue-pink/ Crete, s Greece 288
turrilliana / R / to 16" / deep blue fls / e Balkan Penin. 288
urticifolia / P / 10-28" / sparsely hairy, corolla lilac / s&c Eur. 288
Wormskjoldii / R / to 1' / dark blue fls / N. Am., in mts. 25
Wormskjoldii stelleri - see V. stelleri 20
yezo-alpina - included in V. schmidtiana 200

VERONICASTRUM - Scrophulariaceae
exaltata - included in VERONICA 61
sibericum - scarcely distinguishable from V. virginicum 99
virginicum / P / to 6' / fls white or purplish / e N. Am. 224 / 99

VERTICORDIA - Myrtaceae
forestii Fontanesii / bushy shrub / 3-4'/ fls wht, pink/ s Australia 61
nitens / Sh / to 3' / fls gold-yel / w Australia 25

VESICARIA - Cruciferae
graeca - see ALYSSOIDES graeca 286
ultriculata - see ALYSSOIDES utriculata 286

VESTIA - Solanaceae
lycioides / Sh / 1' / fls grn-yel; evergreen; ill-scented / Chile 64 / 25

VIBURNUM - Caprifoliaceae
acerifolium / Sh / to 6' / frs red to blackish / e N. Am. 139
alnifolium / Sh / to 10' / white fls / e N. Am. 194 / 25

Awabuki - see V. odoratissimum v. Awabuki 167 / 25
burejaeticum / Sh / 15' / white fls; bluish-black frs / Manchuria 25
X Burkwoodii / Sh / 6' / fragrant pinkish fls; evergreen lvs 184 / 25
X Burkwoodii 'Chenault' - fls pale rose,turn wht; autumn lvs bronze 25
x carlcephalum - large fls, white w. red; autumn foliage brilliant 25
Carlesii / Sh / 5' / fragrant white fls / Korea 129 / 25
cassinoides / Sh / 6-10' / red, blue to black frs / e N. Am. 110 / 139
Davidii / HH Sh / to 5' / frs dark blue / China 129
dentatum / Sh / 10' / frs blue-black / e N. Am. (317) / 139
dilitatum / Sh / to 10' / fls white, frs scarlet / Japan 300 / 25
dilitatum v. hizenense / alleged V. d x V. japonicum / nearly glab. 200
ellipticum / Sh / 6'+ / fls whitish; frs blackish / Calif. to Wash. 222
erosum / Sh / 6-10' / white fls; red frs / Japan 167 / 139
erubescens / Sh / to 15' / wht. fls, tinged pnk / w China, Himalayas 25
Farreri / Sh / to 10' / frs red / n China 139
fragrans - see V. farreri 139
furcatum / Sh / 10'+ / frs red to black / Japan, Korea 139
ichangense / HH Sh / to 6' / showy, scarlet-red frs / c & w China 25
X juddii / Sh / 6' / fls white, pink in bud 129
japonicum / HH Sh / to 6' / fragrant wht. fls; evergreen lvs / Japan (628) / 25
Lantana / Sh / 15' / fls white; frs red / Eurasia 210 / 25
Lantana 'Macrophyllum' - larger fls & fruit clusters (318)
Lentago / Sh / 15'+ / frs blue-black / e North America 139
lobophyllum / Sh / to 15' / wht fls, lg cymes; fr br red / China 25
mongolicum / Sh / to 7' / black frs / ne Asia 25
odoratissimum v. awabuki /HH Sh/ to 12'/ lvs glossy, fls white/Japan 25
Opulus / Sh / 10'+ / white fls, red frs / Eurasia 129 / 139
Opulus 'Notcutt's Var.' - larger fls & fruits than type 139
Opulus 'Xanthocarpum' - golden yellow frs 139
phlebotrichum / HH Sh / 6' / fls white or pinkish; frs red / Japan 167 / 25
plicatum / Sh / to 10' / white fls / China, Japan 25
plicatum 'Lanarth' - strong growth, less horizontal than type 139
prunifolium / Sh / 18' / frs blue-black / e North America (317) / 139
recognitum / Sh / 3-9" / the northern Arrowwood / ne N. America 99
rhytidophyllum / Sh / red to black frs / c & w China 139
Sargentii / Sh / 12' / scarlet frs / ne Asia 110 / 25
Sargentii 'Susquehanna' / to 6' / red frs; leathery lvs (615) / (319)
setigerum / Sh / 6-10' / yellow to red frs / c & w China 139
setigerum 'Aurantiacum' - fr. orange yellow 25
Sieboldii / Sh / 15' / pink to blue-black frs / Japan 300 / 139
Tinus / HH Sh / 10' / fls white or pinkish; frs black / Medit. reg. 210 / 25
trilobum / Sh / 10'+ / red frs, persisting / n North America 300 / 139
trilobum 'Wentworth' - matures fruit early (318)
Wrightii / Sh / 6'+ / glistening red frs / Far East (318) / 139

VICIA - Leguminosae
crocea / P / 20-32" / fls bright orange-yellow / Caucasus, Iran 9
onobrychoides / P / 2' / purple fls / s Europe 212 / 206
pyrenaica / R / 2-12" / sol fls, br viol-purp / Spain & s France 287
sativa / A or B / to 3" / fls purp, usually paired / Eur, nat N.A. 25
sepium / P / 1-3'+ / fls 2-6, dull bluish-purple / Europe 287
sylvatica / P / 2-5' / fls white, purple veined / n, c & e Europe 148 / 287
tetrasperma / A / 4-24" / 1-2 fls, pale purple / Eur. to Scand-Rus. 287
unijuga / R / 1'+ / brilliant blue fls / Siberia 165 / 96

VINCETOXICUM - Asclepiadaceae
hirundinaria / P / to 4', twining / fls white or yellow / Europe 148 / 288
nigrum / P / 20-40", twining / dark purple fls / Europe 288
officinale - see V. hirundinaria 288
purpurascens - see CYNANCHUM purpurascens 200

VIOLA - Violaceae
adunca / R / low / violet to deep purple fls / w North America 312 / 107
aetolica / R / 6-20" / yellow fls / Balkans 287
alba / R / 2-6" / fragrant white or violet fls / s Europe 287
albida v. chaerophylloides f. sieboldiana - see V. dissecta v. s. 200
[1] 'Alice Whitter' - white & red eye
alpina / R / see V. magellensis 287
altaica / R / to 5" / variable colors, a pansy ancestor / Altai 96
appalachiensis / R / 2½" / fls pale to deep violet / W. Virginia 313
arborescens / R / 4-8" / fls whitish or pale violet / w Medit. reg. 212 / 287
arenaria - see V. rupestris 64
arenaria 'Rosea' - see V. rupestris 'Rosea' 25
'Arkwright Ruby' / 6" / deep ruby-crimson, black blotch 300 / 268
arvensis / A / to 16" / fls cream, yellow, bluish-violet / Europe 287
aurea - see V. purpurea v. aurea 31
Beckwithii / R / 3" / purple fls / w North America 123
Bertolonii / R / to 1' / fls violet or yellow / Maritime Alps 287 / (445)
Bertolonii corsica - see V. corsica 287
Bertonii X Stojanowii - crossed by donor? 61
betonicifolia / R / 6" / fls blue to purple / Asia, Australia 49 / 64
biflora / R / to 8" / yellow fls / Europe, in mts. 148 / 287
biflora v. crassifolia - see V. crassa 200
bissetii / R / 2-4" / fls lg., pale purple / Japan 200
blanda / R / 2-4" / white fls, purple veins / e North America. 224 / 31
boisieuana v. iwagawae - white fls, small plant, Kyushu 200
boninensis - see V. mandshurica v. boninensis 200
bosniaca - see V. elegantula 25
'Bowles Black' - see V. tricolor 'E. A. Bowles' (418) / 14
brevistipulata / R / to 1' / yellow fls / n & c Japan, in high mts. 200
calaminaria - see V. tricolor ssp. subalpina 287
calcarata / R / 2" / fls violet or yellow / Alps 148 / 287
calcarata alba / P / stemless to 4" / petals white / mts s Europe 25
calcarata ssp. zoysii - yellow fls, w Balkans / se Alps 287
canadensis / R / to 1' / white or violet-tinged fls / n N. Am. 31 / 25
canadensis v. rugulosa - slender, stoloniferous rhizome 25
canina / 4-16" / fls blue / Europe 284 / 287
canina 'Alba' - may be ssp. montana, taller, white or blue fls 287
cazorlensis / R / 4" / fls intense pinkish-purple / se Spain 212 / 287
cenisia / R / 2" / fls bright violet / sw Alps 148 / 287
chaerophylloides - see V. dissecta v. chaerophylloides 200
chaerophylloides sieboldiana -see V. dissecta v. chaerophylloides 200
'Chantreyland' / 6" / apricot fls 129 / 268
chinensis - see V. Patrinii 29
conspersa / R / 3½-6" / fls pale violet to whitish / e N. America 31
cornuta / R / 8-12" / fragrant lilac or violet fls / Pyrenees 148 / 287
cornuta 'Alba' - white fls 306 / 107

---

[1] Rocknoll Catalog, Spring 1986.

cornuta 'Minor' / 2-3" / clear blue fls / Pyrenees 123
corsica / R / 4-8" / violet fls / Corsica, Sardinia 287
cotyledon ssp. lologensis / loose growth / fls wht, deep blue/ S. Am. 64
crassa / R / to 5" / fls deep yellow / alpine reg. on Japan 165 / 200
crassiuscula / R/ to 6"/ fls br viol,pink,wht; ptl gold-yel/s Spain 287
cucullata / R / tufted / blue-violet fls, invasive / e & c N. Am. 22 / 99
cucullata alba - see V. papilionacea f. albiflora 99
cucullata 'Freckles' - see V. papilionacea 'Freckles'
[1] cucullata 'White Czar' - lg white fls, delicate markings
cuneata / R / 5" / white & purple fls / California, Oregon 228
Cunninghamii / R / 6" / white fls / New Zealand 238
Curtisii - see V. tricolor ssp. Curtisii (629) / 287
delphinantha / P-R / 2-4" / fls pink or red-purp / n Greece, Bulg. 287
diffusa v. glabella / lvs coarsely wht; fls wht, purp tinge / China 200
dissecta v. chaerophylloides / to 4" / incised; fls purple / Japan 200
dissecta v. chaerophylloides 'Alba' - white fls *
dissecta v. chaerophylloides f. eizanensis - fls purple 200
dissecta v. chaerophylloides eizanensis alba-see white form V. d. c. 25
dissecta v. sieboldiana / 5" / white fls, purple striated 200
diversifolia / R / 1-6" / hairy; fls frag., viol. / e & c Pyrenees 287
Douglasii / R / 4" / light golden-yellow fls / Calif., Oregon 31 / 229
Dubyana / R / 4" / violet fls / Italian Alps 148 / 287
eizanensis - see V. dissecta v. chaerophylloides f. eizanensis 200
eizanensis 'Alba' - see V. dissecta v. chaerophylloides f. e.'Alba' *
eizanensis chaerophylloides - see V. dissecta var. chaerophylloides 200
elatior / P / to 20" / fls pale blue / c & e Europe 287
elegantula / P,B/ to 1' / fls viol; yel striped spot / sw Eur. 287 / 25
emarginata - natural hybrids V. affinis x V. sagittata 25
eriocarpa - see V. pubescens 25
escondidaensis - South American, densely leafy stems 64
eugeniae / R / compact / fls yellow & violet / Apennines 287
fimbriatula / R / to 4" / fls violet-purple / e & c North America 224 / 25
Flettii / R / to 6" / violet fls / Washington, in mts. 31 / 25
X florariensis - yellow & purple fls (588) / 25
'Freckles' - see V. papilionacea 'Freckles' 300 / (321)
glabella / R / 8-12" / yellow fls / Alaska to n Calif., ne Asia 13 / 25
gracilis / R / to 1' / fls violet or yellow / Balkans 18 / 287
gracilis 'Alba' - white fls 208
gracilis 'Lord Nelson' - fls deeper in color 207
gracilis lutea / P / to 6" / fls yellow / Balkans to Asia Minor 25
gracilis major - Cv. of V. gracilis 25
grisebachiana / R / to 4" / violet fls / Balkan alpine meadows 287
grypoceris / R / to 1' / fls pale violet, scentless / Japan 165 / 200
Hallii / R / 4" / fls purple & cream / California, Oregon (275) / 107
'Hazelmere Seedling' / dwarf / elfin violet fls 83
hederacea / R / 2-3" / purple & white fls / Australia 96
hirsutula / R / small / fls reddish-purple / e U.S. 31 / 99
hirta / R / to 6" / violet fls / most of Europe 287
hirtipes / R / to 6" / fls pale rose-purple / Japan, in mts. 200
hispida / R / to 10" / fls violet or yellowish / nw France 287
hondoensis / R/spreads to 3"/fls viol or wht, lip purp striate/Jap. 200
'Jackanapes' / R / yellow & chocolate fls 61

[1] Wayside Catalog, Spring 1986.

japonica / R / 3" / fls purple / Japan, s Korea, in lowlands	165 /	200
Jooi / R / 4-12" / fragrant reddish fls / Roumania	(64) /	287
Keiskei / R / to 8" / white fls / Japan, in mts.	165 /	200
Keiskei v. okuboi - plant pilose		200
Kitaibeliana v. Rafinesquii / A / 4"+ / bluish-white fls / Eurasia		99
Kiusiana - see V. diffusa v. glabella		200
labradorica / R / 4-6" / lilac fls, purplish lvs / n N. Am.		207
labradorica purpurea - characteristic of the sp.		*
lactiflora - see V. japonica		200
lanceolata / R / small / white fls / North America	224 /	99
Langsdorfii / R / to 8" / fls blue-violet to wht / Siberia, nw N.Am.	13 /	25
lobata / R / 4" / fls deep lemon-yellow / California, Oregon		228
lutea / R / 8"+ / fls parti-colored, yellow & violet / w Europe	148 /	287
lutea ssp. sudentica / usually 6" / mts. of c Europe		287
lyallii / R / 6" / white fls, purple veined / New Zealand		238
macedonica - see V. tricolor ssp. macedonica		287
Macloskeyi / R / to 5" / white fls / Alaska to Calif., in mts.	227 /	31
maculata / R / 8" / golden yellow fls / Falklands	293 /	64
magellensis / R / 2-4" / violet or pink fls / Greece, Albania		287
mandshurica / R / to 8" / fls dark purple or striated white / e Asia	165 /	25
mandshurica 'Alba' - the rare white form		200
mandshurica v. boninensis - seashore phase, thicker lvs		200
minor / lvs dp green, lustrous / fls deep purple, small / Japan		200
montcaunica - like V. cornuta, but fls ½ size		287
X montivago - see V. sororia x V. septentrionalis		99
'nana compacta' - name of no botanical standing		25
nephrophylla / stemless / petals dp violet to 1"+ / North America		25
nigra 'Bowles Black' - see V. tricolor 'E. A. Bowles'		14
Nuttallii / R / 2-4" / deep lemon yellow fls / c North America	229 /	31
Nuttallii v. praemorsa / densely hairy lvs / B. C. to n Calif.		25
Nuttallii v. vallecola /lvs 2"-, glab.-sparsely hairy/B.C.-Rockies		25
ocellata / R / 6-12" / white fls, eyed purple / Calif. Ore.	228 /	31
occidentalis / R / 3-6" / white fls / California, Oregon		31
odorata / R / to 6" / fls dark violet or white / Europe	147 /	287
odorata 'Alba' - white fls		287
odorata 'Rosina' - old-rose fls		300
odorata 'Sulphurea' - sulphur-yellow fls		*
orbiculata / R / clump / yelow fls / B.C. to Oregon & Montana	229 /	227
oreades / R / creeps, 2-3" / fls yel, viol; petals wide / Caucasus		287
orphanidis / P / 8-28"/ fls violet or blue /Balk. to c Greece, Bulg.		287
pallens / R / to 3½" / white fls / North America		31
palmata / R / 3-5" / blue-violet fls / e & c United States	224 /	31
palmata var. sororia - see V. sororia		99
palustris / R / stemless / pale lilac-nearly wht fls / n N. Am., Eu.	210 /	25
papilionacea / R / tufted / violet fls, white centers / ec U.S.	31 /	99
papilionacea f. albiflora - whitish gray with purple veins		99
papilionacea - see V. papilionacea f. albiflora		99
papilionacea 'Freckles' / to 6" / white fls, blue spotted		(321)
papilionacea priceana - see V. papilionacea f. albiflora		99
Patrinii / R / tuft / clear lavender fls / n Asia, Japan		96
pedata / R / 4-5" / deep & light purple fls / e North America	224 /	123
pedata f. alba - white fls		99
pedata bicolor - is the first-named form above, hence the type		99
pedata v. concolor - see V. pedata f. alba		99
pedata f. rosea - fls uniformly the lighter colors		99

pedata v. lineariloba - lvs dissected nearly to base 60 / 99
pedatifida / R / to 10" / fls soft reddish-violet / c N. Am. 227 / 31
pedunculata / P / to 2' / fls orange-yel, veined purp / Cal. & Baja 25
pensylvanica - see V. pubescens v. eriocarpa 224 / 25
phalacrocarpa / R / to 8" / deep rose-purple fls / Far East 200
pinnata / R / 4" / fragrant pale violet fls / Alps 148 / 287
praemorsa / R / 3" / deep lemon-yellow fls / California to B.C. 31
Priceana 'Confederate Violet' - see V. papilionacea f. albiflora 99
primulifolia / R / to 10" / fls white / e North America 224 / 25
'Prince John' - see V. tricolor 'Prince John' 25
pubescens / R / 8-12" / lemon-yellow fls, striped purple / ne N. Am. 224 / 31
pubescens v. eriocarpa / 6" / yellow fls / e North America 25
pumilio / R / 2" / small white fls / Japan, on sunny hills 200
purpurea / R / 2-7" / deep lemon-yellow fls / Calif. to B.C. 228 / 31
purpurea v. aurea - arid lands 4500-5000' as V. purpurea 31
pyrenaica / P/ 3-4"/summer lvs shine; fls viol, wht throat/Pyr.Alps 287
Reichenbachiana / R / 6" / violet fls / s, w & e Europe 287
renifolia / P / stemless to 4" / fls white / Nfld to Minn, s to Pa. 25
Riviniana / R / 8" / blue-violet fls / Europe, n Africa 194 / 25
Riviniana minor / fls and fr smaller / more exposed habitats, w Eur. 287
rosea - a listed name of no botanical standing 25
Rossii / R / to 10" / fls pale rose-purple / Far East, in mts. 200
rostrata / R / pale lavender-violet fls / e North America 31
rotundifolia / R / to 6" / deep lemon-yellow fls / e United States 31
rugulosa - see V. canadensis v. rugulosa 225 / 25
rupestris / R / 4" / fls reddish-violet to white / Europe 148 / 287
rupestris 'Rosea' - selection for pink fls 64
sacchalinensis / R / 4-8" / fls pale violet / n Japan, e Asia 200
sagittata / R / 4" / fls violet-purple / ec United States 31 / 25
saxatilis - see V. tricolor ssp. subalpina 287
saxatilis aetolica - see V. aetolica (418) / 287
saxatilis macedonica - see V. tricolor v. macedonica 287
Selkirkii / R / to 4" / pale violet fls / n North America 224 / 228
sempervirens / R / 8" / lemon-yellow fls / Calif. to Vancouver Is. 228 / 31
sempervirens orbiculata - see V. orbiculata 142
septentrionalis / R / 6" / violet fls, varying lighter / e N. Am. 46 / 31
Sheltonii / R / 6" / lemon-yellow & brown fls / nw N. Am. 227 / 31
sieboldii - glabrous petals, fls white; may be form of V. pumilio 200
sieheana / like V. Riviana but fls larger, pale blue or wht/se Eur. 287
sororia / like V. papilionacea / viol. or lav / e & c N. A. 99
sororia 'Confederate Violet' - see V. papilionacea f. albiflora 99
sororia 'Freckles' - see V. papilionacea 'Freckles' 59
Stojanowii / R / to 4" / yell. fls, tinged violet / Bulgaria, Greece 287
striata / P / 1-2' / fls ivory-white or cream / N.Y., Minn. to Ga. 224 / 31
sudetica - see V. lutea ssp. sudetica 287
takedana / fls rose-purp, lat petals slightly bearded / Honshu 200
tokubuchiana - see V. takedana 200
tricolor / A / 8" / fls blue-violet / Europe 148 / 129
tricolor 'Bowles Black' - see V. tricolor 'E. A. Bowles' 280
tricolor 'Chantryland' - see V. 'Chantryland' *
tricolor ssp. Curtisii / 6" / fls variously colored / w Europe 24 / 287
tricolor 'E. A. Bowles' / B / velvety blackish fls 280
tricolor ssp. macedonica / violet & yellow fls / Balkans 287
tricolor ssp. subalpina / sometimes B / to 1' / s & c Europe in mts. 287
trinervata / R/ tufted, to 6"/fls lt to dk blue, yel base/ WA & OR 31 / 25

vaginata / R / to 8" / pale purple fls / Hokkaido, Honshu 165 / 200
variegata / R / 6" / pale violet fls / Siberia, Korea 70 / 96
variegata v. nipponica / lvs wht; var upper, purp beneath / n Orient 200
verecunda v. yakusimana / R / 1" / white fls / Kyushu, in mts. 200
volcanica / HH R / 3" rosette / small blue fls; fleshy lvs / S. Am. 64
Walteri / R / 5" / fls blue violet / Florida to Kentucky 99
X Wittrockiana -botanical name Pansy/ V. tricolor, lutea & attaica 25
yakusimana - see V. verecunda v. yakusimana 200
yedoensis / stemless / lvs white, hairy; ptls wht, purp lines/ Jap. 200 / 25
yubariana / R / 2" / yellow fls / Hokkkaido 200
Zoysii - see V. calcarata ssp. Zoysii 287

VISCARIA - Caryophyllaceae
alpina - see LYCHNIS alpina 286
atropurpurea - see LYCHNIS viscaria ssp. atropurpurea 286
1 'Blue Angel' (syn. SILENE) - azure blue fls
1 'Rose Angel' - rose pink form of 'Blue Angel'
splendens - see LYCHNIS viscaria 'Splendens' 28
viscosa - see LYCHNIS viscaria 286
viscosa albiflora - see LYCHNIS viscaria 'Alba' 135
viscosa atropurpurea - see LYCHNIS viscaria ssp. atropurpurea 286
vulgaria - see LYCHNIS viscaria 286
vulgaria 'Splendens' - see LYNCHIS viscaria 'Splendens' *

VITALIANA - Primulaceae
primuliflora / R / mats / yellow fls / s & wc Europe to Alps 288
primuliflora ssp. canescens /P/trail /fl yel; lvs flat/sw Alps,Pyr. 288

VITEX - Verbenaceae
Agnus-castus / Sh / 6' / violet fls / Medit. region to c Asia 149 / 139
chinensis - see V. negundo v. incisa 222
Negundo / Sh / 15' / fls lilac or lavender / Africa, Asia 144 / 25
Negundo v. incisa - leaflets nearly pinnatifid; northern hardy form 222

VITIS - Vitaceae
aestivalis / Cl/tall/ lvs soft, 2-8"; fr clusters black/e & sc U.S. 25
riparia / Cl/ frag; lvs 3-7"; fr clust. to 6", black / Nova Scot. & s 25
vulpina / Cl / high / glossy back frs, edible after frost / e U.S. 25

VITTADINIA - Compositae
australis / R / to 1' / white-rayed fls / Australia, New Zealand 238 / 93

WACHENDORFIA - Haemodoraceae
paniculata / Gh / 18" / yellow fls / Cape of Good Hope 206
thyrsiflora / Gh / 2' / yellow fls / South Africa 184 / 25

WAHLENBERGIA - Campanulaceae
albomarginata / R / 6" / fls blue to white / New Zealand 30 / 74
brockiei / R / to 4" / fls pale blue-violet / Canterbury, N. Z. 15
congesta / HH R / mat / blue or white fls / New Zealand 25
consimilis - see W. stricta 109

---

1 Thompson and Morgan Catalog, 1986.

dalmatica - see EDRAIANTHUS dalmaticus *
gloriosa / HH R / to 1' / deep blue to purple fls / Australia 61
gracilis - see W. trichogyna 25
graminifolia - see EDRAIANTHUS graminifolius 81
hederacea / R / prostrate / light blue fls / Europe, England 123
marginata / R / 8-16" / blue fls / Far East 164 / 200
Matthewsii / R / 8" / whitish to pale lilac fls / New Zealand 25
pumilio - see EDRAIANTHUS pumilio 289
pumilio ssp. stenocalyx - see EDRAIANTHUS pumilio ssp. stenocalyx *
pygmaea / HH R / ½" / fls white & pale blue / New Zealand 238 / 15
saxicola / R / 3-4" / bright light blue fls / Tasmania 182 / 96
serpyllifolia - see EDRAIANTHUS serpyllious 289
serpyllifolia alba - see EDRAIANTHUS s. 'Albus' *
stricta / HH P / to 2' / soft blue fls, white throat / Australia 109
tasmanica - see W. saxicola 25
trichogyna / P / 10-24" / blue fls, yellowish centers / New Zealand 238 / 25

WALDSTEINIA - Roseaceae
lobata / P / 4-8" / lvs 3 lobed; fls yel / N.Car. to Ga. 250
sibirica - see W. ternata 200
ternata / R / creeping / white fls / Siberia, Japan 200

WATSONIA - Iridaceae
Beatricis / HH C / 3' / pure white fls / Cape Province 279 / 267
coccinea / HH C / to 1' / blood-red fls / South Africa 267
Fourcadei / HH C / 4½' / deep salmon-red fls / South Africa 267
fulgens / HH C /3' / fls coral-red / South Africa 25
humilis / Cl / fl stem 18"/ 8-10 fl, spike pale magenta/Cape S. Afr. 111
marginata / HH C / 3-4' / rose to lilac pink fls / South africa 267
Meriana / C / 2-3' / pink to rose-red fls / Madagascar & S. Africa 95 / 25
pillansii / HH C / 4' / orange-red fls / e Cape Province 267
pyramidata / C / to 5½' / fls rose pink / South Africa 111 / 25
'Sandford Scarlet' - see W. Standfordiae 'Scarlet Form' 267
socium / HH C / 16-20" / reddish-gold fls / Natal 121
Standfordiae / HH C / to 4' / rose-purple / South Africa 267
Standfordiae 'Scarlet Form' / Gh C / 4' / scarlet fls / South Africa 267
versfeldii / HH C / to 5' /fls rose-pnk, stained crimson/ Cape Prov. 267
Wilmaniae / HH C / to 4' / rose- or lilac-purple fls / Cape Province 267
Wordsworthiana / HH C / 5½' / fls purplish-lilac / South Africa 61

WATTAKAKA
sinensis - see DREGEA corrugata 95 / 61

WEIGELA - Caprifoliaceae
Middendorffiana / Sh / 4' / sulphur-yellow fls / Far East 139

WEINMANNIA- Cunoniaceae
trichosperma / HH T / 18' / white fls / Chile 139

WELDENIA - Commelinaceae
candida / HH R / 6" / white fls / Mexico, Guatamala, in high mts. 129 / 25

WELWITSCHIA - Welwitschiaceae
bainesii - see W. mirablis 93
mirabilis - African desert plant of botanical interest 93

WERNERIA - Compositae
nubigena / R / mats / white daisies, often pink-backed / S. America 64

WISSADULA - Marvaceae
holosericea / HH P / 3-6' / yellow or orange fls / w Texas 308 / 226

WISTERIA - Legiminosae
floribunda 'Macrobotrys' / high Cl/ racemes to 3'+/ fls various col. 25

WITTSTEINIA - Epacridaceae
vacciniacea / Gh / 6-12" / fls greenish to reddish / Australia 109 61

WOODSIA - Polypodiaceae
alpina / F / 2-6" / high mountain fern / England, Wales 148 / 159
glabella / F / to 6" / petioles straw-colored / n N. Am., Europe 205 / 25
ilvensis / F / 2-6" / easily raised from spores / G. Britain in mts. 148 / 159
intermedia / F / 2-4" / tufted fronds / Korea 200
mexicana / F / 3-7" / masses of sori / Chisos Mts., Mexico, Texas 25
obtusa / T / 6-12" / gray-green fronds / North America 159
oregana / F / to 8" / bright green fls / Morth America 25
polystichoides / F / 9" / tufted habit / Japan 61
scopulina / F / to 10" / viscid & aromatic frond / nc North America 99

WOODWARDIA - Polypodiaceae
areolata / F/ fronds to 15"/fertile lvs with dk petioles/ Me to Fla 25
virginica / F / to 2" / lvs 9" wide / Nova Scotia to Florida 25

WULFENIA - Scrophulariaceae
alpina - see BESSEYA alpina 198
Amherstiana / R/ 4-8" / dull blue fls / Himalayas 182 / 93
Baldacci / R / 8" / bright blue fls / n Albania (630) / 288
bullii / R / 4-12" / greenish or yellowish fls / nc United States 99
carinthiaca / P / to 8" / deep purple fls,spike to 2' / e Eur.mts. 148 / 25
carinthiaca 'Alba' - white fls 25
orientalis / P / 7-16" / heliotrope fls / Asia Minor 61

WYETHIA - Compositae
amplexicaulis / P / to 2½' / yellow fls; lvs glossy / nw U.S. 229 / 25
helonioides / P / 8-32" / white or creamy fls / c Oregon, Nevada 229
mollis / P / 16-40" / fls yellow / California, Oregon (331) / 229

XANTHISMA - Compositae
texana / P / 4-30" / lemon yellow fls / Texas 226

XATARDA - Umbelliferae
scabra / R / 4-10" / greenish-yellow fls / e Pyrenees 287

XERANTHEMUM - Compositae
annuum / A / rosy-red, purple or white fls / Caucusus, Europe 149 / 93

XERONEMA - Liliaceae
callistemon / Gh P / 2' / bright red fls / Poor Knights Is., N.Z. 164

XEROPHYLLUM - Liliaceae
asphodeloides / P / to 4½' / showy white fls / New Jersey-Georgia 224 / 99
tenax / P / 2-5' / ivory fls / Rocky Mts. & w-ward 229 / 107

XYRIS - Xyridaceae
caroliniana / P / scapes to 3' / bracts yel, grn ctr / Fla to Me & w 99
montana / P / 18" / yellow fls / e North America 25

YUCCA - Liliaceae
aloifolia / HH P / 12'+ / white fls / North Carolina to West Indies 93
angustissima / Gh / to 7' /fls pale cream, tinged rose/ Ariz., Utah 227 / 25
baccata / HH Sh / 2' / white or cream fls / sw United States 229 / 25
Baileyi / P / 14-48" / green-white fls, tinged purple / Col., N. Mex. 302
brevifolia / HH Sh / 15-36" / several-stemmed / California, Nevada 302
brevifolia v. jaegeriana / tender tall pl. / creamy fls / sw U.S. 139
elata / HH Sh / to 20' / fls white-green, tinged pink / Tex., Ariz. 227 / 25
filamentosa / Sh / 4-12' / white-threaed lvs / se United States 147 / 129
flaccida / Sh / 10'+ / whitish fls; less rigid lvs / N. Car. to Ala. 279 / 25
glauca / P / 3-5' / greenish-white lvs / South Dakota to New Mexico 229 / 25
glauca 'Pink Brilliant' - selection for pink fls *
gloriosa / P / to 8' / fls green-white to red / N.C. to Fla. 25
harrimaniae v. neomexicana / P / nar. lvs; wht fls / Col. & N.Mex. 72
neomexicana - see Y. harrimaniae v. neomexicana 72
radiosa / HH Sh / 15-20" / fls white / Arizona, Mexico 28
rupicola / HH Sh / 20" / Twisted-leaf Yucca / c Texas 93
Schottii / HH P / to 15' / small white fls / sw North America 227 / 25
X thornberi - supposed hybrid between Y. baccata & arizonica 227 / 302
utahensis - see Y. glauca x Y. elata 25
Whipplei / HH Sh / to 10' / fls greenish-white / Calif., Baja Calif. 227 / 268

ZANTEDESCHIA - Araceae
aethiopica / HH Tu / to 3' / white splathes / South Africa 128 / 267
aethiopica 'Crowborough' - hardier form of Calla Lily 129 / 267
albomaculata / Gh Tu / 3' / whitish or pale yellow fls / Africa 25
Elliottiana / Gh Tu / 2' / yellow fls / South Africa (418) / 25
rehmannii / Gh Tu / 2' / rosy-purplish to white fls / South Africa 194 / 25

ZANTHOXYLUM - Rutaceae
americanum / Sh or T / 18'+ / jet-black frs / e North America 139
piperitum / Sh / 20' / fls green, 'Japan Pepper' / Orient 25
simulans / Sh or T / 9-20' / reddish frs; lustrous lvs / China 222

ZAUSCHNERIA - Onagraceae
arizonica - see Z. californica ssp. latifolia 25
californica / HH R / to 1' / scarlet fls / California, Mexico 175 / 123
californica 'Dublin' - selection for early-flowering 24
californica ssp. latifolia - somewhat wider-leaved, more herbaceous 139
californica ssp. latifolia 'Alba' - albino form 175
californica ssp. mexicana / gray-green lanceolate lvs / California 228
cana / HH R / 1' / narrow lvs, scarlet fls / sw California 159
'Dublin' - see Z. californica 'Dublin' 24
latifolia - see Z. californica v. latifolia 28
septentrionalis / R / 8" / bright red fls / nw California 228

ZELKOVA - Ulraceae
serrata / T / to 100' / sharply-toothed lvs / Japan 167 / 25

ZENOBIA - Ericaceae
pulverulenta / Sh / 2-4' / white fls / North Carolina to Florida 194 / 139

ZEPHYRANTHES - Amaryllidaceae
X 'Ajax' / HH Bb /, creamy-sulphur fls / Z. citrina X candida 185
andersonii - see HABRANTHUS andersonii 25
atamasco / HH Bb / 10" / white to pink fls / Missouri to Florida 225 / 267
brazoensis / HH Bb / 4-10" / white fls, tinged red / Texas, New Mex. 308 / 25
candida / HH Bb / to 1' / wht fls tinged rose / LaPlata reg., S. Am. 185 / 25
carinata - see Z. grandiflora 25
citrina / HH Bb / 6" / golden yellow fls in autumn / West Indies 185
concolor / HH Bb / 14"+ / yellow fls / Mexico (52)
drummondii / HH Bb / 5-8" /fragrant wht. fls, tinged red/Texas, Mex. 226 / 25
grandiflora / Gh Bb / to 10" / rose or pink fls / s Mex., Guatemala 267 / 25
macrosiphon / HH Bb / 12" / bright red fls / Mexico (645) / 121
mesochloa / A / lvs 10" / wht or pink with green base / Argentina 25
rosea / Gh Bb / ? / rose-red fls in autumn / Cuba 174 / 25
Smallii / HH Bb / to 6" / yellow fls, tube green / s tip of Texas 226 / 25
Treatiae / Bb / 18"?/ fl white, pink tinge, lvs cyl / Fla. 225 / 25
tubispatha / Bb / includes Z. insularum / fl white / West Indies 25
verecunda / HH Bb / 8" / white fls / Mexico, in mts. (631) / 267

ZIGADENUS - Liliaceae
elegans / Bb / to 3' / fls creamy-white / Alaska to Iowa 312 / 99
Fremontii / Bb / 12-16" / fls greenish-white / w North America 227 / 268
Fremontii 'Minor' - dwarfer form *
glaberrimus / HH P / 2-4' / white fls / Virginia to Florida 225 / 99
glaucus / Bb / to 3' / creamy-white fls / e & c United States 224 / 99
gramineus / Bb / 8-15" / pale yellow fls / c Canada 58
makinoana - see Z. sibiricus 200
micranthus / Bb / 2' / greenish fls / Sask. to California 228
Nuttallii / Bb / 2½'/ fls yellowish-white / Tenn. to Kans & Tex. 25
paniculatus / Bb / to 2' / fls yellowish-white / w United States 73 / 25
sibiricus / Bb / 5-10" / few fls, pale yel-grn or purp/Jap, n Asia 200
venenosus / Bb / to 2' / fls whitish / nw North America 229 / 25
venenosus v. gramineus - stem lvs sheathing 73 / 142
venenosus micrantha - see Z. micranthus 228

ZINNIA - Compositae
elegans pumila - the dwarf annual zinnia 283
grandiflora / R / 4-10" / yellow fls / Colorado, Kansas & s-ward 229 / 25
peruviana / A / 2' / yellow to red fls / Arizona, Mexico, Argentina 194 / 25

ZIZANIA - Gramineae
aquatica / A / 10'/'Annual Wild Rice', panicles 3'+/ Me.to Mich & s 25

ZIZIA - Umbelliferae
aptera / P / 1-2' / small yellow fls / North Dakota to Oklahoma 229
aurea / P / to 2½' / yellow fls / e North America 224 / 25

ZYGOPHYLLUM - Zygophyllaceae
atriplicoides / P / 3' / yellow fls / America 206
Fabago / HH P / 1-4' / yellow fls, coppery at base / Syria 81

ADDITIONS AND CORRECTIONS

ACONITUM
  napellus ssp. napellus / P / to 1' / fls viol.-blue / w & wc Eur. 289

AMMI - Umbelliferae
  Majus / P / to 2½' / small white fls / widely cultivated 25

ANTHEMIS
  carpatica / P / cushion to 1' / ray fls white / mts. s Europe 25

AQUILEGIA
  buergeriana v. oxysepala - with spurs incurved 200
1 'Nora Barlow' / P / to 28" / dbl. fls, comb red, pink, green
2 'Snow Queen' - pure white with spurs

ARABIS
2 blepharophylla 'Pink Pearl' / prostrate / deep "sludgy" pink fls

ARACHNOIDES - should read ARACHNIODES

ARISAEMA
3 utile / Tu / lg lvs over br.-purp or grn spathe / Himalayas

ASPLENIUM - all references marked 142 should read 159.
  billottii - variously spelled billotii, bilotti; for 142, see 286 and 159.
  forisense - should read A. forisiense

ASTER
1 'Alma Potschke' / P / to 3½' / fls salm tng. rose; densely branched

ATHYRIUM
  crenulato-serrulatum - use A. crenulatoserrulatum
4 filix-femina "Bornholmensis' - European lg crested variety

BELLEVALIA - Liliaceae
  spicata - see Strangweia spicata

BENSONIA
  oregana - see BENSONIENELLA oregona

BERBERIS
  canadensis / Sh / 3-6' / spiny br., yel. fls; red fr / Va. & s 61 / 36
  fendleri - see B. canadensis
  franchetiana v. macrobotrys / Sh / 4' / fls large to 3" / Yunnan 61

---

[1] Wayside Garden Catalog, Spring 1985.

[2] Thompson & Morgan Seed Catalog, 1984.

[3] QBAGS v. 52:#3

[4] Judith I. Jones in correspondence, 1986.

BESSEYA
  wyomingensis / P / 8-24" / wht hairy; fls pk-purp / Wyo., Id., Nev. 229

BETULA
  Medwediewii / T / tall / catkins stalked, erect / Transcaucasus 61
  Michauxii / Sh / 4-24" / catkins small / Lab. to Nfld. 99
  Michauxii nana / 3-lobed bracts, longer catkins / Nova Scotia 99
  tatewakiana / Sh / to 3' / catkins erect, 4-6" long / Japan 200

BILLARDIERA
  cymosa / Cl / lvs dull, stalkless, fls pk-mauve / Austr. 109

BLECHNUM
  Germainii / F / sterile fronds 2-3", fert. lgr / Brazil, Chile 61

BORDEREA - Dioscorea
  pyrenaica / Cl / to 8" / lvs dark green; fls small green / Pyr. 290

BOSSIAEA
  prostrata / Sh / prostrate / fls br. yel; lvs tiny / Australia 109

BRACHYCOME
  obovata / P / to 8" / fls small white or lilac / Australia 109

BRODIAEA
1 laxa 'Queen Fabiola' / C / 12" / dp blue fls, light ribs / w N.A.

BUPLEURUM
  trirodiatum / P / 2-4" / basal lvs clasp; fls yel. / Siberia 200

CALCEOLARIA
  Fothergillii / P / creeping /fls sulfur, red spots /Pategonia, Falk. 61

CAMPANULA
2 carpatica 'China Doll' / 8" / lav-blue plate fls

CASSIA
  odorata / Sh / to 6' / fls yel, 6-10 pr. lvs / s Australia 109

CASSIOPE
  Stellerana / Sh / spreading / fls white / Wash, Alaska 25

CHEIRANTHUS
3 'Apricot Delight' / 15" / fls soft apricot
4 cheiri 'Fair Lady Mixed' - frag soft colors lemon to purp to mahog.

---

1 White Flower Farm, The Garden Book, Fall 1983.

2 Wayside Catalog, Spring 1986.

3 Park Catalog, 1986.

4 Thompson and Morgan Catalog, 1986.

CHENOPODIUM
Bonus-Henricus / P/ to 2½'/arrow-lvs to 3"; potherb/Eur., nat. N.A. 25

CHIONODOXA
Siehei / Bb / 8-12" / 15 fls per stalk / Asia Minor, Crete 61

CHRYSANTHEMUM
1 Weyrichii / P / to 1' / white to pink / Japan 25

CIRSIUM
undulatum / B / 8-30" / wht-felt. lvs; fls purp. / Vt. w to Sask. 99

CLEMATIS
Jackmanii 'Purpurea Superba' / Cl / to 10' fls dark violet-purple 25
Jackmanii 'Superba' - see C. Jackmanii 'Purpurea Superba'
integrifolia 'Alba' / C / 3-4' / fls wht; lvs simple, 2½-4"/ s Eur. 61
2 'Prince Phillip' / 8-12' / 8-10" fls rich blue, red stripe
2 'Rouge Cardinal' - see C. 'Red Cardinal'
2 'Red Cardinal' / Cl / 12' / fls 4-6", velvet red

CLEOME
Isomeris / Sh or T / to 9' / fls yel; lfts 3; evergreen / California 25

COMMELINA
communis v. hortensis / A / creeping / lg. blue fls / n Asia 200

COPROSMA
obconica / Sh / to 6' / fls sm. wht; short branchlets / New Zealand 15

COREOPSIS
grandiflora Badengold / 3' / two-tone yellow fls 268

CROCOSMIA
3 'Lucifer' / 3½" / fls red; lvs brown-green, sword shape

CROCUS
chrysanthus 'Snow Bunting' - fls wht, blue feather or stripes 25

CRATAEGUS
durobrivensis / Sh / 10-16' / large wht fls; large red fr / N.A. 139

CYCLAMEN
4 persicum giganteum / 9-12" / ruffled wht petals, red edge and eye

---

1 Rocknoll Spring Catalog, 1986.

2 Park Seed Catalog, 1986.

3 Wayside Catalog, Fall 1985.

4 Thompson and Morgan Catalog, 1986.

DRYOPTERIS
1 affinis - see pseudomas and borreri; orangy-brn scales / England
1 affinis ssp. 'Cristata the King'

ESCHSCHOLZIA
2 californicum 'Double Ballerina Mixed' / 10" / fls to 3", rich colors

FRITILLARIA
bucharica / Bb / to 12"? / fls wht, tint grn / n Afghanistan 185
lutea / Bb / small / fls yel / c Caucasus 40

GERBERA
2 'Dwarf Happipot' - 8-12"

HALIMODENDRON - Leguminosacae
halodendron / Sh / to 6' / fls pale purp, gray-downy lvs / Asia 25

HAMAMELIS
mollis 'Brevipetala' / Sh, T / to 30' / fls dp orange / w China 25

HAPPLOPAPPUS
pygmaeus / R / dwarf to 3" / fls yel / Montana 142

HIPPEASTRUM
roseum / Gr Bb / to 6" / bright red, peritube green / Chile 25

HOMERIA
3 pallida / HH C / ? / fls pale yel / Transvaal

ISOMERIS
arborea - see CLEOME Isomeris 25

LYCHNIS
2 X arkwrightii 'Vesuvius' / 12" / bronze-maroon lvs; orange-red fls

MAZUS
japonicus / R / 2-6" / 5 fld, wht to purp with yel bands / Asia 200
rugosus - see M. japonicus 200

PENSTEMON
4 'Goldie' - P. confertus 'Kittitas' x pink f. P. euglaucus / yellow

---

[1] Judith Jones in correspondence, 1986.

[2] Thompson and Morgan Catalog, 1986.

[3] Sima Eliovson, Wild Flowers of Southern Africa, 1980.

[4] Alpines of the Americas. Report of the First Interim International Rock Garden Plant Conference, 1976.

PHLOX
1 sibirica ssp. pulvinata / R /loose cush./ fls many colored / Rockies

POLYSTICHUM
tsus-sinense - correct to P. tsus-simense

POTENTILLA
davurica 'Mandschurica' / st. to 1½' / lvs silky-hairy 25

THLASPI
avalanum / 3-24"/ frs longer than wht fls / Carpathians & Balk. Pen. 286
kovatsii - see T. avalanum 286

VIOLA
dissecta v. chaerophylloides f. chaerophylloides /sm; lvs finely dis. 200

---

[1] William A. Weber, Rocky Mountain Flora, 1976.

INDEX OF REFERENCES, as keyed

1. Adams, Nancy M. Mountain Flowers of New Zealand, 1965.

2. Ahrendt, Leslie Walter Allam. Berberis and Mahonia -- A Taxonomic Revision, 1961.

3. Akademiia naukSSSR Botanicheskii Institut. Flora of the USSR, Vol. 3, 1935 (trans. 1964).

4. Akademiia naukSSSR Botanicheskii Institut. Flora of the USSR, Vol. 4, 1935 (trans. 1970).

5. Akademiia naukSSSR Botanicheskii Institut. Flora of the USSR, Vol. 7, 1937 (trans. 1970).

6. Akademiia naukSSSR Botanicheskii Institut. Flora of the USSR, Vol. 8, (trans. 1970).

7. Akademiia naukSSSR Botanicheskii Institut. Flora of the USSR, Vol. 10, (trans. 1971).

8. Akademiia naukSSSR Botanicheskii Institut. Flora of the USSR, Vol. 12, 1946 (trans. 1965).

9. Akademiia naukSSSR Botanicheskii Institut. Flora of the USSR, Vol. 13, (trans. 1972).

10. Akademiia naukSSSR Botanicheskii Institut. Flora of the USSR, Vol. 18, 1952 (trans. 1967).

11. Akademiia naukSSSR Botanicheskii Institut. Flora of the USSR, Vol. 19.

12. Akademiia naukSSSR Botanicheskii Institut. Flora of the USSR, Vol. 24, (trans. 1972).

13. The Alaska-Yukon Wild Flowers Guide, n.a., 1974.

14. Allan, Mea. E. A. Bowles and His Garden, 1973.

15. Allan, H. H. Flora of New Zealand, Vol. 1., 1961.

16. Alpenglow Gardens Catalog #47, n.d., n.a.

17. Anderson, E. B. Hardy Bulbs, 1964.

18. Anderson, E. B. Rock Gardens, 1959.

19. Anderson, E. B. Seven Gardens, 1973.

20. Anderson, Jacob Peter. Flora of Alaska and Adjacent Parts of Canada, 1959.

21. Angel, Heather. Photographing Nature: Flowers, 1975.

22. Appalachian Mt. Club. Mountain Flowers of New England, 1964.

23. Ashberry, Anne. Miniature Trees & Shrubs, 1958.

24. Bacon, Lional. Alpines, 1973.

25. Bailey Hortorium Staff. Hortus III, 1976.

26. Bailey, Liberty Hyde, Garden of Larkspurs, 1939.

27. Bailey, Liberty Hyde. The Garden of Pinks, 1938.

28. Bailey, Liberty Hyde. The Standard Cyclopedia of Horticulture, 1935.

29. Bailey, Liberty Hyde and E. Z. Bailey, eds. Hortus II, 1941.

30. Bailey, Liberty Hyde and G. H. M. Lawrence. The Garden of Bellflowers, 1953.

31. Baird, Viola B. Wild Violets of North America, 1942.

32. Baker, Mary Frances. Florida Wild Flowers, 1949.

33. Barker, Frank. The Cream of Alpines, 1958.

34. Barr, Claude A. Prairie Gem Ranch Seed Catalog, 1968.

35. Bartlett, Mary. Gentians, 1975.

36. Bean, William Jackson. Trees and Shrubs Hardy in the British Isles, 8th ed, 4 vols. (1970-80).

37. Bean, William Jackson. Trees and Shrubs Hardy in the British Isles, 7th edition, 1950, 1951.

38. Beard, John Stewart, ed. Descriptive Catalog of West Australian Plants, (196-).

39. Beatley, Janice C. Vascular Plants of Nevada Test Site, 1976.

40. Beck, Christabel. Fritillaries, 1953.

41. Bennett, Ralph. Penstemon Nomenclature, 1960.

42. Bennett, Ralph. Studies in Penstemon I, 2nd ed., 1959.

43. Blackburn, Benjamin C. Trees and Shrubs in Eastern North America, 1952.

44. No entry for this number.

45. Blomberry, Alec M. What Wildflower is That?, 1973.

46. Bloom, Alan. Alpines for Trouble-Free Gardening, 1961.

47. Bloom, Alan. Hardy Plants of Distinction, 1965.

48. Bloom, Alan. Making the Best of Alpines, 1978.

49. Bloom, Alan. Perennials for Your Garden, 1971.

50. Böhm, Cestmir. Rock Garden Flowers: a concise guide in color, 1970.

51. Booth, William Edwin. Flora of Montana, Part I, 1950.

52. Bowles, E. A. A Handbook of Crocus and Colchicum for Gardens, 1952.

53. Braun, E. Lucy. The Monocotyledoneae: Cat-tails to Orchids. Vol. I, Vascular Flora of Ohio, 1967.

54. Braun, E. Lucy. The Woody Plants of Ohio, 1961.

55. Brickell, Chris D. and Brian Mathew. Daphne, 1976.

56. Brickell, Chris D. Lilies and Allied Plants, 1972.

57. Brown, Lauren. Grasses: An Identification Guide, 1979.

58. Budd, A. C. Wild Plants of the Canadian Prairie, 1957.

59. Burpee's Catalog, 1980.

60. Campbell, Carlos C., and others. Smoky Mountains Wildflowers, 1964.

61. Chittenden, Fred J., ed., assisted by specialists The Royal Horticultural Society Dictionary of Gardening, A Practical and Scientific Encyclopaedia of Horticulture, 4 vols., 1951, 1956.

62. Clarke, Joy Harold, ed. Rhododendrons for Your Garden, 1961.

63. Clausen, Robert Theodore. Sedum of North America North of the Mexican Plateau, 1975.

64. Clay, Sampson. The Present-Day Rock Garden, 1954.

65. Clements, Frederick E. and Edith S. Clements. Rocky Mountain Flowers, 3rd ed., 1945.

66. Correll, Donovan Stewart. Native Orchids of North America North of Mexico, 1950 [1951).

67. Cowan, John Macqueen. The Journeys and Plant Introductions of George Forrest, 1952.

68. Cox, Euan Hillhouse Methven and P. A. Cox. Modern Rhododendrons, 1956.

69. Cox, Euan Hillhouse Methven and P. A. Cox. Modern Shrubs, 1958.

70. Crane, Florence Hedleston. Flowers and Folk-Lore from Far Korea, 1931 (1969 reprint).

71. Crawford, Ehrenrich and Kucera. Ozark Range & Wildlife Plants, 1969 (USDA).

72. Cronquist, Arthur and others. Intermountain Flora: Vascular Plants of the Intermountain West, U.S.A., vol. I, 1972.

73. Cronquist, Arthur and others. Intermountain Flora: Vascular Plants of the Intermountain West, U.S.A., vol. 6, 1977.

74. Crook, H. Clifford. Campanulas, Their Cultivation and Classification, 1951 (1977 Theophrastus reprint).

75. Darnell, Anthony William. Unfamiliar Flowers for Your Garden, 1975 (Dover Reprint).

76. Davidson, William. All About House Plants, 1974.

77. Davis, P. H., ed. Flora of Turkey (vols. 1-5, 1965-75).

78. Dayton, William A. Notes on Western Range Forbs, 1960.

79. Den Ouden, P. and B. K. Boom. Manual of Cultivated Conifers, 1965.

80. Dodge, Natt Noyes. 100 Desert Wildflowers in Natural Color, 1963.

81. Don, George. A General System Of Gardening and Botany, 4 vol., 1831-1838.

82. Dorman, Caroline. Flowers Native to the Deep South, 1958.

83. J. Drake Catalog, 1975.

84. Dunn, David B. and John M. Gillett. The Lupines of Canada and Alaska, 1966.

85. Duperrex, Aloys. Orchids of Europe, 1961. Trans. A. J. Huxley.

86. Dykes, William Rickatson. The Genus Iris, 1974 (Dover reprint).

87. Elliott, Roy C. Gardener's Guide to the Androsaces (from March, 1965 Quarterly Bulletin of the Alpine Garden Society).

88. Elliott, Roy C. The Genus Lewisia (from vol. 34, 1966, Quarterly Bulletin of the Alpine Garden Society.)

89. Elliott, Roy C. The Genus Lewisia, 1978, 2nd edition (from vol. 34, 1966, with addendum Quarterly Bulletin of the Alpine Garden Society.)

90. Elliott, Roy C. Portraits of Alpine Plants, n.d.

91. Elliott, Roy C. Rock Garden Plants, 1935.

92. Enari, Leonid. Ornamental Shrubs of California, 1962.

93. Encke, Fritz, ed. Parey's Blumengartneri (2 vol. plus index 1958-61).

94. Evans, Ralph L. A Gardener's Guide to Sedums. An Alpine Garden Society Guide, (197-).

95. Everett, Thomas H. New York Botanical Garden Illustrated Encyclopedia of Horticulture, 1980.

96. Farrer, Reginald. The English Rock Garden, 2 vol., 1928.

97. Favarger, Claude von. Alpenflora, 1958.

98. Feldmeier, Carl. Lilies, 1970 (trans.).

99. Fernald, Merritt Lyndon, ed. Gray's Manual of Botany, 8th ed., 1950.

100. Fletcher, Harold R. A Quest of Flowers, 1976.

101. No entry for this number.

102. No entry for this number.

103. No entry for this number.

104. No entry for this number.

105. No entry for this number.

106. Floraire. Plantes Alpines et Vivaces, n.d.

107. Foster, H. Lincoln. Rock Gardening, 1968.

108. Fox, Robin Lane. Variations on a Garden, 1974.

109. Galbraith, Jean. Collin's Field Guide to the Wild Flowers of South-East Australia, 1977.

110. Gault, Simpson Millar. The Dictionary of Shrubs in Colour, 1976.

111. Genders, Roy. Bulbs, a Complete Handbook, 1973.

112. Genders, Roy. Miniature Bulbs, 1961.

113. Genders, Roy. Scented Flora of the World, 1977.

114. Gillett, John M. The Gentians of Canada, Alaska and Greenland, 1963.

115. Gooding, E. G. B., H. R. Loveless, and G. R. Proctor. Flora of Barbados, 1965.

116. Gould, Frank W. Texas Plants, 1962.

117. Gray, Alec. Miniature Daffodils, 1955.

118. Green, Peter Shaw, ed. Plants Wild and Cultivated, 1973.

119. Green, Roy. Asiatic Primulas, 1976.

120. Greer Gardens Catalog, 1977.

121. Grey, Charles Hervey. Hardy Bulbs, 3 vol., 1938.

122. Grey-Wilson, Christopher. Dionysias: The Genus Dionysias in the Wild and in Cultivation, (1969?).

123. Griffith, Anna N. Collins Guide to Alpines, 1964.

124. Grounds, Roger. Ornamental Grasses, 1979.

125. Hall, A. Daniel, Sir. The Genus Tulipa, 1940.

126. Harding, Winton. Saxifrages, 1970.

127. Harrison, Sydney Gerald. A Handbook of Coniferae & Ginkgoacceae, 4th ed., 1966.

128. Hay, Roy, F. R. McQuown and G. and K. Beckett. The Dictionary of House Plants, 1974.

129. Hay, Roy and Patrick M. Synge. The Color Dictionary of Flowers and Plants for Home and Garden, 1969.

130. Hay, Thomas. Plants for the Connoiseur, 1938.

131. Hayes, Doris W. and George A. Garrison. Key to Important Woody Plants of Eastern Oregon and Washington, 1960.

132. Heath, Royton E. Collector's Alpines, 1964.

133. Heath, Royton E. Miniature Shrubs, 1978.

134. Heath, Royton E. Shrubs for the Rock Garden and Alpine House, 1954.

135. Hebb, Robert S. Low Maintenance Perennials, 1975.

136. Hecker, W. R. Auriculas and Primroses, 1971.

137. Heller, Christine A. Wild Edible & Poisonous Plants of Alaska, 1958.

138. Heywood, Vernon H., ed. Flowering Plants of the World, 1978.

139. Hillier's Manual of Trees and Shrubs, 1972.

140. Hills, Lawrence Donegan. The Propagation of Alpines, 1950.

141. Hitchcock, Albert Spear. Manual of Grasses of the United States, 1935.

142. Hitchcock, C. Leo and Arthur Cronquist. Flora of the Pacific Northwest, 1974.

143. House, Homer Doliver. Annotated List of the Ferns and Flowering Plants of New York State, 1924.

144. Hume, Edward Putnam. Some Ornamental Shrubs for the Tropics (USDA), 1951.

145. Hutchinson, John. Common Wild Flowers, 1948.

146. Hutchinson, John. Genera of Flowering Plants. Vol. 1, 1964, Vol. 2, 1967.

147. Huxley, Anthony. Garden Perennials & Water Plants, 1971.

148. Huxley, Anthony. Mountain Flowers in Color, 1967.

149. Huxley, Anthony and W. Taylor. Flowers of Greece & the Aegean, 1977.

150. Hyams, Edward Solomon. Ornamental Shrubs for Temperate Zone Gardens. Vol. II, 1965.

151. Hylander, Nils. The Genus Hosta in Swedish Gardens, 1954.

152. Ingwersen, Will. Alpines, Without a Rock Garden, 1975.

153. Ingwersen, Will. The Dianthus, 1949.

154. Ingwersen, Will. Manual of Alpine Plants, 1978.

155. Jacobsen, Hermann. A Handbook of Succulent Plants, 3 vol., 1960 (in translation).

156. Jacobsen, Hermann. Lexicon of Succulent Plants, 1974.

157. Jaynes, Richard A. The Laurel Book, 1975.

158. Johnstone, George Horace. Asiatic Magnolias in Cultivation, 1955.

159. Kaye, Reginald. Hardy Ferns, 1968.

160. Kew Handlist of Herbaceous Plants, 2nd ed., 1955.

161. Kidd, Mary Maytham. Wild Flowers of the Cape Peninsula, 1950.

162. Kihara, Hitoshi, ed. Fauna and Flora of Nepal Himalaya. Vol. I of Scientific Results of the Japanese Expeditions to Nepal Himalaya, 1952-1953, 1955.

163. Kimura, Koichi [Koiti]. Japanese Medicinal Plants, Vol. II, 1960 [1962] (Japanese Text).

164. Kitamura, Siro, Gen Murata and Masaru Hori. Colored Illustrations of Herbaceous Plants of Japan, #15, 1966 (Japanese text).

165. Kitamura, Siro and Gen Murata. Colored Illustrations of Herbaceous Plants of Japan, #16, 1964. (Japanese Text)

166. Kitamura, Siro, Gen Murata and Tetsuo Koyama. Coloured Illustrations of Herbaceous Plants of Japan, #17, 1964. (Japanese Text)

167. Kitamura, Siro and Syogo Okamoto. Colored Illustrations of Trees and Shrubs of Japan, Vol. 19, 1958. (Japanese Text)

168. Klaber, Doretta. Gentians for Your Garden, 1964.

169. Klaber, Doretta. Rock Garden Plants, 1959.

170. Knight, Frank P. Heaths and Heathers, 1976.

171. Konczewska, Florence. Argentine Flowers, 1976.

172. Lamb Nursery Catalog, n.d.

173. Lawrence, Elizabeth. The Little Bulbs, 1957.

174. Lemmon, Kenneth. The Golden Age of Plant Hunters, 1968.

175. Lenz, Lee W. Native Plants for California Gardens, 1956.

176. Little, Elbert L., Jr. Checklist of the Native and Naturalized Trees of the U.S., 1953.

177. Little, Elbert L, Jr. and Frank H. Wadsworth. Common Trees of Puerto Rico and the Virgin Islands, 1964-74.

178. McDougall, Walter Byron and Herma A. Baggley. The Plants of Yellowstone Park, 1956.

179. McDougall, Walter Byron and Omer E. Sperry. Plants of Big Bend National Park, 1951.

180. Macoboy, Stirling. Tropical Flowers and Plants, 1974.

181. Maire, Rene. Flore de l'Afrique du Nord, Vol. 5, 1958.

182. Makins, Frederick Kirkwood. Herbaceous Garden Flora, 1957.

183. Marchant, Angela and Brian Mathew. Alphabetical Table: The Genus Iris, 1974.

184. The Marshall Cavendish Illustrated Encyclopedia of Gardening, 1968-70. Peter Hunt, ed.

185. Mathew, Brian. Dwarf Bulbs, 1973.

186. Mathew, Brian. The Larger Bulbs, 1978.

187. Mathias, Mildred E. Color for the Landscape, 1973.

188. Mathias, Mildred Ester and Elizabeth McClintock. A Checklist of Woody and Ornamental Plants of California, 1963.

189. Mayfair Nursery. Pot-grown Rock Plants and Alpines, n.d.

190. Menninger, Edwin Arnold. Flowering Trees of the World for Tropics and Warm Climates, 1962.

191. Mohlenbrock, Robert H. The Illustrated Flora of Illinois - Ferns, 1967.

192. Mohlenbrock, Robert H. The Illustrated Flora of Illinois - Flowering Plants; Lilies to Orchids, 1970.

193. Moore, Lucy B. and J. B. Irwin. Oxford Book of New Zealand Plants, 1978.

194. Morley, Brian D. Wild Flowers of the World, 1970.

195. Muenscher, Walter Conrad Leopold. Flora of Whatcom County, State of Washington; Vascular Plants, 1941.

196. Munz, Philip Alexander. A California Flora, 1959.

197. Nehrling, Arno and Irene. Peonies Outdoors and In, 1960.

198. Nelson, Ruth Ashton. Plants of Rocky Mountain National Park, 1953.

199. New York Botanical Garden. Hardy Ferns and Their Culture, 1940. Carol H. Woodward, ed.

200. Ohwi, Jisaburo. Flora of Japan, 1965.

201. Page, Mary and William T. Stearn. Culinary Herbs, 1974.

202. Park's Flower Book, a catalog, 1973.

203. Park's Seed Catalog, 1980.

204. Parsa, Ahmad. Flora of Iran, vol. 4, 1952.

205. Parsons, Frances T. How to Know the Ferns, 1961 (Dover Reprint).

206. Paxton, Joseph C., Sir. A Pocket Botanical Dictionary; comprising the names, history, and culture of all plants known in Britain. (1968 ed.)

207. Perry, Frances. Collins Guide to Border Plants, 1973.

208. Pizetti, Ippolito and Henry Cocker. Flowers: A Guide for Your Garden, 2 vols., 1975.

209. Polunin, Nicholas. Circumpolar Arctic Flora, 1959.

210. Polunin, Oleg. Pflanzen Europas, 2nd ed., 1977.

211. Polunin, Oleg and A. Huxley. Flowers of the Mediterranean, 1965.

212. Polunin, Oleg and B. E. Smythes. Flowers of Southwest Europe, 1973.

213. Porsild, Alf Erling. Botany of the S. E. Yukon Adjacent to the Canal Road, 1951.

214. Porsild, Alf Erling. Vascular Plants of the Western Canadian Arctic Archipelago, 1955.

215. Proctor, George R. A Preliminary Checklist of Jamaica Pteridophytes, 1953.

216. Radford, Albert E., and others. Guide to the Vascular Flora of the Carolinas, 1964.

217. Randolph, Lowell Fitz, ed. Garden Irises, 1959.

218. Rau, M. A. Illustrations of West Himalayan Flowering Plants, 1963.

219. Rauh, Werner. Alpenpflanzen, 4 vols., 1951-1953.

220. Reed, Clyde F. Selected Weeds of the United States. Agriculture Handbook #366, 1970.

221. Rehder, Alfred. Manual of Cultivated Trees and Shrubs, 1927 (First Edition).

222. Rehder, Alfred. Manual of Cultivated Trees and Shrubs Hardy in North America, 1947, 2nd ed.

223. Rice, Elsie Garrett and Robert Harold Compton. Wild Flowers of the Cape of Good Hope, 1950.

224. Rickett, Harold William. Wild Flowers of the United States, Vol. I, 1966.

225. Rickett, Harold William. Wild Flowers of the United States, Vol. II, 1967.

226. Rickett, Harold William. Wild Flowers of the United States, Vol. III, 1969.

227. Rickett, Harold William. Wild Flowers of the United States, Vol. IV, 1970.

228. Rickett, Harold William. Wild Flowers of the United States, Vol. V, 1971.

229. Rickett, Harold William. Wild Flowers of the United States, Vol. VI, 1973.

230. Rohde, Eleanour Sinclair. Uncommon Vegetables; how to grow and how to cook, 1943.

231. Roi, Jacques. Traite des Plants Medicinales Chinoises, 1955.

232. Roper, Lanning. The Gardens in the Royal Park at Windsor, 1959.

233. Roxburgh, William, Nathaniel Wallich and William Carey, eds. Flora Indica, or Descriptions of Indian Plants, 1824 (reprint 1975).

234. Ryberg, Mans. A Taxonomical Survey of the Genus Corydalis Ventenat: with reference to cultivated species, 1955.

235. Rydberg, Per Axel. Flora of Colorado, 1906.

236. Rydberg, Per Axel. Flora of the Prairies and Plains of Central North America, 1932.

237. Salisbury, Sir Edward. Weeds and Aliens, 1961.

238. Salmon, John Tenison. New Zealand Flowers and Plants in Colour, 1967.

239. Saunders, Doris E. Cyclamen, A Gardener's Guide to the Genus, 1973, 1975.

240. Schacht, Wilhelm. Der Steingarten und seine Welt, 1960.

241. Schenk, George. Rock Gardens, 1965.

242. Schnell, Donald E. Carniverous Plants of U.S. and Canada, 1976.

243. Schröter, Ludwig and C. Schröter. Alpine Flowers: Coloured vade-mecum to the Alpine Flora, [192?].

244. Schultes, Richard E. Hallucinogenic Plants, 1976.

245. Scoggan, H. J. Flora of Canada, 1978-79.

246. Seymour, Frank Conkling. The Flora of New England, 1969.

247. Sharpe, Grant William. 101 Wild Flowers of Mt. Rainier National Park, 1957.

248. Sharpe, Grant William. 101 Wild Flowers of Olympic National Park, 1957.

249. Siskiyou Rare Plant Nursery Catalog, 1970, et seq.

250. Small, John Kunkel. Manual of the Southeastern Flora, 1933.

251. Smith, Geoffrey. Easy Plants for Difficult Places, 1967.

252. Smith, George F. and Duncan B. Lowe. Androsaces. Alpine Garden Society, 1977.

253. Stainton, J. D. A. Forests of Nepal, 1972.

254. Standley, Paul C. Trees and Shrubs of Mexico, 1961.

255. Stern, Frederick C. A Chalk Garden, 1960.

256. Stern, Frederick C. Snowdrops and Snowflakes, 1956.

257. Stevens, William Chase. Kansas Wild Flowers, 1948.

258. Steward, Albert Newton. Manual of Vascular Plants of the Lower Yangtze Valley, China, 1958.

259. Steyermark, Julian A. Flora of Missouri, 1963.

260. Stites, Jerry S. and Robert G. Mower. Rock Gardens. College of Agriculture, Cornell University, 1980.

261. Stoker, Fred. A Gardener's Progress, 1939.

262. Stoyanoff, Nikolai and Boris Stefanoff. Flora of Bulgaria, 1948 (Bulgarian Text).

263. Strasbaugh, P. D. and E. L. Core. Flora of West Virginia, 1952-64 (1st Edition).

264. Street, Frederick. Hardy Rhododendrons, 1954.

265. Sukachev, Vladimir Nikolaevich, ed. Studies on the Flora and Vegetation of High Mountain Areas, 1960. (Translated 1965).

266. Suslov, Sergei Petrovich. Physical Geography of Asiatic Russia, 1961 (trans.).

267. Synge, Patrick M. Collins Guide to Bulbs, 1961.

268. Synge, Patrick M., ed. Royal Horticultural Society Supplement to the Dictionary of Gardening, 2nd edition, Suppl. 2, 1969.

269. Synge, Patrick M. and J. W. D. Platt. Some Good Garden Plants, 1962.

270. Tackholm, Vivi. Student's Flora of Egypt, 1956.

271. Tackholm, Vivi and M. Drar. Flora of Egypt, Vol. III, 1954.

272. Tagawa, Motozi. Colored Illustrations of the Japanese Pteridophytes, 1959 (Japanese text).

273. Takeda, Hisayoshi. Alpine Flora of Japan in Colour, 1960 (Japanese text).

274. Taylor, Albert William. Wild Flowers of the Pyranees, 1971.

275. Taylor, George. The Genus Meconopsis, 1934.

276. Taylor, Roy L. and Bruce MacBride. Vascular Plants of British Columbia, 1977.

277. Thomas, Graham Stuart. Colour in the Garden, 1957.

278. Thomas, Graham Stuart. The Old Shrub Roses, 1971.

279. Thomas, Graham Stuart. Perennial Garden Plants, 1976.

280. Thomas, Graham Stuart. Plants for Ground Cover, 1977.

281. Thomas, Graham Stuart. Shrub Roses of To-day, 1962.

282. Thompson & Morgan Catalog, 1975, et seq.

283. Thompson & Morgan Seedlist Catalog, 1971.

284. Tosco, Uberto. The World of Mountain Flowers, 1973 (trans.).

285. Turrill, William Bertram. The Plant-life of the Balkan Peninsula, 1929.

286. Tutin, Thomas Gaskell, ed. Flora Europaea, Vol. I, 1964.

287. Tutin, Thomas Gaskell, ed. Flora Europaea, Vol. II, 1968.

288. Tutin, Thomas Gaskell, ed. Flora Europaea, Vol. III, 1972.

289. Tutin, Thomas Gaskell, ed. Flora Europaea, Vol. IV, 1976.

290. Tutin, Thomas Gaskell, ed. Flora Europaea, Vol. V, 1980.

291. Underhill, Terry L. Heaths and Heathers, 1972.

292. Vajda, Erno. Wild Flowers of Hungary: the origin and development of plant communities, 1956.

293. Vallentin, Mrs. E. F. and Mrs. E. M. Cotton, descriptions. Illustrations of the Flowering Plants and Ferns of the Falkland Islands, 1921.

294. Van Dersal, William R. Native Woody Plants of the U.S., 1938.

295. Van Laren, A. J. [Laren, A.J. van]. Cactus, 1935.

296. Van Melle, Peter Jacobus. In and About the Rock Garden, n.d.

297. Van Melle, Peter Jacobus. Shrubs and Trees for the Small Place, 1943.

298. Venning, Frank D. Cacti, 1974.

299. Viereck, Leslie A. and Elbert L. Little Jr. Alaska Trees & Shrubs, 1972.

300. Wayside Gardens Spring Catalog 1978, et seq.

301. Webb, David Allardice. An Irish Flora, 1963.

302. Webber, John Milton. Yuccas of the Southwest, 1953.

303. Weintraub, Frances C. Grasses Introduced Into the United States, 1953.

304. Welch, Humphrey James. Dwarf Conifers, A complete guide, 1966.

305. Wherry, Edgar T. The Genus Phlox, 1955.

306. Wilder, Louise (Beebe) Mrs. Pleasures and Problems of a Rock Garden, 1928.

307. Willis, John Christopher. A Dictionary of Flowering Plants and Ferns, 1966.

308. Wills, Mary Motz and Howard S. Irwin. Roadside Flowers of Texas, 1961.

309. Wister, John. Lilacs for America, 1943.

310. Wister, John. The Peonies, 1962.

311. Woodcock, Hubert Drysdale and William Thomas Stearn. Lilies of the World, 1950.

312. Zwinger, Ann H. and Beatrice E. Willard. Land Above the Trees, 1972.

313. Strasbaugh, P. D. and E. L. Core. Flora of West Virginia, 1977 (2nd edition).

314. Grey-Wilson, Christopher. The Alpine Flowers of Britain and Europe, 1979.

SUPPLEMENTAL REFERENCES. In parentheses.

Key to abbreviations in SUPPLEMENTAL REFERENCES.

AHM American Horticultural Magazine

ARGS American Rock Garden Society

BARGS Bulletin of the American Rock Garden Society

GC Gardeners' Chronicle, London

GC-GI Gardeners' Chronicle and Gardening Illustrated, London

GC-NH Gardeners' Chronicle and New Horticulturist

GI Gardening Illustrated, London

JRHS Journal of the Royal Horticultural Society. After volume 100 it became The Garden.

JSRGC Journal of the Scottish Rock Garden Club

NHM National Horticultural Magazine. With volume 39 it became American Horticultural Magazine, with volume 50 it became American Horticulturist.

NYBG New York Botanical Garden Garden Journal.

QBAGS Quarterly Bulletin of the Alpine Garden Society

RHS Royal Horticultural Society

Supplemental references have the same number they had in previous editions of this work; numbers omitted from the following list were not used in this edition.

(1) JRHS 96:6 Brickell (Journal of the Royal Hort. Soc.; after 100:5 became The Garden)
(3) QBAGS 36:3
(4) GC 165:3
(5) NHM 37:4 Mulligan-Creech
(6) GC 1675:24 Grey-Wilson
(7) BARGS 31:3
(9) GC 164:2 Mole
(10) Bull. Nat. Sci. Mus., Tokyo. Nakai
(12) QBAGS 43:1 Schwarz
(14) QBAGS 39:4
(15) JRHS 95:4 Blanchard
(16) RHS Lily Yearbook. 1967. Wendelbo
(18) Taxon. 5:10 Killman
(22) GC-NH 166:5 Beckett (Gardeners' Chronicle & New Horticulturist Vol. 165 thru Vol. 166:14)
(23) Brittonia 27:4 Goldblatt
(26) Herbertia 7. 1940 Goodspeed

(27) QBAGS 45:1 Wilson.
(28) QBAGS 45:4 Elliott
(31) QBAGS 43:4 Elliott
(32) Arnoldia 37:2 Pride
(34) QBAGS 37:3 Grey-Wilson
(36) QBAGS 44:3 Elliot, Watson
(39) Gentes Herbarium 7:1 Munz
(40) BARGS 29:1 Cole
(41) QBAGS 43:2 Elliott
(42) ARGS Yearbook. 1941 G. H. M. Lawrence
(44) GC 163:26 Mathew
(45) Lords and Ladies. 1960 Prime
(46) BARGS 35:1 Bacon
(48) BARGS 27:2 Murray-Lyon
(51) BARGS 24:1
(52) Plant Life 11-1955 Hoover
(53) Plant Life 7-1951 Dyer
(54) JRHS 101:4 Bloom
(55) QBAGS 45:2
(56) Genera of Flowering Plants Vol. 1-1964: Vol. 2-1967. Hutchinson
(58) QBAGS 41:2 Schwarz
(59) QBAGS 37:4 Elliott
(60) GC 161:18 Gorer
(61) QBAGS 35:4 Downward, Saunders
(62) BARGS 32:3 Hatch
(63) Plants & Gardens 33:2
(64) QBAGS 42:4 Elliott
(66) Baileya 19:2 Pringle
(67) RHS Lily Yearbook 1968. Furse
(68) BARGS 33:4 Harkness
(69) NHM 14:1 Spingarn
(70) JRHS 102:3 Annand
(71) JRHS 97:2 Knox-Finlay
(73) JSRGC 12:3 Nutt
(74) BARGS 34:2 Wherry
(75) GC-NH 165:6 McQuown
(76) Baileya 5:2 G. H. M. Lawrence
(78) QBAGS 39:3 Grey-Wilson
(79) QBAGS 40:4 Elliott
(80) NYBG Garden Journal 25:6 Belman
(81) QBAGS 42:1 Elliott
(82) GC-NH 166:14 Mathew
(83) Baileya 18:3 Yeo
(84) GC-GI 8/18/56 Ingwersen
(86) JSRGC 12:4 Watt
(94) GC 166:25 Lawder
(95) QBAGS 42:2 Grey-Wilson
(100) QBAGS 41:1 Hecker
(102) JRHS 100:6 Lloyd
(103) JRHS 101:7 J. Elliott
(105) Flora USSR Vol. 6
(106) GC 163:10 Grey-Wilson
(107) Arnoldia 37:4 Weaver
(108) Baileya 11:3 Ingram
(109) JSRGC 15:1 Woodward & McPhail

(110) BARGS 30:3 Myhr
(111) BARGS 25:2 Peterson
(112) GC 164:18 Mathew
(113) Am. Hort. 53:4 Kondo
(116) The Plantsman 1:2 Hunt & Butterfield
(119) QBAGS 41:3 Bacon
(122) GC 163:18 Beckett
(123) JRHS 89:10 Valder
(125) JRHS 104:11 Synge
(128) GC-NH 165:16 Beckett
(129) BARGS 27:3
(132) QBAGS 39:1 Philipson
(133) Plantes Alpines dans le Jardin. 1956 DeVilmorin
(134) GC-GI 148:3 Haes
(135) GI 2-1952 Anderson
(136) GC 164:3 Beckett
(138) GC 164:17 Lawder
(139) Plant Life Vol. 7 Traub
(141) GC-GI 6/18/60 Thomas
(142) GI 9-1955 Johnson
(143) GC 157:7 Bawden
(145) JRHS 101:8 Hamilton
(147) JRHS 103:3 McClintock
(148) Plant Life Vol. 31 1975 Traub
(149) GC 163:4 Martin & Chapman
(150) Jour. Arnold Arboretum 27:3 Johnston
(152) BARGS 32:4 Mitchell
(153) QBAGS 46:1 Ferns
(154) Bull. Hardy Plant Soc. 1:7 Fish
(156) Am. Hort. Mag. 47:2
(161) Am. Hort. Mag. 50:3 G. Wister
(162) Mulloy in corres.
(164) Baileya 8:2 Dress
(165) BARGS 29:2 Humphrey
(166) The Plantsman 1:4 Robson
(169) JRHS 104:5 Synge
(170) GC-NH 165:25 Beckett
(173) GC 167:9 Luscombe
(174) BARGS 23:1 Klaber
(175) GC 167:8 Luscombe
(176) GC-NH 165:12 Mathew
(177) QBAGS 40:2 Elliott
(178) GC 164:22 Grey-Wilson
(179) Div. of Hort., Ottawa - Progress Report 1934-48 Davis
(180) QBAGS 40:3 Hodgkin
(181) GC-NH 165:26 Beckett
(182) Taxon 17:5 Hunt & Welch
(185) BARGS 25:3 Davidson
(186) JSRGC 14:1 Crosland
(187) GC-NH 165:17
(189) Halley, in corres.
(190) The Herb Grower 3:7 Neugebauer
(191) Plant Life Vol. 34 Ravenna
(192) BARGS 27:4 Wheeler
(193) Baileya 7:1 Dress

(194) GI 139:25 Barber
(195) JRHS 102:7 Scott-James
(196) JSRGC 16:1 Evans
(198) Plant Life Vol. 32-1976 Wilson
(202) GC 164:12 Beckett
(206) Plants & Gardens 25:3 McGourty
(209) GC 165:14
(210) GC-NH 165:13 Grey-Wilson
(211) Bull. Hardy Plant Soc. 1:6 Lyall
(212) JRHS 101:9 Chatto
(213) Contrib. U.S. Nat. Herb 28:4 Plants coll. by R.C. Ching - Walker
(215) NHM 24:2 Banks
(217) NHM 42:4
(218) JSRGC 13:3 Forrest
(219) Lodewick in Corres.
(220) JSRGC 14:3 Philipson
(221) BARGS 28:2
(223) JRHS 103:6 Hobhouse
(224) JRHS June 1963
(225) Baileya 14:2 Dress
(226) GC 163:11 Grey-Wilson
(228) GC-NH 166:10 Anderson
(229) Baileya 16:3 Cullen
(230) BARGS 26:3 Hayward
(231) Plant Life Vol. 35 Garaventa
(236) Penstemon Plant Identifier #5. 1974 Lodewick
(237) Bull. Am. Penstemon Soc. Vol. 30. 1971 Bennett
(238) Penstemon Plant Identified #1. 1970 Lodewick
(239) Bull. Am. Penstemon Soc. Vol. 31. 1972 Bennett
(241) GC 167:18 Ingwersen
(242) Am. Hort. Mag. 52:1 Brown
(243) JSRGC 16:3 Stone
(244) JRHS 86:4 Adey
(245) JRHS 104:3
(246) GC-NH 166:18 Beckett
(247) JRHS 103:11 Hellyer
(248) BARGS 33:1 Drew
(249) JRHS 102:1
(250) JRHS 103:4 Beckett
(252) List of Primula Species in Cultivation. (Preliminary Draft) Edwards and Winstanley.
(253) QBAGS 35:2 Cain
(255) JRHS 101:5 Jermyn
(256) JSRGC 13:2 Robertson
(257) Flora USSR Vol. 20
(258) Horticulture 5/1/41 Hamblin
(259) QBAGS 44:1 Barrett
(261) QBAGS 45:3 Elliott
(262) GC-NH 165:22 Beckett
(263) QBAGS 35:1 Erskine
(265) GC 163:3 Ingwersen
(266) Am. Hort. 51:3 Leach
(267) GC-NH 165:19 Rose
(268) GC-NH 166:23 Anderson
(270) Plant Life Vol. 27 - 1971 Ravenna & Ruppel

(271) Plant Life Vol. 26 - 1970 Ravenna
(272) Am. Hort. 53:5 Balick
(273) QBAGS 46:2 Elliott
(274) GC 168:17 Douglas
(275) QBAGS 39:2 Mathew
(276) GC-NH 168:8 Little
(277) GC 5/7/71 Lancaster
(279) GC-NH 165:21 Drake
(280) BARGS 33:4 Horny, Sojak & Weber
(281) Brittonia 11:4 Calder & Saville
(284) BARGS 25:4 Senior
(285) The World of Rock Plants. 1971 Plant Conf. Report. Furse
(289) GC-GI 2/14/59 Cowley
(290) Brickell, in corres.
(291) Innis, in corres.
(292) BARGS 23:3 Crocker
(293) QBAGS 43:3 Mathew, Elliott
(295) JRHS 102:12 Bloom
(296) NYBG Garden Journal 24:3 White & Gunn
(302) JRHS 102:9 Rose
(304) JRHS 103:5 Burgess
(306) BARGS 19:2 Dowbridge
(307) Omitted
(308) Brittonia 27:1 Freeman
(309) BARGS 26:4 Guppy
(312) JRHS 104:4 Beckett
(313) BARGS 32:2 Arneson
(314) GC 164:14 Corley
(315) QBAGS 44:4 Mathew
(317) NYBG Garden Journal 22:3 Frese
(318) Am. Hort. Mag. 41:4 Egolf
(319) Baileya 14:3 Egolf
(321) BARGS 11:2 Scorgie
(324) Woody Plants of the U. of Wash. Arboretum. 1977 Mulligan
(325) Ill. Flora of the Canadian Arctic Archipelago. 1957 Porsild
(326) Clematis. 1965 Lloyd
(327) Exotic Plants. 1971 (Golden Guide) Morton
(328) JRHS 97:5 Knox-Finlay
(329) JRHS 96:9 Hewer
(330) GC 164:4 Cowley
(331) USDA Range Plant Handbook. 1937
(332) Nat. Hort. Mag. 16:1
(333) Australia's Western Wild Flowers. 1968 Morcombe
(334) JRHS 93:6 Blanchard; Sayers
(336) QBAGS 38:2 Elliott
(337) Nat. Hort. Mag. 17:2 W. B. Wilder, Fox, Wilson
(338) Nat. Hort. Mag. 37:3 W. B. Wilder, Fox
(339) JRHS 92:5 G. Thomas
(340) QBAGS 36:1 Holford
(341) Nat. Hort. Mag. 20:3 W. B. Wilder, Fox
(342) Native Trees of Canada. 1973 Hosie
(343) GC 165:8 Seth-Smith
(344) QBAGS 36:4 Downward
(345) GC 166:13 Mathew
(346) JRHS 98:5 Mathew

(347) JRHS 92:9 Gorynski & Siphes
(348) JSRGC 14:4 Mitchell
(349) Nat. Hort. Mag. 13:3
(350) QBAGS 36:2 Elliott
(352) Nat. Hort. Mag 22:1 Marriage
(353) GC 164:9 Mathew
(354) QBAGS 35:3 Furse
(355) Nat. Hort. Mag. 18:1
(356) GC 165:11 Anderson
(357) Nat. Hort. Mag. 22:2
(360) BARGS 37:1 LeComte
(361) Nat. Hort. Mag. 12:1 Jones
(362) BARGS 29:3 Woodward
(365) Flowering Plants of Jamaica. 1972 Adams
(366) GC-GI 6/30/56
(368) JSRGC 16:4 McKelvie
(370) Arnoldia 36:3 Weaver
(371) Baileya 9:4 Hume
(378) QBAGS 46:4 Elliott
(393) Proceedings Int. Lilac Soc. 1978 Pringle
(395) Bull. Am. Penstemon Soc. 38:1 Wayt
(398) QBAGS 47:2 Norman
(399) GC 163:13 Beckett
(401) QBAGS 38:1 Tjaden
(402) Flora USSR Vol. 14
(403) The Plantsman 1:1 R. Elliott
(404) Garden (Formerly Garden Journal, NYBG) 1:3. Foster
(405) Plants & Gardens 30:1 Orton
(408) Am Lily Year Book 1946 Harkness
(410) Horticulture August 1959. Delkin
(412) Florist's Exchange & H. T. W. 3/14/36 Van Melle
(413) GC 157:20 Wada
(414) Am. Hort. Mag. 41:1 C. May
(416) JRHS 104:9
(417) GC 152:25
(418) Flowers of the World. 1972 F. Perry
(419) QBAGS 47:3
(424) Fern Grower's Manual. 1979 Hoshizaki
(426) Nat. Hort. Mag. 16:3 Morrison
(427) BARGS 28:1 Feridi
(428) Nat. Hort. Mag. 14:2 Marriage
(429) Nat. Hort. Mag. 12:4 Miner
(430) BARGS 30:2 Vasak
(431) BARGS 30:4 Baggett
(432) Am. Hort. Mag. 46:3
(433) Nat. Hort. Mag. 34:4 Morrison
(434) Nat. Hort. Mag. 19:2
(435) BARGS 8:4 Senior
(436) Nat. Hort. Mag. 14:2 Jones
(438) BARGS 12:1 Epstein
(439) Nat. Hort. Mag. 13:1 Spingarn
(440) Am. Hort. Mag. 47:1 Senior
(441) GC 162:12 Knowles
(442) GC 162:17 Doerflinger
(443) GC 162:18 Fish

(444) QBAGS 47:4 H. Smith
(445) QBAGS 47:1
(446) BARGS 1:6 Walker
(447) BARGS 23:4 Lackschewitz
(448) Nat. Hort. Mag. 9:3 Morrison & Guernsey
(449) Nat. Hort. Mag. 8:3 Morrison & Guernsey
(450) Nat. Hort. Mag. 8:2 Morrison & Guernsey
(451) Nat. Hort. Mag. 9:2 Morrison & Guernsey
(452) Nat. Hort. Mag. 9:1 Morrison & Guernsey
(453) Am. Hort. Mag. 44:1 Rogers & Hope
(454) Nat. Hort. Mag. 29:2 Hope
(455) Nat. Hort. Mag. 31:3 Blasdale
(456) BARGS 10:3 Nearing
(457) Am. Hort. Mag. 45:4 Minninger
(458) BARGS 3:3 Van Melle
(459) Nat. Hort. Mag. 21:1 Hope
(460) Nat. Hort. Mag. 38:1 Morrison
(461) Am. Hort. Mag. 49:1 Armstrong
(462) BARGS 3:5 Richardson
(463) BARGS 12:2 Klaber
(464) Am. Hort. Mag. 48:1 Baumgardt
(465) Nat. Hort. Mag. 18:3 Morrison & Guernsey
(466) BARGS 26:2 Vasak
(467) Am. Hort. Mag. 47:2
(468) Nat. Hort. Mag. 16:4 Morrison & Guernsey
(469) Nat. Hort. Mag. 38:3 Plaisted & Lighty
(470) Am. Hort. Mag. 49:4 Galle
(471) Nat. Hort. Mag 36:1 Hu
(472) BARGS 17:4 Marshall
(473) Nat. Hort. Mag. 11:3 Morrison & Guernsey
(474) Nat. Hort. Mag. 20:3 Sherrard
(475) Nat. Hort. Mag. 38:2
(476) BARGS 3:6 Wherry
(477) BARGS 18:2 Worth
(478) BARGS 26:3 Davidson
(479) Nat. Hort. Mag. 15:1 Jones
(480) Am. Hort. Mag. 44:4 Coe
(481) Nat. Hort. Mag. 12:1 Newcomer
(482) Nat. Hort. Mag. 12:2 Fox
(483) Am. Hort. Mag. 48:3 De Graaf
(484) Nat. Hort. Mag. 16:2 Fox
(485) QBAGS v. 12 p. 91
(486) QBAGS v. 1 p. 69
(487) QBAGS v. 3 p. 206
(488) QBAGS v. 14 p. 216
(489) QBAGS v. 4 p. 7
(490) QBAGS v. 14 p. 34
(491) QBAGS v. 6 p. 291
(492) Brittonia 16:1 Hardin
(493) QBAGS v. 4 p. 145
(494) QBAGS v. 11 p. 92
(495) QBAGS v. 11 p. 95
(496) QBAGS v. 10 p. 139
(497) QBAGS v. 7 p. 79
(498) QBAGS v. 10 p. 55

(499) QBAGS v. 2 p. 214
(500) QBAGS v. 15 p. 209
(501) QBAGS v. 10 p. 130
(502) QBAGS v. 3 p. 241
(503) QBAGS v. 14 p. 196
(504) QBAGS v. 14 p. 220
(505) Flore Laurentienne. 1964 Fr. Marie-Victorin
(506) BARGS 2:2 Henry
(507) QBAGS v. 2 p. 302
(508) QBAGS v. 1 p. 41
(509) QBAGS v. 15 p. 247
(510) QBAGS v. 14 p. 232
(511) QBAGS v. 14 p. 242
(512) QBAGS v. 7 p. 7
(513) QBAGS v. 4 p. 255
(514) QBAGS v. 1 p. 12
(515) QBAGS v. 7 p. 208
(516) QBAGS v. 2 p. 36
(517) QBAGS v. 7 p. 288
(518) QBAGS v. 10 p. 207
(519) QBAGS v. 12 p. 35
(520) GC 161:15
(521) GC 161:18
(522) GC 161:23
(523) QBAGS v. 13 p. 172
(524) QBAGS v. 9 p. 23
(525) QBAGS v. 5 p. 178
(526) QBAGS v. 4 p. 151
(527) QBAGS v. 11 p. 162
(528) QBAGS v. 7 p. 135
(529) QBAGS v. 12 p. 165
(530) Am. Hort. Mag. 45:2 Jefferson
(531) QBAGS v. 12 p. 142
(532) QBAGS v. 1 p. 121
(533) Am. Hort. Mag. 40:2 Menninger
(534) Am. Hort. Mag. 39:4 Hope
(535) Nat. Hort. Mag. 17:2 Hope
(536) QBAGS v. 1 p. 45
(537) QBAGS v. 5 p. 323
(538) QBAGS v. 6 p. 254
(539) Am. Hort. Mag. 45:1
(540) Nat. Hort. Mag. 28:4 Synge
(541) Nat. Hort. Mag. 24:3 Coombs
(542) BARGS 9:5 Walker
(543) QBAGS v. 11 p. 18
(544) QBAGS v. 4 p. 145
(545) BARGS 8:2 Barr
(546) Nat. Hort. Mag. 30:1
(547) Nat. Hort. Mag. 26:2
(548) BARGS 2:1 Wherry
(549) Nat. Hort. Mag. 11:2 Berry
(550) BARGS 31:2 Halda
(551) BARGS 6:2
(552) BARGS 17:2 Mitchell
(553) QBAGS v. 4 p. 345

(554) QBAGS v. 2 p. 26
(555) QBAGS v. 4 p. 280
(556) QBAGS v. 15 p. 6
(557) QBAGS v. 12 p. 36
(558) QBAGS v. 4 p. 273
(559) Am. Hort. Mag. 48:4 Egolf
(560) BARGS 6:3 Fox
(561) BARGS 31:4 Halda
(562) Nat. Hort. Mag. 30:4 Mulligan
(563) Am. Hort. Mag. 46:1 Galle
(564) Am. Hort. Mag. 39:2 Davidian
(565) Nat. Hort. Mag. 36:3 Sersanous
(566) Nat. Hort. Mag. 14:3 Bowers
(567) Nat. Hort. Mag. 7:3 Morrison & Guernsey
(568) Nat. Hort. Mag. 31:1 (Azalea Handbook).
(569) QBAGS v. 14 p. 231
(570) QBAGS v. 1 p. 107
(571) QBAGS v. 8 p. 127
(572) QBAGS v. 14 p. 39
(573) Am. Hort. Mag. 40:1 - Palm Issue.
(574) QBAGS v. 3 p. 3
(575) Am. Hort. Mag. 41:1 Thomas
(576) QBAGS v. 15 p. 170
(577) Nat. Hort. Mag. 12:3 Marriage
(578) BARGS 25:1 Plestil
(579) QBAGS v. 2 p. 294
(580) QBAGS v. 15 p. 243
(581) QBAGS v. 9 p. 304
(582) QBAGS v. 9 p. 135
(583) QBAGS v. 7 p. 366
(584) QBAGS v. 15 p. 95
(585) QBAGS v. 9 p. 303
(586) QBAGS v. 8 p. 30
(587) QBAGS v. 3 p. 221
(588) Nat. Hort. Mag. 17:3
(589) Nat. Hort. Mag. 10:2
(590) Nat. Hort. Mag. 9:4
(591) QBAGS v. 9 p. 118
(592) QBAGS v. 3 p. 267
(593) QBAGS v. 8 p. 201
(594) QBAGS v. 9 p. 114
(595) QBAGS v. 3 p. 268
(596) QBAGS v. 3 p. 280
(597) QBAGS v. 8 p. 216
(598) QBAGS v. 8 p. 302
(599) QBAGS b. 3 p. 259
(600) QBAGS v. 3 p. 260
(601) QBAGS v. 9 p. 107
(602) QBAGS v. 10 p. 102
(603) QBAGS v. 3 p. 279
(604) QBAGS v. 8 p. 212
(605) JRHS 105:2 Bloom
(606) BARGS 29:2 Davidson
(607) JRHS 103:8 Schilling
(608) BARGS 9:6 Ingwersen

(609) Nat. Hort. Mag. 23:2 Fox & Wilder
(610) Am. Hort. Mag. 43:4 Wyman
(611) Am. Hort. Mag. 44:3 Green
(612) Nat. Hort. Mag. 37:1 Creech
(613) QBAGS v. 6 p. 165
(614) Nat. Hort. Mag. 36:4 Coe
(615) Am. Hort. Mag. 47:4 Egolf
(606) BARGS 30:1 Davidson
(617) BARGS 8:6 Barneby
(618) QBAGS v. 11 p. 102
(619) BARGS 22:4 Hutmire
(620) Nat. Hort. Mag. 35:1 Fox
(621) QBAGS v. 15 p. 205
(622) Nat. Hort. Mag. 18:4 Morrison & Guernsey
(623) QBAGS v. 14 p. 102
(624) Nat. Hort. Mag. 15:3 Morrison
(626) QBAGS v. 9 p. 289
(627) Nat. Hort. Mag. 13:4 Morrison & Carron
(628) Am. Hort. Mag. 42:1 Egolf
(629) QBAGS v. 15 p. 186
(630) QBAGS v. 1 p. 40
(631) Nat. Hort. Mag. 21:3 E. K. Balls
(639) JRHS 103:5 Bloom
(640) JRHS 104:6 J. Elliott
(644) JSRGC 17:1 Evans
(645) The Plantsman 2:1 Herklots